£49.50

Springer Series in
Surface Sciences

22

Editor: Robert Gomer

Springer Series in **Surface Sciences**

Volume 1: **Physisorption Kinetics** By H. J. Kreuzer, Z. W. Gortel

Volume 2: **The Structure of Surfaces** Editors: M. A. Van Hove, S. Y. Tong

Volume 3: **Dynamical Phenomena at Surfaces, Interfaces and Superlattices**
Editors: F. Nizzoli, K.-H. Rieder, R. F. Willis

Volume 4: **Desorption Induced by Electronic Transitions, DIET II**
Editors: W. Brenig, D. Menzel

Volume 5: **Chemistry and Physics of Solid Surfaces VI**
Editors: R. Vanselow, R. Howe

Volume 6: **Low-Energy Electron Diffraction**
Experiment, Theory and Surface Structure Determination
By M. A. Van Hove, W. H. Weinberg, C.-M. Chan

Volume 7: **Electronic Phenomena in Adsorption and Catalysis**
By V. F. Kiselev, O. V. Krylov

Volume 8: **Kinetics of Interface Reactions** Editors: M. Grunze, H. J. Kreuzer

Volume 9: **Adsorption and Catalysis on Transition Metals and Their Oxides**
By V. F. Kiselev, O. V. Krylov

Volume 10: **Chemistry and Physics of Solid Surfaces VII**
Editors: R. Vanselow, R. Howe

Volume 11: **The Structure of Surfaces II**
Editors: J. F. van der Veen, M. A. Van Hove

Volume 12: **Diffusion at Interfaces: Microsopic Concepts**
Editors: M. Grunze, H. J. Kreuzer, J. J. Weimer

Volume 13: **Desorption Induced by Electronic Transitions, DIET III**
Editors: R. H. Stulen, M. L. Knotek

Volume 14: **Solvay Conference on Surface Science** Editor: F. W. de Wette

Volume 15: **Surfaces and Interfaces of Solids** By H. Lüth

Volume 16: **Theory of the Atomic and Electronic Structure of Surfaces**
By M. Lannoo, P. Friedel

Volume 17: **Adhesion and Friction** Editors: M. Grunze and H. J. Kreuzer

Volume 18: **Auger Spectroscopy and Electronic Structure**
Editors: G. Cubiotti, G. Mondio, K. Wandelt

Volume 19: **Desorption Induced by Electronic Transitions, DIET IV**
Editors: G. Betz, P. Varga

Volume 20: **Semiconductor Surfaces** By W. Mönch

Volume 21: **Surface Phonons** Editors: W. Kress, F. de Wette

Volume 22: **Chemistry and Physics of Solid Surfaces VIII**
Editors: R. Vanselow, R. Howe

R. Vanselow R. Howe (Eds.)

Chemistry and Physics of Solid Surfaces VIII

With 234 Figures

Springer-Verlag
Berlin Heidelberg New York London
Paris Tokyo Hong Kong Barcelona

Professor Ralf Vanselow, Ph.D.

Department of Chemistry and Laboratory for Surface Studies,
The University of Wisconsin-Milwaukee,
Milwaukee, WI 53201, USA

Dr. Russell Howe

Department of Chemistry, University of Auckland,
Private Bag, Auckland, New Zealand

Series Editors

Professor Dr. Gerhard Ertl

Fritz-Haber-Institut der Max-Planck-Gesellschaft, Faradayweg 4–6,
D-1000 Berlin 33

Professor Robert Gomer, Ph. D.

The James Franck Institute, The University of Chicago, 5640 Ellis Avenue,
Chicago, IL 60637, USA

Managing Editor: Dr. Helmut K.V. Lotsch

Springer-Verlag, Tiergartenstrasse 17,
D-6900 Heidelberg, Fed. Rep. of Germany

ISBN 3-540-52679-X Springer-Verlag Berlin Heidelberg New York
ISBN 0-387-52679-X Springer-Verlag New York Berlin Heidelberg

2154/3150-543210 – Printed on acid-free paper

Preface

This volume contains review articles written by the invited speakers at the ninth International Summer Institute in Surface Science (ISISS 1989), held at the University of Wisconsin-Milwaukee in August of 1989.

During the course of ISISS, invited speakers, all internationally recognized experts in the various fields of surface science, present tutorial review lectures. In addition, these experts are asked to write review articles on their lecture topic. Former ISISS speakers serve as advisors concerning the selection of speakers and lecture topics. Emphasis is given to those areas which have not been covered in depth by recent Summer Institutes, as well as to areas which have recently gained in significance and in which important progress has been made.

Because of space limitations, no individual volume of Chemistry and Physics of Solid Surfaces can possibly cover the whole area of modern surface science, or even give a complete survey of recent progress in this field. However, an attempt is made to present a balanced overview in the series as a whole. With its comprehensive literature references and extensive subject indices, this series has become a valuable resource for experts and students alike. The collected articles, which stress particularly the gas–solid interface, have been published under the following titles:

Surface Science: Recent Progress and Perspectives, Crit. Rev. Solid State Sci. 4, 125–559 (1974); *Chemistry and Physics of Solid Surfaces*, Vols. I, II, and III (CRC Press, Boca Raton, FL 1976, 1979, 1982); Vols. IV, V, VI and VII: Springer Ser. Chem. Phys., Vols. 20 and 35, and Springer Ser. Surf. Sci., Vols. 5 and 10 (Springer, Berlin, Heidelberg 1982, 1984, 1986, 1988).

Each of the past volumes contains one chapter in which a selected surface scientist reviews in a historical framework his own research achievements. Distinguished surface scientists such as P.H. Emmet, E.W. Müller, F.C. Tompkins, G.-M. Schwab, W.E. Spicer, G.A. Somorjai and H.D. Hagstrum have provided their services in the past.

This volume begins with a historical review by *G. Ertl* on the reactivity of surfaces. He outlines the historical development of the surface science approach to heterogeneous catalysis, in which his own research has played a pivotal role, and illustrates it with recent examples. The theme of surface science in catalysis is continued by *Ceyer*, who discusses new mechanisms for the activation of molecules at surfaces, a subject of particular relevance in the present search for more economic natural gas conversion processes. *White* reviews his own work in the area of photochemistry, emphasizing the use of photons to break bonds within molecules. The closely related topic of desorption induced by electronic transitions is reviewed by *Madey* et al.; in this case, both photon-stimulated and electron-stimulated effects are considered. Most industrial metal catalysts consist of small metal particles rather than well-defined single crystals. The modeling of such industrial catalysts with well-defined metal clusters is a growing area of research. *Sachtler* reviews the use of zeolites as porous hosts in which to prepare, isolate and study the surface reactivity of small metal clusters. Nonlinear optical effects at surfaces and interfaces are attracting increasing attention as probes of surface structure and reactivity. Techniques such as second harmonic generation (SHG) and infrared visible sum frequency generation (SFG) allow the observation of solid–liquid and solid–solid interfaces as well as those in solid–gas systems. Pulsed laser methods will in principle permit picosecond time resolution. These exciting new methods are described by *Hall* et al. High electric fields can be present in a variety of chemical systems. For example, within the double layer at an electrolyte/electrode interface, the static electric field strength can reach 10^7 V/cm and around localized charges in zeolites values of 10^8 V/cm have been estimated. *Kreuzer* deals with the impact of such high electric fields on surface physics and surface chemistry. In 1989, R. Pool posed the question: "Is it necessary for me to learn what is going on in chaos and is there anything that I would do or think differently if I was to avail myself of modern (chaotic) thinking?" *Gadzuk* provides some guidance to surface scientists so that they might better answer such a question for themselves. He brings out a number of possibly relevant aspects of nonlinear studies featuring some of the "paradigms of chaos", and then shows a few of the already existing examples in surface science, which have drawn, in one way or another, upon developments from the world of chaos.

The remaining reviews deal with a variety of techniques used for surface characterization. *Canter* et al., after providing a brief overview of early low-energy positron diffraction (LEPD) experiments (1979–1983), discuss experimental advances leading to the recent use of LEPD for surface structure determinations. The surface atomic geometries of CdSe$10\bar{1}0$ and CdSe$11\bar{2}0$ are identified as an example. The authors critically discuss the differences between low-energy electron diffraction (LEED) and LEPD in such structure determinations. *Rabalais* et al. describe time-of-flight scattering and recoiling spectrometry (TOF-SARS), a nondestructive technique for surface analysis, which collects both neutrals and ions in a multichannel TOF mode. It is sensitive to all elements, including hydrogen. The authors compare it to LEED and AES and demonstrate its application in a structural analysis of clean W{211} and of O_2 and H_2 on W{211}. *Unguris* et al. review scanning electron microscopy with polarization analysis (SEMPA). This technique combines the finely focused beam of the scanning electron microscope (SEM) with secondary electron spin polarization analysis to obtain high-resolution images of the surface magnetic microstructure of ferromagnetic materials. As examples, the authors discuss the analyses of iron and cobalt crystals, ferromagnetic glasses, domain walls and magnetic storage media. Low energy electron microscopy (LEEM) has become operational only very recently. However, because of its high lateral resolution, it has proven its usefulness in a number of areas very rapidly. Its inventor, *Bauer*, discusses in detail the fundamentals (such as resolution, intensity and contrast mechanism) and the powerful combination of this technique with LEED and photoemission electron microscopy (PEEM). The application is demonstrated in studies of surface imperfections, epitaxy, segregation, adsorption induced surface modifications and surface layers. The surfaces investigated include Si{111}, Si{100}, Mo, W, Pb{110} and Au{100}. The adsorbates/deposits include C, Cu, Au and Co. Photoelectron spectroscopy of adsorbed xenon (PAX), can be considered as a reversible, non-destructive decoration and titration technique, which provides information about the number density of specific kinds of surfaces and, in particular, about the local surface potentials of these sites. It can provide potential differences on an atomic scale and seems to complement scanning tunneling microscopy (STM) in a unique way. *Wandelt* discusses in depth the physical principles of PAX, deals with the technique and addresses some methodical aspects. The application concentrates on metal surfaces (including metal powders) and deals with surface steps (vicinal surfaces), initial stages of film growth and the thermal stability of metal/metal interfaces. The last group of surface characterization techniques is devoted to the presently most popular and most rapidly developing microscopy, the STM and its offsprings. *Tersoff* reviews the interpretation of STM images based on recent theoretical work. He points out that STM images tend to fall into two classes: i) "simple" and noble metals and ii) semiconductors. He also discusses mechanical interactions between STM-tip and sample. *Murday* et al. provide a general overview on STM and its relatives under the title "Proximal Probes: Techniques for Measuring at the Nanometer Scale". It deals in detail with STM, ballistic electron emission microscopy (BEEM), atomic force microscopy (AFM), near-field optical microscopy (NFOM), and, as much as necessary for the understanding of the parent technique, with field electron emission. They look into the research opportunities at the nanometer scale and explore briefly nanometer scale fabrication using proximal probes. Two chapters deal with specific applications of STM, STS and AFM. *Avouris* et al. study semiconductor surface chemistry, i.e., the topography and valence electronic structure of surfaces, with atomic resolution. They demonstrate that STM and atom-resolved tunneling spectroscopy (ARTS) can, in favorable cases, go a long way to achieving useful results. Because of the linkage with *Avouris*'s topic, we inserted at this point the chapter by *Tong* et al. on "Bonding and Structure on Semiconductor Surfaces". Finally, *McClelland* et al. present recent experimental and theoretical developments involving tribology at the atomic scale. After a brief summary of key concepts in classical tribology, they review experimental results obtained by instruments such as the field ion microscope (FIM), the AFM and the surface force apparatus (SFA). At the end, they discuss theoretical models in which the motion of individual atoms is followed.

As in previous volumes, a thorough subject index is provided, together with extensive lists of references.

We would like to thank the sponsors of ISISS: The Office of Naval Research (Grant No. N00014-89-J-1682) as well as the Graduate School, the College of Letters and Science and the Laboratory for Surface Studies at the University of Wisconsin-Milwaukee. Their support made both the conference and the publication of this volume possible. The cooperation of the authors and the publisher is gratefully acknowledged.

Milwaukee, Auckland
January 1990

Ralf Vanselow
Russell Howe

Contents

1. Reactivity of Surfaces

G. Ertl

Fritz-Haber-Institut der Max-Planck-Gesellschaft, Faradayweg 4–6,
D-1000 Berlin 33, West Germany

On July 29, 1823, J.W. Doebereiner, professor of chemistry at the University of Jena, informed his minister, J.W. Goethe, about his observation that finely divided platinum causes hydrogen to react with oxygen "by mere contact", whereby the platinum even starts to glow due to the heat evolved in this process [1.1]. This discovery was a sensation in the scientific world (perhaps comparable to the turmoil about recent reports about the so-called "cold fusion", with the essential difference that Doebereiner's effect was real!) and prompted many researchers to further investigations in this field, for which somewhat later Berzelius coined the term "catalysis" [1.2]. A tentative explanation was offered by *Faraday* [1.3] who speculated:

> "dependent upon the natural conditions of gaseous elasticity combined with the exertion of that attractive force possessed by many bodies ...; by which they are drawn into association more or less close, ... and which occasionally leads to the combination of bodies simultaneously subjected to this attraction".

It was, however, only during the first third of our century that the principles underlying the reactivity of solid surfaces started to become truly explored. Around 1900, *W. Ostwald* [1.4] formulated catalysis as a kinetic phenomenon in a manner which is still valid. Most remarkable intuition was demonstrated by Langmuir who, in the course of experiments with a Pt wire interacting with oxygen and carbon monoxide, wrote on February 2, 1915, into his laboratory notebook [1.5]:

> "The CO molecules striking the clean Pt surface adhere to it (perhaps not all but a fraction) probably largely because of the unsaturated character of their carbon atoms".

And somewhat later, on April 21, 1915 [1.6]:

> "1. The surface of a metal contains spaces according to a surface lattice.
> 2. Adsorption films consist of atoms or molecules held to the atoms forming the surface lattice by chemical forces".

This view was in strong contrast to the then more or less generally accepted theories of adsorption in terms of simple condensation of gases in relatively thick layers due to unspecified long-range forces. Langmuir was certainly inspired by the recently recognized atomic structure of crystals based on the discovery of x-ray diffraction and was probably triggered by a lecture given by W.H. Bragg in his laboratory in late 1914.

Langmuir formulated his novel concepts in a monumental paper published in 1916 [1.7], from which only two short statements will be quoted:

- "Adsorption is very frequently the result of the strongest kind of chemical union between the atoms of the adsorbed substance and the atoms of the solid".
- "The atoms in the surface of a crystal must tend to arrange themselves so that the total energy will be a minimum. In general, this will involve a shifting of the positions of the atoms with respect to one another".

This latter conclusion was only 'rediscovered' during the last two decades and is currently the subject of intense experimental and theoretical investigation.

In 1915, Langmuir was essentially a 'nobody', while a few years later he was the leading scientist invited to give the Opening Address on "Chemical Reactions on Surfaces" at a Discussion Meeting of the Faraday Society which took place on September 28, 1921 [1.8]. Let me quote again a short section from this paper:

> "Most finely divided catalysts must have structures of great complexity. In order to simplify our theoretical consideration of reactions at surfaces, let us confine our attention to reactions on plane surfaces. If the principles in this case are well understood, it should then be possible to extend the theory to the case of porous bodies. In general, we should look upon the surface as consisting of a checkerboard ..."

This strategy precisely outlines the now widely adopted "surface science" approach whose experimental realisation had to wait, however, for nearly half a century. Nowadays, the reactivity of surfaces is studied in great detail by the use of clean single crystal surfaces and the arsenal of modern surface physics, and great progress in our understanding of the underlying elementary processes has been made in recent years.

I was informed that it is the aim of this introductory 'Historical Review' to sketch this development and to illustrate it by means of selected examples originating mainly from the author's own achievements. While I am very much honored by this invitation on the one hand, it might, on the other, suggest that relevant contributions from this person in the future are no longer to be expected. In order to escape from this dilemma I therefore decided to combine this retrospective analysis with several results of our present activities.

1.1 Chemisorbed Phases – Alterations of the Substrate Bonds

Volume 1 of the journal "Surface Science" appeared in 1964 which marked the beginning of the renaissance of low energy electron diffraction (LEED) and hence this was essentially the only technique during these years to provide microscopic information about properties of extended single crystal surfaces. (Field emission and field ion microscopy require thin tips as objects of investigation). I published my first paper in this journal in 1967 [1.9]. It concerned (among others) the investigation of the interaction between oxygen and a Cu{110} surface. Comparison of data resulting from interaction with either O_2 or N_2O led to the conclusion that oxygen is present on the surface in atomic form. From the relative intensities of the LEED 'extra' spots it was further concluded that these are essentially not due to scattering at the O atoms, but to alterations of the unit cell by displacements of the Cu substrate atoms ("reconstruction"). A series of LEED patterns reflecting the structural transformations of this surface with increasing O coverage is reproduced in Fig. 1.1: Low O_2 exposures at first cause the appearance of sharp streaks between the substrate lattice spots (Fig. 1.1), reflecting the formation of highly anisotropic nuclei of the adsorption phase which are narrow in the [110] direction and fairly long in the [001] direction. With increasing exposure, these streaks contract to spots of a 2×1 structure which is completed at $\theta = 0.5$ (Fig. 1.1b). Beyond that, nucleation of a new $c6 \times 2$ phase starts which at first coexists in the form of islands with the 2×1 phase, as reflected by the superposition of diffraction spots from both structures (Fig. 1.1c), and finally covers the whole surface (Fig. 1.1d).

These conclusions drawn from observations in reciprocal space are nicely confirmed by current investigations in real space by means of scanning tunneling microscopy (STM) [1.10]. Figure 1.2a shows the anisotropic domains of the growing 2×1 phase which are mainly formed via homogeneous nucleation. The completed 2×1 phase is shown in Fig. 1.2b with even higher magnification and atomic resolution, whereby it is, however, still unclear whether in this case O or Cu atoms are actually seen. Figure 1.2c is an image from the coexisting 2×1 (long rows) and $c6 \times 2$ patches, while in Fig. 1.2d the surface has been largely converted into the $c6 \times 2$ phase. (The varying grey scale reflects the differing heights of terraces separated from each other by atomic steps).

Numerous investigations have been performed on the structural properties of the O/Cu{110} system over the past two decades [1.11–1.31]. While for the 2×1 phase general consensus exists that the substrate surface is reconstructed and the O atoms are located in the 'long bridge' sites, the type of reconstruction is still under debate: There is strong experimental evidence for a 'missing row' structure, but in a quite recent STM study this model has been discarded in favor of the 'buckled-row' model, in which every other [001] row is lifted [1.32].

We nevertheless believe that the O/Cu{110}-2×1 structure is of the same type as the analogous structure formed by O on the Ni{110} surface for which in a recent LEED investigation the 'missing row' structure could be clearly established [1.33] (Fig 1.3). This analysis (yielding excellent R factors) was performed by using a recently developed optimisation scheme [1.34] which, in the present case, permitted simultaneous refinement of 8 structural parameters. Apart from the absence of every second Ni row in the topmost layer, the interaction with the adsorbate causes alterations of the substrate structure even in deeper layers, e.g. some 'buckling' (by ± 0.05 Å) of the third layer Ni atoms. While the spacing

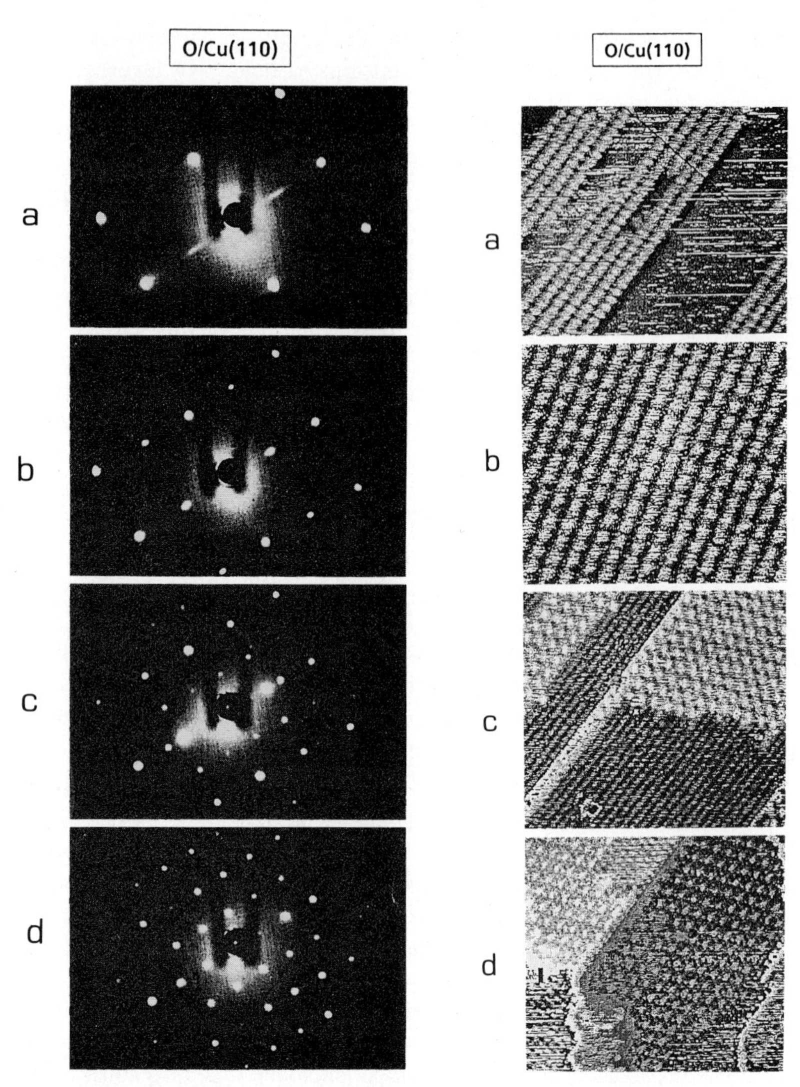

Fig. 1.1. Series of LEED patterns from Cu{110} with increasing oxygen exposure [1.9]. (**a**) 'Streaking' 2×1 pattern; (**b**) 2×1 structure; (**c**) coexistence of 2×1- and $c\,6 \times 2$; (**d**) $c\,6 \times 2$ structure

Fig. 1.2. STM images from the O/Cu{110} system corresponding to the conditions of Fig. 1.1 [1.10]. a) Anisotropic 2×1 domains besides patches of the clean surface; b) 2×1 structure; c) coexistence of 2×1- and $c\,6 \times 2$ patches; d) predominantly $c\,6 \times 2$ domains (varying length scales)

between the first and second Ni layer of the clean surface is contracted by 9% (with respect to the bulk value) it is expanded by 4% in the presence of the adsorbate.

The original conception of *Langmuir* [1.8] where a surface is to be considered as a 'checkerboard' of adsorption sites formed by the essentially rigid substrate lattice (– in spite of his own speculation about possible alterations

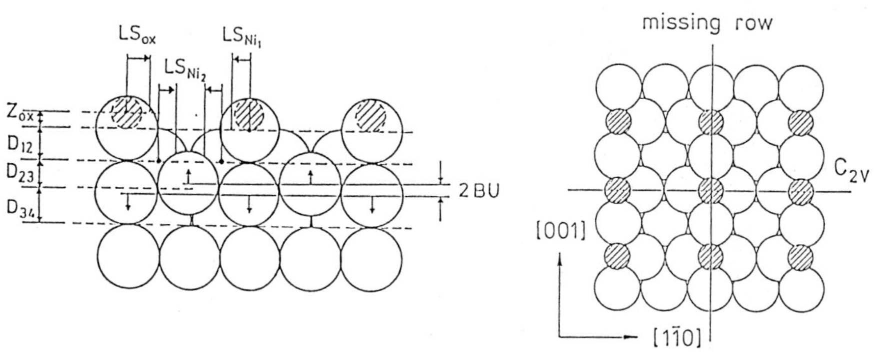

Fig. 1.3. Structure of the 2×1 O-Ni{110} phase as derived from LEED intensity analysis [1.33]. Structural parameters: $z_{ox} = 0.20$ Å; $D_{12} = 1.30$ Å; $D_{23} = 1.23$ Å; $D_{34} = 1.25$ Å; $LS_{ox} \simeq 0.1$ Å; $LS_1 = LS_2 = 0$; $BU = \pm 0.05$ Å

of the atomic positions quoted above –) is obviously valid only in exceptional cases. Even if, in cases of adsorbates interacting only weakly with the substrate, the surface remains undistorted by the adsorption process, lattice gas systems are not necessarily formed, but rather incommensurate overlayer phases in which the interactions between the adsorbed particles are dominating their mutual configurations. The systems H/Pd{100} [1.35] and H/Fe{110} [1.36], for example, belong to the category of lattice gas systems almost in the Langmuirian sense, whereby, however, the operation of mutual interactions betwen the adsorbed particles causes ordering phenomena at low temperatures in contrast with Langmuir's random distribution of the checkerboard sites. For chemisorption systems the adsorbate-surface bond energies are usually of similar magnitude to the cohesive energies between surface atoms, and hence the formation of the adsorption bond is frequently associated with displacements of the substrate atoms. The driving force has to be sought in the energy gain through the more favorable atomic configuration offered to the adsorbate which overcompensates the energy necessary for altering the geometry for the clean surface. The heat of chemisorption of CO and Pt{100}, for example, differs by about 10 kcal mol^{-1} between the two possible surface modifications [1.37]. In this case the clean surface is already reconstructed ('hex') [1.38] and becomes transformed into the 'normal' 1×1 structure as soon as the CO coverage exceeds a critical value. The adsorbed CO molecules then form islands of a $c2 \times 2$ phase with local coverage $\theta = 0.5$. After thermal desorption, the 1×1 structure is metastable and transforms back into the 'hex' structure with an activation energy of about 25 kcal mol^{-1} [1.39]. The energetics of these transformations are depicted schematically in Fig. [1.4].

Similarly, the clean Pt{110} surface is reconstructed into a 'missing row' 1×2 structure which is again transformed into the 1×1 phase by adsorption of CO [1.40–1.47]. The associated variation of atomic density within the substrate unit cell is achieved by surface diffusion, which may be restricted to very short

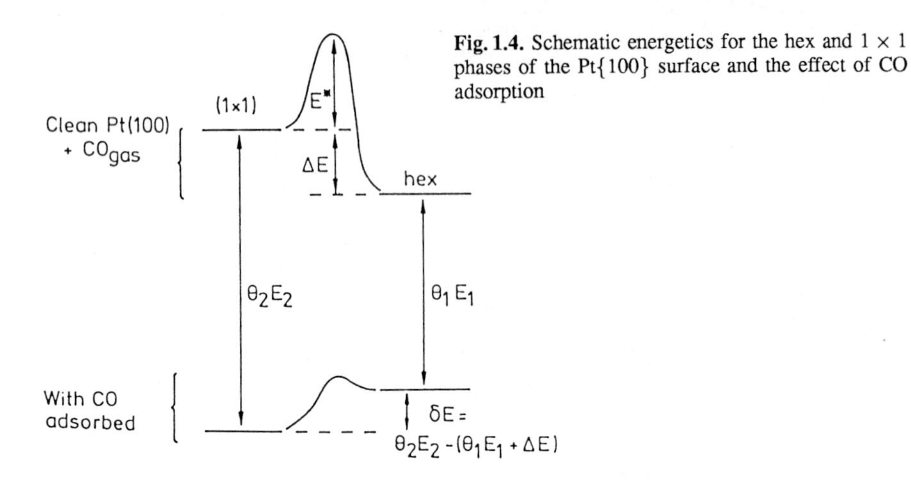

Fig. 1.4. Schematic energetics for the hex and 1 × 1 phases of the Pt{100} surface and the effect of CO adsorption

distances (or even suppressed) if the temperature is not high enough. Figure 1.5 shows STM images from the Pt{110} surface on which CO adsorption at room temperature causes homogeneous nucleation (through local adsorbate density fluctuations) of very small 1 × 1 patches simply by jumps of surface atoms over 1–2 lattice spacings [1.48]. The first nuclei are formed after a critical exposure has been reached, and further increase of the coverage causes the formation of additional nuclei rather than growth of already existing ones. Figure 1.6 shows the variation of the density of 1 × 1 patches with CO exposure (until at saturation the surface looks like a 'swiss cheese'), together with the change of the half-order beam intensity of the LEED pattern signaling the loss of long-range 1 × 2 order. At higher temperatures, on the other hand, the enhanced mobility of Pt surface atoms enables the formation of extended 1 × 1 regions.

With the O/Cu{110} system discussed above the mobility of surface Cu atoms is already very high at room temperature so that transport of matter over long distances as associated with the formation of the 'missing row' 2 × 1 phase is easily feasible.

1.2 Chemical Transformation – Alteration of the Adsorbate Bonds

The strong molecule-surface interaction will, of course, also affect the bonds within the adsorbate, even leading to bond dissociation or to the formation of new ones. This forms in fact the basis of heterogeneous catalysis. While the analogy to normal chemical transformations had already been recognized by Langmuir, more quantitative concepts were first developed by *M. Polanyi* [1.49] and by *H.S. Taylor* [1.50]. The latter had observed that adsorption processes may be rather slow and dependent on temperature, a phenomenon for which he proposed the term 'activated adsorption'. (A review about the development of this field was presented by *G. Ehrlich* [1.51] at the last ISISS conference). Rationalisation of the (activated) dissociative adsorption of a diatomic molecule in terms of a schematic

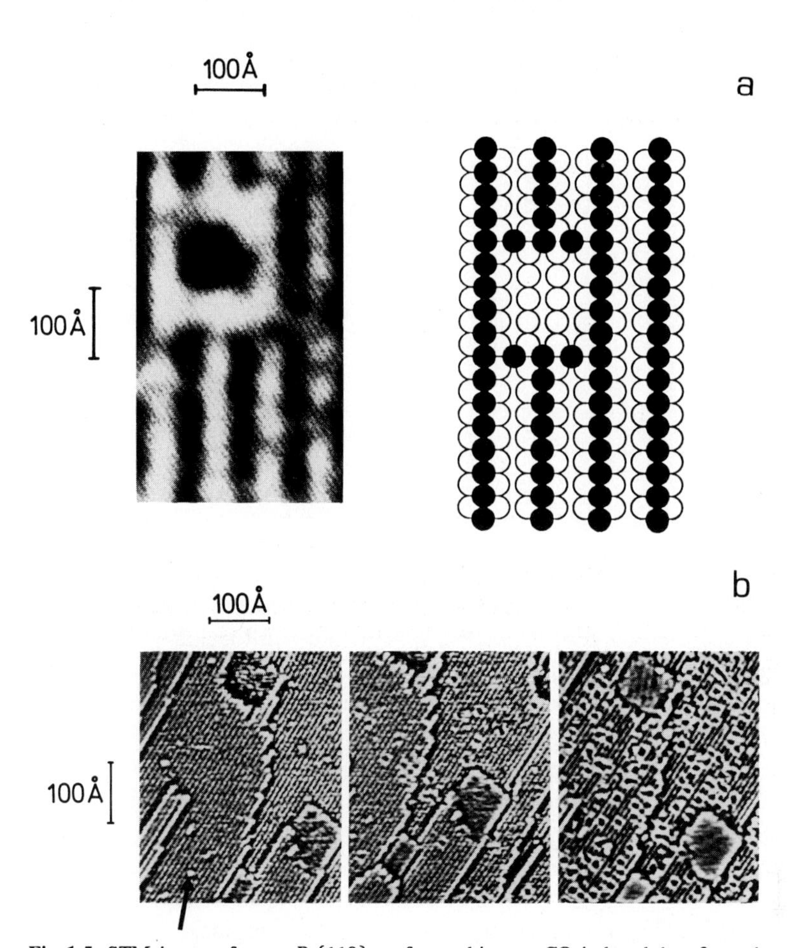

Fig. 1.5. STM images from a Pt{110} surface subject to CO induced $1 \times 2 \rightarrow 1 \times 1$ structural transformation at 300 K. a) Single nucleus and corresponding ball model; b) increasing density of nuclei with increasing CO exposure

potential energy diagram was performed in 1931 by Lennard-Jones [1.25] at a Faraday discussion meeting, and this type of diagram (Fig. 1.7b) is still familiar to everbody working in the field of surface chemistry. The Lennard-Jones potential diagram may, however, be misleading, and it seems to be more reasonable to plot the potential energy as a function of *two* coordinates, the distance x of the molecule from the surface and the interatomic distance y within the molecule (Fig. 1.7a). The shallow minimum at large x then corresponds to the molecular 'precursor', the transition state marked by $\neq$ to the crossing of the two curves in the Lennard-Jones diagram, and the valley at small x extending to increasing y to the dissociated state A_{ad}. The dashed line in Fig. 1.7a marks the 'reaction coordinate ϱ, i.e. the path of the reaction with minimum energy requirement. The variation of the potential along ϱ is plotted in Fig. 1.7c.

The theoretical basis for this view of the progress of chemical reactions in general was established in 1931 in a fundamental paper by *Polanyi* and *Eyring*

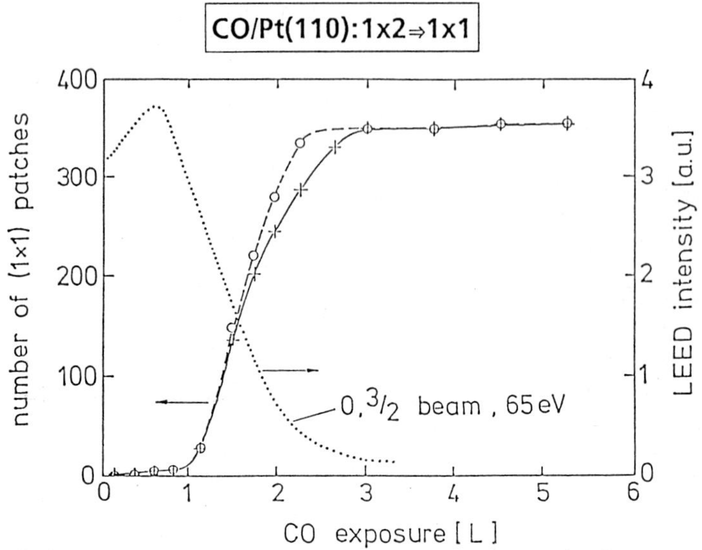

Fig. 1.6. CO induced $1 \times 2 \rightarrow 1 \times 1$ transformation of a Pt{110} surface at 300 K. *Full line*: Variation of the number of 1×1 nuclei as derived from STM with CO exposure. *Broken line*: Same data determined several minutes after termination of the respective gas exposure, indicating further slow nucleation due to local coverage fluctuations. *Dotted line*: Variation of the intensity of a half-order LEED spot with CO exposure. The initial increase is caused by a change in electron reflectivity due to build-up of the adsorbed phase

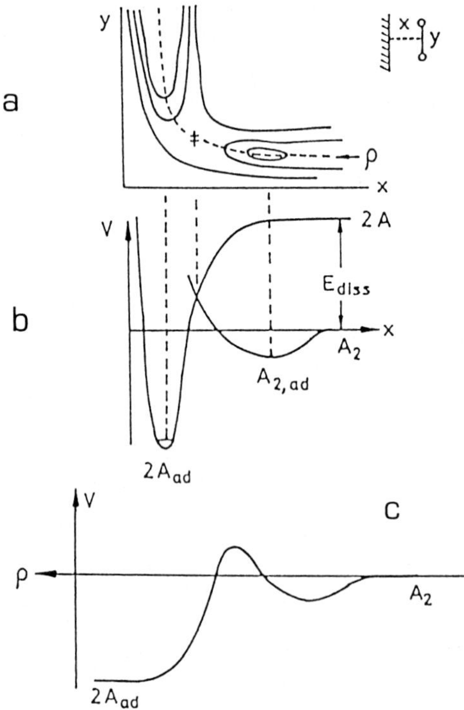

Fig. 1.7a–c. Schematic potential diagram for dissociative adsorption of a diatomic molecule. (**a**) Two-dimensional contour plot; (**b**) one-dimensional Lennard-Jones diagram; **c**) variation of the potential along the reaction coordinate ϱ

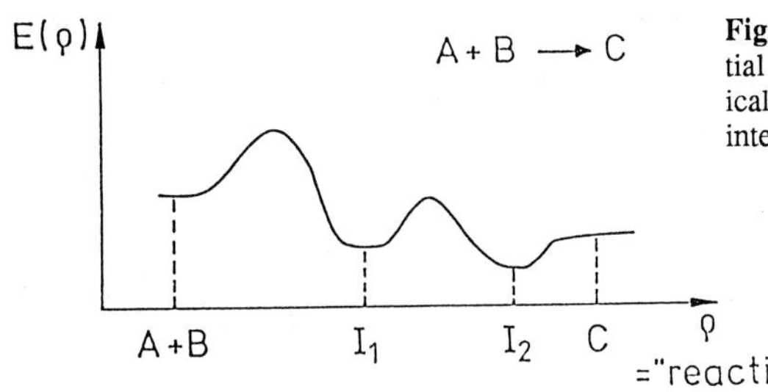

Fig. 1.8. Schematic one-dimensional potential along the reaction coordinate for a chemical reaction $A + B \rightarrow C$ passing through intermediates I_1 and I_2

[1.53], shortly after the advent of quantum mechanics and its application to the problem of the chemical bond. Solution of the Schrödinger equation for a system with nuclear coordinates R_j and electronic coordinates r_i within the Born-Oppenheimer approximation will yield a multi-dimensional potential surface $E(R_j)$, whose minimum energy trajectory represents the reaction coordinate ϱ. Figure 1.8 shows a schematic plot of $E(\varrho)$ for an arbitrary reaction: The local minima mark the nuclear configurations (= structures) of the reaction intermediates I_1 and I_2 as well as their stabilities, while the local maxima represent the transition states and their activation barriers. Passage of the system over this potential surface will take place through a large manifold of trajectories and will be associated with continuous transformation of energy between the various degrees of freedom. Experimental investigation of these energy transformation processes in molecule-surface interaction may be achieved by the application of molecular beam and laser spectroscopic techniques [1.54], and the contribution by *S. Ceyer* at this conference covers certain aspects of this dynamics (see Chap. 2). Full theoretical evaluation of the potential surface is still only feasible for the most simple reactions. For more complex processes (including reactions at surfaces) one will usually be content with the experimental determination of the essential features of the $E(\varrho)$ trace, namely identification of the reaction intermediates and the activation energies (or even better the individual rate constants, including the preexponential factors), which then establishes the 'reaction mechanism'. In conjunction with knowledge about the concentrations of the reacting species this will then enable evaluation of the overall reaction rate.

With surface reactions a complication arises, in that the rate is not simply determined by the concentrations (partial pressures) of the gaseous reactants but rather by those of the surface species or, more precisely, by their mutual configurations on the surface which may be sensitively affected by adparticle interactions (leading e.g. to island formation). Any conclusions about the actual reaction mechanism based solely on the dependence of the rate on the partial pressures, which are the external control parameters, may be quite misleading, as will be shown by the following example.

"The reaction which takes place at the surface of a catalyst may occur by interaction between molecules or atoms adsorbed in adjacent spaces on the surface, ... or it may take place directly as a result of a collision between a gas molecule and adsorbed molecule or atom on the surface".

With these words, *Langmuir* [1.8] formulated the two basic types of bimolecular reactions at surfaces, which are now generally classified either as Langmuir-Hinshelwood (LH, i.e. $A_{ad} + B_{ad} \rightarrow P$) or Eley-Rideal (ER, i.e. $A_{ad} + B_{gas} \rightarrow P$) mechanisms. Langmuir had investigated the catalytic oxidation of carbon monoxide, $CO+1/2\,O_2 \rightarrow CO_2$, over a Pt wire as early as 1915, but published a full account only a few years later [1.55]. He came (ironically!) to the conclusion that this reaction proceeds via the ER mechanism by collision of gaseous CO molecules with adsorbed O atoms because the rate of CO_2 formation was found to depend (under his conditions) linearly on O_2 pressure and inversely on the CO pressure, $r \alpha p_{O_2}/p_{CO}$. This puzzling rate law, whereafter the rate decreases with the increase of the concentration of one of the reactants, had in fact already been found previously [1.56] and had prompted *Bodenstein* and *Fink* [1.57] to a completely different view of the nature of the adsorbed layer: CO condenses on the surface, and the higher the p_{CO}, the thicker will be this layer. If the reaction takes place at the solid surface itself, diffusion through this condensed layer will be delayed and hence the inhibiting effect of CO. According to Langmuir, however, this is due to the increasing occupation of adsorption sites by adsorbed CO which inhibit adsorption of oxygen. Since, on the other hand, the rate law rules out a similar blocking of CO adsorption by adsorbed oxygen, the natural conclusion was to assume reaction via the ER mechanism outlined before.

The first report on the 'surface science' approach to a catalytic process was published in 1969 for the same reaction taking place at a Pd{110} surface [1.58]. The state of the surface was monitored by LEED while simultaneously the stationary reaction rate r was determined. Figure 1.9a shows a plot of r versus temperature for a fixed composition of the gas phase. Below 100 °C the LEED pattern exhibits spots from the CO adlayer signaling high CO coverages. With increasing temperature this coverage becomes continuously lower due to the increasing rate of desorption, oxygen may adsorb and hence the rate increases. Beyond 300 °C the oxygen coverage (as monitored by the O derived LEED structures) decreases and therefore the rate also decays. At fixed p_{CO}, T, on the other hand (Fig. 1.9b), the rate increases at first linearly with p_{O_2} (as in Langmuir's experiments) until the appearance of the oxygen LEED patterns signals the presence of substantial O coverage, but there is no indication for a decrease of the rate which would have to take place, if adsorbed O atoms inhibited the adsorption of CO and if the reaction proceeded along the LH mechanism. For this reason too, (erroneously!) operation of the ER mechanism was concluded.

In fact, mutual inhibition of adsorption is asymmetric with this system [1.59]: While CO tends to form densely packed layers which inhibit dissociative oxygen adsorption beyond a critical coverage, adsorbed O atoms, on the other hand, form relatively open surface structures, into which CO adsorption still may take place with high probability – hence the reaction rate does not decrease, even if the surface is saturated with O. A clear distinction between both mechanisms under these circumstances, only becomes possible on the basis of time resolved experiments: With the ER mechanism the product is formed during the collision, i.e. unmeasurably fast, while with the LH mechanism the CO molecule is

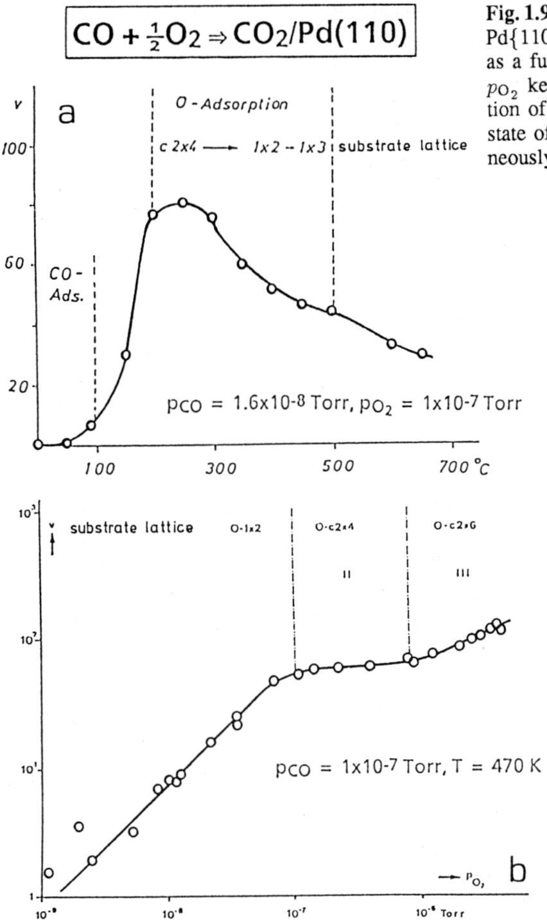

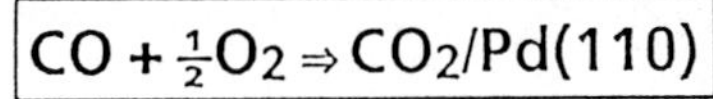

Fig. 1.9. Catalytic oxidation of CO on a Pd{110} surface [1.58]. a) Stationary rate as a function of temperature with p_{CO} and p_{O_2} kept fixed; b) stationary rate as a function of p_{O_2} with p_{CO} and T constant. The state of the surface was monitored simultaneously by LEED as indicated

first adsorbed and is expected to have a finite lifetime before it is transformed into CO_2. Measurements with a modulated molecular beam relaxation technique clearly demonstrated that the latter is indeed the case [1.60]. The results of a typical experiment are reproduced in Fig. 1.10: At $t = 0$, an oxygen covered Pt{111} surface is exposed to a periodically modulated CO beam. The oxygen coverage θ_0 decays continuously, while the reaction rate R remains constant over most of the time and decreases only towards the completion of O_{ad} consumption. Most important, the relaxation time τ between CO impact and CO_2 formation is of the order 5×10^{-4} s and thereby definitely rules out a direct collision-type reaction mechanism.

The schematic potential diagram illustrating the progress of this simple catalytic reaction is reproduced in Fig. 1.11 [1.61]. Most of the exothermicity is released to the solid with the adsorption steps. Both the adsorption energies and the activation energy for the LH reaction depend on the partial coverages, so that formulation of a simple rate law in the sense of Langmuir kinetics cover-

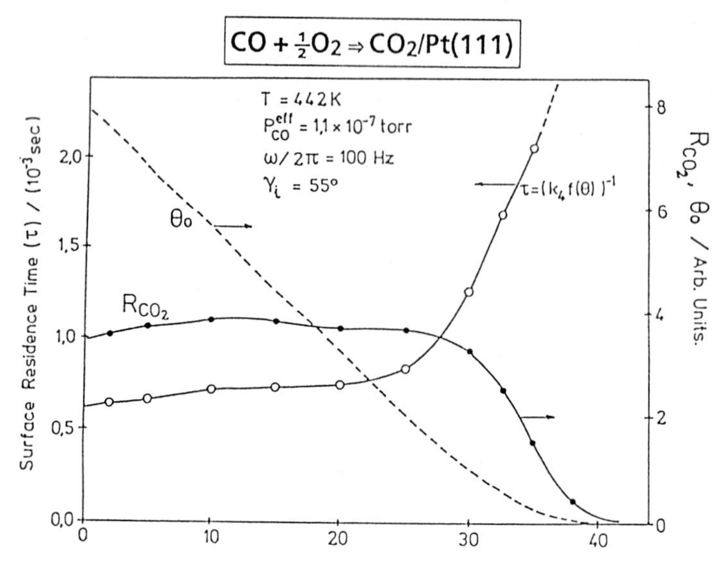

Fig. 1.10. Modulated molecular beam experiment on the mechanism of CO oxidation at a Pt{111} surface [1.60] (explanation see text)

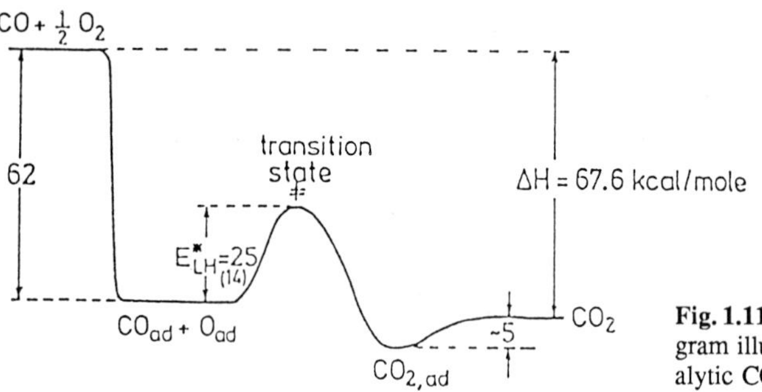

Fig. 1.11. Schematic potential diagram illustrating the progress of catalytic CO oxidation on platinum

ing all conditions is not feasible. Nevertheless, the steady-state kinetics, even under high pressure conditions, could be successfully modelled mathematically with parameters on the elementary steps derived from UHV single crystal studies [1.62]. (Similarly, the kinetic parameters for the elementary steps of ammonia synthesis as determined with Fe single crystal surfaces [1.63] served to develop a kinetic model which correctly predicts the yields of industrial reactors under conditions of 'real' catalysis [1.64].) Under conditions of high CO coverage, the reaction rate indeed closely follows the expression $r \alpha p_{O_2}/p_{CO}$, that means it is independent of total pressure which has recently been verified for total pressure variations over 6 orders of magnitude [1.65]. This is simply a consequence of the competition for adsorption sites between CO and O_2, whereby oxygen adsorption is rate-determining.

Figure 1.12 shows the variation of the stationary rate of CO_2 formation on a flat Pt{110} surface (full curve) as a function of CO partial pressure for fixed

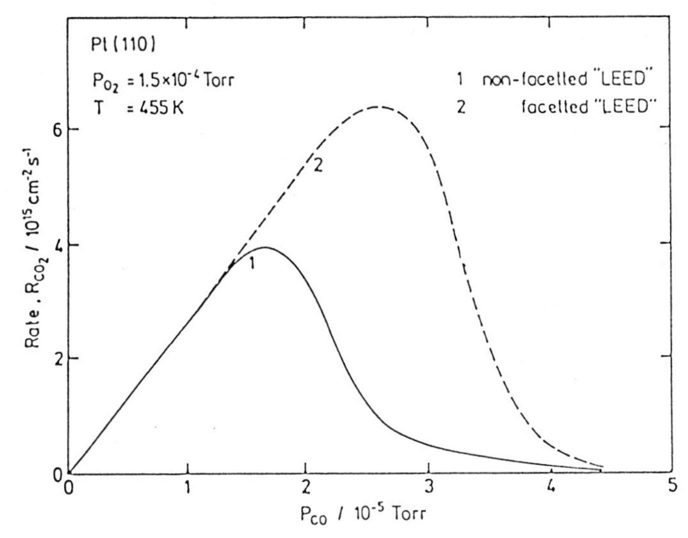

Fig. 1.12. Steady-state rate of CO_2 formation on Pt{110} as a function of p_{CO} for $p_{O_2} = 1.5 \times 10^{-4}$ Torr and $T = 455$ K. *Full curve*: 'flat' surface. *Broken curve*: 'facetted' surface

temperature and O_2 pressure. At low CO pressures, the surface is largely covered by adsorbed oxygen, and the rate increases linearly with p_{CO}, since now CO adsorption is rate-limiting. With increasing p_{CO}, however, the rate passes through a maximum, since adsorbed CO increasingly inhibits oxygen adsorption and now $r \sim 1/p_{CO}$ and oxygen adsorption becomes rate-limiting. The dashed curve in Fig. 1.12 was recorded with a Pt{110} surface containing a substantial concentration of steps. The presence of these defects has no noticeable effect on the CO sticking coefficient (which is always close to unity), but on the other hand, substantially increases the sticking coefficient for dissociative oxygen chemisorption. As a consequence, the steady-state rate of CO_2 formation is the same for both surfaces in the low p_{CO} region where CO adsorption is rate-limiting, but markedly differs in the high p_{CO} region where oxygen adsorption is rate-determining. This is a striking example of a catalytic reaction which may be structure-insensitive as well as structure-sensitive, depending on the applied conditions.

1.3 Nonlinear Dynamics in Surface Reactions

The data for the flat Pt{110} surface shown in Fig. 1.12 are in fact not true steady-state values. If the system is kept under constant conditions in the high CO-coverage regime (i.e. beyond the rate maximum) at not too high temperatures, the rate continuously increases with time (Fig. 1.13) [1.66]. Parallel to this increase the LEED spots split and continuously move apart, as reflected by the sequence of beam profiles reproduced in Fig. 1.14. This is due to the development of a periodic step and terrace structure (facetting) approaching inclined planes with a (210) orientation. These facets exhibit higher oxygen sticking coefficients and

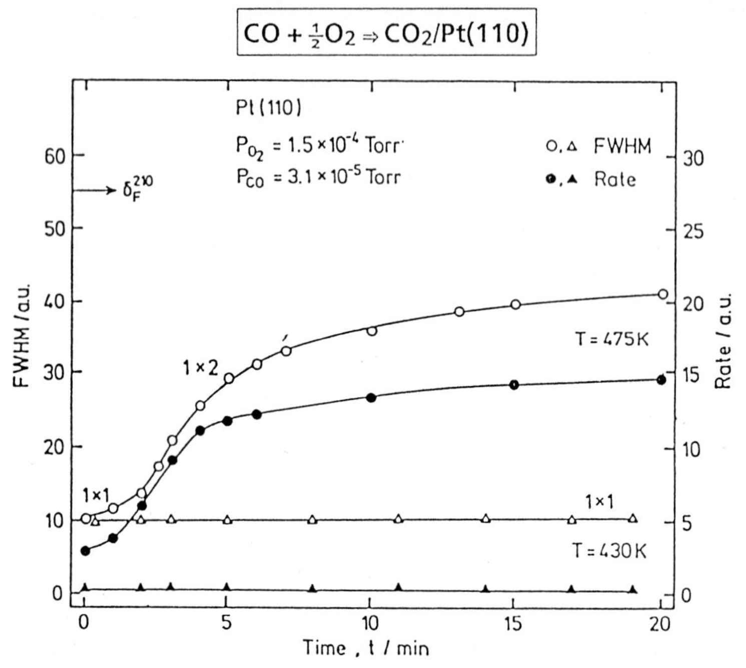

Fig. 1.13. Catalytic CO oxidation on a Pt{110} surface at fixed control parameters (p_{CO}, p_{O_2}, T) [1.66]. At T = 475 K the reaction rate is finite and increases continuously with time (*dark circles*), while simultaneously the LEED spots split and move apart (*open circles*). At 430 K the reaction rate is negligibly small and does not vary with time (*dark triangles*), and the surface structure remains stable (*open triangles*)

hence the continuous increase of the reaction rate. This effect does not take place by keeping the sample in an atmosphere of either CO or O_2 alone, but requires a continuously ongoing reaction. If the gas flow is switched off, the facets are thermally annealed and the initial flat surface is restored, provided the temperature is high enough to enable sufficient surface mobility. If the temperature is even higher, the rate of annealing will dominate over that for facetting, and the surface remains flat.

Variations of the topography of a catalyst surface during the reaction are observed quite frequently [1.67], and in fact even *Langmuir* [1.8] had already reported about such an effect with CO oxidation at a Pt wire:

> "Closer examination shows that the values for k tended to increase steadily, indicating that the filament was undergoing a progressive change in the direction of becoming a better catalyst .... There is good evidence that the effect is caused by changes in the structure of the surface itself, brought about by the reaction. After the wire has been used, the surface becomes very rough".

This effect cannot be caused by a thermodynamic principle, namely the tendency for lowering the surface free energy by the creation of new crystal faces,

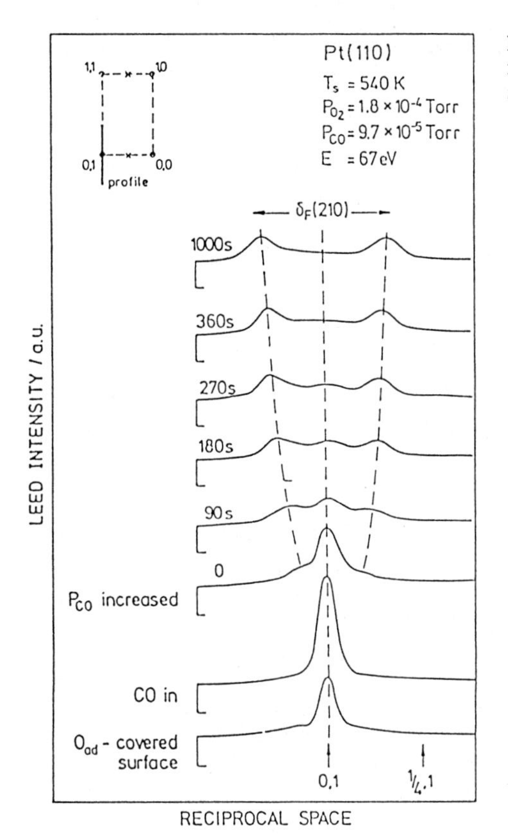

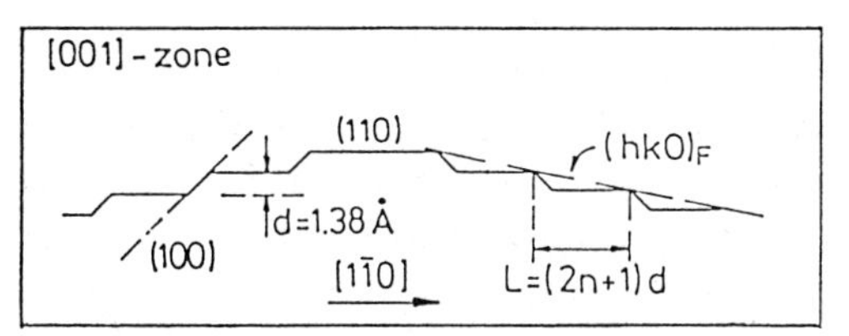

Fig. 1.14. A series of LEED beam profiles reflecting the progress of facetting of Pt{110} surface during catalytic CO oxidation [1.66]

since it is only observed while the catalytic reaction is going on, i.e. under conditions of non-equilibrium. Microscopically, it can be traced back to the lifting of the 1×2 reconstruction of the Pt{110} surface by CO adsorption coupled to the creation of defect sites with enhanced oxygen sticking coefficient, so that the adsorbed CO is reacted off again and the surface tends to restore its 1×1 structure and thus is continuously 'ploughed'. Recent computer simulations [1.68] demonstrated that such a sequence of elementary steps may indeed cause an initially flat surface to become periodically roughened.

More generally speaking, the observed phenomena are a consequence of nonlinear dynamics governing the behavior of the system under consideration. We are dealing with an open system, far from equilibrium, whose temporal behavior – i.e. the time dependence of the concentrations of the various surface species x_i – can be described by a set of coupled nonlinear differential equations of the form

$$\frac{dx_i}{dt} = F_i(x_j, p_k) ,$$

where p_k are the external control parameters, such as temperature and partial pressures. If the state variables x_i are, in addition, allowed to be spatially dependent, these equations expand into

$$\frac{\delta x_i}{\delta t} = D_i \Delta x_i + F_j(x_j, p_k) ,$$

where D_i are the diffusion coefficients and Δ the Laplace operator.

While near-equilibrium processes exhibit linear relations between fluxes and driving forces (the Onsager relations) and tend to minimum entropy production, completely new phenomena – such as the continuous *increase* of the entropy production associated with progressing facetting and increase of the reaction rate with the present system – may come into play with systems obeying the nonlinear laws outlined. The fundamental consequences for chemical systems were explored in detail by *Prigogine* [1.70], although some of the basic principles of spatial pattern formation had already been discovered previously by *Turing* [1.71]. According to Turing, an initially spatially homogeneous system subject to reaction and diffusion of its constituents, may undergo symmetry breaking triggered by local fluctuations and leading to stationary pattern formation. It is believed that the formation of regular facets on a Pt{110} surface during catalytic CO oxidation belongs to this category of Turing instabilities.

Even more interesting consequences are observed with respect to the temporal behavior. While linear systems operated off-equilibrium approach a stationary state if the external control parameters are kept constant (called a fixed point), the behavior of nonlinear systems may be much more complex, viz. it may lead to various types of temporal oscillations including transitions to (deterministic) chaos.

As an example, Fig. 1.15 shows the temporal evolution of the system $CO + 1/2\,O_2 \rightarrow CO_2/\,Pt\{110\}$ as monitored through the work function variation $\Delta\phi$, if one of the control parameters (p_{O_2}) is altered at the point marked by an arrow to a new constant value [1.72]. (Note that the conditions were chosen in a way such that no interference with the effect of facetting described above comes into play.) In this case, $\Delta\phi$ is directly proportional to the rate of CO_2 formation and does not reach a new stationary value, but rather develops into sustained temporal oscillations. Phenomena of this type have been observed for the first time with heterogeneously catalyzed reactions in *Wicke*'s laboratory [1.73] and have since then been investigated extensively with polycrystalline catalyst samples [1.74] as well as, more recently, with single crystal surfaces [1.69, 75, 76]. Oscillating reactions have been explored even more widely with homogeneous systems, in particular with the famous Belousov-Zhabotinskij reaction [1.77]. With the present system the occurrence of these oscillations is

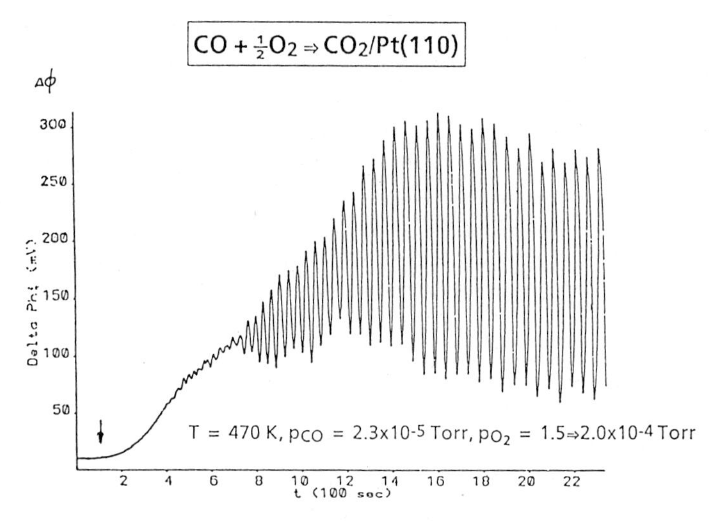

Fig. 1.15. Evolution of kinetic oscillations with the CO oxidation on Pt{110} [1.72]

restricted to a rather narrow range in parameter space and is coupled to a periodic transformation of the surface structure between the reconstructed 1×2 and the non-reconstructed 1×1 modification caused by varying CO coverages. These two structural modifications are characterized by different oxygen sticking coefficients, and hence the surface switches periodically between states of different reactivity. A similar mechanism had previously been found to underly kinetic oscillations with the same reaction taking place on a Pt{100} surface, as was directly demonstrated by simultaneously recording the work function and the intensities of LEED spots monitoring the periodic hex $\leftrightarrow$ 1×1 transformations [1.78]. In this latter case the oxygen sticking coefficient on the two surface modifications differs by more than two orders of magnitude, and as a consequence the existence region for oscillations in the parameter space is considerably broader [1.79, 80]. Generally, the kinetic oscillations under discussion will be coupled to a periodic switching of the surface between two states of differing reactivity. Apart from adsorbate induced surface structural transformations, this may also be caused by the build-up and removal of a 'subsurface' oxygen phase, as recently verified for Pd{110} [1.81] (oxide model) or even of a carbon deposit [1.82].

Successful mathematical modelling of the kinetic oscillations occurring with the CO oxidation on Pt{100} could be achieved by solution of a set of four coupled nonlinear differential equations, describing the temporal variation of the CO coverage on both surface phases (hex and 1×1), that of the O coverage on the 1×1 phase (the sticking coefficient on hex is negligible), and of the hex $\leftrightarrows$ 1×1 transformations depending on the respective CO coverages [1.83]. This can be further simplified to only three equations for Pt{110} where the difference between the 1×1 and 1×2 phases with respect to CO adsorption is neglected, albeit the $1 \times 2 \leftrightarrow 1 \times 1$ transformation induced by CO has to be included [1.84].

The temporal oscillations of Pt{100} are frequently irregular, while they are more regular with Pt{110}. (This is connected with the different mode of spatial self-organisation [1.80].) In the latter case slight variations of one of the external parameters may lead to a transition from simple periodic behavior to chaos via a sequence of period doublings (Feigenbaum scenario) [1.85]. The corresponding bifurcation diagram is reproduced in Fig. 1.16. It shows the $\Delta\phi$ amplitudes upon decreasing p_{CO}, starting from a simple stationary state (at $p_{CO} = 3.9 \times 10^{-5}$ Torr), passing via a Hopf bifurcation into the regime of simple harmonic oscillations and then through two period-doubling steps into the chaotic regime with irregular amplitudes.

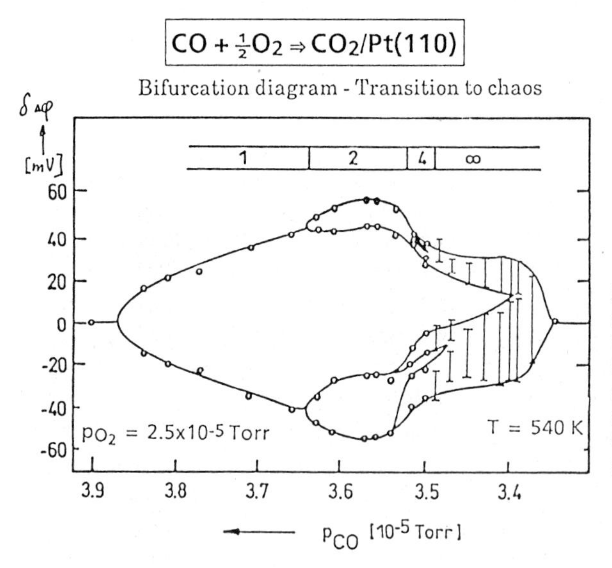

Fig. 1.16. Bifurcation diagram for the oscillatory CO oxidation on Pt{110} [1.85]. Positive and negative amplitudes (with respect to the respective average $\Delta\phi$ value) as a function of the control parameter p_{CO}

Other temporal effects observed include the rich phenomena of externally forced oscillations [1.86], as well as the occurrence of damped oscillations with the reaction $NO + CO \rightarrow 1/2\,N_2 + CO_2$ on Pt{100} [1.87].

In order to observe macroscopic variations of the reactivity there must exist efficient coupling between the various parts of the catalyst's surface. Under high pressure conditions and with supported catalysts this will usually take place through heat conductance, while with the low pressure single crystal systems under discussion here, non-isothermal effects can clearly be ruled out. With the CO oxidation on Pt{110} the oscillatory range is very narrow and the small partial pressure modulations ($\lesssim 1\%$) associated with the oscillations are sufficient to synchronize the system almost instantaneously [1.80]. This is quite different with the same reaction taking place on Pt{100}. Here, coupling between reaction and surface diffusion leads to a wavelike propagation of the structural transformation across the surface area as was demonstrated by scanning LEED experiments [1.88]. The spatial pattern formation occurring with this system can be imaged in even more detail by use of the recently developed technique of scanning photo-emission microscopy (SPM) [1.89]: Monochromatized UV light is focused through the objective of a micoscope onto a small spot ($\sim 3\mu$) of a surface and the total yield of electrons emitted is recorded by a channeltron. This yield depends on the local work function, and by rapidly scanning the position of the irradiated spot, a map of the work function distribution across the surface – reflecting differences in the state of the surface such as O or CO covered etc. – is constructed. Figure 1.17 shows a series of SPM images recorded from a $1 \times 1\,mm^2$ section of a Pt{100} surface at intervals of 1 min during the occurrence of temporal oscillations [1.90]. The formation of 'dissipative structures' by symmetry breaking of an initially spatially homogeneous system is clearly evident. These spontaneously forming spatial inhomogeneities propagate as waves with typical velocities of the order of 1 mm/min, and excitation of such 'chemical waves' may also be performed under non-oscillatory conditions by local external

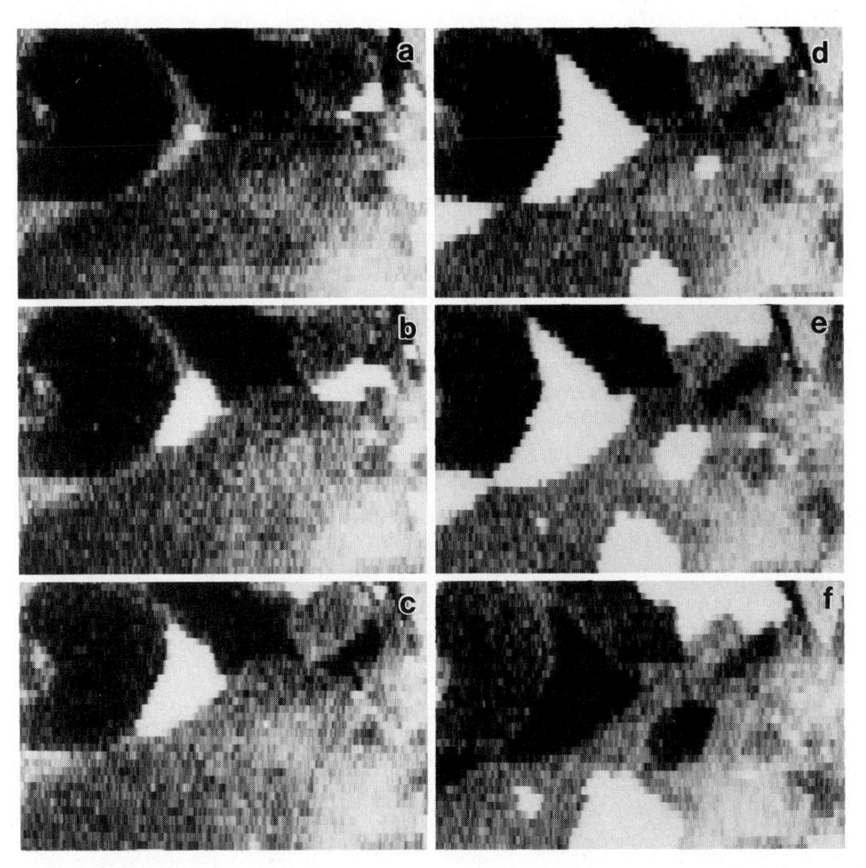

Fig. 1.17. Spatial pattern formation associated with the oscillatory CO oxidaton on Pt{100}. SPM images from a $1 \times 1\,mm^2$ surface area recorded in intervals of 1 min and reflecting the spatial differences of the work function and their variation with time [1.90]

perturbations via laser-induced thermal desorption of CO from a small spot on the surface [1.91].

Propagating 'chemical waves' in excitable media had already been discovered at the beginning of this century by *Luther* [1.92], who proposed these phenomena as models for nerve conductance.

1.4 Conclusion

This excursion had necessarily to be fragmentary and is characterized by the subjective view of the author. It appears remarkable that most of the basic principles of surface reactivity had already been proposed many years ago, but are only now becoming accessible to detailed investigation. The field of nonlinear dynamics, on the other hand, has been established more recently and is at present in a stage of rapid development. Much progress is also still to be expected from research in state-selective surface chemistry including photochemistry, areas which were completely excluded from this review and whose further exploration will hopefully bring us the answer to the question of how surface reactions really occur.

References

1.1 A. Mittasch: *Döbereiner, Goethe und die Katalyse*, (Hippokrates, Stuttgart 1951)
1.2 J. Berzelius: Jber. Chem. **15**, 1936 (237)
1.3 Quoted from H.S. Taylor: Trans. Faraday Soc. **28** (1932), 131
1.4 W. Ostwald: Physik. Z. **3** (1902), 313
1.5 I. Langmuir: Laboratory Notebook February 2, 1915. Quoted from G.L. Gaines, G. Wise: In *Heterogeneous Catalysis-Selected American Histories*, ed. by B.H. Davis, W.P. Hettinger, American Chem. Society (1983), p.13
1.6 ibid., April 21, 1915
1.7 I. Langmuir: J. Am. Chem. Soc. **38** (1916), 2221
1.8 I. Langmuir: Trans. Faraday Soc. **17** (1922), 607
1.9 G. Ertl: Surface Sci. **6** (1967), 208
1.10 D. Coulman, J. Wintterlin, R.J. Behm, G. Ertl: Phys. Rev. Lett. **64** (1990), 1761 and in preparation
1.11 G.W. Simmons, D.F. Mitchell, K.R. Lawless: Surface Sci. **8** (1967), 130
1.12 R.P.N. Bronckers, A.G.J. de Wit: Surface Sci. **12** (1981), 133
1.13 J.A. Yarmoff, D.M. Cyr, J.H. Huang, R.S. Williams: Phys. Rev. **B 33** (1986), 3856
1.14 E. van de Riet, J.B.J. Smeeb, J.M. Fluit, A. Niehaus: Surface Sci. **214** (1989), 111
1.15 R.N. Niehus, G. Comsa: Surface Sci. **140** (1984), 18
1.16 A. Spitzer, H. Lüth: Surface Sci. **118** (1982), 121
1.17 I.E. Wachs, R.J. Madix: Appl. Surface Sci. **40** (1973), 545
1.18 F.H.P.M. Habraken, G.A. Bootsma: Surface Sci. **87** (1979), 333
1.19 F.H.P.M. Habraken, G.A. Bootsma, P. Hofmann, S. Hachicha: Surface Sci. **88** (1979), 285
1.20 J.F. Wendelken: Surface Sci. **108** (1981), 605
1.21 J. Lapujoulade, Y. Le Cruer, M. Lefort, Y. Lejay, E. Maurel: Surface Sci. **118** (1982), 103
1.22 A.G.J. de Wit, R.P.N. Bronckers, J.M. Fluit: Surface Sci. **82** (1979), 177; **90** (1979), 676; **104** (1981), 384
1.23 M. Bader, A. Puschmann, C. Ocal, J. Haase: Phys. Rev. Lett. **57** (1986), 3273
1.24 R.A. Di Dio, D.M. Zehner, E.W. Plummer: J. Vac. Sci. Technol. **A 2** (1984), 852
1.25 R. Courths, B. Cord, H. Wern, H. Saalfeld, S. Hüfner: Solid State Commun. **63** (1987), 619
1.26 A.W. Robinson, J.S. Somers, D.E. Ricken, A.M. Bradshaw, A.L.D. Kilcoyne, D.P. Woodruff: to be published

1.27 U. Döbler, K. Baberschke, J. Haase, A. Puschmann: Phys. Rev. Lett. **52** (1984), 1437
1.28 R. Feidenhans'l, I. Stensgard: Surface Sci. **133** (1983), 453
1.29 K.S. Liang, P.H. Fuoss, G.J. Hughes, P. Eisenberger: in *The Structure of Surfaces*, ed. by M.A. Van Hove, S.Y. Tong (Springer, Berlin, Heidelberg 1985) p.245
1.30 J.A. Stroscio, M. Persson, W. Ho: Phys. Rev. **B 33** (1986), 6758
1.31 B. Hillert, L. Becker, M. Pedio, J. Haase, to be published
1.32 F.M. Chua, Y. Kuk, P.J. Silvermann: Phys. Rev. Lett. **63** (1989), 386
1.33 G. Kleinle, J. Wintterlin, G. Ertl, R.J. Behm, F. Jona, W. Moritz: Surface Sci. **225** (1990), 171
1.34 G. Kleinle, W. Moritz, D.L. Adams, G. Ertl: Surface Sci. **219** (1989), L637
1.35 R.J. Behm, K. Christmann, G. Ertl: Surface Sci. **99** (1980), 320;
K. Binder, D.P. Landau: Surface Sci. **108** (1981), 503
1.36 R. Imbihl, R.J. Behm, K. Christmann, G. Ertl, T. Matsushima: Surface Sci. **117** (1982), 257;
W. Kinzel, W. Selke, K. Binder: Surface Sci. **121** (1982), 13; **125** (1983), 74;
W. Moritz, R. Imbihl, R.J. Behm, G. Ertl, T. Matsushima: J. Chem. Phys. **83** (1985), 1959
1.37 R.J. Behm, P.A. Thiel, P.R. Norton, G. Ertl: J. Chem. Phys. **78** (1983), 7438; 7448
1.38 P. Heilmann, K. Heinz, K. Müller: Surface Sci. **83** (1979), 487;
M.A. Van Hove, R.J. Koestner, P.C. Stair, J.P. Biberian, L.K. Kesmodel, I. Bartos, G.A. Somorjai: Surface Sci. **103** (1981), 189
1.39 K. Heinz, E. Lang, K. Strauss, K. Müller: Appl. Surface Sci. **11/12** (1982), 611;
P.R. Norton, J.A. Davies, D.K. Creber, C.W. Sitter, T.E. Jackman: Surface Sci. **108** (1981), 205
1.40 C.M. Comrie, R.M. Lambert: J. Chem. Soc. Farad. I **72** (1986), 1659
1.41 T.E. Jackman, J.A. Davies, D.P. Jackson, W.N. Unertland, P.R. Norton: Surface Sci. **120** (1982), 389
1.42 H.P. Bonzel, S. Ferrer: Surface Sci. **118** (1982), L263
1.43 R. Imbihl, S. Ladas, G. Ertl: Surface Sci. **206** (1988), L903
1.44 P. Hofmann, S.R. Bare, D.A. King: Surface Sci. **117** (1982), 245; **144** (1984), 347
1.45 B.E. Hayden, A.W. Robinson, P.M. Tucker: Surface Sci. **192** (1987), 163
1.46 P. Fenter, T. Gustaffson: Phys. Rev. **B 38** (1988), 10197
1.47 N. Freyer, M. Kiskinova, G. Pirug, H.P. Bonzel: Appl. Phys. **A 39** (1986), 209
1.48 T. Gritsch, D. Coulman, R.J. Behm, G. Ertl: Phys. Rev. Lett. **63** (1989), 1086
1.49 M. Polanyi: Z. Elektrochem. **27** (1921), 142; **35** (1929), 561
1.50 H.S. Taylor: J. Am. Chem. Soc. **53** (1931), 578
1.51 G. Ehrlich: In *Chemistry and Physics of Solid Surfaces VII*, ed. by R. Vanselow, R.F. Howe (Springer, Berlin, Heidelberg 1988) p.1
1.52 J.E. Lennard-Jones: Trans. Faraday Soc. **28** (1932), 333
1.53 M. Polanyi, H. Eyring: Z. Phys. Chem. **12 B** (1931), 279
1.54 See e.g. G. Ertl: Proc. Solvay Conf. on Surface Science, ed. by F.W. De Wette (Springer, Berlin, Heidelberg 1988), p. 322
1.55 I. Langmuir, Trans. Faraday Soc. **17** (1922), 621
1.56 M. Bodenstein, F. Ohlmer: Z. Phys. Chem. **53** (1905), 166
1.57 M. Bodenstein, C.G. Fink: Z. Phys. Chem. **60** (1907), 1
1.58 G. Ertl, P. Rau: Surface Sci. **15** (1969), 443
1.59 T. Engel, G. Ertl: Adv. Catalysis **28** (1979), 1
1.60 T. Engel, G. Ertl: J. Chem. Phys. **69** (1978), 1267;
C.T. Campbell, G. Ertl, H. Kuipers, J. Segner: J. Chem. Phys. **73** (1980), 5862
1.61 G. Ertl: In *Chemistry and Physics of Solid Surfaces III*, ed. by R. Vanselow, W. England (CRC Press 1982), p.19
1.62 S.H. Oh, G.B. Fisher, J.E. Carpenter, D.W. Goodman: J. Catal. **100** (1986), 360;
A.G. Sault, D.W. Goodman: Adv. Chem. Phys. **76** (1989), 153
1.63 G. Ertl: In *Catalysis-Science and Technology*, ed. by J.R. Anderson, M. Boudart, Vol.4 (Springer, Berlin, Heidelberg 1983), p.209
1.64 P. Stoltze, J.K. Nørskov: Phys. Rev. Lett. **55** (1985), 2502; Surface Sci. **197** (1988), L230;
M. Bowker, I. Parker, K. Waugh: Surface Sci. **197** (1988), L223;
J.A. Dumesic, A.A. Trevino, to be published
1.65 M. Boudart, to be published
1.66 S. Ladas, R. Imbihl, G. Ertl: Surface Sci. **197** (1988), 153
1.67 M. Flytzani-Stephanopoulos, L.D. Schmidt: Progr. Surf. Sci. **9** (1979), 83

1.68 R. Imbihl, A.E. Reynolds, D. Kaletta, to be published
1.69 G. Ertl, Adv. Catalysis **37** (1990), in press
1.70 G. Nicolis, I. Prigogine: *Self-Organisation in Nonequilibrium Systems* (Wiley, New York 1977)
1.71 A.M. Turing: Phil. Trans. Roy. Soc. (London) **B 237** (1952), 37
1.72 M. Eiswirth, G. Ertl: Surf. Sci. **177** (1986), 90
1.73 P. Hugo: Ber. Bunsenges. Phys. Chem. **74** (1970), 121;
H. Beusch, P. Fieguth, E. Wicke: Chem. Ing. Techn. **44** (1972), 445
1.74 For recent review see L.F. Razón, R.A. Schmitz: Catal. Rev. **28** (1986), 89; Chem. Eng. Sci. **42** (1987), 1005
1.75 G. Ertl, P.R. Norton, J. Rüstig: Phys. Rev. Lett. **49** (1982), 177
1.76 R. Imbihl, to be published (review)
1.77 O. Gurel, D. Gurel: *Oscillations in Chemical Reactions* (Springer, Berlin, Heidelberg 1983)
1.78 M.P. Cox, G. Ertl, R. Imbihl, J. Rüstig: Surface Sci. **134** (1983), L517;
R. Imbihl, M.P. Cox, G. Ertl: J. Chem. Phys. **84** (1986), 3519
1.79 M. Eiswirth, R. Schwankner, G. Ertl: Z. Phys. Chem. N. F. **144** (1985), 59
1.80 M. Eiswirth, P. Möller, K. Wetzl, R. Imbihl, G. Ertl: J. Chem. Phys. **90** (1989), 510
1.81 R. Imbihl, S. Ladas, G. Ertl: Surface Sci. **219** (1989), 88
M. Ehsasi, C. Seidel, H. Ruppender, W. Drachsel, J.H. Block, K. Christmann: Surface Sci. **210** (1989), L198
1.82 N.A. Collins, S. Sundaresan, Y.J. Chabal: Surface Sci. **180** (1987), 136
1.83 R. Imbihl, M.P. Cox, G. Ertl, H. Müller, W. Brenig: J. Chem. Phys. **83** (1985), 1578
1.84 K. Krischer: thesis, in preparation
1.85 M. Eiswirth, K. Krischer, G. Ertl: Surface Sci. **202** (1988), 565
1.86 M. Eiswirth, G. Ertl: Phys. Rev. Lett. **60** (1988), 1526;
M. Eiswirth, P. Möller, G. Ertl: Surface Sci. **208** (1989), 13
1.87 S.B. Schwartz, L.D. Schmidt: Surface Sci. **183** (1987) L269; **206** (1988), 169;
M.R. Bassett, J.P. Dath, T. Fink, R. Imbihl, G. Ertl: to be published
1.88 M.P. Cox, G. Ertl, R. Imbihl: Phys. Rev. Lett. **54** (1985), 1725
1.89 H.H. Rotermund, G. Ertl, W. Sesselmann: Surface Sci. **217** (1989), L383
1.90 H.H. Rotermund, S. Jakubith, A. von Oertzen, S. Kubala, G. Ertl: J. Chem. Phys. **91** (1989), 4942
1.91 T. Fink, R. Imbihl, G. Ertl: J. Chem. Phys. **91** (1989), 5002
1.92 R. Luther: Z. Elektrochem. **12** (1906), 596

2. New Mechanisms for the Activation and Desorption of Molecules at Surfaces

S.T. Ceyer

Department of Chemistry, Massachusetts Institute of Technology, Cambridge, MA 02139, USA

2.1 Translational Activation of CH_4

Many chemical reactions occurring on the surfaces of solid materials appear to proceed only under high pressures of the gaseous reactants but not at low pressures ($< 10^{-4}$ Torr), despite favorable thermodynamics. This lack of reactivity at the low pressures where ultrahigh vacuum (UHV) surface science techniques are operable is known loosely as the pressure gap in the reactivity in heterogeneous catalysis [2.1, 2]. Our group proposed that an origin of the pressure gap is the presence of a barrier to dissociative chemisorption of at least one of the reactants upon collision with the surface [2.3–6]. Since it is the translational or internal energy of the incident molecule that is important in surmounting this barrier and not the surface temperature, the rate of the reaction is limited by the flux of incident molecules with energies above the energy of the barrier. High pressures simply increase the absolute number of high energy molecules, thereby increasing the reaction rate sufficiently for the products to be detected.

To verify this hypothesis, we systematically increased the energy of the incoming molecule while monitoring for the onset of dissociation in an ultrahigh vacuum-molecular beam apparatus which combines high resolution electron energy loss spectroscopy with molecular beam techniques. The molecular beam provides a convenient source of monoenergetic molecules at low pressures and electron energy loss spectroscopy, a vibrational spectroscopy for adsorbed species, is a sensitive and chemically specific detector of the adsorbed products of the dissociative chemisorption event. We probed the dissociative chemisorption of CH_4 on Ni{111}, since the steam reforming of CH_4 over a Ni catalyst ($CH_4 + H_2O \rightarrow CO + 3H_2$) is an example of a reaction which appears to proceed only at high pressures. Steam reforming of natural gas is the commercial process for H_2 production. Measurements of the dissociation probability as a function of the CH_4 incident energy showed that there is indeed a barrier to the dissociation of CH_4. Our low-pressure, dissociation probability measurements as a function of energy were found to agree very well with the rates of CH_4 decomposition on a Ni{111} crystal measured under high-pressure conditions as a function of reactor temperature [2.7]. The agreement between the low and high-pressure experiments carried out in different laboratories establishes the presence of a barrier along the reaction coordinate as an origin of the pressure gap in heterogeneous catalysis.

Besides providing a link between UHV surface science and high-pressure catalysis, these studies of the dynamics of the dissociative chemisorption of CH_4 have provided a detailed microscopic picture of the mechanism for the C–H bond breaking process [2.5]. We have shown that a deformation model explains the role of translational and vibrational energy in promoting dissociative chemisorption and suggests that tunneling is the final step in the C–H bond cleavage. Specifically, the probability for dissociative chemisorption of CH_4 on Ni{111} is observed to scale exponentially with the CH_4 translational energy in the direction normal to the surface. Vibrational energy in the CH_4 bending modes is found to be as effective as translational energy but surface temperature has no effect on the dissociation probability. These results are interpreted in terms of a barrier to CH_4 dissociation largely associated with the energy required to deform CH_4. The deformation is necessary to distort CH_4 away from its spherical shape, so that the hydrogen atoms, which effectively bury the carbon atom, are pushed sufficiently out of the way to allow a strong attractive interaction and bond formation between the surface Ni atoms and the carbon atom. The formation of a Ni–C bond is necessary along with Ni–H bond formation in order for there to be sufficient energy release to break the 100 kcal/mol C–H bond. Higher translational energy in the direction normal to the surface results in greater deformation of the molecule upon impact and, hence, a higher probability for dissociative chemisorption. This process is known as translational activation.

2.2 Collision Induced Dissociative Chemisorption and Collision Induced Desorption of CH_4

Additional corroboration of the deformation model arises from the observation of a new kind of mechanism for dissociative chemisorption, collision induced dissociative chemisorption. If the barrier to dissociation of CH_4 is largely the energy required to distort CH_4, then it should be possible to supply this deformation energy to CH_4 physisorbed on Ni{111} by collision with an inert gas atom. The impact of the inert gas atom is predicted to pound the molecularly adsorbed CH_4 into the distorted shape of the transition state that leads to dissociation. We showed that this mechanism does occur by monitoring the dissociation rate of CH_4 physisorbed on Ni{111} at 47 K induced by the impact of an incident Ar or Ne atom beam [2.3, 8, 9]. The absolute cross section for collision induced dissociation, which is proportional to the dissociation rate, is measured over a wide range of kinetic energies (28–52 kcal/mol) and angles of incidence of a Ne or Ar atom beam. Unlike the translational activation of CH_4 which exhibits strict normal energy scaling, the collision induced dissociation cross section displays a complex dependence on the energy of the impinging inert gas atoms, uncharacteristic of normal energy scaling.

A two-step, dynamical model for the mechanism of collision induced dissociation is shown to provide excellent agreement with the energy and angular dependence of the cross section for dissociation [2.9]. The model depicts the initial collision between the Ar or Ne and the physisorbed CH_4 to be impulsive and bimolecular. The energy transferred to CH_4 is therefore described by the classical mechanics of the collision of two hard spheres. However, the energy transferred to CH_4 in the normal direction only is important because, as shown by the translational activation results, only the normal kinetic energy is effective in promoting dissociation. The magnitude of the energy transferred to CH_4 in the normal direction is dictated not only by the energy and incident angle of the impinging Ar or Ne atom but by the impact parameter of the collision. It is this dependence of the energy transfer on impact parameter that leads to the breakdown of normal energy scaling in the Ar or Ne kinetic energy. Once this collision and energy transfer occurs, the Ar or Ne no longer participates in the dissociation process. The CH_4 molecule is accelerated into the surface by its newly acquired energy, is deformed upon impact with the surface, and dissociates. The probability for CH_4 dissociation at the value of the normal energy acquired by CH_4 after its collision with Ar is given by the previous translational activation results [2.5]. In this way, the model calculations allow the translational activation data to be mapped onto the cross sections for collision induced dissociation. Therefore, translational activation and collision induced activation are shown to be completely consistent. They are simply different ways to provide the energy to deform the CH_4 molecule but, once deformed, the mechanism for the dissociation is the same.

In competition with collision-induced disssociation, we have observed another process, collision induced desorption [2.3, 8, 10]. The absolute cross section for desorption increases with the incident angle of the Ar atoms at high total kinetic energy and remains approximately constant at low kinetic energy. Classical trajectory simulations indicate that desorption is predominantly the result of direct collisions of Ar with CH_4. The complex angle and energy dependence is shown to arise from the competition between the decrease in the energy available in the normal direction and the increase in the geometrical cross section for the collision as the incident angle increases. Multiple collisions do not play a significant role because the barrier to motion of physisorbed CH_4 tangential to the smooth surface is very small. Surface mediated processes or through-surface effects such as hot spots do not contribute measurably to desorption. The collision cross section and normal energy transfer upon a single collision are the two sole important components in determining the dependences of the cross section for collision induced desorption on energy and incident angle.

But perhaps more important than the physics behind collision induced dissociation and desorption is the importance of the very observation of these processes for understanding the chemistry occurring on surfaces under high-pressure conditions. In a high-pressure environment, a catalyst surface is covered with adsorbate and the adsorbate-covered catalyst is continually bombarded by a large flux of high energy molecules. Therefore, having shown that collision induced processes occur, we suggest that no mechanism for surface reactions under high-pressure conditions can now be considered complete without an assessment of the importance of collision induced chemistry and desorption as major reaction steps. In fact, in the catalytic literature, there are a dozen papers which note effects of the quantity and kind of inert gas on the rates of heterogeneous catalytic reactions [2.11]. It is now important to investigate these reactions in light of the knowledge that these processes do occur. Collision induced chemistry and desorption are additional explanations for the origin of the pressure gap in heterogeneous catalysis. They are additional reasons why surface chemistry in a high-pressure environment is often very different from the chemistry in UHV environments.

2.3 New Methods for Activation: New Syntheses

Having established this link between high-pressure catalysis and UHV surface science, we now know how to bypass the high-pressure requirement simply by raising the energy of the incident molecule (translational activation) or collisionally inducing dissociation (collision induced activation). We have used both methods to synthesize and identify spectroscopically, by high resolution electron energy loss spectroscopy, an adsorbed CH_3 radical under low pressure, ultrahigh vacuum conditions. This was accomplished originally by measuring the vibrational spectrum of methane after deposition on the surface at 140 K with a translational energy of 17 kcal/mol. The crystal temperature is maintained at a low value in order to trap the nascent product of the dissociative chemisorption event rather than a species produced by thermal decomposition of the nascent product. The dissociation products are identified as an adsorbed methyl radical and adsorbed H atom [2.4, 5].

Adsorbed methyl radicals have long been invoked as reaction intermediates in a wide variety of hydrocarbon-surface reactions carried out both in an ultrahigh vacuum environment and under high-pressure conditions. Despite their importance as proposed reaction intermediates, this is the first synthesis and unambiguous identification of them on a single crystal metal surface. Adsorbed methyl radicals have not been produced previously because there is no simple way to synthesize them. Methane and ethane, natural candidates for the clean production of methyl radicals, are completely unreactive with most metal surfaces under low-pressure conditions of the adsorbing gas. However, it is now clear both why methane is unreactive at low pressures and how to activate it. The variability of the collision energy afforded by molecular beams makes them a tool with which novel adsorbates can be synthesized.

With adsorbed CH_3 species conveniently and cleanly synthesized, their stability and reactivity have been probed by monitoring the vibrational spectrum as a function of surface temperature [2.12–14]. The methyl radicals adsorbed on Ni{111} are stable below surface temperatures of 140 K. Above this temperature, the methyl radicals dissociate to form CH and the CH species recombine at 230 K to form C_2H_2, a rehybridized form of acetylene. At temperatures above 440 K, the C_2H_2 species dissociates into adsorbed carbon and hydrogen.

The ability to activate CH_4 with molecular beam techniques has allowed gas phase benzene to be synthesized from methane over a Ni{111} crystal [2.15]. The dissociation of methane is collisionally induced by impact of Kr atoms on a layer of CH_4 physisorbed on Ni{111} at 47 K. The crystal is then heated to 395 K to desorb H_2 and to form adsorbed C_2H_2. Methane is again physisorbed on the Ni surface partially covered with C_2H_2 and the Kr atom bombardment and annealing are repeated. Repetition of this procedure several times increases the C_2H_2 coverage to about 0.25 monolayer. At this coverage, heating the surface to 395 K causes the C_2H_2 to trimerize to form benzene. The presence of benzene is verified by the vibrational spectrum. Upon raising the surface temperature to 430 K, benzene desorbs. The gas phase benzene is identified as such by mass spectroscopy. The benzene remaining on the surface dissociates to form adsorbed carbon and hydrogen. This procedure represents the first synthesis of benzene from methane over a single catalyst and establishes molecular beams as a synthetic tool. The production of benzene observed here may provide an important first step for the direct formation of liquid fuels from natural gas.

Acknowledgments. This work is supported by the National Science Foundation (CHE-8508734) and the Petroleum Research Fund

References

2.1. G.A. Somorjai: *Chemistry in Two Dimensions: Surfaces* (Cornell University Press, Ithaca 1981)
2.2. P. Stoltze, J.K. Norskov: *Phys. Rev. Lett.* **55**, 2502 (1985)
2.3. S.T. Ceyer: *Ann. Rev. Phys. Chem.* **39**, 479 (1988)
2.4. S.T. Ceyer, J.D. Beckerle, M.B. Lee, S.L. Tang, Q.Y. Yang, M.A. Hines: *J. Vac. Sci Technol.* **A5**, 501 (1987)
2.5. M.B. Lee, Q.Y. Yang, S.T. Ceyer: *J. Chem. Phys.* **87**, 2724 (1987)
2.6. S.T. Ceyer: *Langmuir* **6**, 82 (1990)
2.7. T.P. Beebe, Jr., D.W. Goodman, B.D. Kay, J.T. Yates, Jr.: *J. Chem. Phys.* **87**, 2305 (1987)
2.8. J.D. Beckerle, Q.Y. Yang, A.D. Johnson, S.T. Ceyer: *J. Chem. Phys.* **86**, 7236 (1987)
2.9. J.D. Beckerle, A.D. Johnson, Q.Y. Yang, S.T. Ceyer: *J. Chem. Phys.* **91**, 5756 (1989)
2.10. J.D. Beckerle, A.D. Johnson, S.T. Ceyer: *Phys. Rev. Lett.* **62**, 685 (1989)
2.11. R.R. Hudgins, P.L. Silveston: *Catal. Rev. Sci. Eng.* **11**, 167 (1975)
2.12. S.T. Ceyer, M.B. Lee, Q.Y. Yang, J.D. Beckerle, A.D. Johnson: *Methane Conversion* ed. by D. Bibby, C. Chang, R. Howe, S. Yurchak, (Elsevier, Amsterdam 1988), p. 51
2.13. Q.Y. Yang, S.T. Ceyer: *J. Vac. Sci. Technol.* **A6**, 851 (1988)
2.14. Q.Y. Yang, A.D. Johnson, S.T. Ceyer: in preparation
2.15. Q.Y. Yang, A.D. Johnson, K.J. Maynard, S.T. Ceyer: *J. Am. Chem. Soc.* **111**, 8748 (1989)

3. Photochemistry at Adsorbate-Metal Interfaces: Intra-adsorbate Bond Breaking

J.M. White

Department of Chemistry, University of Texas,
Austin, TX 78712 USA

While photochemistry in gases, liquids and solids has a long and well-documented record [3.1], and photochemistry, including photoelectrochemistry, at semiconductor-adsorbate interfaces is well known [3.2], photochemistry at adsorbate-metal interfaces is a newly emerging and rapidly expanding area of surface chemistry. There are also a number of recent interesting experiments involving photochemistry of molecules adsorbed at insulator surfaces [3.3]. In this paper we focus on *bond cleavage* within the first monolayer of adsorbate rather than on adsorbate-metal bond cleavage or photochemistry in multilayers. The latter topic has been the subject of a number of investigations and reviews [3.4–6]. Table 3.1 lists a number of recent successful photochemical intraadsorbate bond cleavage experiments [3.7–27]. An interesting comparison is given in Table 3.2, which lists experiments where no photochemistry was observed [3.28–31].

Surface photochemistry has obvious relations to electron- [3.32], ion- [3.33], and particle- [3.34] stimulated chemistry. It is more general, in the sense of focusing on both desorbed and retained species, than DIET, desorption induced by electronic transitions, a subject of wide interest [3.35]. From a practical perspective, the possibility of sub-micron spatial control of chemistry at surfaces is of great interest, obviously so in the electronic materials community. In general, one can argue that photons, through localized resonant absorption involving specific electronic degrees of freedom, have greater potential for mode-specific chemistry than do particle excitations. While this view may be overly optimistic, since many electronic excitations, particularly in solids, are delocalized, and since the pathway to the desired products may have to compete with a number of paths with no particular mode-specificity, nevertheless, the photochemistry of electronically excited species at surfaces offers attractive conceptual possibilities. From a scientific perspective, it is a great challenge to make measurements, and to build and test models that describe these systems.

3.1 General Considerations

Defining the system and process components. For the purpose of organization, we arbitrarily divide the *system* into three parts: bulk substrate, adsorbate-substrate complex, and bulk (multilayer) adsorbate. We also arbitrarily divide the *process* into three sequential components: (1) excitation, (2) temporal evolution

Table 3.1. Systems showing photochemical intra-adsorbate bond cleavage

Adsorbate	Metal	Ref.	Comments on Observed Photochemistry
O_2	Pt{111}	[3.7]	Multiple wavelength and angle-dependent processes
	Pd{111}	[3.8]	Multiple wavelength dependent processes
	Pd{110}	[3.9]	Oxygen photochemistry found on reconstructed Pt{110}-(1 × 2)
H_2O	Pd{111}	[3.10]	Formation OH(a) and desorption of parent
CH_2CO	Pt{111}	[3.11]	Postirradiation TPD and SSIMS evidence for breaking H_2C-CO bond
$Mo(CO)_6$	Rh{100}	[3.12]	Cleavage of multiple, but not all, Mo-CO bonds
$Fe(CO)_5$	Ag{110}	[3.13]	Cleavage of Fe-CO bonds
NO_2	NO/Pd{111}	[3.14]	Formation of NO and its desorption dynamics; polarized light
CH_3Cl	Pt{111}	[3.15]	Wavelength dependent cleavage of C-Cl bond; coadsorption with CD_3Br
	Ag{111}	[3.16]	Wavelength dependent cleavage of C-Cl bond
	Ni{111}	[3.17]	C-Cl dissociation and dynamics of CH_3 desorption in multilayers
CH_3Br	Pt{111}	[3.18]	Threshold for C-Br dissociation, HREELS of CH_3(a), and effect of C
	Ag{111}	[3.19]	C-Br dissociation threshold, multilayer vs. monolayer response
	Br/Ni{111}	[3.20]	Dynamics of CH_3 desorption, particularly in multilayers
	Ru{001}	[3.21]	C-Br bond cleavage, HREELS identification of CH_3(a)
	Cu/Ru{001}	[3.21]	C-Br bond cleavage
CH_3I	Pt{111}	[3.22]	By XPS, slow C-I bond cleavage compared to C-Br in CH_3Br
	Ag{111}	[3.19]	C-I bond cleavage
CH_2I_2	Al (poly)	[3.23]	Desorption of CH_2 and C_2H_4 during laser photolysis
Cl_2CO	Ag{111}	[3.19]	Desorption of CO during photolysis, retention of all Cl
COS	Ag{111}	[3.19]	Desorption of CO, retention of S
cis-ClC_2H_2Cl	Pt{111}	[3.24]	*cis-trans* interconversion and C_2H_2(a) formation
	Pd{111}	[3.24]	*cis-trans* interconversion and C_2H_2(a) formation
trans-ClC_2H_2Cl	Pt{111}	[3.24]	*cis-trans* interconversion and C_2H_2(a) formation
	Pd{111}	[3.24]	*cis-trans* interconversion and C_2H_2(a) formation
C_2H_5Cl	Pt{111}	[3.25]	C-Cl bond cleavage, accumulation of C_2H_5(a) and Cl(a)
	Ag{111}	[3.19]	C.Cl bond cleavage threshold
C_6H_5Cl	Ag{111}	[3.26]	C-Cl bond cleavage, accumulation of adsorbed phenyl
ClC_2H_4Br	Pt{111}	[3.27]	Both C-Cl and C-Br bonds break, for λ where C_2H_5Cl does not
	Ag{111}	[3.19]	Both C-Cl and C-Br bonds break, for λ where C_2H_5Cl does not
CH_2I_2	Ag{110}	[3.28]	For monolayer coverages, no desorption accompanying laser pulses. No evidence concerning retained species

Table 3.2. No intra-adsorbate bond cleavage

Adsorbate	Metal	Ref.	Comments
H_2O	Ag{111}	[3.29]	Only unchanged molecular water desorption in TPD
NO	Pt{111}	[3.30]	Postirradiation TPD indistinguishable from unirradiated case. No N(a) or O(a) retained
C_6H_6	Ag{111}	[3.31]	Only unchanged molecular desorption in TPD and no residual C by XPS and AES
CH_3OH	Ag{111}	[3.29]	Only unchanged molecular methanol desorption in TPD

and electronic quenching of the excited state, and (3) thermal equilibration after quenching. Note that we are considering the adsorbate-substrate complex as an entity distinguishable from the bulk substrate and the bulk adsorbate. This complex is arbitrarily defined to include the first layer of metal and the first layer of the adsorbate. In many cases, the bulk adsorbate is not present by construction of the experiment and we will only need to discuss two of the three parts. We can excite one or more of the three parts of the system but for intra-adsorbate bond breaking, at least a part of the initial excitation must become localized in the adsorbate-substrate complex. An important question concerns the relative contributions to the observed bond breaking process, of two kinds of electronic excitation – excitation of the adsorbate-substrate complex and bulk substrate electronic excitation.

Excitation, temporal evolution and temperature. When a material system, at a well-defined temperature, absorbs a photon, a specific quantum transition between two states of that system is involved. The final state is, at birth, set apart from the thermal distribution of the rest of the system but, with time, the system relaxes to a new equilibrium position. In any case of bond cleavage following an electronic excitation, the nuclei involved in the bond of interest must separate, i.e., the system must convert electronic energy into specific nuclear motion. The more rapidly this can be done, the more competitive the bond dissociation becomes; thermal relaxation processes can not be so fast that they overwhelm the mode-specific steps of interest. The electronic degree of freedom initially receiving the photon energy is coupled to all the other degrees of freedom of the system – some couplings may be weak while others are strong – and it is the interplay of these couplings that will determine whether mode-specific non-thermal chemistry can be realized.

Suppose we want to break a chemical bond in this kind of system. Assuming that thermal activation, by itself, is not effective, then following excitation the bond breaking must occur on a time scale that is rapid compared to the time required to reach the new thermal equilibrium position. In the simplest such process, bond breaking occurs directly from the initially prepared electronically excited state. In more complex pathways, the initial excitation is converted, on some competitive time scale, into other modes of motion which are still highly excited, compared to the thermal distribution, and these lead to the desired bond cleavage. In this context, it is important to realize that surface photochemical processes, like ordinary thermal ones, typically involve a series of elementary steps; only one, most often the initiating step, is required to be photochemical. Thus, surface photochemistry can be temperature dependent; for example, a photochemically prepared key intermediate may react thermally in a subsequent elementary step(s) to produce the detected product. The critical factor is getting to the key intermediate in competition with all competing relaxation channels.

Optical excitation of the adsorbate-substrate complex. Photons interacting with the adsorbate-metal complex will, as in the gas phase, cause localized excitations. For submonolayer and monolayer systems, the adsorbate-substrate complex is optically thin; the optical absorption coeffcient is such that nearly all the incident flux passes through the layer either to be absorbed in or reflected from the metal. The small fraction of the total that is absorbed by the adsorbate-substrate complex produces electronically excited states which can lead to interesting chemistry. We refer to this as direct adsorbate-substrate excitation. As in the gas phase, it will be wavelength dependent and its probability will depend upon the effective absorption coefficient, a parameter which is unknown.

Since the molecules adsorbed at an interface typically are oriented, and since optical absorption is governed by the quantum mechanical matrix element $\langle \boldsymbol{\mu} \cdot \boldsymbol{E} \rangle^2$, where $\boldsymbol{\mu}$ is the transition dipole vector and $\boldsymbol{E}$ is the electric field vector, we expect to find interesting polarization effects at the surface. Examples are discussed in Sect. 3.3. The metal, itself, makes interesting optical contributions, and we now examine these.

Optical response and excitation of the metal. It is well known that metals reflect, absorb and alter the polarization of incident light [3.36]. The bulk optical properties of a metal are characterized by a wavelength-dependent complex number, which is reported in two forms – either the refractive index, $n(\omega)$ and $k(\omega)$, or the dielectric function, $\varepsilon'(\omega)$ and $\varepsilon''(\omega)$.

One key property is the penetration depth of the incident light. As for molecules this can be described in terms of the optical absorption coefficient, α, which is related to $k(\omega)$ by the relation:

$$\alpha = \frac{4\pi k}{\lambda}, \tag{3.1}$$

where λ is the wavelength of the incident light. For two typical metals, Pt and Ag, at two wavelengths spanning the range of typical interest, 230 nm (5.4 eV) to 500 nm (2.5 eV), Table 3.3 gives the characteristics. The characteristic absorption length, $1/\alpha$ (which is useful in a Beer's Law formula, $I/I_0 = \exp(-\alpha d)$, where d is the penetration length), does not change very much over this range and corresponds to 30 or 40 layers of metal.

The optical electric field at the surface of a metal is very different from that of a wave propagating in a uniform medium. At the interface between the vacuum and the metal, light is reflected and interferes, constructively and destructively,

Table 3.3. Absorption of light by Pt and Ag

Wavelength [nm]	Pt k	Pt α	Pt $1/\alpha$ [nm]	Ag k	Ag α	Ag $1/\alpha$ [nm]
230	1.75	0.0955	10.5	1.35	0.0737	13.6
500	3.9	0.0979	10.2	3.12	0.0784	12.8

with the incoming light to give an angularly dependent optical response. In the UV region, transition metals also absorb significant amounts of light; the fraction depends upon the angle of incidence and the polarization.

Any light wave can be described by two mutually orthogonal electric field components which are perpendicular to the propagation direction. For light incident on metals, these two components are referenced to the surface of the metal. For the first, p-polarization, the electric vector is perpendicular to the plane defined by the surface normal and the direction of incidence (called the plane of incidence). For the second, s-polarization, the electric vector is parallel to the plane of the metal and perpendicular to the plane of incidence.

For Pt at 5 eV, Fig. 3.1 shows five curves describing reflection and absorption of s- and p-polarized light. For s-polarized light, the electric vector is always parallel to the surface, cannot cause excitations that involve μ vectors perpendicular to the surface, and decays away monotonically as the angle of incidence increases. For p-polarized light, there is an electric field component parallel to the surface which behaves qualitatively like s-polarized light. There is also a com-

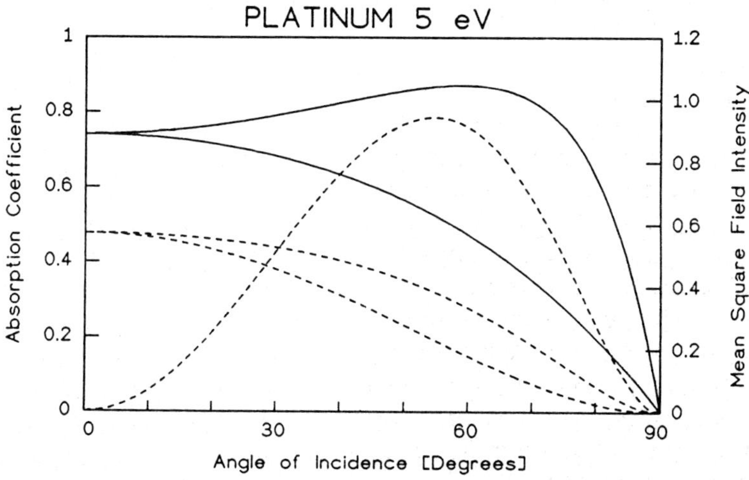

Fig. 3.1. Absorption (*solid curves*) and mean square electric field intensities (*dashed curves*) for 5 eV photons incident on Pt. The upper solid curve describes absorption of p-polarized light, the lower curve, s-polarized. The dashed curve that maximizes at 57° describes the mean square electric field perpendicular to the Pt surface when p-polarized light is incident. At 60°, the intermediate dashed curve describes the mean square electric field parallel to the surface when p-polarized light is incident and the lowest dashed curve when s-polarized light is used. For normal incidence (0°) there is no difference between s- and p-polarized light and there is no component perpendicular to the surface

ponent perpendicular to the surface, which will excite transitions with μ vectors perpendicular to the surface, and which, with increasing angle of incidence, rises to about 57° before dropping sharply.

Polarization dependence of absorption. Absorption, which will govern the excitation of metal electrons and, thereby, the production of photoelectrons (above the vacuum level) and hot electrons (below the vacuum level) also has a polarization dependence. For s-polarized light, the absorption drops with incident angle, qualitatively like the reflection of s-polarized light. For p-polarized light, which is equivalent to s-polarized light at normal incidence, the absorption rises with increasing angle of incidence out to about 60°, and then falls.

While hot electrons will always be produced, the production of photoelectrons will be determined by the work function of the metal-adsorbate system. To realize intra-adsorbate bond breaking, the energy deposited in these electrons must migrate to, and become localized in, the adsorbate-substrate complex. This is analogous to dissociative electron attachment in the gas phase [3.37]. Unlike the resulting temporary negative ions in the gas phase, these surface excitations are subject to strong attractive image charge forces and quenching by charge transfer to the metal. At high optical powers, e.g. high-power pulsed lasers, non-linear optical processes will produce photoelectrons even when the photon energy is less than the work function.

Transport of electrons to adsorbate-substrate complex. Once formed, the excited electrons will undergo inelastic scattering events with the lattice (phonons) and the surrounding electron density. The scattering lengths depend strongly on the excitation energy, measured with respect to the Fermi level. For example, according to one calculation, at 5 eV above the Fermi level the electron-electron scattering length is 5 nm, at 3 eV it is 15 nm and at 1 eV, 90 nm [3.38]. For electrons excited 1 eV above the Fermi level, the inelastic scattering length ($\approx$ 90 nm) exceeds the optical absorption length ($\approx$ 15 nm) by a large factor so that the majority of the electrons formed by both normal and grazing incidence will arrive at the surface without electron-electron scattering. On the other hand, for electrons excited near the vacuum level in Ag or Pt, say 5 eV above E_f, the scattering length ($\approx$ 5 nm) is much less than the optical absorption length ($\approx$ 10 nm). Thus, at grazing incidence the average energy of an electron arriving at the surface will be higher, just as it is in x-ray and ultraviolet photoelectron spectroscopy. This tends to be offset by the lower absorption at grazing incidence (Fig. 3.1).

For very low energy electrons, where the electron-electron scattering length is greater than 100 nm, electron-phonon scattering will become the dominant scattering mechanism. The characteristic scattering lengths remain long and have little influence on the energy distribution of hot electrons arriving at the adsorbate-substrate complex.

This discussion is limited to excited electrons; similar considerations apply to holes.

Quenching of electronically excited states. The electronically excited states, formed either by direct optical excitation or by transport of substrate-excited electrons, have finite lifetimes. These lifetimes, compared with the time required for the nuclear motion needed to break a bond, will determine the probability of bond cleavage. The excited state initially prepared in the direct excitation model, will have a valence band hole and an occupied excited state antibonding orbital. The state formed by substrate excited electrons will be different; there is no valence band hole and, while the same antibonding orbital may be occupied, its energy will be significantly higher. Thus, the quenching rates will be different for these two cases because the key energies and orbitals involved are different. For example, given a direct adsorbate-substrate interaction, quenching must involve refilling the hole. The hole lifetime is, thus, a crucial consideration in bond breaking. This process is not important for the substrate-mediated electron excitation process.

For all electronic quenching, one important consideration is the resonance, or lack of it, between the substrate and adsorbate orbitals involved. Resonance interactions allow energy and charge exchange on a much faster time scale than non-resonant ones because the latter always require multi-electron excitations. Reasonable time scales for resonant electronic relaxation processes lie in the 1 to 10 fs (10^{-15} s) range, whereas non-resonant electronic quenching is roughly 10^2 slower and in the range of 100 fs [3.5, 39].

Bond breaking times. How do these times compare to the time required to break a chemical bond? One kind of limit can be obtained by considering a molecule or atom moving with room temperature (300 K) average thermal speed. For a light fragment, say CH_3, the average thermal speed is about 780 ms^{-1} or 1.3×10^{-13} s Å^{-1} (130 fs Å^{-1}). Another estimate is realized by giving the departing fragment a translational energy of 2 eV (5 eV photon energy −3 eV C–Br bond energy); this will reduce the time scale by roughly one order of magnitude to about 10 fs Å^{-1}. Making the reasonable assumption that stretching a chemical bond by 1 Å in an adsorbate is sufficient to activate it with respect to dissociative interactions with a supporting metal substrate, we come to the conclusion that, unlike fluorescence which cannot compete with quenching because the fluorescence lifetime is on the order of 10^{-9} s (1 ns), bond dissociation on repulsive potential surfaces will be competitive with quenching, particularly when non-resonant processes control the quenching rate.

A more detailed calculation, that gives further insight, is based on the following hypothesis and questions. Following the schematic of Fig. 3.2, suppose methyl bromide is photodissociated in the presence of a quencher that does not alter the gas phase electronic potential energy curves; the quencher simply causes a Franck-Condon transition from the excited state to the ground state. How far would the C–Br bond have to lengthen so that when the molecule quenched, sufficient kinetic energy would have accumulated in the bond coordinate to allow bond breaking? How long would the excited state have to live to give this separation?

Based on analysis of the isolated molecule CH_3Br potential energy curves [3.40], the C–Br bond would have to stretch only 0.6–0.7 Å beyond its equilib-

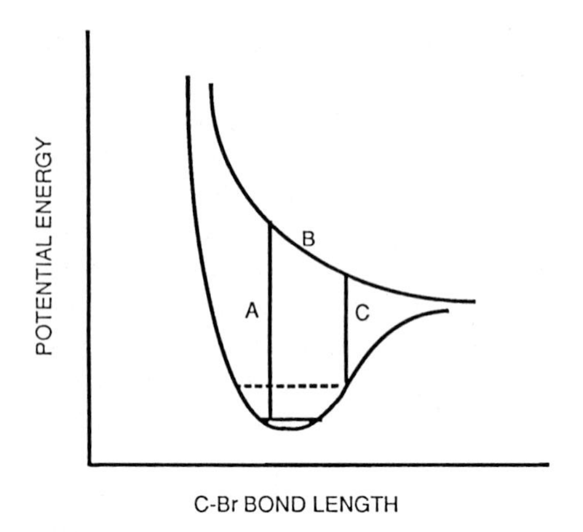

Fig. 3.2. Schematic of photodissociation of C–Br bond in gas phase CH_3Br. The following sequence of events is shown, (*A*) a Franck-Condon transition from the ground electronic and vibrational state to the repulsive excited state, (*B*) C–Br bond extension during the excited state lifetime, and (*C*) a Franck-Condon quenching transition to a vibrationally excited ground state

rium distance to gain the required energy. The excited states are very repulsive, $-3.5\,\mathrm{eV/Å^{-1}}$, in this region and only 15–20 fs are required to accomplish this movement. This rough calculation simply demonstrates that bond breaking along a simple repulsive excited state potential energy curve will compete with any quenching process that takes longer, on the average, than 10 fs. While many photochemical bond breaking processes are more complex, requiring say, the rearrangment of many atoms, or intramolecular energy transfer, or electronic state curve crossing, many others are prompt and, therefore, are excellent candidates for surface photochemistry.

Electronic-to-vibrational energy transfer. In the foregoing, we have introduced, without saying so, electronic-to-vibrational energy transfer. When the excited state quenches in Fig. 3.2, bond breaking still occurs because the C–Br bond is vibrationally excited. Except in special circumstances, quenching will always return the molecule to a vibrationally and rotationally excited region of the ground state potential energy surface. This will be the case regardless of the attractive or repulsive character of the excited state. Thermalization of the vibrationally excited state will be slower than quenching of excited electronic states, and particularly when in contact with an active metal substrate, may live long enough to undergo efficient bond cleavage. Considering the metal as a catalyst, we should be encouraged to look for photochemically driven bond breaking processes in molecules that are either not readily photodissociated in the gas phase or not readily decomposed by electron attachment.

3.2 Experimental Considerations

The instrumentation used for the study of photochemistry at adsorbate-metal interfaces falls into two classes, determined, to some extent, by the goal of the experiment. In the first class, the focus is on dynamics of the desorbing species.

This requires a pulsed UV light source (for example, an excimer laser operating at a relatively low frequency (several Hz), and a time resolved detection system (for example, a gated mass spectrometer, laser-induced fluourescence, or multiphoton ionization system). In the second class, the focus is on products retained at the surface. The light source can be a conventional Hg or Xe arc lamp which provides many more wavelengths than typical laser-based systems and is simpler to add to existing surface analysis systems. The analysis is typically done after photolysis using conventional tools of surface science such as TPD, HREELS, XPS, $\Delta\Phi$ and UPS. For both kinds of experiments, conventional ultra-high vacuum technology and procedures are required.

Characterization of the surface species before and after irradiation is certainly required before any adequate interpretation of dynamics data can be given. On the other hand, dynamics informs us of how products form. Thus, the two kinds of experiments complement each other. Taken together, they still fall short of a reasonably detailed picture of the dynamics because temporal evolution after photon absorption and before product appearance is not followed. This will require dynamical spectroscopic studies, on a femtosecond time scale, of the surface itself.

In our lab, work has thus far focused on the second class and has proven very valuable. An example of the first class is given in Sect. 3.3.3.

3.3 Examples

3.3.1 Methyl Bromide on Pt$\{111\}$ and Ag$\{111\}$

Background surface science. As indicated in Table 3.1, CH_3Br has been widely studied in our laboratory as well as in others and with qualitatively the same results. We focus here on Pt$\{111\}$ and Ag$\{111\}$ and first review the thermal chemistry. When adsorbed on either Pt$\{111\}$ [3.18] or Ag$\{111\}$ [3.19] at 100 K, CH_3Br adsorbs molecularly and reversibly; there is negligible thermal decomposition. On Pt, and likely on Ag, the vibrational spectrum, as monitored by HREELS, has frequencies that are perturbed very little from gas phase values. Specular and off-specular HREELS data, when combined with a lower work function and a low thermal desorption temperature (165 K), point to an adsorption geometry where the Br is toward the surface, the C–Br bond lies off the surface normal with a broad distribution of polar angles, and the Br–Pt interaction is very weak.

Photochemistry on Pt$\{111\}$. As indicated in Fig. 3.3, irradiation with a filtered 100 W Hg arc leads to readily distinguishable effects in post-irradiation TPD. The appearance of CH_4 is strongly wavelength dependent, even when the light source output is increased to offset the loss of high energy contributions as cut-off filters are added. The simple interpretation of the appearance of products retained by the surface is that optical excitation leads to cleavage of the C–Br bond and, with significant probability, retention of CH_3. During subsequent TPD, these

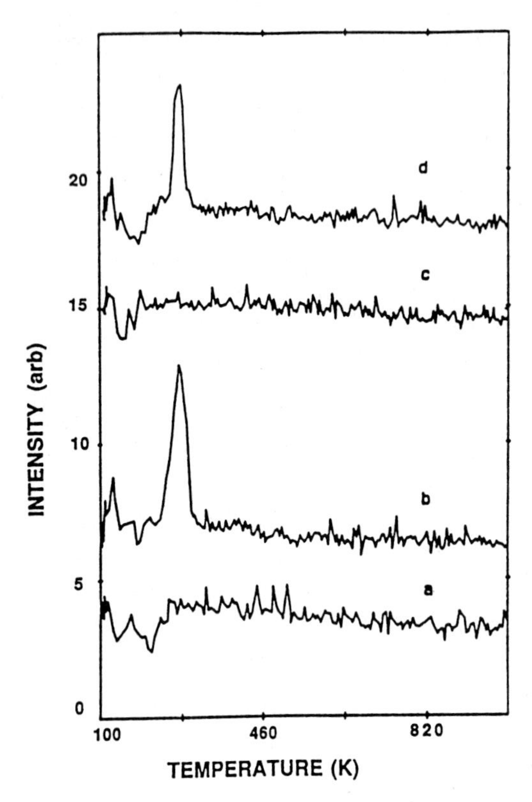

Fig. 3.3. Photochemistry of 1 monolayer of CH_3Br on Pt{111} at 100 K. Mass 16 (CH_4^+) is followed in TPD after photolysis. (*a*) No light but the monolayer held in vacuum for the photolysis time (30 min). (*b*) Full arc for 30 min. (*c*) Irradiation through a 420 nm cut-off filter. (*d*) Irradiation through a 300 nm cut-off filter. From [3.18]

methyl fragments are hydrogenated using hydrogen either accumulated from the background or from a modest decomposition of CH_3 itself. There is an interesting contrast with Ag{111}; the exclusive C- and H-containing product is ethane.

Armed with these results, we substantiated this interpretation through: (1) HREELS measurements which identify the CH_3 group bound to the Pt through a C–Pt bond, the presence of a strong Br–Pt bond, and the absence of C–Br bonds, when the parent molecules are desorbed and (2) XPS measurements which distinguish the Br in CH_3Br from that in Pt-Br on the basis of a 1 eV core level binding energy shift, and which demonstrate that some parent is thermally desorbed, not photo-desorbed.

We conclude, as anticipated based on the bond breaking time requirements outlined above, that photochemistry is competing with quenching to a thermalized ground state. The cross section at 250 nm for the loss of CH_3Br, either through desorption or dissociation, is lower, but by only a factor of two, than the gas phase optical absorption cross section (10^{-20} cm^2). However, in passing to 300 nm, the gas phase cross section drops by 10^4 [3.41] but the surface photochemical cross section drops by less than a factor of 10. The C–Br bond breaking is observed to at least 360 nm but is negligible at 420 nm.

The origin of this red-shifted optical response is an interesting problem. It can be ascribed to one or both of the following: (1) hot electrons excited above the Fermi level but below the vacuum level or (2) direct excitation from a weakly interacting (with the surface) ground state to a strongly interacting, and thereby stabilized with respect to the gas phase, excited state. Some caution must be exercised here. Because of strong interactions with the substrate, the characteristics of these excited states are not representable in terms of a perturbed electronic state of the isolated molecule.

CD_3Br *Coadsorbed with* CH_3Cl. To test for molecular specificity, the photochemistry of CH_3Cl was compared with CH_3Br; coadsorption with CD_3Br was of particular interest [3.15]. The results can be summarized as follows: (1) by itself, adsorbed CH_3Cl shows no photoactivity for wavelengths longer than 300 nm; (2) in coadsorbed monolayers, for wavelengths longer than 300 nm, only the C–Br bond is broken; and (3) for wavelengths around 250 nm, both C–Br and C–Cl bonds are broken.

Consistent with CH_3Br, the CH_3Cl photochemistry is red-shifted. Of even greater significance, these results insist that any interpretation account for molecule specificity. The latter is automatically built into any direct excitation model that rationalizes the red shift because the non-bonding electrons on the halogen atoms are more tightly held in the case of Cl. It is also possible to explain molecular specificity within the framework of a hot electron model. Within coadsorbed layers, the local rather than the average potential controls the probability that hot electrons excite a given species. Since this local potential will depend upon the adsorbate, adsorbate specific wavelength dependences are likely.

The case for CH_3I. Unlike CH_3Br and CH_3Cl, CH_3I thermally decomposes on Pt{111}, and leads to CH_4 with a TPD peak (280 K) within a few degrees of that found in the photolysis of the other methyl halides [3.42]. While confirming the interpretation of the photochemical results, this thermal chemistry complicates post irradiation TPD analysis of photolyzed CH_3I. XPS analysis, which can be done without changing the substrate temperature, circumvents this difficulty and easily distinguishes between I strongly bound to C and I strongly bound to Pt [3.22]. Compared to CH_3Br, the unfiltered arc photolysis of CH_3I is about 10-fold slower, even though the optical absorption coefficients favor CH_3I by a factor of $10^{2.5}$ [3.41]. This interesting result points out the deficiency of any direct excitation model which relies solely on red-shifted gas phase excitation spectra to predict the wavelength dependence for a series of similar molecules. Substrate excitation models also have difficulty with the fact that both CH_3Cl and CH_3I photolyze more slowly than CH_3Br. For the three methyl halides there seem to be two opposing trends, one related to excitation, the other to quenching. We expect methyl iodide to be easiest to excite and decompose but if it also is the most rapidly quenched, then it becomes possible to explain the observed reversal of behavior. In this regard, two other interesting observations should be made. First, CH_2I_2 on Ag is not very active photochemically [3.28], but CH_3I is [3.19]. While these results need to be confirmed by measurements in the same system under the same conditions, it seems clear that the same kinds of substrate and adsorbate specificity we have come to expect in thermal surface chemistry, will be even richer for excited states.

Local thermal effects. To test for local thermal effects that are not measured by the bulk temperature, we examined the photolysis of coadsorbed CH_3Cl and Xe [3.15]. With the bulk temperature at 65 K, the photochemistry is not inhibited by coadsorption of Xe and there is no detectable desorption of Xe during photolysis of CH_3Cl. We conclude that local thermal excitation can be neglected. If the excitation of the adsorbate-substrate complex involved transient local surface heating, then desorption of Xe would occur if the local temperature rose above 80 K.

Influence of a Carbon Spacer Layer. To examine the photochemistry of CH_3Br not bonded directly to the metal substrate, we studied C/Pt{111} [3.18] and learned that: (1) a spacer layer of C is sufficient to blue-shift the wavelength response of the system, (2) the blue-shift brings the photochemistry into the same region as the gas phase optical absorption of CH_3Br and (3) with C, a rough estimate of the CH_3Br losses due to photochemistry have a cross section (4×10^{-21} cm^2) at 254 nm that is lower, but not strongly so, than both the reaction cross section on C-free P{111} (6×10^{-21} cm^2) and the gas phase optical absorption cross section (10^{-20} cm^2). But see discussion of C_2H_5Cl on Ag{111} below.

Unlike C-free Pt{111}, the C-covered surface showed no distinguishable monolayer CH_3Br peak in TPD in the presence of 1 ML of carbidic C. Regardless of the exposure, the only peak was at 124 K, equal to that for a multilayer on clean Pt. HREELS data confirm that adsorbed CH_3Br is only slightly perturbed from gas phase values and that the first layer is disordered. The work function of the C-covered surface is 4.5 eV ($\Delta\Phi = 1.3$ eV), compared to 4.3 eV ($\Delta\Phi = 1.5$ eV) for 1 ML of CH_3Br on Pt{111}.

Unlike the unfiltered arc results, photolysis with a 315 nm (3.9 eV) cut-off filter leads to no detectable photochemistry. There is some loss of CH_3Br, as measured in postirradiation TPD, but this desorption is fully attributable to thermal heating of the substrate. Compared to the C-free surface, the photolysis rate here is more than 50 times slower and the cross section lies below 10^{-22} cm^2, our detection limit. In the absence of C, there is significant photochemistry for a 350 nm cut-off (3.5 eV); to get the same rate with C, requires the full arc.

Assuming for both cases, an optically thin adsorbate-substrate complex and a given cut-off filter, the metal absorption properties, and the resulting excited electron distributions, will be the same. Thus, the adsorbate-substrate complex, including the C layer, will experience a common energy distribution (with respect to the Fermi level) of excited electrons. Within the adsorbate-substrate complex, however, the local potentials experienced by an electron will likely be quite different even though the lateral average, which determines the work function, is nearly the same. If hot electrons are responsible for the photochemistry on C-free Pt{111} at 315 and 360 nm, then in the presence of C, these same electrons either cannot reach the adsorbed methyl bromide or, if they do, the excited states no longer dissociate. These results are consistent with the direct excitation model as well; in the presence of C, which decouples the excited state from the metal, CH_3Br does not experience red-shifted excitation.

CH_3Br *and* C_2H_5Cl *on* Ag{111}. The above results point to special characteristics of the first monolayer, adsorbate-metal interactions. With regard to red-shifted photochemistry, the first monolayer is distinctive for all of the halogenated molecules we have studied. To emphasize this point and to make direct comparisons between two active adsorbates, we undertook the investigation of CH_3Br and C_2H_5Cl photochemistry on Ag{111} [3.19]. Ethyl, rather than methyl, chloride was chosen because multilayers of methyl chloride cannot be prepared at 100 K. From this comparison, we learn: (1) that the chemisorbed layer is special and can be photolyzed at much longer wavelengths than the physisorbed second layer, (2) that the monolayer photochemistry is much faster than the multilayer even when both are active, (3) that the rate attributable to the presence of the second layer is greater than would be expected for direct optical absorption alone, and (4) that the excess rate is attributable to photoelectron-induced chemistry.

Some of the relevant results are summarized in Fig. 3.4 for methyl bromide and in Fig. 3.5 for ethyl chloride. The experiments were carried out by reproducibly dosing a clean Ag{111} surface with either 1 or 2 ML of the parent molecule, photolyzing the surface for 30 min., and then analyzing the surface by TPD for the loss of parent and the formation of AgBr or AgCl product. The wavelength distribution was changed using cut-off filters whose energies are

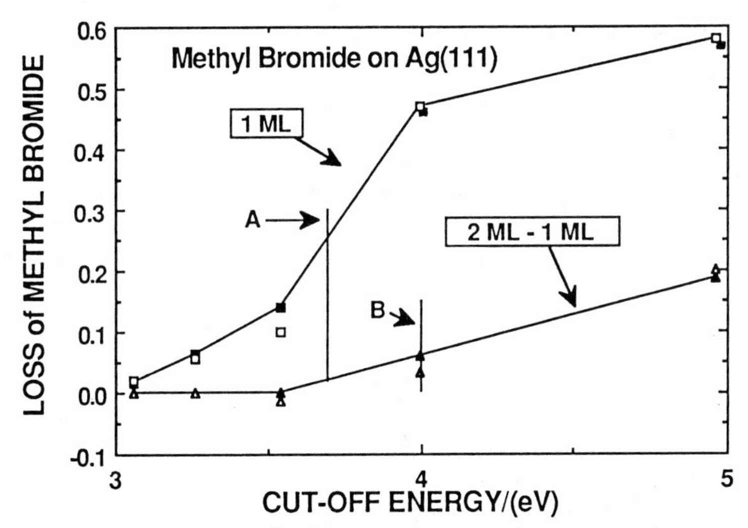

Fig. 3.4. Dependence, on Ag{111}, of the photodissociation of methyl bromide on the cut-off energy. The cut-off energy was controlled with cut-off filters. The photolysis was for thirty minutes with a focused 100 W Hg arc lamp. The 1 ML curve shows both loss of parent methyl bromide TPD peak area (*closed squares*) and gain of high temperature AgBr TPD (*open squares*). The lower curve is the difference between the results measured for 2 ML and 1 ML photolyses and is attributed to the second layer. The AgBr and methyl bromide loss peak areas are normalized at 4.9 eV for the 1 ML case. Data from [3.19]

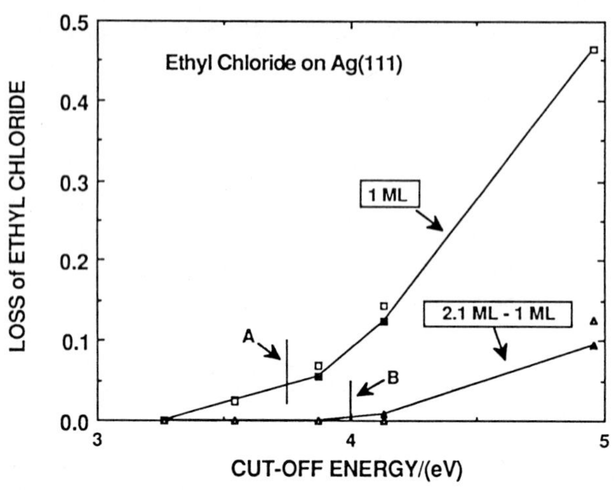

Fig. 3.5. Dependence of the photodissociation of ethyl chloride on the cut-off energy. The cut-off energy was controlled with cut-off filters. The photolysis was for thirty minutes with a focused 100 W Hg arc lamp. The 1 ML curve shows both loss of parent ethyl chloride TPD peak area (*closed squares*) and gain of high temperature AgCl TPD (*open squares*). The lower curve is the difference between the results measured for 2 ML and 1 ML photolyses and is attributed to the second layer. The AgCl and ethyl chloride loss peak areas are normalized at 4.9 eV for the 1 ML case. Data from [3.19]

used to construct the abscissa of the figures. The desorption peak temperatures for monolayer and multilayer amounts of these two molecules are distinctly different, so it is a simple matter to determine the extent to which each is photolyzed. Each of the graphs shows two curves; the loss of parent for an initial coverage of 1 ML and the *increment* for the second monolayer (2 ML–1 ML). Each of the curves has two sets of points associated with it – the TPD-determined loss of parent molecule desorption (which is calibrated) and the TPD peak area of either AgCl or AgBr (which is scaled to the loss-of-parent peak at 4.9 eV, 1 ML). The two vertical bars are the work functions, determined by UPS, for the two initial conditions.

For all conditions. The loss of parent and the gain of AgBr or AgCl have precisely the same dependence on the cut-off energy; this demonstrates the internal consistency of our procedures.

For the 1 ML *photochemistry*: (1) The cut-off energies are different for the 1 ML cases; 3 eV for CH_3Br and about 3.3 eV for C_2H_5Cl. (2) While both curves for 1 ML have the same shape, i.e. slowly increasing rates at low energies and much more rapidly increasing rates at higher energies, the break is about 0.4 eV higher for ethyl chloride. (3) Unlike the rate for C_2H_5Cl, the rate for 1 ML of CH_3Br levels off between 4 and 5 eV; the CH_3Br gas phase absorption cross section rises by 10^{+3} over the same region and, when combined with estimates of the incremental number of photons added in passing between these two cut-offs, we estimate that the rate should increase by at least a factor of 10^2.

(4) The gas phase photodissociation cross section for C_2H_5Cl, which is equal to the absorption cross section, is less than 10^{-24} cm^2 throughout the energy region of Fig. 3.5 [3.41]; there can be no measurable monolayer photochemistry due to direct excitation of a species with such a gas phase cross-section. (5) In both 1 ML systems, photochemistry extends to energies well below the work function cut-off for photoelectrons and, while the work functions are the same to within 0.05 eV (3.7 eV), the photochemistry cut-offs differ by at least 0.3 eV. (6) At 4.9 eV, the measured rate ratio is 1.25 in favor of CH_3Br whereas the gas phase ratio of optical extinction coefficients exceeds 10^4.

For the multilayer systems: (1) For both adsorbates, the incremental rate due to the second ML is a strong function of the cut-off filter energy. For C_2H_5Cl the second layer adds no photochemistry until the cut-off energy exceeds 4 eV. For CH_3Br, the second layer adds nothing until the cut-off energy is above 3.5 eV. (2) The work functions for these two multilayer systems are the same (4.0 eV) but, while the photochemistry in the second layer drops to zero at this value for C_2H_5Cl, it does not for CH_3Br. Thus, at least for methyl bromide, there is photochemistry in the second layer even when photoelectrons cannot be produced. (3) For ethyl chloride at 4.9 eV, the photolysis rate in the 2nd ML is more than 10^2 faster than can be accounted for by direct optical absorption; the extrapolated optical absorption cross section is below 10^{-24} cm^2 [3.41] and the reaction cross section is greater than 10^{-22} cm^2. The same holds for the 2nd ML of CH_3Br at 4 eV. (4) Comparing the changes in yield between 4 and 5 eV, and assigning the changes to 4.9 eV photons, the rate of photolysis for CH_3Br is only 1.5 times that of C_2H_5Cl whereas the ratio of the optical extinction coefficients exceeds 10^4. (5) Between 4 and 5 eV, the yield of the 2nd layer of methyl bromide increases by a factor of 4, whereas the optical extinction coefficient increases by a factor of 10^3.

These results lead us to the following conclusions. First, for monolayer C_2H_5Cl, and for both monolayer and multilayer CH_3Br, photochemistry occurs under conditions where there are no photoelectrons; thus, while photoelectrons may contribute for photon energies above the work function, they are not required. Second, for both adsorbates at 1 ML coverage, the photochemistry is red-shifted but by different amounts; thus, the photochemistry is molecule specific but is not characterizable using gas phase optical absorption coefficients or even a red-shifted version of them. Third, the photochemistry for ethyl chloride in the 2nd layer is still red-shifted and occurs only for wavelengths where there are photoelectrons. However, based on measurements in the CH_3Br/Pt{111} system, [3.18] we expect the number of photoelectrons to increase by about 10^4 within 0.2 eV of the threshold and then increase by less than a factor of 10 over the next 1 eV. Thus, regardless of the exact placement of the work function of the multilayer in Fig. 3.4, the yield due to the 2nd ML does not track the *number* of photoelectrons. Perhaps, the cross sections depend strongly on the energies of the hot and photo electrons. Fourth, the photochemistry of the 2nd layer of methyl bromide is also red-shifted; it cannot be accounted for on the basis of gas phase extinction coefficient variations or a 1 eV shifted version of them; neither can it be accounted for on the basis of photoelectrons. There must be some coupling to the metal.

Summarizing remarks. A few summarizing remarks are in order to close this section. First, photochemistry, as distinct from thermal chemistry is well-established; the alkyl chloride and bromide systems have no thermal chemistry under the conditions of our experiments but show strong photochemistry. Second, the photochemistry depends on the wavelength and the coverages of reactants, products and coadsorbates. There is a particular distinction between the first and second layers and, while we have not discussed it here, there is also a dependence on coverage within the first layer. Third, the photochemistry cannot be described in terms of the gas phase photochemical properties of the adsorbate; even a systematically red-shifted version of the absorption spectrum is inadequate. Fourth, while the relative contributions of direct excitation of the adsorbate-substrate complex and of indirect excitation via excited substrate electrons, remain to be established, it is clear that photoelectrons are not required. Most likely, both hot electrons and direct excitation make a contribution. It is reasonable to assume that the adsorbate-substrate complex has allowed electronic excitations in the wavelength region of interest. Since these involve a valence band hole which may lie below the valence band of the metal, they could have a fairly long lifetime, as discussed in Sect. 3.1. Even if the hole is in resonance with the valence band there is no reason to expect the quenching rate for such a state to be orders of magnitude faster than states formed by hot electron excitations. However, the adsorbate-substrate complex is optically thin; most of the absorbed photons are absorbed in the metal. Thus, even if only a small fraction of these delocalized substrate excitations become localized at the surface, they can still compete with the small number of localized excitations formed by direct excitation. Finally, regardless of the physics controlling the photochemistry, it is exciting to consider the surface chemical synthesis prospects offered by surface photochemistry. Two examples will suffice; we have prepared large coverages of ethyl fragments bound to Pt{111} by photolyzing C_2H_5Cl [3.25] and phenyl fragments on Ag{111} by the photolysis of C_6H_5Cl [3.26].

3.3.2 Oxygen on Pt{111}

Introduction to Oxygen. The alkyl halides, as well as other halogen-containing molecules listed in Table 3.1, are generally characterizable as having very weak interactions with the substrate when adsorbed at low temperatures, and no thermal dissociation in subsequent TPD. In our early work, this was very helpful in distinguishing between photo and thermal chemistry, but later it became of interest to look for systems where the adsorbate-substrate interactions were much stronger. We first found photochemistry in the CH_2CO-Pt{111} system [3.11], but with complex thermal chemistry. There was no detectable dissociation of NO on Pt{111} [3.30], consistent with pulsed laser desorption data [3.6], and no dissociation of C_6H_6 on Ag{111} [3.31]. We do find dissociation of COS on Ag{111} [3.19] and, of great interest, multifacetted photochemistry of O_2 on Pt{111} [3.7]. Very comparable results have been obtained independently for O_2–Pd{111} [3.8] and O_2–Pt{110} [3.9]. If TPD is any guide, the Pd{111} system is even richer than Pt{111}. Before describing the photochemistry of O_2 on Pt{111}, we briefly review the background surface science.

Background surface science. The adsorption of O_2 on Pt{111} has been thoroughly characterized. At 100 K, peroxo (O_2^{2-}) is formed and gives an intense O–O stretch (870 cm^{-1}) in HREELS and an increased work function (+0.8 eV) from the clean surface value (5.8 eV) [3.43, 44]. The saturation coverage of O_2 at 100 K is 0.44 ML and, upon heating, 0.34 ML of this desorbs as molecular oxygen, with no isotope exchange, at 150 K. The remainder dissociates, also near 150 K, to give 0.25 ML of O(a). The atomic oxygen, which forms a (2×2) LEED structure and is characterized by a single HREELS band at 490 cm^{-1}, recombines and desorbs quantitatively above 650 K [3.43–45]. This desorption calibrates the oxygen coverages.

Photochemistry. Figure 3.6 summarizes the HREELS data showing the development of the 475 cm^{-1} band with irradiation time using the unfiltered arc; the inset shows, on a semi-logarithmic scale, the loss of the molecular O_2 HREELS signal with time. A more detailed analysis [3.7] shows that the 650 cm^{-1} band,

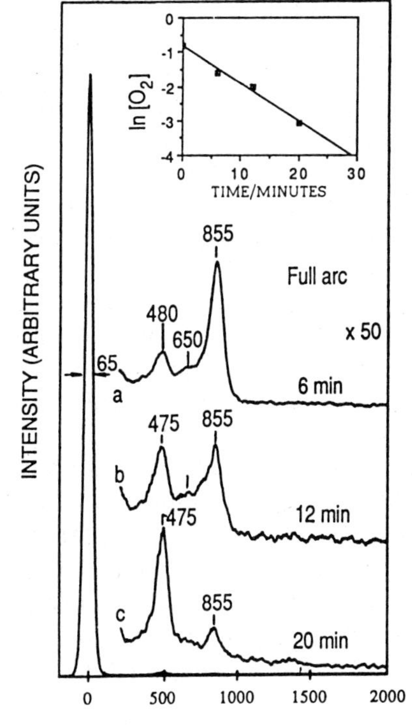

Fig. 3.6. HREELS data for the photolysis of 0.45 ML O_2 after irradiation with an unfiltered Hg arc spectrum (0.05 ± 0.01 W/cm^2) for various lengths of time. The inset shows a semilog plot of the concentration of O_2 based on the HREELS intensity at 865 cm^{-1}. From [3.7]

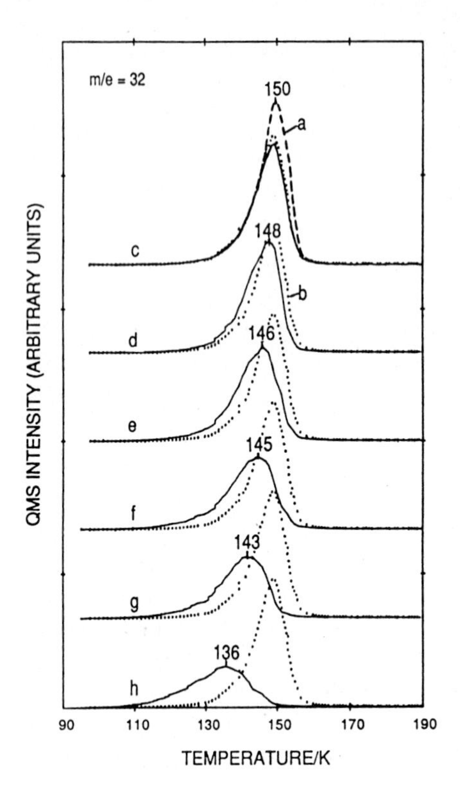

Fig. 3.7. Low temperature part of post irradiation TPD of molecular oxygen. In each case 0.45 ML of O_2 was dosed at 95 K. The curves represent the following conditions: (*a*) 0.45 ML dosed and desorbed immediately, (*b*) heat to 97 K for 6 min then desorb, (*c–h*) irradiate for 6 min with 460, 420, 350, 315, 295, and no filter, respectively. Spectrum (*b*) is reproduced with each of the other curves. From [3.7]

which we tentatively ascribe to a bridged bonded peroxo species [3.44], rises in the first 6 min but then falls and drops below the detection limits by 20 min.

Figure 3.7 shows: (1) the low temperature desorption of molecular oxygen after irradiation using various cut-off filters (solid curves), (2) the corresponding curves for no irradiation but a temperature-time course equivalent to that measured with illumination (dotted curves), and (3) the O_2 desorption with no illumination and no thermal processing (dashed curve). In each case, the energy flux to the surface was $0.07 \pm 0.01\ \mathrm{W\,cm^{-2}}$, O_2 was dosed at 95 K, and the surface temperature rose to 97 K during illumination. As the cut-off wavelength drops, more energetic photons are available, and the area beneath the solid curves and the peak temperature both decrease. This area, combined with the area under the high temperature recombination peak, was used in a difference calculation of the desorption occuring during irradiation. Separate mass spectrometer measurements confirm that O_2 is desorbed during illumination.

Figure 3.8 summarizes desorption, R_{des}, and dissociation, R_{diss}, rates versus cut-off wavelength, λ_0, where R_i is the *average* rate, measured over a 6 min photolysis interval. Desorption, monitored by the decay of the total O_2 TPD, and dissociation, monitored by the growth of the atomic oxygen HREELS signal, were measured in separate experiments. For each experiment, the initial O_2 coverage was 0.45 ML and the angle of incidence was 57° with respect to the surface normal. With no cut-off filter (230 nm cut-

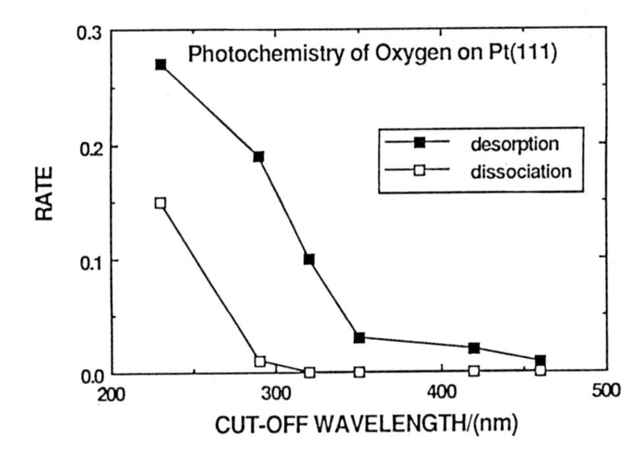

Fig. 3.8. For O_2 on Pt{111}, the variation of rate of desorption (*closed squares*) and dissociation (*open squares*) with cut-off wavelength. Data from [3.7]

off), the desorption rate ($0.27\,\mathrm{ML\,min^{-1}/(W\,cm^{-2})}$) exceeds the dissociation rate ($0.15\,\mathrm{ML\,min^{-1}/(W\,cm^{-2})}$) by about a factor of 2. As the photon energy distribution is shifted to lower energies by successively higher wavelength cut-off filters, both rates drop rapidly, but the dissociation rate, with a threshold at about 300 nm, has a much different wavelength dependence than the desorption rate, which has a threshold above 480 nm. The desorption rate appears to be the sum of two components, one controlling between 350 and 460 nm, the other dominating below 350 nm. While the qualitative aspects are reliably revealed, the quantitative values of these *average* rates must be interpreted carefully, since they depend on coverage (of reactants and products) and extent of photolysis. We only note here that the *average* cross sections for the full arc are $5.7 \times 10^{-20}\,\mathrm{cm}^2$ for dissociation and $1.2 \times 10^{-19}\,\mathrm{cm}^2$ for desorption. From work presently underway, we know that the initial cross sections are higher by as much as a factor of 3, but with different fractional increases for the two channels.

Figure 3.9 shows the variation of $R_{\mathrm{diss}}/R_{\mathrm{des}}$ with the incident angle of the unfiltered arc, measured with respect to the surface normal. This angular variation changes the orientation of the plane of polarization with respect to the surface. At normal incidence this plane is parallel to the surface, whereas at grazing incidence it is nearly normal to it. Interestingly, and as discussed in detail elsewhere [3.7], the angular dependences of dissociation and desorption are quite different; as the incident angle increases out to 75°, the dissociation rate drops while the desorption rate increases.

Possible pathways. From these results, we make the following observations: (1) Molecular oxygen, chemisorbed in the peroxo form on Pt{111} at 100 K, is photoactive; photochemistry is not overwhelmed by rapid electronic quenching processes that convert localized excitations into bulk heat. (2) Two channels, dissociation and desorption, are evident and have very different optical responses; the excitations responsible for these two paths, and possibly the two components of the O_2 desorption as well, must differ. A third path may be involved; the

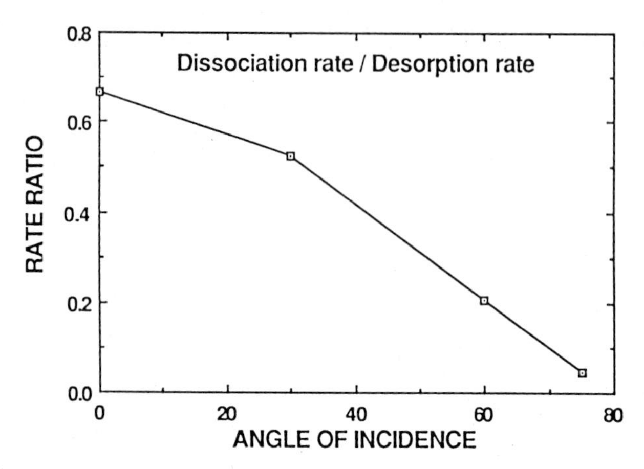

Fig. 3.9. For O_2 on Pt{111}, variation of the rate ratio, dissociation/desorption, as a function of the angle of incidence of the light. Data from [3.7]

rise and fall of the vibrational band near $650\,cm^{-1}$, Fig. 3.1, points to the photochemical conversion of peroxo oxygen into a second form of adsorbed molecular oxygen. Three very similar channels, dissociation, desorption and surface rearrangement, have been found on Pd{111} [3.8].

We are interested in how the optical excitation is linked to the observed photochemistry and these results provide some guidance. First, the chemistry must be the result of direct adsorbate-substrate excitation and/or hot electrons, and not photoelectrons, since the work function (6.6 eV) exceeds the maximum photon energy (5.3 eV).

Second, the very strong angular dependence of R_{diss}/R_{des} suggests that at least one channel is controlled by direct excitation. Assuming that the hot electron distributions are not strongly dependent upon the angle of incidence, there would be little, if any angular variation with angle of the rate ratio if they were solely responsible. But see discussion below.

Direct excitation pathways to dissociation, analogous to peroxide photochemistry [3.41], and to desorption, analogous to organometallic photochemistry [3.46], are consistent with our obvservations. The O–O bond in hydrogen peroxide dissociates upon photolysis below 300 nm and the transition is described as an excitation from the non-bonding electrons associated with O into a repulsive σ^* orbital that is antibonding with respect to the O–O bond. Fortuitously perhaps, the wavelength response of hydrogen peroxide and the peroxo species on Pt are very nearly the same and, assuming the transition dipole vector lies in the O–O bond axis, the angular dependence of the dissociation rate is also nicely explained.

A schematic of the proposed model is shown in Fig. 3.10. The focus is on the Pt valence d-band and the frontier orbitals of oxygen. UPS results indicate that the width of the Pt d-band is about 5 eV and that the $1\pi_g^*$ level is located near its bottom [3.47]. Upon adsorption, the degeneracy of $1\pi_g^*$ is lifted; the orbital

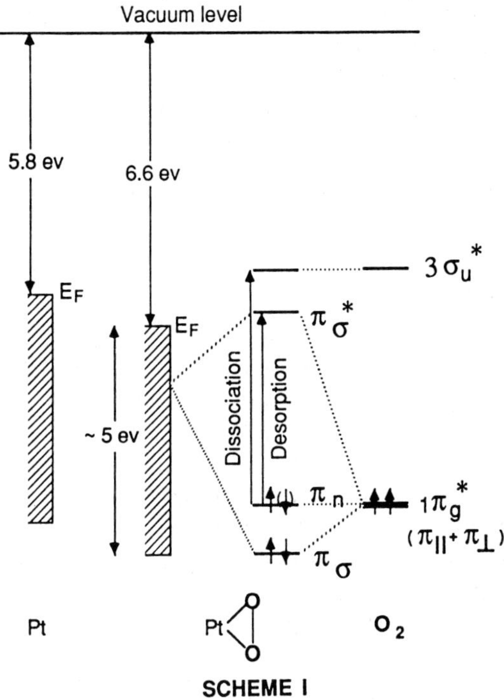

Fig. 3.10. Schematic model of the interaction of O_2 with Pt{111}, indicating possible dissociation and desorption channels. From [3.7]

parallel to the surface is largely non-bonding, π_n, with respect to Pt, while the one perpendicular to the surface interacts strongly with the d-band (π_σ: d_{xz} and d_{yz}). The coupling of the latter results in a bonding, π_σ, and an antibonding, π_σ^*, orbital (or band). The former is located mainly on oxygen and the latter mainly on Pt. The result is O_2^{2-} or O_2^- (filled or partially filled π_n orbital), as in organometallic peroxo compounds [3.48]. Direct excitation from π_n to σ_u^*, as in hydrogen peroxide, can account for the dissociation.

Turning to molecular desorption, and keeping in mind that it appears to be the sum of two components, we consider direct excitation of the Pt–O bond as one of these. In Fig. 3.10, excitation from π_n to π_σ^* would decrease the bonding between O and Pt and could lead to desorption. Interestingly, the photodissociation of $[P(C_6H_5O_2]_2PtO_2$, which generates singlet O_2, starts at 450 nm and extends to lower wavelengths [3.46]. While the transition dipole vector is not known, a component along the Pt–O bond direction, perpendicular to the surface for the O_2–Pt{111} case, is likely. This would be consistent with the observed increased desorption as the angle of incidence increases.

Hot electron tunneling into orbitals that are antibonding with respect to the O–Pt (π_σ^*) and O–O (σ_u^*) bonds could also lead to desorption and dissociation, respectively. In the diagram of Fig. 3.5, excitation of electrons in the substrate at energies between the Fermi level and the vacuum level could populate the π_σ^* and/or the σ_u^* orbitals and lead to desorption and dissociation. In this context, it is important to remember that the energetic position of these orbitals will be significantly higher in the absence of a valence band hole. For *all* of the observed chemistry to be mediated by hot electrons requires an explanation of the strong angular dependence of the rate ratio (Fig. 3.9). If the cross section for hot electron processes is very sensitive to the electron energy distribution, and if the inelastic electron scattering length is comparable to or shorter than the characteristic optical absorption depth, then increasing the angle of incidence will increase the average energy of the hot electrons that reach the surface [3.38]. We consider an explanation in terms of contributions from both direct excitation and hot electron excitation as more likely.

Regardless of the initial excitation mechanism, it is important to bear in mind that electronic quenching of the excited states can lead to a vibrationally excited ground state and, subsequently, depending on the modes excited, to dissociation and desorption. This path may be particularly relevant for O_2 /Pt{111} since there is only a small thermal barrier (150 K) to both desorption and dissociation.

Summarizing remarks. These results on Pt{111} compare very favorably with those obtained on Pd{111} [3.8] and on Pt{110} [3.9] suggesting that the photochemistry is not strongly dependent on the substrate. The presence of at least three processes, identifiable with photoexcitation, and having different wavelength and angle-of-incidence dependences, points to the participation of different excited states and different excitation channels. Experiments using s- and p-polarized excitation as a function of angle-of-incidence are underway in an effort to help determine the relative contributions of direct and indirect excitation to each of the product channels. Particularly when the molecules are oriented with respect to the surface, this procedure should provide considerable insight. An example is discussed in the next section.

3.3.3 Nitrogen Dioxide on NO-Covered Pd{111}

Recently, using the methods of molecular dynamics, the photochemistry of dimerized NO_2 on NO-covered Pd{111} has been studied [3.14]. This system, excited with pulsed excimer laser radiation (four energies between 3.5 and 6.4 eV), showed photochemical NO production and two NO desorption channels, one characterized by distinctly non-thermal state populations, the other showing thermal accomodation to the surface. Desorbing NO was detected using time-of-flight (TOF), laser-induced fluorescence (LIF) and TPD methods. For the fast channel, the yield increased strongly above 4 eV, the mean translational energy was independent of the photon energy, there was a high degree of rotational excitation, the translational and rotational energies were positively correlated, and the angular distribution was sharply peaked in the direction of the surface normal. For the slow channel, both the translational and rotational degrees of freedom were accomodated to the surface temperature and there was a dependence on sample temperature supporting the thermal desorption of physisorbed NO.

As part of this study, a method, based on polarized excitation, was devised to distinguish between direct excitation of the adsorbate-substrate complex and

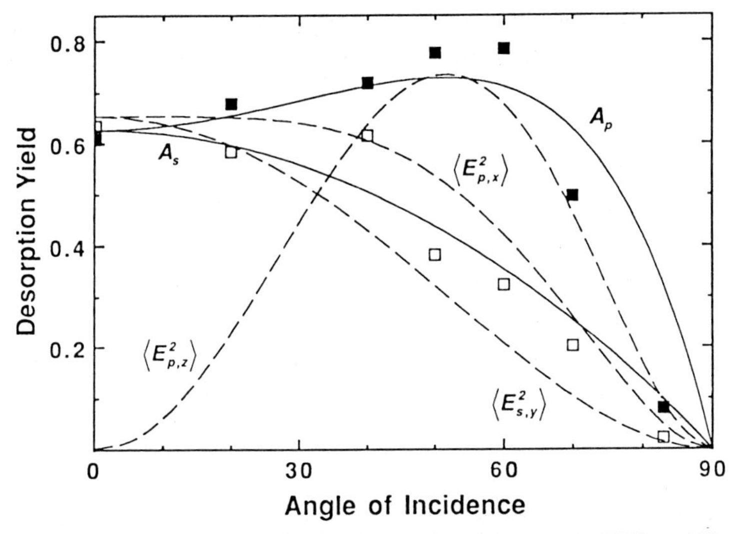

Fig. 3.11. Variation of the desorption yield of NO, during photolysis of NO_2 on NO-covered Pd{111}, with angle of incidence using polarized 6.4 eV irradiation. Rate data for s-polarized (*open squares*) and p-polarized (*closed squares*) are shown. Solid curves are absorption curves for the metal and the dashed curves are relative mean square electric strengths at the surface. From [3.14]

substrate excitation followed by hot electron-induced chemistry. Using the known bulk optical constants of Pd and the Fresnel equations, one can predict, for s- and p-polarized light, the angular variation of the absorption, A_i ($i = s, p$), and of the square of the surface electric field strength, $\langle E^2_{i,k}\rangle$ ($k = x, y, z$) (The z-direction is defined as the surface normal; y and z, together, define the plane of incidence; and x is thus perpendicular to the plane of incidence and in the plane of the surface.), as described in Sect. 3.1. If the yields are dominated by *substrate* absorption, they should follow directly the calculated optical absorption curves; if they are dominated by *direct absorption*, they should follow $\langle \boldsymbol{\mu} \cdot \boldsymbol{E}\rangle^2$. The computed curves are shown, along with the experimental NO yield data points, in Fig. 3.11. The experimental data (6.4 eV) for s-polarized light (open squares) has its maximum at normal incidence and decays monotonically as the angle of incidence increases. The experimental data for p-polarized light (filled squares) increases out to 60° and then drops sharply. Compared to the computed curves, it is clear that the data track the calculated absorption curves for s- and p-polarized light except, perhaps, at large angles. It is also possible to choose a dipole moment orientation such that the direct absorption model will qualitatively fit this data. A reasonable fit to the p-polarization data is obtained by assigning a weight of 0.33 to the perpendicular, $\langle E^2_{p,z}\rangle$, and 1.0 to the parallel, $\langle E^2_{p,x}\rangle$, components. Using the same weight on $\langle E^2_{s,y}\rangle$ as on $\langle E^2_{p,x}\rangle$ gives and adequate fit to the s-polarization data. Some systematic error is removed, and a more stringent test is realized, by comparing yield ratios, Y_p/Y_s, as a function of incident angle, with variations predicted by the direct absorption and substrate excitation models. As shown in Fig. 3.12, for both 6.4 and 5 eV, the substrate excitation model gives a

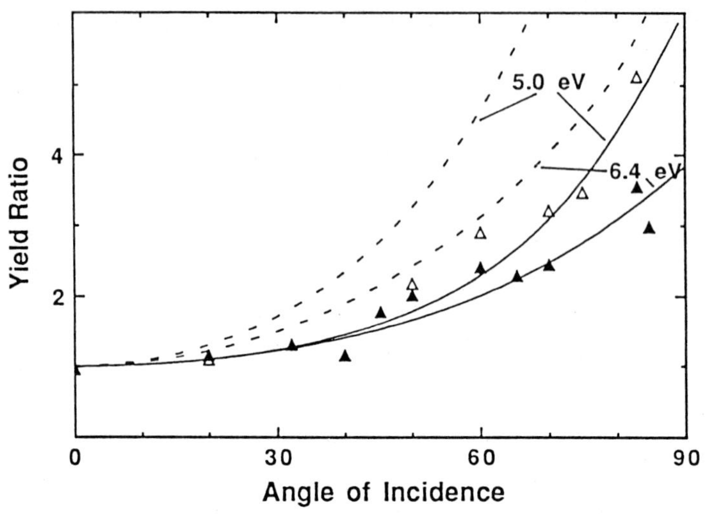

Fig. 3.12. Ratio of yields, p-polarized/s-polarized, of NO formed photochemically using 6.4 (*closed triangles*) and 5 eV (*open triangles*) irradiation. Solid and dashed curves are predictions based on absorption by the metal and absorption by the molecules at the surface, respectively. From [3.14]

very nice fit while the substrate excitation model is much too high at large angles. Thus, for this case, the substrate excitation model provides the best description.

Summarizing remarks. This study is particularly noteworthy because because it is the first to measure the polarization dependence of a surface photochemical reaction involving intra-adsorbate bond cleavage. When combined with fundamental surface science and dynamics measurements, this kind of measurement will provide a very nice description of photochemistry at adsorbate-metal interfaces.

3.4 Prospects

For each of the above examples, opportunities for further work have been noted. Taking a broader perspective, there are obvious general questions to be resolved regarding the pathways by which the photon energy is used to break intra-adsorbate bonds. In addition, the surface spectroscopy of active species and their temporal evolution is a very challenging problem requiring femtosecond time resolution. The relation between the adsorbate structure and the photochemical cleavage of intra-adsorbate and adsorbate-substrate bonds, as in the O_2–Pt$\{111\}$ example, is not yet established. The connections between electron-induced chemistry and photo-induced chemistry need to be established, particularly in those cases where substrate-mediated hot electrons are making a large contribution.

From a chemical reactions and synthesis perspective, there are exciting opportunities. Many intermediates have been proposed, but few observed, in heterogeneous catalysis because at operating temperatures an active intermediate cannot be stable enough to accumulate in large concentrations. Photochemistry of adsorbate-substrate systems offers the opportunity to prepare some of these at very low surface temperatures and then to study their reactivity. Aside from these connections with existing literature, photochemistry of adsorbate-substrate species offers the prospect of steering reactants along selected pathways that are not important in either the gas phase or in thermally driven surface chemistry.

Acknowledgements. I am indebted to numerous people for many discussions and for their good advice. The contributions from my own research group are obvious and their patience with my insistence about "shining light on it" is deeply appreciated. The stimulation and collaboration of the Campion and White groups has been rewarding. Support by the National Science Foundation and by the Army Research Office is gratefully acknowledged. The pleasant hospitality of Professor G. Ertl and his department at the Fritz-Haber-Institut in Berlin, and the support of the Alexander von Humboldt-Stiftung while this manuscript was being prepared, is greatly appreciated.

References

3.1 See for example, *Advances in Photochemistry*, ed. by J.N. Pitts, Jr., G.S. Hammond, W.A. Noyes, Jr., Vols. 1–8 (Academic, New York); M.N.R. Ashford, J.E. Baggatt: *Molecular Photodissociation Dynmaics* (Royal Society of Chemistry, London 1987)
3.2 J.R. Creighton: J. Vac. Sci. Technol. **A4**, 669 (1986);
W. Ho: Comments Cond. Mater. Phys. **13**, 293 (1988);
J.R. Swanson, C.M. Friend, Y.J. Chabal: J. Chem. Phys. **87**, 5028 (1987);
Z. Ying, W.Ho: Phys. Rev. Lett. **60**, 57 (1988)
3.3 E.B.D. Bourdon, P. Das, I. Harrison, J.C. Polanyi, J. Segner, C.D. Stanners, P.A. Young: Faraday Disc. Chem. Soc. **82**, 343 (1986);
F.L. Tabares, E.P. Marsh, G.A. Bach, J.P. Cowin: J. Chem. Phys. **86**, 738 (1987);
C.-C. Cho, J.C. Polanyi, C.D. Stanners: J. Chem. Phys. **90**, 598 (1989);
I. Harrison, J.C. Polanyi, P.A. Young: J. Chem. Phys. **89**, 1475 and 1498 (1988);
K. Domen, T.J. Chuang: J. Chem. Phys. **90**, 3318 (1989)
3.4 T.J. Chuang: Surf. Sci. Rep. **3**, 1 (1983)
3.5 P. Avouris, R.E. Walkup: Ann. Rev. Phys. Chem. **40**
3.6 D.S. King, R.R. Cavanagh: Adv. Chem. Phys. **76**, 45 (1989)
3.7 X.-Y. Zhu, S.R. Hatch, A. Campion, J.M. White: J. Chem. Phys. **91**, 5011 (1989)
3.8 X. Guo, L. Hanley, J.T. Yates, Jr.: J. Chem. Phys. **90**, 5200 (1989)
3.9 T. Matsushima: private communcation
3.10 G. Ertl: private communication
3.11 B. Roop, S.A. Costello, C.M. Greenlief, J.M. White: Chem. Phys. Lett. **143**, 38 (1988)
3.12 T.A. Germer, W. Ho: J. Chem. Phys. **89**, 562 (1989)
3.13 F.G. Celii, P.M. Whitmore, K.C. Janda: Chem. Phys. Lett. **138**, 257 (1987)
3.14 E. Hasselbrink, S. Jakubith, S. Nettesheim, M. Wolf, A. Cassuto, G. Ertl: J. Chem. Phys. **92**, 1509 (1990)
3.15 B. Roop, K.G. Lloyd, S.A. Costello, A. Campion, J.M. White: J. Chem. Phys. **91**, 5103 (1989)
3.16 X.-L. Zhou, J.M. White: Surface Sci. (in press)
3.17 E.P. Marsh, T.L. Gilton, W. Meier, M.R. Schneider, J.P. Cowin: Phys. Rev. Lett. **61**, 2725 (1988);
T.L. Gilton, C.P. Dehnbostel, J.P. Cowin: J. Chem. Phys. **91**, 1937 (1989)
3.18 S.A. Costello, B. Roop, Z.-M. Liu, J.M. White: J. Phys. Chem. **92**, 1019 (1988);
B. Roop, S.A. Costello, Z.-M. Liu, J.M. White: Springer Series in Surface Sci. **14**, 343 (1988);
S. Hatch, X.-Y. Zhu, A. Campion, J.M. White: in preparation

3.19 X.-L. Zhou, J.M. White: Surface Sci. (in press)
3.20 E.P. Marsh, M.R. Schneider, T.L. Gilton, F.L. Tabares, W. Meier, J.P. Cowin: Phys. Rev. Lett. **60**, 2551 (1988)
3.21 B. Roop, Y. Zhou, Z.-M. Liu, M.A. Henderson, K.G. Lloyd, A. Campion, J.M. White: J. Vac. Sci. Technol. **A7**, 2121 (1989)
3.22 Z.-M. Liu, S. Akhter, B. Roop, J.M. White: J. Am. Chem. Soc. **110**, 8708 (1988)
3.23 A. Modl, K. Domen, T.J. Chuang: Chem. Phys. Lett. **154**, 187 (1989)
3.24 V.H. Grassian, G.C. Pimentel: J. Chem. Phys. **88**, 4484 (1988)
3.25 K.G. Lloyd, A. Campion, J.M. White: Catal. Lett. **2**, 105 (1989;
K.G. Lloyd, B. Roop, A. Campion, J.M. White: Surface Sci. **214**, 227 (1989)
3.26 X.-L. Zhou, J.M. White: J. Chem. Phys. (in press)
3.27 S.K. Jo, J.M. White: to be published
3.28 K. Domen, T.J. Chuang: J. Chem. Phys. **90**, 3332 (1989)
3.29 S.R. Coon, L. Schoenecker, J.M. White: unpublished
3.30 X.-Y. Zhu, J.M. White: unpublished
3.31 X.-L. Zhou, J.M. White: to be published
3.32 T.E. Madey: in *Inelastic Particle-Surface Collisions*, ed. by W. Heiland, E. Taglauer, Springer Ser. in Chem. Phys., Vol. 17 (Springer, Berlin, Heidelberg 1981)
3.33 N. Winograd: in *Chemistry and Physics of Solid Surfaces V*, ed. by R. Howe, R. Vanselow, Springer Ser. in Chem. Phys., Vol. 35 (Springer, Berlin, Heidelberg 1981)
3.34 N.H. Tolk, R.G. Albridge, A.V. Barnes, R.F. Haglund, Jr., L.T. Hudson, M.H. Mendenhall, D.P. Russell, J. Sarnthein, P.M. Savundararaj, P.W. Wang: in *Desorption Induced by Electronic Transitions, DIET III*, ed. by R.H. Stulen, M.L. Knotek, Springer Ser. in Surf. sci., Vol. 13 (Springer, Berlin, Heidelberg 1988)
3.35 *Desorption Induced by Electronic Transitions, DIET I*, ed. by N. Tolk, M. Traum, J. Tully, T. Madey, Springer Ser. in Chem. Phys., Vol. 24 (Springer, Berlin 1983);
Desorption Induced by Electronic Transitions, DIET II, ed. by W. Brenig, D. Menzel, Springer Ser. in Surf. Sci., Vol. 4 (Springer, Berlin 1985);
Desorption Induced by Electronic Transitions, DIET III, ed. by R.H. Stulen, M.L. Knotek, Springer Ser. in Surf. Sci., Vol. 13 (Springer, Berlin, Heidelberg 1988)
3.36 D.M. Kolb: in *Spectroelectrochemsitry: Theory and Practice*, ed. by R.J. Gale (Plenum, New York 1988), p. 87
3.37 L.G. Christophoru (ed.): *Electron-Molecule Interactions and Their Applications*, Vol. 1 and 2 (Academic, New York 1988)
3.38 G.F. Derbenwick, D.T. Pierce, W.E. Spicer: Methods of Experimental Physics, **11**, 67 (1974)
3.39 D.R. Jennison, E.B. Stechel, A.R. Burns: Springer Ser. Surf. Sci., Vol. 13 (1988), p. 167
3.40 G.N.A. van Veen, T. Baller, A.E. deVries: Chem. Phys. **92**, 59 (1985)
3.41 H. Okabe: *Photochemistry of Small Molecules* (Wiley, New York 1978), p. 282
3.42 M.A. Henderson, G.E. Mitchell, J.M. White: Surface Sci. **184**, L325 (1987)
3.43 J.L. Gland, B.A. Sexton, G.B. Fisher: Surface Sci. **95**, 587 (1980)
3.44 H. Steiniger, S. Lehwald, H. Ibach: Surface Sci. **123**, 1 (1982);
N.R. Avery: Chem. Phys. Lett. **96**, 371 (1983)
3.45 J.L. Gland: Surface Sci. **93**, 487 (1980)
3.46 A. Volger, H. Kunkely: J. Am. Chem. Soc. **103**, 6222 (1981)
3.47 I. Panas, P. Siegbahn: Chem. Phys. Lett. **153**, 458 (1988);
W. Ranke: Surf. Sci. **209**, 57 (1989
3.48 M.H. Gubelman, A.F. Williams: in *Structure and Bonding*, Vol. 55, ed. by M.J. Clarke et al. (Springer, Berlin, Heidelberg 1983)

4. Desorption Induced by Electronic Transitions

Theodore E. Madey [1], *S.A. Joyce** [2] *and J.A. Yarmoff* [†] [2]

[1] Department of Physics and Astronomy and Laboratory for Surface Modification
Rutgers, The State University of New Jersey, Piscataway, NJ 08855-0849, USA
[2] National Institute of Standards and Technology, Gaithersburg, MD 20899, USA

Bombardment of a surface by electrons or photons can cause rupture of surface bonds and desorption from the surface, by inducing transitions to repulsive electronic states. The phenomenon of desorption induced by electronic transitions (DIET) includes both electron stimulated desorption (ESD) and photon stimulated desorption (PSD) [4.1–3]. DIET processes are of widespread importance in many areas of science and technology, including beam damage in surface analysis using x-rays or electrons, in electron and photon beam lithography, and in radiation physics of interstellar space, to name a few.

The dynamics of DIET processes have attracted a great deal of attention from experimentalists and theorists in recent years [4.1–9]. It is the purpose of this chapter to summarize briefly what is known about the mechanisms of DIET processes, and to illustrate the use of a DIET-based measurement (electron stimulated desorption ion angular distributions – ESDIAD [4.4, 8, 9]) in studies of molecular structure and dynamics at surfaces. We also discuss beam-induced damage, and the use of synchrotron radiation methods to initiate and characterize DIET.

In Sect. 4.1, we outline the mechanisms of electron and photon stimulated desorption of neutrals, positive ions and negative ions, emphasizing recent findings. Section 4.2 contains a discussion of experimental methods in ESDIAD, and Section 4.3 includes a description of positive and negative ion ESDIAD, and synchrotron radiation studies of adsorbed fluorine-containing molecules.

4.1 Mechanisms of DIET

4.1.1 Background; Desorption of Neutrals

DIET processes are generally initiated by energetic electrons or photons (typically a few eV to >1000 eV) incident on a surface containing adsorbed monolayers of atoms or molecules, or terminal bulk atoms which cause electronic excitations in

* NAS/NRC postdoctoral Research Associate
Present Adress: Sandia National Laboratory Albuquerque, NM 87185
† Present Adress: Department of Physics, University of California
Riverside, California 92521

the surface species. These excitations can result in desorption of ions, ground state neutral species, or metastable species from the surface. Recently, there has been a great deal of experimental and theoretical interest in ESD and PSD processes. Most experimental studies have relied on measurement of ESD or PSD ions, due to their relative ease of detection, although many studies of ground state and excited neutrals are now available. The principles and many applications of ESD and PSD have been summarized in several books and review articles [4.1–9].

Various models to describe desorption from covalent or ionic surfaces following valence or core electronic excitations have been formulated [4.1–12]. Although the models differ in detail, they have much in common. A fast ($\sim 10^{-16}$ s) initial electronic excitation is followed by a rapid electronic rearrangement ($\sim 10^{-15}$ s) to a repulsive electronic state having a lifetime of $\sim 10^{-14}$ s. In this state, repulsive electronic energy is converted to nuclear motion, and desorption is initiated. The initial force is generally along the ground state bond vector. However, as the desorbing species starts to leave the surface, its energy, charge state and trajectory can be influenced by reneutralization (de-excitation) processes, image force effects, etc. [4.4, 13, 14].

Much of the available information concerning excitation mechanisms in DIET comes from measurements of threshold energies for desorption processes; these have been reviewed extensively recently by *Avouris* and *Walkup* [4.5]. The threshold energies for desorption of neutral species are as low as 5 eV or less; the fundamental processes which cause neutral desorption include direct valence and shallow excitations of surface species, as well as substrate excitations ("hot" electron effects [4.15]). Desorption of excited neutrals or metastables requires higher energies [4.5].

4.1.2 Positive Ions

Positive ion desorption is initiated by valence and shallow core excitations that have thresholds of 15 eV or higher. Deep core excitations (e.g., C1s at 280 eV and O1s at 530 eV) also correlate with positive ion desorption thresholds as new desorption channels become available. A particularly useful model [4.11], in which positive ion desorption is initiated by Auger decay following core-hole ionization, was formulated by *Knotek* and *Feibelman* (KF). In this model, localized holes produced by Auger decay in the final (excited) state of a surface bond cause a repulsive Coulomb interaction that leads to desorption. The KF mechanism was originally invoked to explain O^+ desorption from maximal valency metal oxides, such as WO_3, TiO_2 etc. However, the mechanism has been generalized to covalent systems as well, and is often referred to as Auger-stimulated desorption [4.12].

Multiply charged positive ions (C^{++}, O^{++}) are seen at energies above deep-core ionization thresholds for adsorbed atoms on metal surfaces [4.7]. In recent work, *Baragiola*, *Madey* and *Lanzillotto* [4.16] report the ESD of multiply charged Si ions (Si^+, Si^{++}, Si^{+++} from SiO_2 under electron bombardment. The emission of multiply charged substrate ions is surprising, since numerous pathways for the dissipation of electronic excitations are expected to occur due to close interactions between the many neighboring atoms. The electron energy dependence of the ion yields show structure related to the Si 2p and O 1s thresholds, but are delayed in energy (see Fig.4.1). This suggests that ions are emitted as a result of an Auger process. Desorption from highly repulsive multiply charged ion states is also manifested in high energies of the ejected ions, which for O^+ reach at least 25 eV. Although the ESD yields are clearly related to core excitations, the yields are *not* directly proportional to the respective core hole ionization cross sections for Si 2p and O 1s. The authors conclude that multiple electronic excitations, i.e., excitations in shallow electronic states such as O 2s in addition to deeper core excitations, are necessary to cause desorption of multiply-charged Si. Other evidence for multiple electronic excitations in DIET is seen above the O 1s threshold in PSD of singly charged O^+ from CO adsorbed on metals [4.7, 17].

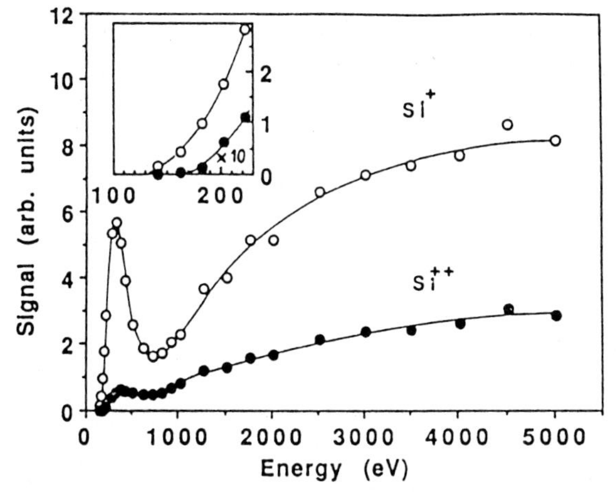

Fig. 4.1. ESD yields of singly and multiply-charged ions (Si^+, Si^{++}) as a function of electron energy during bombardment of a 65 Å thick SiO_2 film. The inset shows a magnified view of the threshold region. (From Baragiola, Madey, Lanzillotto [4.16], with permission)

4.3 Negative Ions

Desorption of negative ions from surfaces may be initiated by dissociative attachment (DA) or dipolar dissociation (DD) processes. For an adsorbed diatomic molecule AB, these processes are represented by

$$e^- + AB \rightarrow AB^{-*} \rightarrow A^- + B \qquad \text{(DA)}$$

$$e^- + AB \rightarrow AB^* \rightarrow A^- + B^+ + e^- \qquad \text{(DD)}$$

where $*$ indicates an electronically excited species. In DA, the electron is initially captured by a resonant process. If the transient anion formed in this process

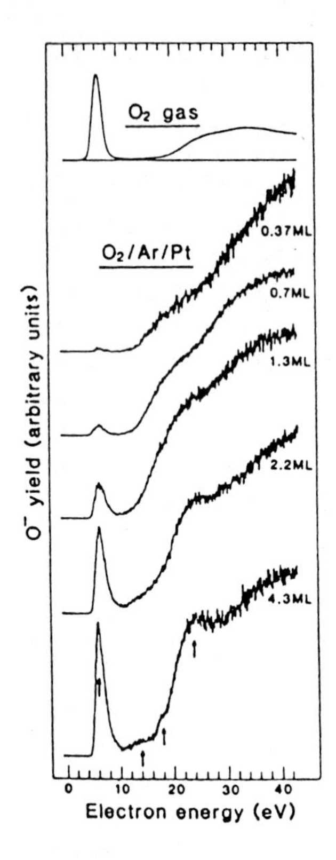

Fig. 4.2. ESD of a negative ion, O^-, as a function of electron energy. The O^- yield from O_2 on Pt is affected by isolating the O_2 molecules from the metal substrate by 1.3, 2.2 and 4.3 ML (monolayer) of an Argon spacer layer; this is shown in the lowest three curves. For all curves except the upper one, which represents the gas phase cross section for O^- production, the O_2 coverage is 0.1 ML. The curves labeled 0.37 and 0.7 ML are the O^- yields for coadsorption of these amounts of Ar with 0.1 ML O_2 (from L. Sanche [4.18], with permission)

cannot transfer energy to the substrate, it can undergo dissociation and eject a negative ion. Thresholds for ESD of negative ions via DA processes can be quite low, less than 5 eV. Thresholds for DD are higher, typically $\gtrsim$ 15 eV.

Sanche and his colleagues [4.18] have pioneered the ESD of negative ions; an example of their work is shown in Fig. 4.2, where the O^- desorption yield from O_2 on an Ar-covered Pt surface is plotted vs. electron energy [4.19]. For low Ar coverages, when the O_2 interacts directly with the Pt surface, the DA cross section is low due to quenching of the excited state by the substrate and the dominant desorption mechanism is DD. As the Ar insulating layer increases in thickness, desorption initiated by DA becomes more probable.

Dissociative attachment has recently been identified with low energy electron induced dissociation and desorption of neutral fragments, e.g., dissociation of Cl_2CO on Ag{111} at electron energies < 2 eV [4.20]!

4.1.4 Photon Stimulated Desorption; Surface Photochemistry

In discussing DIET processes, it is important to recognize that ESD and PSD are generally thought to be initiated by the same elementary excitation of the surface [4.6]. There are differences, however, in the shapes of ESD spectral yield curves (ion yield vs. photon energy) and in the magnitudes of the excitation cross sections for electron and photon excitation. The origin of the spectral shapes and the relevant excitation physics are discussed in the literature [4.6].

Historically, PSD processes have been identified mainly with ionizing radiation, i.e., photons with energies > 15–$20\,\mathrm{eV}$, such as vacuum ultraviolet or synchrotron radiation. In the past several years there has been an enormous interest in the interaction of intense low energy photons ($< 5\,\mathrm{eV}$) with surfaces (lasers, Hg arc lamps, etc.). Dissociation and desorption of molecules induced by low energy photons can occur with high cross sections [4.15, 20, 21], and have been identified with such processes as "hot electrons" excited in the substrate, and dissociative attachment. Processes induced by low energy ($\lesssim 5\,\mathrm{eV}$) photon sources are generally referred to as "surface photochemistry" (see Chap. 3 by J.M. White).

It is becoming clear in many cases that surface photochemistry represents another class of DIET processes, so that the distinction between DIET and surface photochemistry is largely one of semantics.

4.1.5 Use of ESD in Surface Structure Determination: ESDIAD

ESD has been particularly useful in surface structure determination due to the fact that ESD ions desorb in discrete cones of emission, in directions determined by the orientation of the surface bonds that are ruptured by the excitations. For example, ESD of CO bound in a "standing up" configuration on a surface will result in desorption of O^+ in a direction perpendicular to the terrace plane. ESD of H^+ from "inclined" OH, or from NH_3 bonded via the N atom, will occur in directions away from the surface normal. Measurements of the electron stimulated desorption ion angular distribution (ESDIAD) patterns can thus provide direct information about the structure of molecules oriented on surfaces [4.4, 8, 9, 22]. Angle-resolved PSD is more difficult experimentally, but provides similar information [4.23].

An important issue in ESDIAD of positive and negative ions from surfaces concerns the influence of final-state effects on the trajectories of desorbed ions. As mentioned above, the initial repulsive forces propelling ESD ions from surfaces often originate from 2 hole repulsive states which act along the original chemical bond direction [4.4, 8, 9]. However, final state effects, including image field effects, reneutralization, and coadsorbed impurity ions can influence the measured ion desorption angles by distorting the ion trajectories and thus the angular distributions. Several recent papers have described classical trajectory calculations in which the image force and reneutralization [4.13] were included. The calculations indicate that for desorption angles $\lesssim 30°$, there is little difference between the bond angle and ion desorption angle.

Dong, *Nordlander* and *Madey* [4.14] have performed classical trajectory calculations which illustrate the influence of coadsorbed impurity ions on the trajectories of ESD ions from surface molecules. The results indicate that care must be exercised in interpreting ESDIAD data: ion trajectories can be strongly influenced by adsorbed impurity atoms, particularly for fully ionized impurities, for low ion energy, and for large desorption angle (measured with respect to the surface normal).

4.2 Experimental Procedures for Positive and Negative Ion ESDIAD

Our ESDIAD experiments are performed in a stainless-steel ion-pumped ultrahigh vacuum (uhv) chamber containing LEED/ESDIAD optics (Fig. 4.3), a cylindrical mirror analyzer for Auger Electron Spectroscopy, a quadrupole mass spectrometer (QMS), an ion sputter gun, and an ancillary gas dosing system [4.4, 8, 9]. Samples are mounted on a XYZ rotary manipulator, and their temperatures are controlled from $\sim$ 100 to $\sim$ 1500 K. Surfaces are cleaned by Ar^+ ion bombardment and annealing; surface purity is monitored by AES and surface order is checked using low energy electron diffraction (LEED).

The ion angular distributions are imaged using a dual microchannel plate electron multiplier and phosphor assembly (Fig. 3) that allows detection of single ions. Ion masses are measured using the QMS. The ESDIAD data are collected and analyzed using a video data acquisition system [4.24], built around a high sensitivity video camera and a video frame grabber in a microcomputer. The

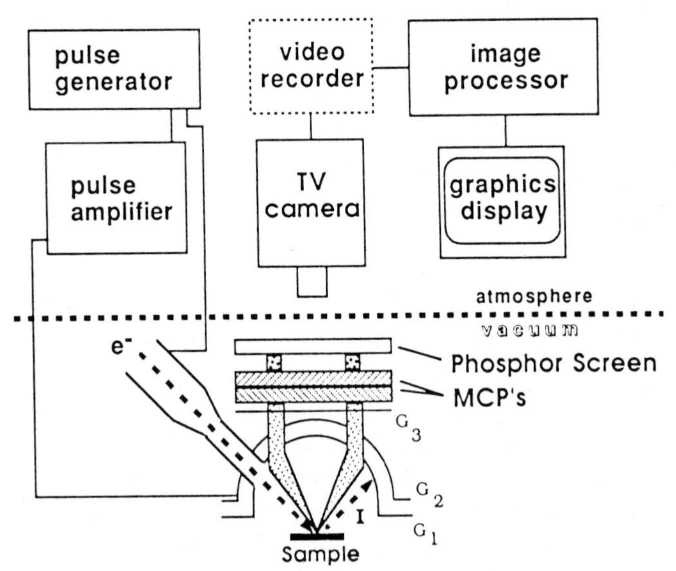

Fig. 4.3. ESDIAD apparatus used for measuring ESD ion angular distributions. The grids labeled $G_1G_2G_3$ are used to control electric fields. The ion signal is amplified using the microchannel plates (MCPs) and the secondary electrons from the MCPs are accelerated to the phosphor screen to produce light pulses. The resultant ESDIAD pattern is detected with a video camera. The images are digitized and processed using a computer system. The pulse electronics are used for time-of-flight measurements, particularly for detection of negative ions

video signal is digitized to 8 bits (256 levels) of gray scale. The images can be obtained directly from the phosphor using the video camera (in real time or off-line using video tape data storage), or from previously-obtained photographs. The digitization of the data allows it to be displayed in various forms, including line contour and perspective plots, in black-and-white or false color.

Measurements of *positive* ion ESDIAD can be made in a *dc* mode, by appropriate biasing of the grids and microchannel plate entrance to reject secondary electrons. The major technical challenge in *negative* ion ESDIAD is to separate the weak negative ion signal from the much larger (by $\sim 10^6$) secondary electron signal. This is accomplished using time-of-flight techniques [4.25], by pulsing the electron beam and "gating" a grid to suppress electrons, as indicated in Fig. 4.3. In our apparatus, the electron flight time is ~ 30 ns and the flight time for an F^- ion is ~ 900 ns.

Experiments using synchrotron radiation are performed on beamline UV-8*b* at the National Synchrotron Light Source at Brookhaven National Laboratory [4.26]. The samples are cleaned by cycles of argon ion bombardment and oxidation/annealing treatments. Annealing temperatures are measured with an optical pyrometer. Sample cleanliness is determined by the absence of carbon and oxygen signal in the valence band photoemission spectra. The sample is exposed to gases in a separate preparation chamber and transferred under UHV to the analyzer chamber. All photoemission and PSD measurements are made with the sample at or near room temperature. The combined resolution of the monochromatic light and the ellipsoidal mirror analyzer (EMA) is ~ 0.1 eV for the photoemission studies and ~ 0.4 eV for PSD.

4.3 DIET Studies of a Model System: PF_3 on Ru (0001)

4.3.1 ESDIAD of Positive and Negative Ions: Background

There are many examples in which ESDIAD has been used to identify bonding structures of adsorbed molecules (see examples in Refs. 4.4, 8, 9); many of these structures have been confirmed using other surface-sensitive methods. Examples include "standing up" CO and NO on metals, "inclined" OH and CO on surfaces, NH_3 and H_2O adsorbed on many surfaces, and adsorbed hydrocarbons. Of particular interest are the many observations of impurity-induced ordering of H_2O, NH_3 and CO observed using positive-ion ESDIAD. Studies have been made for oxide surfaces of Ti and W and for the bonding of small molecules to Si surfaces.

Recently, the first measurements of negative ion ESDIAD were reported [4.28]. These studies included ESDIAD from several fluorine-containing molecules adsorbed on Ru (0001) at 110 K, for which high yields of both F^- and F^+ ions are seen. The negative ions originate mainly from parent molecular adsorbates, while the positive ion yield is higher for adsorbed dissociation fragments. There is striking angular anisotropy observed in negative ion ESDIAD, provided structural information complementary to that derived from positive ion ESDIAD. The comparison of ESDIAD data for F^+ and F^- also provides insights into the ion formation mechanisms.

In the following paragraphs, we describe DIET studies of an interesting adsorbed molecule, PF_3 on Ru (0001), to illustrate a range of complementary phenomena: ESDIAD of positive and negative ions, as well as synchrotron radiation studies of electron- and photon-stimulated dissociation.

4.3.2 DIET of PF_3

Why study DIET of adsorbed PF_3? There are several good reasons. First, the bonding of PF_3 to Ru and other metals is well understood: PF_3 bonds to a Ru (0001) surface exclusively in a singly coordinated manner through the phosphorous atom at the Ru atop site [4.28]. There are striking similarities between the coordination chemistry of PF_3 with metal atoms and the bonding of singly coordinated CO with metal atoms. As in the case of CO on Ru (0001), PF_3 adsorbs and desorbs reversibly in uhv, without dissociation [4.29]. Second, adsorbed PF_3 is known to yield both positive and negative ions in ESD [4.25, 27, 30], so that it is a good prototype molecule for comparing positive and negative ion ESDIAD. Finally, adsorbed PF_3 is very susceptible to beam damage by electrons and photons, and several dissociation fragments are formed with interesting structural and dynamical properties.

For all PF_3 coverages of less than ~ 0.8 of saturation, the positive ion ESDIAD patterns are characterized by a "halo" of ion emission (a representative image from such a coverage is shown in Sect. 4.3.3 in Fig. 4.6a). The predominant ion desorbing is F^+ as determined both by the use of time-of-flight and a quadrupole mass spectrometer. There are no structural changes in the positive ion ESDIAD patterns for these low coverages as a function of temperature, with the "halo" remaining up to desorption of PF_3, above which there is no ion emission. For the low PF_3 coverages, there is no reproducible negative ion emission. These results are consistent with a model of PF_3 bonded via the P atom to Ru, with the F atoms inclined away from the surface. The halo is suggestive of free rotation of PF_3 about the Ru-P bond [4.29, 30].

For coverages at or near saturation at 300 K, where LEED indicates the formation of a long range $\sqrt{3} \times \sqrt{3}$ R 30° ordered overlayer, there are dramatic differences in the ESDIAD images. As shown in Fig. 4.4, there is significant azimuthal ordering of the positive ion emission and there is also significant negative ion emission. The positive ion ESDIAD pattern contains six off-normal beams whose azimuthal orientation is between atom rows of the Ru (0001) substrate. In addition, there is intense emission along the surface normal direction. The negative ion desorption pattern is azimuthally ordered as well, with off-normal emission again between the substrate atom rows. Unlike the positive

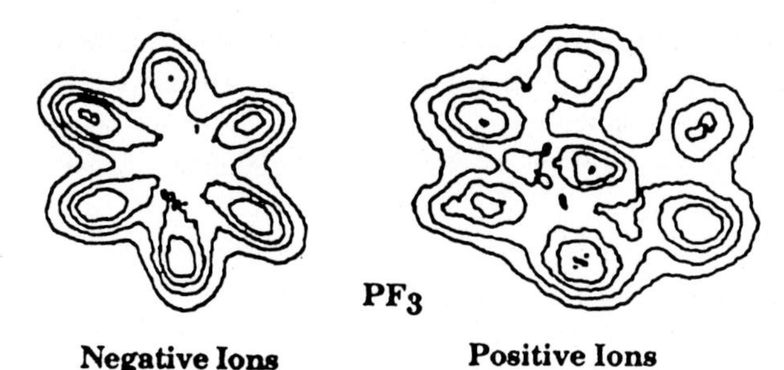

Fig. 4.4. Negative and positive ion ESDIAD contour plots for PF_3 adsorbed on Ru(0001). Images acquired at a substrate temperature of 120 K after annealing to ~ 300 K. Positive ions are F^+, negative ions are predominantly F^- [4.27]

ion pattern, there is a minimum for the desorption along the surface normal for the negative ions. TOF measurements show that the off-normal emission is exclusively F^-, while the emission along the surface normal is composed of both F^- and a species with an atomic mass of 50 ± 5, which we ascribe to PF^-. Again, as in the case of low coverages, F^+ is the only positive ion observed at saturation coverage.

The six emission lobes of the F^+ and F^- ESDIAD patterns confirm that PF_3 is azimuthally oriented on the hexagonal Ru substrate. The molecules, which appear to rotate freely about the Ru-P bond at lower coverages, are contrained azimuthally at saturation coverage in a close packed $\sqrt{3} \times \sqrt{3}$ R 30° array, as shown schematically in Fig. 4.5. Just as interlocking gears in a triangular array

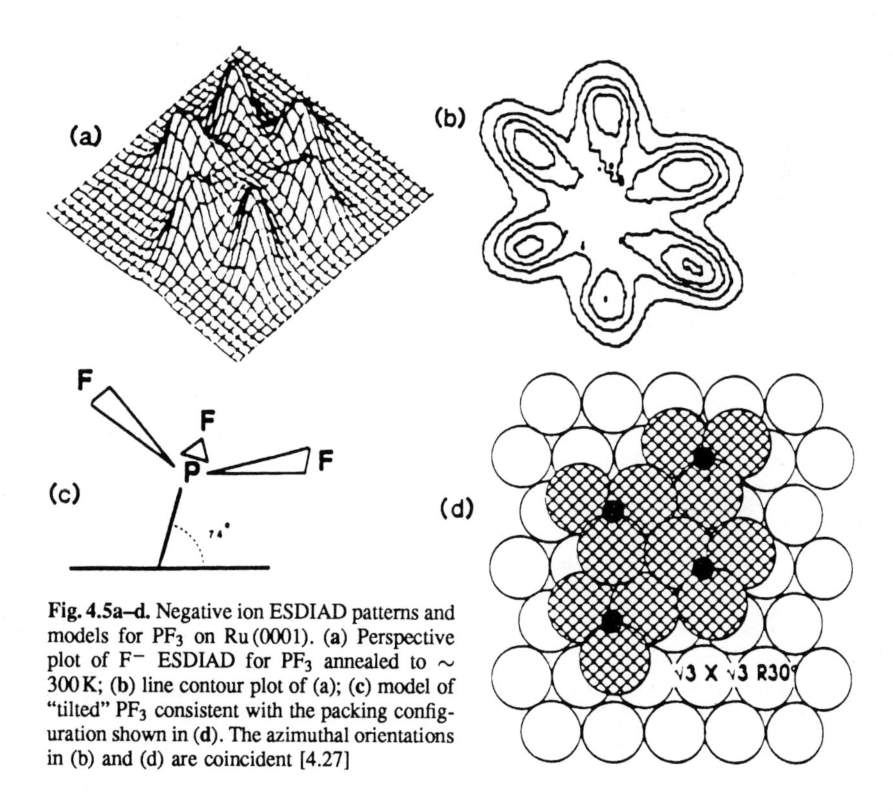

Fig. 4.5a–d. Negative ion ESDIAD patterns and models for PF_3 on Ru(0001). (**a**) Perspective plot of F^- ESDIAD for PF_3 annealed to ~ 300 K; (**b**) line contour plot of (a); (**c**) model of "tilted" PF_3 consistent with the packing configuration shown in (**d**). The azimuthal orientations in (b) and (d) are coincident [4.27]

are unable to rotate, so the PF_3 molecules are locked into the configuration shown in (Fig. 4.5d). Although the molecules cannot rotate freely, they are able to "tilt" to relieve crowding; this may result in the azimuthal elongation of the off-normal F^- beams (i.e., the beams are petal-like rather than roughly circular).

4.3.3 Electron- and Photon-Induced Damage to PF_3

The extreme sensitivity of PF_3 to energetic radiation has been observed previously in a number of studies. We observe dramatic changes in the F^+ ion angular distributions for subsaturation coverages of PF_3 as a function of electron beam exposure as shown in Fig. 4.6. The halo of F^+ emission seen for the undamaged surface is converted, after an electron beam exposure of 6×10^{-3} C/cm^2, to six azimuthally ordered off-normal beams and a beam directed along the surface normal. A comparison of the azimuthal direction of the off-normal ESDIAD beams with the LEED pattern indicates that the off-normal emission is directed between the rows of atoms on the Ru (0001) surface. Studies of the initial changes in the angle resolved intensities of the various features of the F^+ ESDIAD image (i.e., the normal emission beam, the off-normal emission between the atom rows and the off-normal emission along the atoms rows (representing the remaining halo emission)) show that the normal emission increases most rapidly, the off-normal emission increases at a rate approximately a third to a fourth more slowly, and the halo emission decreases at an even smaller rate. Figure 4.6 shows both ESDIAD data and schematic ESDIAD patterns illustrating the electron-induced changes (halo to hexagon to normal emission).

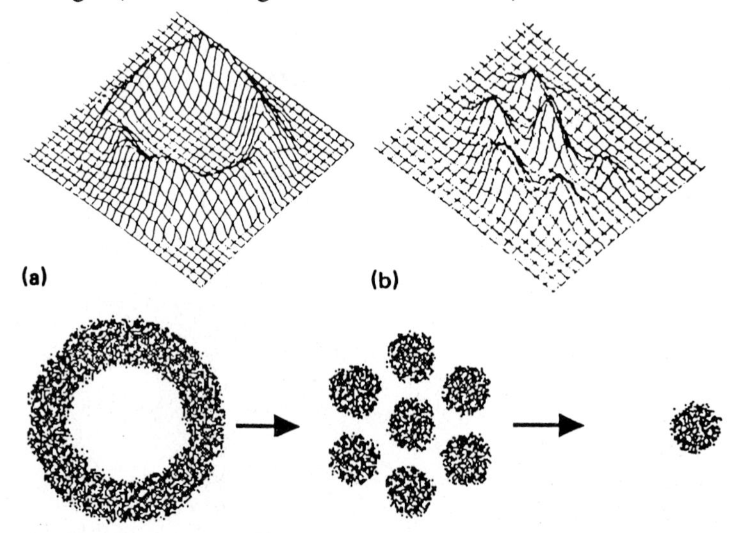

Fig. 4.6a,b. Electron beam-induced changes in F^+ ESDIAD patterns for a half-saturation coverage of PF_3 on Ru (0001). **(a)** Before extensive electron-bombardment; **(b)** after an electron exposure of 6×10^{-3} C/cm^2. The schematic patterns in the lower portion illustrate the beam-induced changes in the F^+ ESDIAD [4.25]

At saturation coverage at 300 K where both positive and negative ions are observed, beam damage experiments show no striking changes in the symmetry of the images as function of electron exposure. There are, however, differences in overall intensity changes between the positive and negative ions. For the positive ions, the overall intensity *increases* as a function of exposure to the electron beam similar to the behavior seen for subsaturation coverages. The negative ion intensity, on the other hand, *decreases* with exposure. The ESD cross section for depletion of the species giving rise to the F^- is $2.5 \times 10^{-16}\,cm^2$ [4.25].

Even more dramatic evidence for the radiation-induced dissociation of PF_3 can be seen after exposing the surface to intense synchrotron light or moderate fluxes of energetic electrons. Soft x-ray photoemission spectra illustrating this point are shown in Fig. 4.7. Spectrum (a) shows P $2p_{3/2,1/2}$ photoemission from the molecular PF_3 on an as-exposed surface. Spectrum (b) is from the same surface after direct exposure for 1 minute to the "white" synchrotron light which comes through the monochromator at zero-order. There are four well resolved components arising from the three phosphorous fluorides ($PF_x, x = 1$–3) and adsorbed P. Very similar results are obtained from a surface exposed to energetic electrons as shown in spectrum (c). The electron exposure occurred while measuring the overlayer diffraction using a conventional LEED apparatus. The primary electron energy was 100 eV, the beam current on the order of $1\,mA/cm^2$, and while the total exposure time varied across the surface, no spot was illuminated for longer than $\sim$ 1 minute. The near-equivalence of photon and electron stimulated processes has been noted previously [4.6].

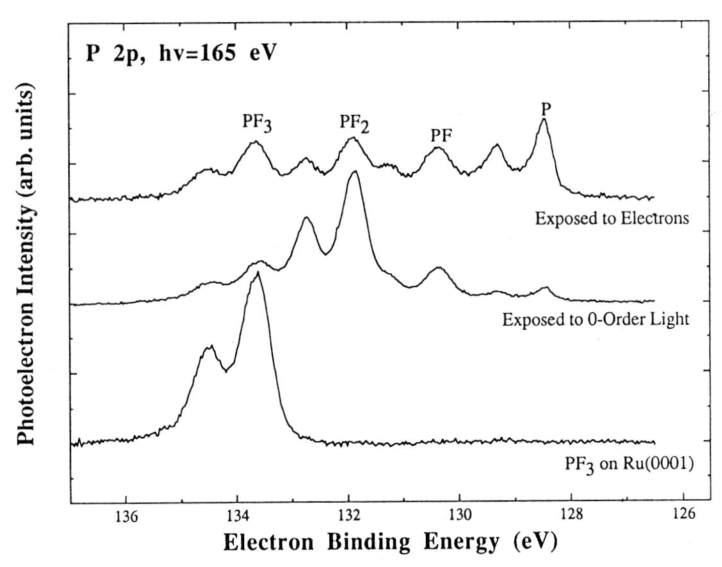

Fig. 4.7. Photoemission from phosphorous $2p$ region illustrating the effect of electron and photon induced damage to a saturation coverage of PF_2 on Ru (0001). (*Bottom*): $P2p_{3/2,1/2}$ before beam damage. (*Middle*): following exposure to zero-order synchrotron radiation. (*Top*): following electron beam exposure. See text for details

While the formation of the subfluorides by energetic radiation is clear, the fate of the fluorine which is removed cannot be determined by examination of the P 2p level spectra. The substrate Ru core levels (the 4s and 4p) showed no changes resulting from PF_3 dissociation. These levels, however, are rather broad and therefore any Ru-F bonding cannot be ruled out. Examination of the F 2s shallow core level spectra, not shown, before and after exposure to zero-order light shows a substantial decrease ($\sim$ 30%) in intensity indicating that fluorine has desorbed. As the absolute flux of photons at the surface is not known, a total desorption cross section cannot be calculated. However, this result clearly demonstrates the extreme sensitivity of PF_3 to energetic radiation which leads to the dissociation of the parent and retention of the phosphorous subfluorides on the surface.

The emergence of the subfluorides of phosphorous during irradiation as seen in the SXPS results suggests that the changes seen in the F^+ ESDIAD images are indeed due to these fragmentation products. We associate the hexagonal F^+ ESDIAD pattern (Fig. 4.6) with three domains of PF_2 species bonded to Ru via the P atom, and the normal emission pattern with PF species bonded perpendicular to the Ru surface via the P atom.

The detailed kinetics of the electron- and photon-induced fragmentation of PF_3 are currently under investigation. Recent experiments demonstrate that the PSD F^+ desorption probability near the P 2p edge is highly sensitive to the nature of the adsorbed fragment (PF_3, PF_2, PF) [4.31]. We are particularly interested in examining the electron and photon energy dependence of the dissociation processes, e.g., above and below the P 2p edge. The details of the parent-daughter relationships in the PF_3 fragmentation will also be examined.

4.4 Summary and Conclusions

We have summarized briefly some of the basic physics and chemistry of DIET, with emphasis on the use of ESDIAD in determining surface molecular structure. In particular, we have compared positive and negative ion ESDIAD in studies of PF_3 adsorbed PF_3 on the Ru (0001) surface.

We find for adsorbed PF_3 that positive and negative ion ESDIAD provide complementary information. Whereas the negative ion signal is dominated by emission from molecularly adsorbed PF_3, the F^+ ESDIAD signal also contains contributions from dissociation products (PF_2, PF).

Insights into beam damage processes in adsorbed PF_3 are provided by parallel ESDIAD and synchrotron radiation studies. Soft x-ray photoemission demonstrates clearly the chemical nature of the dissociation products, and the ESDIAD measurements provide a direct view of the bonding structure of the PF_3 and fragments (PF_2, PF).

Acknowledgements. The authors acknowledge, with pleasure, the contributions of colleagues to the research and the ideas discussed herein, especially A.L. Johnson and R.A. Baragiola. The non-synchrotron portion of this work was also supported in part by the U.S. Department of Energy, Office of Basic Energy Sciences. This research was carried out in part at the National Synchrotron Light Source, Brookhaven National Laboratory, which is supported by the U.S. Department of Energy (Division of Materials Sciences and Division of Chemical Sciences) under contract No. DE-AC02-76CH00016.

References

4.1 N.H. Tolk, M.M. Traum, J.C. Tully, T.E. Madey, (eds.): Springer Ser. in Chem. Phys. Vol. **24** (Springer, Berlin, Heidelberg 1983) p. 1
4.2 W. Brenig, D. Menzel, (eds.): Springer Ser. in Surf. Sci. Vol. **4** (Springer, Berlin, Heidelberg 1985) p. 1
4.3 R.H. Stulen, M.L. Knotek, (eds.): Springer Ser. Surf. Sci. Vol. 13 (Springer, Berlin, Heidelberg 1988) p. 1
4.4 T.E. Madey: Science **234**, 316 (1986); T.E. Madey, D.E. Ramaker, R.L. Stockbauer: Ann. Rev. Phys. Chem. **35**, 215 (1984)
4.5 Ph. Avouris, R. Walkup: Ann. Rev. Phys. Chem. **40**, 173 (1989)
4.6 M.L. Knotek: Physica Scripta **T6**, 94 (1983); Rept. Prog. Phys. **47**, 1499 (1984); Ref. [4.1] p. 139
4.7 D. Menzel: J. Vac. Sci. Technol. **20**, 538 (1982); Nucl. Instr. Meth. **B13**, 507 (1986)
4.8 T.E. Madey, A.L. Johnson, S.A. Joyce: Vacuum **38**, 579 (1988)
4.9 T.E. Madey, S.A. Joyce, A.L. Johnson: Proceedings of Spring College in Condensed Matter Physics 1988, ICTP (Trieste) (in press)
4.10 D. Menzel, R. Gomer: J. Chem. Phys. **41**, 3311 (1964); P.A. Redhead: Can. J. Phys. **42**, 886 (1964)
4.11 M.L. Knotek, P.J. Feibelman: Phys. Rev. Lett. **40**, 964 (1978); Surf. Sci. **90**, 78 (1979)
4.12 D.E. Ramaker: Springer Ser. Chem. Phys. **24**, 70 (1983); Springer Ser. Surf. Sci. Vol. **4** (Springer, Berlin, Heidelberg 1985)p. 10; J. Vac. Sci. Technol. **A1**, 1137 (1983)
4.13 Z. Miskovic, J. Vukanic, T.E. Madey: Surf. Sci. **169**, 405 (1986); Surf. Sci. **141**, 285 (1984)
4.14 C.Z. Dong, P. Nordlander, T.E. Madey: In G. Betz, P. Varga (eds.): Springer Ser. in Surf. Sci. Vol. 19 (1990)
4.15 S.A. Buntin, L.J. Richter, R.R. Cavanagh, D.S. King: Phys. Rev. Lett. **61**, 1312 (1988); E.P. Marsh, M.R. Schneider, T.L. Gilton, F.L. Tabares, W. Meier, J.P. Cowin: Phys. Rev. Lett. **60**, 2551 (1988); Phys. Rev. Lett. **61**, 2535 (1988)
4.16 R.A. Baragiola, T.E. Madey, A.-M. Lanzillotto: Phys. Rev. B **41**, 9541 (1990)
4.17 R. Jaeger, J. Stohr, R. Treichler, K. Baberschke: Phys. Rev. Lett. **47**, 1300 (1981)
4.18 For a review of negative ion electron stimulated desorption, see L. Sanche: In *DIET III*, ed. by M. Knotek, R. Stulen, Springer Ser. in Surf. Sci., Vol. **13** (Springer, Berlin, Heidelberg 1988) p. 78. Also, see E. Bauer, In Proc. of DIET I, ed. by N.H. Tolk, M.M. Traum, J.C. Tully, T.E. Madey: Springer Ser. in Chem. Phys., Vol. **24** (Springer, Berlin, Heidelberg 1988), p. 104; D. Lichtman, ibid, p. 117
4.19 H. Sambe, D.E. Ramaker, L. Parenteau, L. Sanche: Phys. Rev. Lett. **59**, 236 (1987); **59**, 505 (1987)
4.20 X.-L. Zhou, J.M. White: J. Chem. Phys. **92**, 1504 (1990)
4.21 Z. Ying, W. Ho: Phys. Rev. Lett. **60**, 57 (1988)
4.22 J.J. Czyzewski, T.E. Madey, J.T. Yates Jr.: Phys. Rev. Lett. **32**, 777 (1974)
4.23 T.E. Madey, R.L. Stockbauer, F. van der Veen, D.E. Eastman: Phys. Rev. Lett. **45**, 187 (1980)
4.24 A.L. Johnson, R. Stockbauer, D. Barak, T.E. Madey, in Ref. 3, op. cit., p. 130
4.25 S.A. Joyce, A.L. Johnson, T.E. Madey: J. Vac. Sci. Technol. **A7**, 2221 (1989)
4.26 D.E. Eastman, J.J. Donelon, N.C. Hein, F.J. Himpsel: Nucl. Instr. and Methods **172**, 327 (1980); F.J. Himpsel et al.: Nucl. Instr. and Methods **222**, 107 (1984); J.A. Yarmoff, S.A. Joyce: Phys. Rev. **B40**, 3143 (1989)
4.27 A.L. Johnson, S.A. Joyce, T.E. Madey: Phys. Rev. Lett. **61**, 2578 (1988)
4.28 F. Nitschke, J. Küppers, G. Ertl: J. Chem. Phys. **74**, 5911 (1981)

4.29 S.A. Joyce, A.L. Johnson, J.A. Yarmoff, T.E. Madey, in preparation
4.30 M.D. Alvey, J.T. Yates Jr., K.J. Uram: J. Chem. Phys. **87**, 7221 (1987); M.D. Alvey, J.T. Yates Jr.: J. Chem. Soc. **110**, 1782 (1988)
4.31 S.A. Joyce, J.A. Yarmoff, T.E. Madey, in preparation

5. Transition Metal Clusters and Isolated Atoms in Zeolite Cages

W.M.H. Sachtler

Center for Catalysis and Surface Science, Northwestern University, Evanston, Illinois, 60208, USA

Surface science research has made much use of metal single crystals. Results have been obtained of high relevance to chemisorption and catalysis, in particular with respect to the nature of the chemisorption bonds, their dipole moments and their crystal face specificity. The coverage dependence of specific parameters and the important phenomenon of surface reconstruction have also been studied quantitatively. Among the methods used, only field ion and field electron microscopy make use of highly curved surfaces which expose a variety of crystal faces of high and low Miller indices, so that the surface migration can be studied across face boundaries [5.1–3]. Other surface science methods use macroscopic single crystals and focus on *one* large crystal face, usually of rather high atom density.

The overwhelming majority of industrial metal catalysts consists of very small particles of a transition metal, often with diameters in the order of 1 nm, supported on some oxidic carrier. Such particles do not expose well developed crystal faces. The active sites consist either of single atoms or of ensembles of several atoms, usually of high coordinative unsaturation. Atomic reconstructions have low activation energies and enable the particle to respond readily to changes in the chemical environment. Rhodium particles, for instance, have been claimed to be disrupted by chemisorption of carbon monoxide [5.4], forming $Rh(I)(CO)_2$ [5.5]. In bimetallic catalysts "alloy" particles are often formed even for those systems whose macroscopic phase diagrams display wide miscibility gaps, as geometric constraints are virtually absent for very small particles; when chemical bonds are favorable, large and small metal atoms easily combine to bimetallic clusters. Such clusters are also formed with compositions for which no stable macroscopic crystal lattices exist [5.6].

Extrapolations of experimental results obtained with macroscopic single crystal faces to small, highly dynamic metal particles are, therefore, subject to severe limitations. Supported metal particles often display a wide variety of sizes and shapes unless they have been entrapped in well defined cages of a zeolite. Zeolite encaged metal particles are comparable to metal clusters formed in molecular beams, because the chemical interaction of such particles with the zeolite matrix is small in comparison to the metal-metal bond strength. First, very small "primary" particles are formed, which are mobile even at fairly low temperatures; they coalesce and form "secondary" particles with sizes and shapes that depend on the zeolite geometry. In this respect two basic types of zeolites have to be distinguished: those with linear channels and those in which "cages" can be defined, which are separated from each other by narrow "windows". Migration and repeated coalescence of metal particles is facile in linear channels, leading ultimately to large crystallites. In zeolites that have relatively large cages with smaller windows, however, growing metal particles become entrapped: once the size of the particle exceeds the diameter of the cage window, particle migration is impossible, unless the zeolite lattice is destroyed at elevated temperatures. This consequence of channel geometry on particle stability is illustrated by comparing platinum in two different zeolites: in mordenite with straight channels, heavy metal sintering is observed at 500°C, whereas platinum particles remain stable up to 800°C in faujasites [5.7], which contain 13 Å wide "supercages" with 7.5 Å windows as well as smaller "sodalite cages" and very small "hexagonal prisms". The present chapter will, therefore, focus on research done with metal loaded faujasites, in particular the synthetic zeolite NaY, the structure of which is shown in Fig. 5.1.

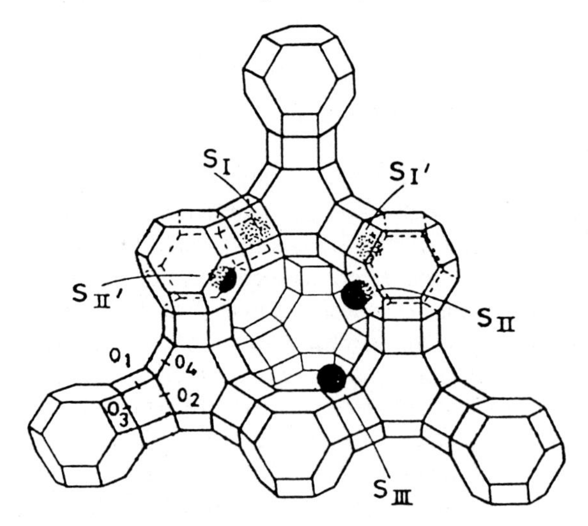

Fig. 5.1. Cages and preferred ion positions in zeolite Y

A large arsenal of experimental methods is available and has been used for the characterization of small metal clusters in zeolite cages. Excellent reviews have been published [5.8–18] on research in this field. The conclusions are based on data combining the following techniques:

1. Transmission electron microscopy of ultramicrotome cuts [5.7,19,21]
2. X-ray diffraction of high load metal/zeolites [5.22, 23]
3. Small Angle X-ray Scattering (SAXS) [5.24, 25]
4. Near Edge X-ray Absorption [5.26]
5. Extended X-Ray Absorption Fine Structure (EXAFS) [5.27, 28]
6. Radial Electron Distribution (RED) [5.29, 30]
7. ^{129}Xe NMR [5.31]
8. Electron Spin Resonance (ESR) [5.32]
9. Spin Echo Ferromagnetic Nuclear Resonance [5.33]
10. Mössbauer spectroscopy [5.34]

11. X-Ray Photoelectron Spectroscopy (XPS) [5.35–39]
12. Chemisorption of hydrogen and other probe molecules [5.40]
13. IR spectroscopy of chemisorbed CO and NO [5.41–43]
14. Temperature programmed reduction (TPR) [5.44–46]
15. Temperature programmed oxidation (TPO) [5.47]
16. Rate and selectivity of catalytic probe reactions [5.48–50]. Of these, the first 13 are static, i.e. describe the result of some process of particle generation, the last three methods are dynamic.

5.1 Preparation of Encaged Particles

Basically, two procedures are being used to obtain metal particles inside zeolite cages:

1. Decomposition of volatile complexes which are brought into the zeolite by sublimation or by deposition from solution;
2. Ion exchange, followed by calcination and reduction.

Each method has its advantages. The ion exchange method is usually preferred for elements which readily form positive ions in aqueous solution; however, the protons which are formed during reduction of the metal ions act as sites of high Brønsted acidity, which are not always desirable. For metals such as molybdenum, tungsten or rhenium, which are most ubiquitous as complexed *negative* ions in aqueous solution, ion exchange against zeolitic cations is impossible. In such cases, metal complexes (e.g. metal carbonyls) are the preferred metal precursors. If the metal atom in this complex is zerovalent, no protons are formed when the complex decomposes and metal particles are formed.

Zeolite encaged metal carbonyl complexes of transition metals, including Ni, Fe and Rh and Ir, have been studied by various authors. *Herron* et al.[5.51] prepared mixed Ni carbonyl phosphines in zeolite X by a ship-in-a-bottle technique, using $Ni(CO)_4$ and trimethyl- or triethylphosphine. *Bein* et al. [5.52] prepared mixed Ni complexes containing Ni^0 and Ni^{2+} with CO and phosphine ligands and characterized them by EXAFS, Solid State NMR and FTIR. For Rh complexes in zeolites we refer to the work of *Gelin* et al. [5.53] and *Bergeret* et al. [5.5]. *Mantovani* et al. [5.41] report transformation of $Rh_4(CO)_{12}$ into $Rh_6(CO)_{16}$. *Gelin* et al. state that $Ir_4(CO)_{12}$ is formed under much milder conditions in NaY than in solution, suggesting that zeolite cages are thermodynamically and/or kinetically favoring the formation of such complexes. *Bergeret* et al [5.54] prepared $Rh_6(CO)_{16}$ clusters by reacting $CO+H_2O$ with Rh^{3+} in NaY, and $Ir_6(CO)_{16}$ was prepared [5.55] from Ir^{3+} by reacting with $CO+H_2O$. Recently, formation of palladium carbonyl clusters has been observed in supercages of NaY[5.56]. This finding illustrates the favorable conditions for clusters inside zeolite cages, as formation of palladium carbonyl adducts outside a zeolite has only been observed in molecular beams. [5.57]

The volatile metal carbonyl $Mo(CO)_6$ was used as a precursor for zeolite encaged molybdenum particles [5.58], and rhenium has been introduced into zeolites by decomposing the carbonyl $Re_2(CO)_{10}$ either in an inert atmosphere [5.58] or in hydrogen [5.59]. $Re_2(CO)_{10}$ is strongly physisorbed on NaY, due to the high electric field inside zeolite cages. The mechanism of $Re_2(CO)_{10}$ decomposition depends on the acidity of the samples, as follows from a comparison of NaY and its acidic counterpart NaHY [5.60], where the majority of the Na^+ ions has been exchanges against protons. Decomposition of organometallic complexes in zeolites is done either thermally or by chemical interaction in a gas flow or by microwave discharge [5.61].

5.2 Thermodynamic "Driving Forces" Favoring Locations and Particle Morphology

For multivalent transition metal cations, a location in cages of high negative charge density is energetically most favorable, provided that the cage size is compatible with the size of the ion. For zeolites Y with a Si/Al ratio of 2.49, Monte Carlo calculations show that 99% of the hexagonal prisms contain two or more Al ions, i.e. their negative charge density is extremely high [5.62]. Electrostatic forces prevail in determining the relative energies of a given ion at various locations, but ligand field effects are also important, as the oxygen ions of the cage wall act as ligands to the transition metal ion. Ions such as Ni^{2+} or Cr^{3+} can be accommodated in hexagonal prisms, but the larger Pt^{2+} ions do not fit in these small cavities, they are most favorably accommodated in sodalite cages. Only very few organic molecules can enter these cages; catalysis therefore occurs in the supercages. Cation migration processes such as:

$$Pt^{2+}(\text{superc.}) + 2Na^{+}(\text{sodal.}) \longrightarrow Pt^{2+}(\text{sodal.}) + 2Na^{+}(\text{superc.}) \quad (5.1)$$

and

$$Ni^{2+}(\text{superc.}) + 2Na^{+}(\text{hexpr.}) \longrightarrow Ni^{2+}(\text{hexpr.}) + 2Na^{+}(\text{superc.}) \quad (5.2)$$

are exothermic reactions for faujasites with high Al/Si ratios. The kinetics of these cation migration processes is largely governed by steric factors. Cations that are coordinated to ligands such as NH_3 groups, easily enter the supercages by ion exchange; they are, however, unable to migrate to smaller cages without first shedding part or all of their ligands.

Reduction of encaged metal ions with hydrogen:

$$n\, M^{2+} + H_2 \longrightarrow 2H^{+} + (M^{0})_n \quad (5.3)$$

results in the formation of metal atoms or particles and protons. These become attached to $O^{=}$ ions of the cage wall, thus forming Brønsted acid sites of high acidity. The energetically most favorable location of a neutral transition metal atom is at bonding distance with other transition metal atoms. Assuming that the interaction with the zeolite framework is weak, the energy gain due to formation of multiatomic particles from isolated metal atoms will be in the order of the heat of sublimation, (90 kcal/mol for Pd, 102 kcal/mol for Ni, and 135 kcal/mol for Pt). For geometric reasons, multiatomic metal particles are only possible in supercages, voids, along dislocation lines, or at the external surface of the zeolite. It follows that the direction of the thermodynamic driving force for atom relocations is different for multivalent cations and for neutral atoms or particles of transition metals; i.e. the equilibrium:

$$X_n(\text{supercage}) \longrightarrow nX(\text{smaller cage}) \qquad (5.4)$$

is shifted to the right for X = metal cation, but to the left for X = metal atom. The thermodynamic driving force for ion reduction thus depends significantly on the atomic environment: isolated metal ions in a small zeolite cage are first reduced to isolated atoms, but ions that are already in contact with an existing metal nucleus are reduced to atoms that form part of a metal lattice. The former type of reduction can be endothermic, the latter is strongly exothermic for elements of Groups VIIIb and VIIIc. Reduction of ions in a supercage can immediately result in growth of a metal particle, but an activated migration through a narrow cage window is required before an isolated atom in a small cage can be attached to a metal particle in a larger cavity.

Due to their weak interaction with the zeolite matrix, small metal particles move freely through the channel system, unless mobility is controlled by steric factors. Metal particles can coalesce with each other in voids, or at the external surface, but in ordinary channels and cages, particle migration and growth is limited by the dimensions of the host lattice. Electron microscopic evidence of *Bergeret* and *Gallezot* [5.7, 20] and *Kleine* et al. [5.63] show that metal particles inside faujasite zeolites are roughly spherical and of dimensions comparable to those of the supercages. Radial electron distribution data are in good agreement with these electron microscopic findings, for Pt best fit has been obtained for the model of a truncated fcc tetrahedron containing 40 atoms. The hypothesis of plate-like or "raft" shapes has been categorically eliminated by the RED results.

The smallest platinum particle which would be too large to pass through the 7.5 Å window of a faujasite supercage would be an icosahedron consisting of 13 Pt atoms, one in the center and 12 equivalent "surface" atoms. This model has been proposed by *Burton*[5.64] and *Gordon* et al. [5.65] as a stable configuration of a Pt_{13} particle, but experimental evidence for its existence inside supercages is lacking, all SAXS and XRD data show that the symmetry of small zeolite entrapped metal particles is of the fcc type [5.30, 66]. A regular Pt_{13} cluster with fcc structure has also one central atom and 12 "surface" atoms, but their coordination number is lower than that of the Pt_{13} icosahedron. The shortest metal-metal distances in the smallest zeolite entrapped particles are slightly shortened with respect to the macroscopic fcc lattice. This shortening is relaxed when hydrogen is adsorbed on the particle.

Once an encaged particle is significantly larger than the window of its cage, it is "trapped" and can grow in a non-destructive way only by attachment of smaller entities to it. A state of considerable stability is achieved when all metal atoms inside a zeolite are organized in such encaged particles. Further growth requires either local collapse, recrystallization of the zeolite matrix, or detachment of metal atoms from one particle and attachment to another particle i.e. "Ostwald ripening". It appears that these phenomena occur only at rather high temperatures; careful studies by *Bergeret* et al. [5.48] show that the crystallinity of the faujasite matrix is little affected by calcination at 800 K, followed by reduction at 750 K, while the sizes of the metal particles are rather uniform.

The identification of cage filling palatinum and palladium particles of fcc structure shows that the morphology of zeolite encaged metal particles can be rationalized to certain extent by considering the geometric relation of particle and cage. Other experimental results require, however, a more detailed knowledge of the mechanisms of the processes occurring during formation of the particles. In particular, a wealth of evidence shows that the temperatures of calcination and reduction of ion exchanged samples affect the size and location of the reduced metal particles in zeolites. Careful control of these parameters can lead to predominance of any of the following metal morphologies: (1) isolated metal atoms, (2) particles which are smaller than their cage, but larger than its window, (3) cage filling particles, (4) contiguous particles filling several adjacent supercages, without, however, destroying the zeolite matrix, (5) larger particles in voids or at the external surface.

5.3 Mechanisms of Metal Particle Formation

Figure 5.2 shows schematically the parallel and sequential reactions during calcination and reduction of platinum or palladium ions, which have been introduced as ligated, e.g. tetrammine ions in the supercages of NaY or its acidic form NaHY. Mechanistic details of these processes have been unravelled by means of temperature programmed reduction (TPR), temperature programmed desorption (TPD) and extended x-ray absorption fine structure (EXAFS) [5.27]. There are significant differences between platinum and palladium, although the ions had been introduced in both cases by ion exchange as $M(NH_3)_4^{2+}$ ions. The main purpose of the calcination program is to oxidize the NH_3 groups in order to prevent autoreduction, in addition to removing water left in the zeolite after ion exchange. With platinum, a calcination temperature T_C of 360°C could be defined, where deammination is virtually complete, while Pt ions are still in the supercages. In this respect platinum is unique: for other Group VIII metals, ligands removal from the ion and ion migration to smaller cages cannot be separated as clearly. Calcination at higher temperature induces ion migration also for platinum. This, and the fact that the platinum ion is too large to enter the hexagonal prisms, provides an opportunity for studying two rather well defined situations of Pt ions in zeolite Y, viz.:

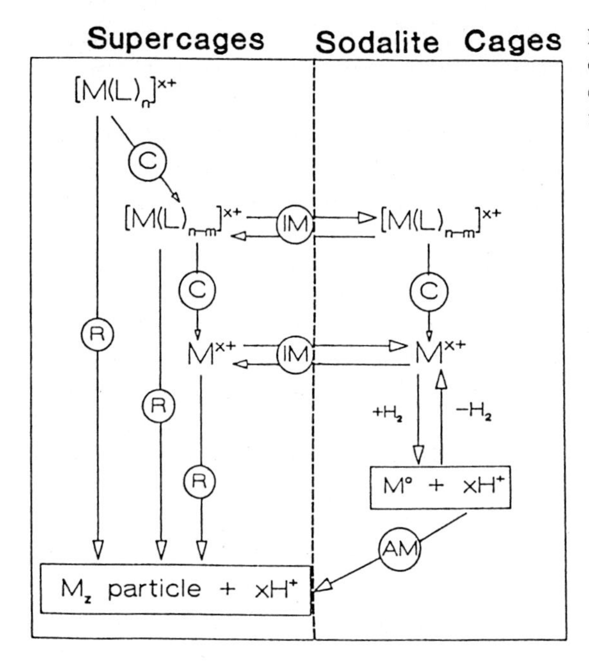

Fig. 5.2. Parallel and sequential processes during calcination and reduction of ion exchanged transition metals in zeolite Y

a) After calcination at 550°C: virtually all Pt ions are bare and located in SI' sites in sodalite cages as found by *Gallezot* et al. by means of X-ray diffraction [5.7]
b) After calcination at 360°C: virtually all Pt ions are bare and located in supercages;

For the unreduced samples, EXAFS showed marked differences of (a) and (b); the Fourier transform peak characteristic of the interaction between the absorber Pt atom and its second neighbor was much more intense for (a) than for (b). This indicates that Pt ions in SI' sites have more Si or Al ions in their second neighbor shell than Pt ions in supercages, but it is also possible that some sodalite cages contain two Pt ions. As scattering of emitted electrons due to Pt is much more intense than due to Si or Al, a small fraction of Pt dimers in sodalite cages would suffice to explain the higher peak intensity. Migration of Pt ions from supercages to sodalite cages results in higher stabilization; and thus the characteristic TPR peak shifts from ~ 20°C to ~ 250°C.

A dramatic difference between situations (a) and (b) becomes apparent after reduction: if all Pt^{2+} ions are located in supercages, very small Pt particles are formed, the coordination number (CN) of the Pt atoms, detected by EXAFS is 2.6. If the Pt^{2+} ions are chased to the sodalite cages, reduction results in large particles and a coordination number CN = 8.0 is found with EXAFS. The dispersion, measured by hydrogen adsorption, is different: H/Pt = 0.3 in case (a), but H/Pt = 1.10 in (b). The large Pt particles that are formed after reducing the samples that were calcined at 550°C, are clearly visible by electron microscopy at the external zeolite surface. Also the Pt–Pt distances, derived from the k^1

weighted EXAFS functions are clearly different: 2.65 Å for the small Pt particles in case (b), but 2.74 Å in case (a). The latter value differs only slightly from that of bulk Pt (2.76 Å).

The presence of co-exchanged multivalent cations, e.g. Fe^{2+}, which can effectively block sodalite cages and hexagonal prisms, thus forcing Pt^{2+} ions to stay in supercages, can prevent the formation of large particles on the external surface of zeolites even at high calcination temperatures. This follows from the EXAFS data for the coordination number of Pt atoms and the distance to first shell Pt neighbors: Fe^{2+} containing samples that were calcined at 550°C have the same characteristics as Fe free samples that had been calcined at 360°C [5.27].

Retention of the "naked" Pd ions in supercages of NaY after complete oxidation of the ammine ligands is not possible. Mass spectrometric analysis of oxygen consumption and nitrogen formation shows that the destruction of the third ammine ligand of the original $Pd(NH_3)_4^{2+}$ ion gives rise to an extremely sharp peak in the temperature programmed oxidation (TPO) plot; this event is accompanied by migration of the mono-ammine ion, $Pd(NH_3)^{2+}$, to the sodalite cage, where the last ammine ligand is finally destroyed [5.47].

The reduction mechanism of Pd is, of course, totally different for ions in supercages and for isolated ions in sodalite cages. When $Pd(NH_3)_x^{2+}$ ions in the channel system of the supercages are reduced, some metal nuclei are formed first, which grow by trapping migrating metal ions. Under reducing conditions each trapped ion becomes part of a growing particle. Primary particles will migrate and coalesce with each other. The growing particles ultimately bulge through cage windows into neighboring supercages, i.e. grape-shaped particles are formed, filling several adjacent supercages, as has been observed by electron microscopy [5.7]. The situation is entirely different if isolated atoms have first been formed in smaller cages. When these atoms migrate into supercages, they become not only trapped by Pd particles, but *formation of new nuclei* is frequent and a high metal dispersion results. The number of metal particles inside supercages therefore, increases with increasing reduction temperature.

Experimental evidence supporting this model has been obtained by hydrogen chemisorption measurements. The dissociative adsorption of dihydrogen requires at least two adjacent metal atoms and is impossible on an isolated metal atom. Since H_2 is able to pass through the O_6 window and reduce metal ions which are located in sodalite cages, adsorption of H_2 is a sensitive probe for metal particles containing more than one atom. The observed H/Pd^0 ratio is shown in Fig. 5.3 as a function of the reduction temperature, T_R, for the two cases of interest: For samples which have been calcined at 500°C, most Pd atoms are isolated in the sodalite cages after reduction at low temperature and thus are unable to adsorb H_2. With increasing reduction temperature an increasing number of Pd atoms escape to supercages where they form small particles capable of chemisorbing H_2. This is the evidence shown in Fig.5.3a: in this case *hydrogen adsorption initially increases* with T_R. Therefore, for very high T_C the H/Pd^0 value passes through a maximum and further decreases due to particle growth, as is usually observed.

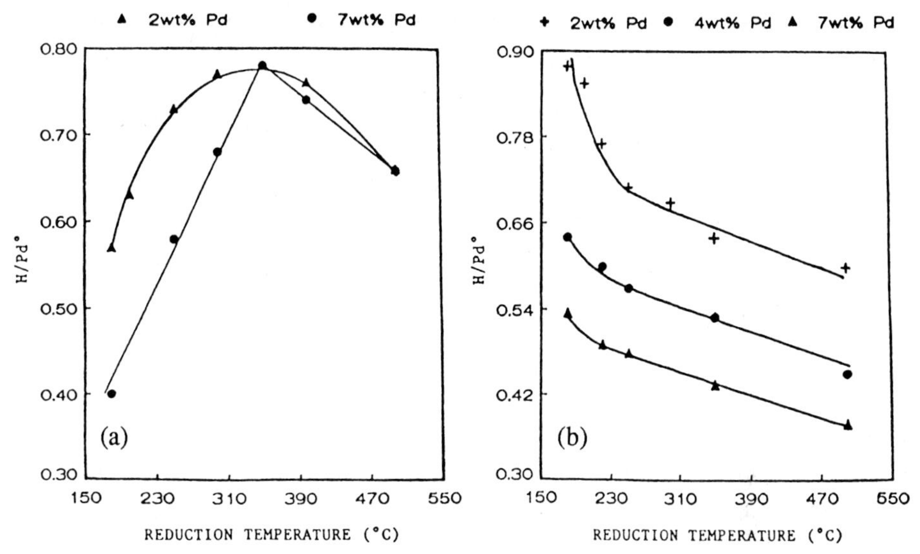

Fig. 5.3. Dispersion H/Pd^0 as a function of reduction temperature for 2 wt%, 4 wt% and 7 wt% Pd in NaY; (**a**) after calcination at 500°C; (**b**) after calcination at 250°C

Very similar results are obtained by FTIR of adsorbed CO [5.67]. The spectra show that the intensity of the CO bands in the region characteristic for multiply bonded CO strongly increases with T_R and passes through a maximum at the same value of TR where the H/Pd ratio attains its maximum value. Both experimental criteria, H/Pd and IR bands of multiply bonded CO, thus identify the presence of accessible Pd ensembles at the surface of Pd particles inside supercages.

After calcination at 250°C, however, the majority of the Pd ions are retained in the supercages. In this case, *dispersion decreases* immediately with increasing reduction temperature, since primary particles that are able to pass through 7.5 Å windows coalesce and trap more ions (see Fig. 5.3b). A semi-stable situation is attained when all ions and primary particles able to migrate through cage windows, are used up. At still higher temperatures secondary particle growth sets in, presumably by Ostwald ripening or by partial lattice collapse of the zeolite matrix. The break between these stages depends on the metal loading in the predictable manner. The differences between Figs. 5.3a and 5.3b and the concomittant IR data illustrate the potential of controlling catalyst behavior by locating ions in predetermined positions.

Since adsorption of CO is non-dissociative on Pd at room temperature, CO in the linear mode can be adsorbed on isolated metal atoms, even inside sodalite cages. Accordingly, we found that the total amount of CO adsorption decreases with reduction temperature for samples which had been calcined at 500°C, or at 250°C.

The differences between the dispersion characteristics of Pt and Pd, resulting from the different activation energies for migration through O_6 windows are remarkable: the highest dispersion of Pt is obtained by retaining the ions inside supercages, whereas with Pd, high dispersion is achieved after reduction in sodalite cages and initiating a "shower" of such atoms into supercages, where additional nuclei are formed.

5.4 Identification of Isolated Atoms and Electron Deficient Particles

Isolated metal atoms in zeolites have been reported for Pt [5.12, 24, 68] and Pd [5.12, 22, 69]. Experimental data for the existence of isolated metal atoms include: (1) re-oxidation by protons of Pt atoms to Pt^{2+} ions and gaseous H_2, and the absence of this phenomenon when all Pt atoms are located in supercages [5.27]; (2) inability of such metal atoms to dissociatively chemisorb H_2 or adsorb CO in the bridging mode; (3) inaccessibility of these atoms to larger molecules in catalytic processes, such as hydrogenation of benzene [5.48]. These data do not discriminate between models which differ only with respect to the location of the protons which are formed during metal reduction. It is, therefore, still possible to speculate that some protons become attached to cage walls, while others remain bonded to metal atoms. For the reduction of Pd^{2+} the following equations can be written:

$$\mathrm{Pd}^{2+}_{\mathrm{sodal}} + \mathrm{H}_2 + 2\mathrm{O}^{2-}_{\mathrm{wall}} \longrightarrow Pd^{0}_{\mathrm{sodal}} + 2\mathrm{H} - \mathrm{O}^{-}_{\mathrm{wall}} \tag{5.5a}$$

$$\mathrm{Pd}^{2+}_{\mathrm{sodal}} + \mathrm{H}_2 + \mathrm{O}^{2-}_{\mathrm{wall}} \longrightarrow [\mathrm{Pd} - \mathrm{H}]^{+}_{\mathrm{sodal}} + \mathrm{H} - \mathrm{O}^{-}_{\mathrm{wall}} \,. \tag{5.5b}$$

An identification of $[Pd\text{-}H]^+$ should be possible by using IR spectroscopy of adsorbed carbon monoxide. It is known that the vibrational frequency of CO is higher on positively charged sites than on neutral sites. It has indeed been found that on mildly reduced Pd/NaY, CO adsorption causes a band at 2120 cm^{-1} ascribed to CO adsorbed on a $[Pd\text{-}H]^+$ entity[5.70]. If this assignment is correct, the $[Pd\text{-}H]^+$ would be a representation of the "isolated, electron-deficient" Pd atom which has often been postulated in the literature. The attachment of the CO molecule to this species will then take place through the O_6 ring between supercage and sodalite cage. The concentration of $[Pd\text{-}H]^+$ or $[Pt\text{-}H]^+$ should depend on the concentration of protons in the zeolite. This agrees with data reported by *Tri* et al. [5.71], who report for CO adsorbed on Pt/NaHY, that the stretching frequency shifts from $2090\,cm^{-1}$ to $2068\,cm^{-1}$, if the Pt/NaHY sample is neutralized with NaOH. Migration of the $[Pd\text{-}H]^+$ complex through the O_6 ring into a supercage and release of the proton will enable the Pd atom to form new nuclei with other atoms or become attached to a growing Pd article.

Electron deficiency of Pd or Pt in zeolites is, however, not limited to isolated atoms. There is substantial evidence for the existence of electron deficient multiatomic particles of these elements in faujasite supercages. For instance, the XPS data of *Foger* and *Anderson* [5.72], show a clear shift in bond energy for platinum in Pt/NaY and Pt/LaY, indicating a positive charge on the metal particle. Such electron deficient metal particles have been found to display extraordinary catalytic activities. *Dalla Betta* and *Boudart* [5.73] found that the rate of neopentane hydrogenolysis on 1 nm Pt aggregates in Y zeolites is 40 times larger than on Pt supported on conventional supports. Recently, we tested reduced Pd/NaY catalysts for this reaction and found that the catalytic activity per exposed and accessible Pd atom was almost two orders of magnitude higher than that of Pd/SiO_2 catalysts tested in the same apparatus under the same conditions. Neopentane conversion is exceptionally appropriate for determining the intrinsic activity of metal/zeolite catalysts as this reaction, in contrast to most other hydrodcarbon conversions, does not cause catalyst deactivation, because the neopentane molecule is not capable of forming olefins or carbenium ions. Markedly enhanced catalytic activity of positively charged palladium particles in neopentane conversion [5.74] and hydrogenation of carbon monoxide to methanol [5.75] has been reported also for Al_2O_3 supported metals, where independent data (IR band of $Pd^{\delta+}$-CO, and abstraction of Pd^{n+} ions with acetyl acetone) prove that the metal particle carries a positive charge. *Gallezot* et al. [5.76] showed that the high resistance of Pt/NaY to sulphur is due to electron deficiency. They correlated empirical parameters such as the ratio of the adsorption coefficient of benzene and toluene to electron deficiency and observed that the positive charge on the metal increases with the acidity of the zeolite. Interestingly, *Bezoukhanova* et al. [5.77] conclude from IR data of adsorbed CO and NO that Pt in basic zeolites carries a negative charge. The phenomenon of "electron deficiency", responsible for extraordinary high catalytic activity and high resistance against sulphur, is of fundamental importance to the catalysis of metals [5.78], in particular in zeolites.

In the case of amorphous supports, such as alumina, it has often been presumed that Lewis acid sites act as electron acceptors, however, in the case of zeolite supported metals, the nature of electron acceptor sites is not obvious. One model assumes that electron deficient particles are actually adducts of protons and metal clusters $[Pd_n\text{-}H_x]^{x+}$. In the case of metals in NaY, these protons are formed during metal reduction. The positive charge will be distributed over the metal atoms of the cluster and the hydrogen atom(s), this distribution being controlled by electron affinities (ionization potentials, work functions) and electrostatic forces which favor, of course, location of the positive charge on the metal cluster at the shortest possible distance from the negative charge, localized in the cage wall.

This model is supported by recent spectroscopic evidence of palladium carbonyl clusters in NaY. With Pd/NaY samples which have been reduced at a low temperature, e.g. 200°C, admission of CO gives rise to highly structured FTIR spectra with very sharp bands, markedly differing from the conventional spectra of CO on supported Pd [5.79]. The new spectra have been attributed to a palladium carbonyl cluster [5.43]. Figure 5.4 shows a set of these spectra at various stages of CO release. The position of the major bands agrees with those reported by *Bradshaw* and *Hoffmann* [5.80] for CO adsorption on Pd single crystal faces; additional bands in the terminal C–O stretching region around 2125 cm^{-1} indicate CO adsorbed on Pd^+ and $Pd^{\delta+}$.

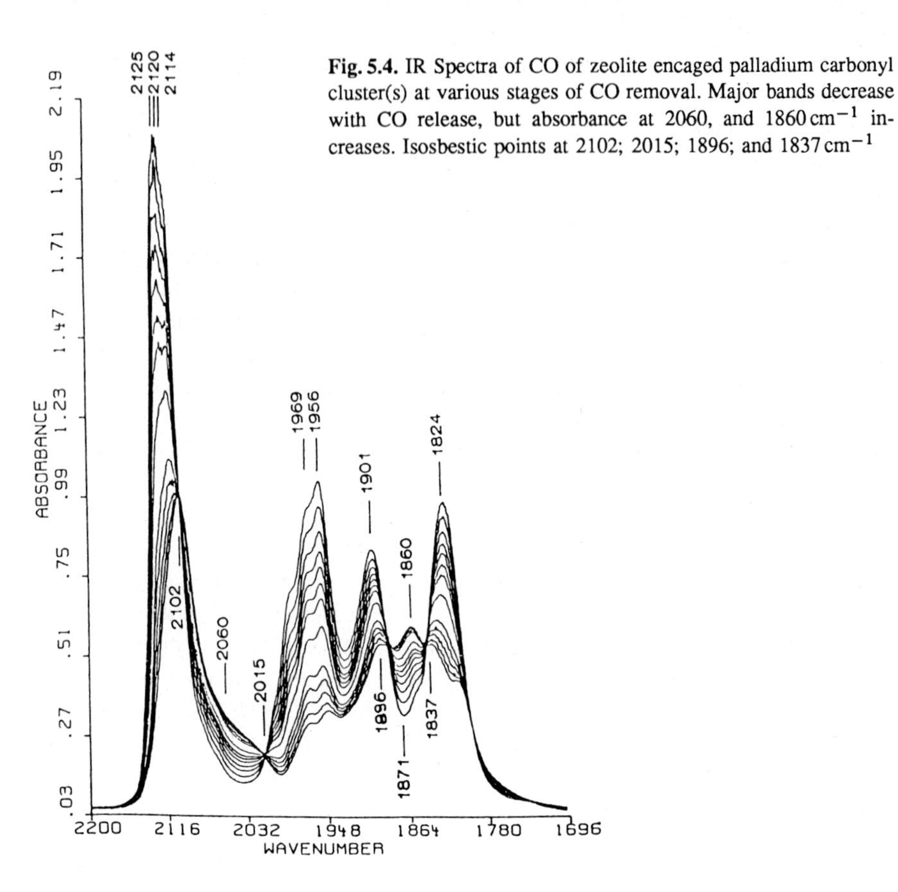

Fig. 5.4. IR Spectra of CO of zeolite encaged palladium carbonyl cluster(s) at various stages of CO removal. Major bands decrease with CO release, but absorbance at 2060, and 1860 cm^{-1} increases. Isosbestic points at 2102; 2015; 1896; and 1837 cm^{-1}

Of particular relevance is the change of the spectra upon removing part of CO by purging in Ar and upon readmitting CO [5.70]. The data indicate that CO release from the cluster is not a simple thermal desorption process, but that the carbonyl *clusters react with zeolite protons*, resulting in a positively charged $[Pd_x$-$(CO)_y$-$H_z]^{z+}$ complex. This has been confirmed by independently monitoring the IR band of the zeolite O–H vibration. It is found that the intensity of this band decreases, when CO is removed from the Pd carbonyl, and it increases again, when CO is admitted and adsorbed on the cluster. This shows that an equilibrium of the type:

$$Pd_n(CO)_a + xH^+ \longrightarrow [H_x - Pd_n(CO)_z]^{x+} + (a - z)CO_{gas} \qquad (5.6)$$

is established between zeolitic protons and the Pd carbonyl cluster [5.70]. The model of the proton-palladium cluster adduct is a straightforward generalization of this conclusion; it assumes that zeolitic protons interact with Pd particles, both in the presence and in the absence of carbonyl ligands.

5.5 Formation of Bimetallic Particles

The principles which govern the formation of metal particles in zeolite cages can be utilized for the design of bimetallic particles of predetermined composition and morphology. The reducibility of zeolite encaged ions is often found to be strongly affected by the presence of a second element. Different causes can contribute to this: the more noble element is reduced first, providing sites where ions of the second element can be attached, but also sites for the dissociative adsorption of hydrogen. In this case "cherry" type particles will be expected with the less noble element surrounding the core of the more noble element. Some authors have also speculated about hydrogen spill-over. In this model, H atoms that have been adsorbed on the reduced noble metal would have to migrate through the zeolite and reduce ions of the less noble element even at considerable distance. It is difficult to imagine an energetically favorable path for the migration of hydrogen atoms, i.e. protons *and* electrons over an electrically insulating surface; and no experimental data are known for this to occur in zeolites.

Other potential mechanisms for mutual effects of metal on each other's reducibility include site blocking, i.e. preventing the escape of ions from supercages, and cation-cation bonding.

NaY encaged pairs of RuNi and RuCu were studied by *Elliott* and *Lunsford* [5.81], and PtCu, IrCu, RuCu, and RhCu by *Tebassi* et al. [5.82]. The results can be rationalized in terms of the "cherry" model. Recently, *Moretti* and *Sachtler* [5.50] studied the reduction kinetics in PtCu/NaY and PtCu/HNaY. In the absence of Pt they found that reduction of Cu occurs in two steps, $Cu^{2+} \longrightarrow Cu^{+}$ and $Cu^{+} \longrightarrow Cu^{0}$. An equilibrium of the type

$$2Cu^{+} + H_2 \longrightarrow 2H^{+} + 2Cu^{0} \qquad (5.7)$$

is established. Therefore, the second reduction step is absent in NaHY; in NaY this step is reversible, i.e. H_2 is evolved at elevated temperature under the conditions where TPD is conventionally measured. Pt strongly enhances the reduction of Cu^{2+} to Cu^{0}; reduction is always complete and the TPR peaks for Pt and Cu merge into one broad peak. Hydrogen adsorption and catalytic tests of the reduced samples are consistent with formation of bimetallic clusters with surfaces enriched in Cu. The samples also show a stereoselective catalysis which suggests that supercages become completely filled.

A second method to prepare bimetallic, zeolite encaged catalysts makes use of the decomposition of an organometallic complex of one metal on the surface of the reduced particles of another metal. This method is, in particular, advantageous if only one metal is able to form positive ions in aqueous solution, whereas the other element forms complexed negative ions at the pH compatible with the stability of the zeolite. It has been used with Y zeolites for PtMo by *Tri* et al. [5.71, 83, 84] and for PtRe by *Tsang* et al. [5.59] and *Dossi* et al. [5.60, 85]. It has been found that the presence of reduced platinum in the zeolite, prior to adsorption and decomposition of $Re_2(CO)_{10}$, changes the kinetics of the decomposition markedly, indicating the formation of bimetallic PtRe clusters, i.e. Re is selectively deposited on the Pt particles in the supercages, although only a minority of these cages contains platinum particles. Moreover, the mechanisms of reductive decomposition are different: in the absence of platinum a Re hydrido carbonyl intermediate is formed, whereas in the presence of Pt no hydrogen consumption is registered at low temperature. Methane formation from hydrogen and residual CO ligands occurs at a lower temperature in the presence of Pt than in its absence, indicating formation of bimetallic PtRe clusters. This is confirmed by the TPR profile of the samples after oxidation [5.59] and by the specific catalytic activity of the reduced catalyst. The chemistry of metal/zeolite systems and the elementary steps of metal particle formation have thus been unravelled to the extent that formation by design of zeolite encaged bimetallic particles has become possible.

5.6 Summary and Conclusions

Zeolites provide a unique opportunity for studying the chemistry and physics of very small transition metal clusters and isolated metal atoms. Supercages of faujasite type zeolites are preferred locations for metal clusters which can be characterized by combining physical techniques, including EXAFS, SAXS, FTIR, and diffuse reflectance spectroscopy, with transient chemical techniques, in particular TPR, TPD and catalysis.

Isolated atoms of transion metals are formed when metal ions that were brought into supercages by ion exchange and stripped of their ligands migrate into smaller cages and then are reduced. Such atoms can be oxidized by zeolite protons to metal ions and H_2. Isolated atoms are unable to dissociatively adsorb H_2, though this adsorption is rapid on multiatomic clusters. Particles that are formed via isolated atoms differ from those formed by reduction of ions in supercages.

Attaching CO ligands to small primary particles of e.g. palladium gives rise to palladium carbonyl clusters in supercages. Adducts of metal clusters rise to palladium carbonyl clusters in supercages. Adducts of metal clusters and zeolite protons are models for "electron-deficient metal" particles that are known to display extremely high catalytic activity and high resistance against sulphur poisoning.

Bimetallic clusters are formed either by reducing a zeolite in which ions of both metals have been incorporated, e.g. Pt^{2+} and Cu^{2+} ions, or by decomposing a physisorbed complex of one metal on the reduced cluster of the other metal, e.g. $Re_2(CO)_{10}$ on Pt/NaY. Adsorption and catalysis of such bimetallic clusters differ significantly from those of the monometallic counterparts. In some cases, formation of monometallic and bimetallic clusters with predetermined size and adsorption characteristics has been achieved.

Acknowledgement Support from the U.S. Department of Energy under Contract DE-FG02-87ERA3654 and grants-in aid of the Engelhard Corporation and the Exxon Education Foundation are gratefully acknowledged.

References

5.1. W.J.M. Rootsaert, L.L. van Reijen, W.M.H. Sachtler: J. Catal. **1**, 416 (1962)
5.2. V.V. Gorodetskii, B.E. Nieuwenhuys, W.M.H. Sachtler, G.K. Boreskov: Appl. Surf. Sci. **7**, 355 (1981)
5.3. V.V. Gorodetskii, B.E. Nieuwenhuys, W.M.H. Sachtler, G.K. Boreskov: Surface Science **108**, 225 (1981)
5.4. D.C. Koningsberger, H.F.J. Van Blik, J.B.A. Van Zon, R. Prins: Int. Congr. Catal. [Proc] 8th; Vol. **5**, p. 123–134 (1984)
5.5. G. Bergeret, P. Gallezot, P. Gelin, Y. Ben Taârit, L. Lefèbvre, C. Naccache, R.D. Shannon: J. Catal. **104**, 279 (1987)
5.6. J.H. Sinfelt: J. Catal. **29**, 308 (1973)
5.7. P. Gallezot, G. Bergeret: In *Catalyst Deactivation*, ed. by E.E. Petersen, A.T. Bell (M. Dekker, New York 1987) 263–296
5.8. P.A. Jacobs, W. De Wilde, R.A. Schoonheydt, J.B. Uytterhoeven: J.C.S. Faraday Trans I, **72**, 1221 (1976)
5.9. J.B. Uytterhoeven: Acta Phys. Chem.: Szeged **24**, 53 (1987)
5.10. P.A. Jacobs: In *Metal Clusters in Catalysis*, ed. by B.C. Gates et al. (Elsevier, Amsterdam 1986), 357
5.11. P. Gallezot: Catal. Rev. Sci. Eng. **20**, 121 (1979)
5.12. P. Gallezot, G. Bergeret: In *Metal Microstructure in Zeolites*, ed. by P.A. Jacobs et al. (Elsevier, Amsterdam 1982), p. 167
5.13. I.E. Maxwell: Adv. Catal. **31**, 1 (1982)
5.14. P. Gallezot: Proc. 6th Int. Zeolite Conf., ed. by A. Bisi, D.H. Olson (Butterworth 1984), 352–367
5.15. P. Gallezot: *Metal Clusters in Zeolites* in "Metal Clusters", ed. by M. Moskovits (Wiley, New York 1986), 219–247
5.16. Kh.M. Minachev, E.S. Shpiro: Kin. i Kataliz. **27**, 824 (1986); Engl.: Kinetics and Catal., 712 (1987)
5.17. D.J. Delafosse: Chim. Phys. **83**, 791 (1986)
5.18. K. Klier: Langmuir **4**, 13 (1988)
5.19. P. Gallezot, I. Mutin, G. Dalmai-Imelik, B.J. Imelik: Microsc. Spectr. Electr. **1**, 1 (1976)
5.20. G. Bergeret, P. Gallezot, B.J. Imelik: Phys. Chem. **85**, 411 (1981)
5.21. A. Kleine, P.L. Ryder, N. Jaeger, G. Schulz-Ekloff: J. Chem. Soc. Faraday Trans. I, **82**, 205 (1986)
5.22. P. Gallezot, B. Imelik: Adv. Chem. Ser. **121**, 66 (1973)
5.23. Y. Kim, K.J. Seff: Am. Chem. Soc. **99**, 7055 (1977)
5.24. P. Gallezot, A. Alarcon-Diaz, J.A. Dalmon, A.J. Renouprez, B.J. Imelik: Catal. **39**, 334 (1975)
5.25. A. Barcicka, S. Pikus: Proc. Conf. Appl. Crystallography, Kozubnik (Poland) (Editor and Publisher unknown), 543 (1978)
5.26. P. Gallezot, R. Weber, R.A. Dalla Betta, M. Boudart: Z. f. Naturforschung **34a**, 40 (1979)
5.27. M.S. Tzou, B.K. Teo, W.M.H. Sachtler: J. Catal. **113**, 220 (1988)
5.28. M.S. Tzou, B.K. Teo, W.M.H. Sachtler: Langmuir **2**, 773 (1986)
5.29. P. Ratnasamy, A.J. Leonard: Catal. Rev. **6**, 293 (1972)
5.30. P. Gallezot, A. Bienenstock, M.N. Boudart: Nouveau J. Chimie **2**, 263 (1978)
5.31. J. Fraissard, T. Ito, L.C. de Menorval, M.A. Springuel-Huet: In *Metal Microstructures in Zeolites*, ed. by P.A. Jacobs et al. (Elsevier, Amsterdam 1982), 179
5.32. C. Naccache, M. Primet, M.V. Mathieu: Adv. Chem. Ser. **121**, 266 (1973)
5.33. Z. Zhang, S.L. Suib, Y.D. Zhang, W.A. Hines, J.I. Budnick: J. Am. Chem. Soc. **110**, 5569 (1988)
5.34. V.R. Balse, J.A. Dumesic, W.M.H. Sachtler: Catal. Letter **1**, 275 (1988) and references therein
5.35. L.A. Pedersen, J.H. Lunsford: J. Catal. **61**, 11 (1980)

5.36. G.V. Antoshin, E.S. Shpiro, O.P. Tkachenko, J.B. Nikishenko, M.A. Ryashentseva, V.I. Avaev, Kh. Minachev: In Proc. 7th Int. Congr. Catal., ed. by T. Seyama, K. Tanabe (Elsevier, Amsterdam 1981), p. 302
5.37. S.L.T. Anderson, M.S. Scurrell: J. Catal. **71**, 233 (1981)
5.38. K. Foger, J. Anderson: J. Catal. **54**, 318 (1978)
5.39. J.C. Védrine, M. Dufaux, C. Naccache, B. Imelik: Chem. Soc. Faraday I, **74**, 440 (1978)
5.40. T.M. Tri, J. Massardier, P. Gallezot, B. Imelik: In *Metal Support and Metal Additives Effects in Catalysis*, ed. by B. Imelik et al. (Elsevier, Amsterdam 1982), 141
5.41. E. Mantovani, N. Palladino, A. Zanobi: J. Mol. Catal. **3**, 285 (1977–78)
5.42. C. Besoukhanova, J. Guidot, D. Barthomeuf, M. Breysse, J.R. Bernard: J. Chem. Soc. Faraday I, **77**, 1595 (1981)
5.43. L.L. Sheu, H. Knözinger, W.M.H. Sachtler: Catal. Letters **2**, 129 (1989)
5.44. D. Exner, N. Jaeger. K. Moeller, G.J. Schulz-Ekloff: Chem. Soc. Faraday Trans I, **78**, 3537 (1982)
5.45. S.H. Park, M. S. Tzou, W.M.H. Sachtler: Appl. Catal. **24**, 85 (1986)
5.46. M.S. Tzou, H.J. Jiang, W.M.H. Sachtler: React. Kin Catal. Lett. **35**, 207 (1987)
5.47. S.T. Homeyer, W.M.H. Sachtler: J. Catalysis **117**, 91 (1989)
5.48. G. Bergeret, P. Gallezot: J. Chem. Phys. **87**, 1160 (1983)
5.49. S.M. Augustine, M.S. Nacheff, C.M. Tsang, J.B. Butt, W.M.H. Sachtler: Catalysis: Theory to Practice, Proc. Intern. Congr. Catalysis, 9th, Calgary, Canada, ed. by M.J. Phillips, M. Ternan (Chem. Inst. Canada, Ottawa, CDN 1988), Vol. 3, 1190–1197 (1988)
5.50. G. Moretti, W.M.H. Sachtler: J. Catal. **115**, 205 (1989)
5.51. N. Herron, G.D. Stuckey, C.A. Tolman: Inorganica Chimica Acta **100**, 135 (1985)
5.52. Th. Bein, S.J. Mc Lain, D.R. Corbin, R.D. Farley, K. Moller, G.D. Stuckey, G. Woolery, D. Sayers: J. Am. Chem. Soc. **110**, 1801 (1988)
5.53. P. Gelin, F. Lefèbvre, B. Elleuch, C. Naccache, Y. Ben Taârit: In *Intrazeolite Chemistry*, ed. by D. Stucky, F.G. Dwyer (American Chemical Society, Washington, DC, ACS Symposium Series **218**, 455, 1983)
5.54. G. Bergeret, P. Gallezot, P. Gelin, Y. Ben Taârit, F. Lefèbvre, R.D. Shannon: Zeolites **6**, 392 (1987)
5.55. G. Bergeret, F. Lefèbvre, P. Gallezot: In New Developments in Zeolite Science and Technology: Proc. 7th Int. Zeolite Conference, Tokyo, ed. by Y. Murakami, A. Ijima, J. Eard (1986), 401
5.56. L.L. Sheu, H. Knözinger, W.M.H. Sachtler: Catal. Lett. **2**, 129 (1989)
5.57. D.M. Cox, K.C. Reichmann, D.J. Trevor, A. Kaldor: J. Chem. Phys. **88**, 111 (1988)
5.58. P. Gallezot, G. Coudurier, M. Primet, B. Imelik: In *Molecular Sieves II*, ed. by J.R. Katzer, Am. Chem. Soc. Washington, D.C., ACS Symp. Ser. **40**, 144 (1977)
5.59. C.M. Tsang, S.M. Augustine, J.B. Butt, W.M.H. Sachtler: Appl. Catal. **46**, 45–56 (1989)
5.60. C. Dossi, J. Schaefer, W.M.H. Sachtler: J. Molec. Catal. **52**, 193 (1989)
5.61. R.P. Zerger, K.C. McMahon, M.D. Seltzer, R.G. Michel, S.L. Suib: J. Catal. **99**, 498 (1986)
5.62. L.M. Aparicio, J.A. Dumesic, S.M. Fang, M.A. Long, M.A. Ulla, W.S. Millman, W.K. Hall: J. Catal. **104**, 381 (1987)
5.63. A. Kleine, P.L. Ryder, N. Jaeger, G. Schulz-Ekloff: J. Chem. Soc. Faraday Trans. I, **82**, 205 (1986)
5.64. J. Burton: J. Catal. Rev. Sci-Eng. **9** (2), 209 (1974)
5.65. M.B. Gordon, M.C. Cryot-Lackmann, Desjonquères: Surf. Sci. **80**, 159 (1979)
5.66. M. Boudart, M.G. Samant, R. Ryoo: Ultramicroscopy **20**, 125–134 (1986)
5.67. L.L. Sheu, H. Knözinger, W.M.H. Sachtler: J. Molec. Catalysis **57**, 62 (1989)
5.68. T. Kubo, H. Arai, H. Tominaga, T. Kunugi: Nippon Kagaku Kaishi **7**, 1199 (1974)
5.69. S.T. Homeyer, W.M.H. Sachtler: J. Catal. **118**, 266 (1989)
5.70. L.L. Sheu, H. Knözinger, W.M.H. Sachtler: J. Am. Chem. Soc. **111**, 8125 (1989)
5.71. T.M. Tri, J.P. Candy, P. Gallezot, J. Massardier, M. Primet, J.C. Védrine, B. Imelik: J. Catal. **79**, 396 (1983)
5.72. K. Foger, J.R. Anderson: J. Catal. **54**, 318 (1978)
5.73. R.A. Dalla Betta, M. Boudart: In "Proc. 5th Int. Congr. Catalysis", ed. by H. Hightower (North Holland, Amsterdam 1973), 1329
5.74. W. Juszczyk, Z. Karpinski, I. Ratajczykowa, Z. Stanasiuk, J. Zielinski, L.L. Sheu, W.M.H. Sachtler: J. Catalysis **120**, 68 (1989)
5.75. J.M. Driessen, E.K. Poels, J.P. Hindermann, V. Ponec: J. Catal. **82**, 26 (1983)

5.76. P. Gallezot, J. Datka, J. Massardier, M. Primet, B. Imelik: Proc. 6th Intern. Congr. Catal. **2**, 696 (1976)
5.77. C. Besoukhanova, J. Guidot, D. Barthomeuf, M. Breysse, J.R. Bernard: J. Chem. Soc. Faraday I, **77**, 1595 (1981)
5.78. R.C. Baetzold: J. Chem. Phys. **55**, 4383 (1971)
5.79. R.P. Eischens, W.A. Pliskin, S.A. Francis: J. Chem. Phys. **22**, 1786 (1954)
5.80. A.M. Bradshaw, F. Hoffmann: Surface Science **52**, 449 (1975) and **72**, 513 (1978)
5.81. D.J. Elliott, J.H. Lunsford: J. Catal. **57**, 11 (1979)
5.82. L. Tebassi, A. Sayari, A. Ghorbel, M. Dufaux, C. Naccache: J. Mol. Catal. **25**, 397 (1984)
5.83. T.M. Tri, J. Massardier, P. Gallezot, B. Imelik: J. Catal. **85**, 244 (1984)
5.84. T.M. Tri, J. Massardier, P. Gallezot, B. Imelik: J. Mol. Catal. **25**, 151 (1984)
5.85. C. Dossi, C.M. Tsang, W.M.H. Sachtler, R. Psaro, R. Ugo: "Energy and Fuels" **3**, 468 (1989)

6. Studies of Bonding and Reaction on Metal Surfaces Using Second-Harmonic and Sum-Frequency Generation

R.B. Hall, J.N. Russell, J. Miragliotta, P.R. Rabinowitz

Exxon Research and Engineering, Rt. 22E Annandale, NJ 08801, USA

Over the past 10 years there has been a resurgence in efforts to develop nonlinear optical techniques for studies of surfaces and interfaces. Chief among these techniques are second-harmonic generation (SHG) and ir-visible sum-frequency generation (SFG). These second-order nonlinear processes offer several advantages. They are inherently sensitive only to the interfacial region between centrosymmetric media. This is because, in the electric dipole approximation, these processes are forbidden in the bulk of such media but are allowed at the interface where the symmetry is broken. Thus, so-called buried interfaces, i.e. surfaces under liquids, high pressure gases, and even other solids, may be investigated. In addition, with sufficiently short-duration laser excitation, these techniques have the potential to provide sub-picosecond time resolution.

This resurgence has been motivated in part by improvements in laser systems, and in part by the number of recent investigations of adsorption and chemical reaction on clean, well characterized surfaces. Recent results have pointed out inaccuracies in the original assumptions about the origin of the nonlinear signals, and have indicated that there is a rich variety of information to be obtained. The purpose of this review is to summarize progress in this area. The emphasis will be on studies of single-crystal metal surfaces under ultra-high-vacuum conditions. Some studies of single-crystal semiconductor surfaces will be included where the results are central to the discussion. The reader is referred to reviews by *Richmond* et al. [6.1] and by *Shen* et al. [6.2], for summaries of studies of nonmetallic surfaces in air, or of metal and nonmetal surfaces under liquids.

6.1 Second-Harmonic Generation

6.1.1 Background

The initial studies of surface second-harmonic generation were carried out over 20 years ago [6.3–10]. *Bloembergen* and co-workers solved Maxwell's equations for a nonlinear dielectric at the interface between a linear and a nonlinear medium [6.3], and subsequently demonstrated second-harmonic generation in reflection from a semiconductor mirror [6.4]. *Brown* et al. reported SHG from a silver mirror shortly thereafter, the first demonstration of surface SHG from a centrosymmetric substrate [6.5].

Jha related, theoretically, the nonlinear polarization of metal surfaces to the conduction electrons but suggested that the surface contribution could originate from the discontinuity of the normal component of the electric field at the surface [6.6]. The discontinuity of the electric field is especially large at a high-dielectric-constant material, such as a metal. The discontinuity is described by a quadrupolar surface term, in apparent agreement with polarization studies [6.5] which suggested that the second-harmonic signals originated from an electric-quadrupole type source. *Bloembergen* and *Shen* suggested that the dominant contribution to SHG from metallic Ag arises from bound electrons in the ion cores at the surface rather than from free conduction electrons [6.7]. Later work by *Bloembergen* et al. on the wavelength dependence of SHG from Ag and Au surfaces indicated that interband transitions can play a role [6.8]. Similarly, *Sonnenberg* and *Heffner* showed that a simple free electron model was not sufficient to account for the dependence of SHG on angle of incidence [6.9]. Finally, relatively weak SHG signals were detected by *Wang* and *Duminiski* from non-metal surfaces [6.10], and by *Wang* from air-liquid interfaces [6.11].

The status at the end of the first decade, then, was as follows. It was found that in a medium with inversion symmetry, surface contributions to SHG could be comparable with or greater than the bulk contributions [6.5, 8–10]. The discontinuity of the electric field at the interface was assumed to be a critical contribution [6.6, 12], but contributions from other effects such as the nonlinear polarization of free electrons, or bound electrons, or interband resonances could be important. Experimental results on the intensity and polarization of surface SHG as a function of angle of incidence and polarization of the incident field could be accounted for by the macroscopic formalism that had been developed, as summarized by *Bloembergen* et al. [6.12]. On this basis, one expected that surface SHG from a high-refractive-index substrate should be particularly strong and nearly independent of surface contaminants. In addition, for cubic crystals, the SHG signal should be independent of surface orientation. Both of these expectations were shown later to be incorrect.

More recently, experiments with better control of surface cleanliness have shown that although metals do generally produce strong surface SHG signals, adsorbates can, and in general do, have a pronounced effect. As early as 1969, *Brown* and *Matsuoka* reported that adsorption dramatically reduces SHG signals from atomically clean Ag [6.13]. Later experiments showed that SHG intensities can be affected by submonolayer coverages [6.14, 15]. With regard to the second expectation, several studies reported that different crystal faces can produce very different SHG signals [6.16–18]. These experiments indicated that SHG had the potential to reveal new information about bonding and structure at surfaces that was inaccessible by other techniques. They also pointed out the inadequacy of the existing theory and the need to perform experiments under controlled, well characterized, conditions in order to obtain data needed to advance the theory. Considerable progress has been made in the theory, especially for free-electron-like metals. These will be discussed in the next section. There has also been considerable expansion of the experimental data base. It is now clear that the second-harmonic response contains information on the electronic properties of metal surfaces, including free- and bound-electrons, surface states, interband resonances and how all of these are affected or participate in surface bonding and reaction, as well as on the in-plane symmetry of the surface layer and the orientation of adsorbtes. The early research emphasis was on the origins of nonlinear polarization. More recently, the emphasis has turned to applications of SHG for investigating surface phenomena including bonding and vibrational relaxation, adsorbate diffusion, and surface reaction kinetics. These applications are discussed below.

6.1.2 Theory

(a) Macroscopic Formalism. The theory of second-harmonic generation has been reviewed in some detail recently by *Richmond* et al. [6.1] and *Shen* [6.2]. We present here an abbreviated summary of the macroscopic equations. This is followed by a discussion of current theories of the microscopic origin of the nonlinear polarization.

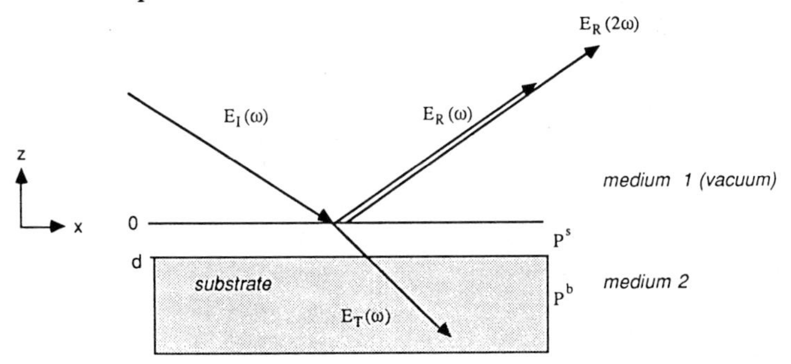

Fig. 6.1. Three-layer dielectric model for the nonlinear response of a vacuum/solid interface. The incident pump field, $E_{\rm I}(\omega)$, induces a linear and a nonlinear polarization, $P^{\rm s}$, in the surface region of medium 2. The induced polarizations gives rise to both a specularly reflected field, $E_{\rm R}(\omega)$, at frequency ω, and an SH field, $E_{\rm R}(2\omega)$, at frequency 2ω. Due to the transmitted fundamental field, $E_{\rm T}(\omega)$, a bulk nonlinear polarization, $P^{\rm b}$, will also contribute to the total reflected SH signal (see text). After [6.2]

A simple physical picture of SHG from an interface between a vacuum and a solid dielectric material is given in Fig. 6.1. A three layer model is used so that a surface layer of microscopic thickness, d, can be included explicitly. This is done to allow for the fact that the outermost layer(s) of a solid typically have electronic, optical and structural properties that are distinct from the bulk. The bulk is characterized by a dielectric constant, $\varepsilon_{\rm b}$, and the surface by a dielectric constant, $\varepsilon_{\rm s}$. These are often taken to be the same. From the vacuum side, an electromagnetic field, $\boldsymbol{E}_{\rm I}(\omega)$, with frequency ω is incident on the surface of the solid. This field induces both linear and nonlinear source polarizations in the surface layer and in the bulk of medium 2. The linear source polarization radiates from the boundary in the reflected direction at frequency ω, $\boldsymbol{E}_{\rm R}(\omega)$. The nonlinear source polarization radiates in the same direction as the reflected beam, but at frequency 2ω. The electric field vector at the second-harmonic frequency is denoted $\boldsymbol{E}_{\mathrm{R}}(2\omega)$. In general, there are also transmitted fields at both ω and 2ω.

The nonlinear polariaztion, $\boldsymbol{P}^{(2)}(2\omega)$, which is responsible for the SHG, can be written as [6.2]:

$$\boldsymbol{P}^{(2)}(2\omega) = \boldsymbol{P}_{\mathrm{s}}(2\omega)\delta(z) + \boldsymbol{P}_{\mathrm{b}}(2\omega) , \tag{6.1}$$

where $\boldsymbol{P}_{\mathrm{s}}(2\omega)$ is the second-order polarizability of the surface, $\delta(z)$ is a delta function at $z = 0$,

$$\boldsymbol{P}_{\mathrm{s}}(2\omega) = \overleftrightarrow{\chi}_{\mathrm{s}}^{(2)} : \boldsymbol{E}_{\mathrm{I}}(\omega)\boldsymbol{E}_{\mathrm{I}}(\omega) , \tag{6.2a}$$

and $\boldsymbol{P}_{\mathrm{b}}(2\omega)$ is the second-order polarizability of the bulk

$$\boldsymbol{P}_{\mathrm{b}}(2\omega) = \overleftrightarrow{\chi}_{\mathrm{b}}^{(2)} : \boldsymbol{E}_{\mathrm{I}}(\omega)\boldsymbol{E}_{\mathrm{I}}(\omega) . \tag{6.2b}$$

The bulk second-order nonlinear susceptibility, $\chi_{\mathrm{b}}^{(2)}$, vanishes in the bulk of medium 2 because of inversion symmetry under the electric dipole approximation. However, the second-order nonlinearity does not vanish in the bulk if the higher order magnetic dipole and electric quadrupole source terms are considered. These contributions are typically small and will not be discussed here. The reader is referred to references [6.1–3] for more information.

At the interface, the inversion symmetry is broken and because of this spatial discontinuity there is an electric dipole contribution to the surface second-order nonlinearity, $\chi_{\mathrm{s}}^{(2)}$. It is typical to treat the nonlinear polarization as a dipole sheet, placed just below the surface (at $z = 0$). Once the form of the nonlinear polarization, $\boldsymbol{P}^{(2)}(2\omega)$, is established, the radiated SH fields are found by solving the wave equation [6.12], given the boundary conditions.

The intensity of the reflected second harmonic signal from the polarization induced by $\boldsymbol{E}_{\mathrm{I}}(\omega)$ is then [6.1,2]:

$$\boldsymbol{I}(2\omega) = 32\pi^3\omega^2c^{-3}\sec^2\Theta_{2\omega}|e(2\omega)\chi_{\mathrm{s,eff}}^{(2)} : e(\omega)e(\omega)|^2 I^2(\omega) . \tag{6.3}$$

In this equation, $\Theta_{2\omega}$ represents the angle between the propagation direction of the radiated SH light and the surface normal, $I(\omega)$ is the pump intensity, and $e(2\omega)$ is the unit polarization vector, in cartesian coordinates, at the SH frequency. The vectors $e(\omega)$ and $e(2\omega)$ are related to the unit polarization vectors in medium 2 by Fresnel coefficients. The effective surface nonlinear susceptibility, $\chi_{\mathrm{s,eff}}^{(2)}$, is the sum of the surface nonlinear susceptibility, $\chi_{\mathrm{s}}^{(2)}$, and the bulk magnetic dipole contributions to the nonlinearity. Symmetry dictates the number of independent nonzero tensor elements in $\chi_{\mathrm{s,eff}}^{(2)}$.

In considering a system consisting of the surface and an adsorbed layer, the net second-order susceptibility is generally treated as the sum of the separate components plus an interaction term. This interaction term accounts for nonadditivity caused, for example, by changes in the individual contributions due to adsorbate/surface bonding. That is, $\chi^{(2)}_{s,\mathrm{eff}}$ is taken to be: [6.2, 19]

$$\chi^{(2)}_{s,\mathrm{eff}} = \chi^{(2)}_{ss} + \chi^{(2)}_{A} + \chi^{(2)}_{I} , \tag{6.4}$$

where $\chi^{(2)}_{ss}$ is the nonlinear susceptibility of the bare substrate surface, $\chi^{(2)}_{A}$ denotes the nonlinear susceptibility of the adlayer when far away from the surface, and $\chi^{(2)}_{I}$ a component resulting from the interaction of the adlayer with the surface. In the case of metals and semiconductors, $\chi^{(2)}_{ss}$ is relatively large. As will be discussed below, this stems largely from the second-order polarizability of the free electrons and extended-band-state electrons in metals. The electrons in molecular, covalent bonds, however, are much less polarizable and, in the absence of resonance enhancement, $\chi^{(2)}_{A}$ is generally more than an order of magnitude smaller than $\chi^{(2)}_{ss}$. At the same time, the formation of a surface chemical bond tends to localize the metal free-electrons, leading to a significant change in the hyperpolarizability of the metal surface. This results in appreciable values for $\chi^{(2)}_{I}$. In general, on metal surfaces, adsorption or surface reaction is detectable because of changes in the interaction term, and not because of the contribution of $\chi^{(2)}_{A}$ to the total susceptibility.

(b) Microscopic Theory. The above equations were derived from macroscopic formalisms which require, and contain, no information on the microscopic origin of the nonlinear response. *Bloembergen* and *Pershan* [6.3], for example, simply assumed a generalized dipole source at the interface.

There have been a number of theoretical investigations of relatively simple, free-electron-like surfaces. Early on, *Rudnick* and *Stern* calculated the quantum mechanical second-order response of an electron gas [6.20]. They calculated the surface and bulk currents which are the source of the second-harmonic response. The surface current, a consequence of the structural discontinuity at the surface, is localized to within a few atomic layers and has components parallel and perpendicular to the surface. Importantly, they found that the surface currents were as important as the field discontinuity in producing SHG [6.21]. A free-electron gas, admittedly, is an extremely simplified model of a metal surface, but it contains some of the essential characteristics of free-electron metals and provides some key insights. It shows, for example, that the perpendicular surface current is an important factor in the nonlinear response. This might help explain the pronounced effects of submonolayer adsorbates, since it is the perpendicular surface currents that are most likely to be influenced by surface bonding. At the same time, this simple model suffers from a number of limitations. It assumes the electrons are noninteracting and hence ignores the effects of collective electron motion. It assumes a sharp discontinuity of electron density at the surface and hence the contribution from the tail of the electron distribution out from the surface is not included. Finally, it does not include any effects of surface states, bound electrons, or interband resonances.

The first of these limitations was addressed by several groups who used a hydrodynamic electron-gas model to account for collective motion of the electrons

[6.23, 24]. *Sipe* and coworkers showed that the hydrodynamic model gives similar results for the parallel and perpendicular surface currents as the free-electron-gas model [6.23]. These calculations, however, kept the unrealistic truncation of the electron distribution at the surface.

In more recent calculations, *Weber* and *Liebsch* [6.25, 26] used a Lang-Kohn potential to approximate the exponential falloff of the electron density away from the surface. Time-dependent, density-functional theory was used to calculate the second-order polarizability for jellium metals of various bulk charge densities. They calculated how the electron density as a function of distance from the surface varied in response to an applied electromagnetic field. The expression derived for the response of the electron density to applied fields, in the low frequency limit, is:

$$\eta(z,t) = \eta_{\mathrm{DC}}(z) + \eta_{\omega}(z)\sin(\omega t) + \eta_{2\omega}(z)\cos(2\omega t) + \dots , \tag{6.5}$$

where, $\eta_{\mathrm{DC}}(z)$ is the spatial distribution of the electron density induced by a DC field, η_{ω} is the current induced by an AC field at ω, and $\eta_{2\omega}$ is the current at 2ω. From this, the second-order polarizability is calculated,

$$P_2(z) = \int_{-\infty}^{z} dz' \, \eta_{2\omega}(z') \, . \tag{6.6}$$

An example of the results obtained, for a jellium metal with a bulk electron density equivalent to that of aluminum, is shown in Fig. 6.2. For this calculation, there is no in-plane surface current, but a substantial perpendicualr current is obtained. Note especially the dramatic effect of including the exponential fall-off of the bulk electron density into the vacuum region ($z > 0$). (Results for a step

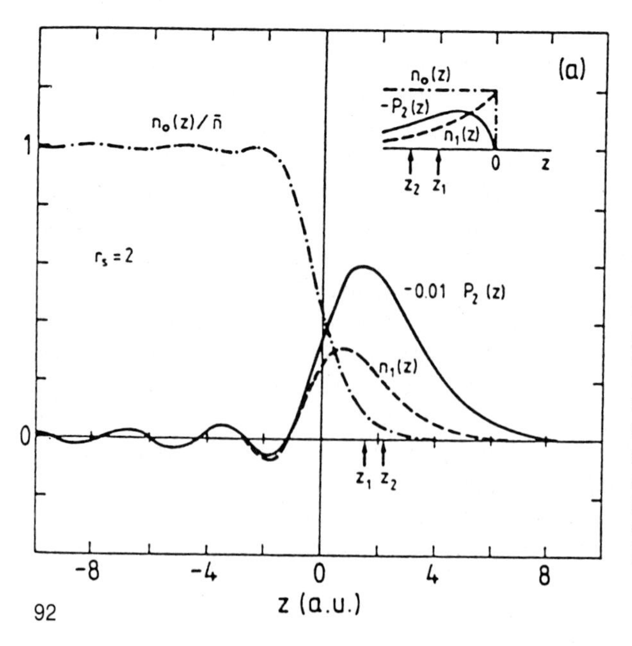

Fig. 6.2. Second-order polarization $P_2(z)$ (*solid line*) and the first-order density $n_1(z)$ (*dashed line*) induced at a metal surface by a static uniform electric field. The normalized equilibrium profile is shown by the dot-dashed line. The arrows indicate the centroid positions, z_1 and z_2, of $n_1(z)$ and $P_2(z)$, respectively. (*Inset*) qualitative form of corresponding quantities in the hydrodynamic model. From [6.25]

function distribution are shown in the inset.) The dominant contribution to the second-order polarizability arises from this tail.

The perpendicular surface current calculated by Weber and Leibsch is more than an order of magnitude greater than that calculated by Sipe et al. However, similarly large values for the perpendicular second-order polarizability were obtained by *Chizmeshya* and *Zaremba*, who used a generalized Thomas-Fermi model, to calculate a quantum mechanical nonlinear response [6.27]. Again, the difference with Sipe's results is attributed to the step function distribution used in the hydrodynamic calculations.

Experimental data on which these theories might be tested are just starting to become available. Data on "simple" metal surfaces under well controlled conditions are required. Recent results by *Plummer* and coworkers [6.28] on Al{100} and Al{111} agree reasonably well with the predictions of the density functional calculations of Weber and Leibsch, although lower frequencies for the driving laser field are required for a more rigorous test. Available microscopic theories do not include the effects of bound electrons, interband transitions and surface states. Some early efforts [6.7, 29] attempted to address some of these issues, but no theory yet available is sufficiently complete to calculate the response of transition metals. As will be discussed, much of the experimental data on transition metals suggest that these effects are significant, if not dominant.

6.1.3 Experimental Methods

A schematic illustrating a generalized experimental configuration is given in Fig. 6.3. Linearly polarized pulsed laser light of frequency ω strikes the surface at an angle α from the surface normal. Light of frequency 2ω is generated at the surface and the reflected component of the SH light comes off at the specular angle, along with reflected light at frequency ω. The propagation properties of the two light fields are similar. The reflected fundamental beam can be separated from the SH beam by a dispersing prism, by color filters, by a monochromator, or by some combination thereof. A monochromator is useful to verify the monochromaticity of the measured light, and to discriminate against luminescence or fluorescence generated at windows. In general, it is important to analyze the polarization of the SH light. Care should be taken that the polarization dependence of the components in the optical train (windows, monochromators, filters, mirrors) are properly taken into account. The SH photons are detected with a high gain photomultiplier. The signal intensity from metal surfaces is generally sufficiently high that analog electronics can be used, photon counting is not required. The signal pulse from the phototube typically is measured with a gated integrator circuit, to discriminate against fluorescence and uncorrellated photons. In most cases, with multimode lasers, there is significant fluctuation in the SH signal due to fluctuations in the pump laser intensity resulting from mode beating. In such circumstances, the signal to noise ratio can be improved by normalizing the SH signal to a signal from a nonlinear reference medium, excited

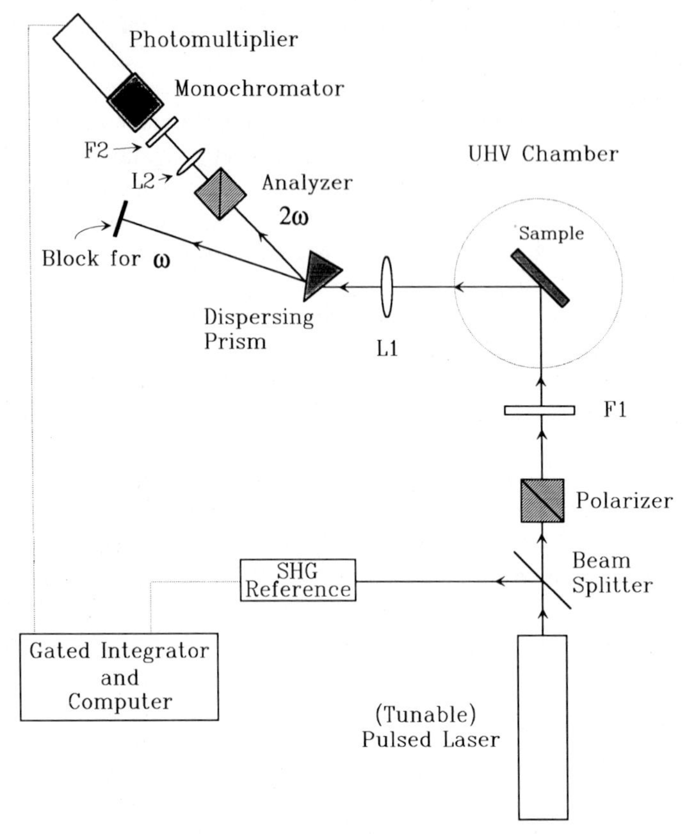

Fig. 6.3. Experimental setup for a typical SHG experimental setup. Linearly polarized light of frequency ω is incident on a sample in an ultra-high-vacuum chamber. The generated SH signal (2ω) is separated from the fundamental by a dispersing prism. The polarization of the SH beam is selected by a polarization analyzer, and the polarized beam is focussed by lens L2, filtered of scattered visible light by filter F2 and/or monochromator, and is detected by a photomultiplier. The SH signal is measured by a gated integrator and normalized to a reference SH signal as discussed in the text

by a portion of the pump laser beam. In general, it is important to confirm that the SH intensity exhibits a quadratic dependence on the intensity of the pump laser.

In addition to measuring the polarized SH intensity as a function of polarization of the pump beam, it is generally desirable to be able to measure the phase of the SH signal relative to the pump beam, the dependence of the SH response on angle of incidence and on azimuthal angle (with respect to some characteristic direction in the plane of the surface), and the time dependence of the SH signal following a perturbation. The experimental apparatus generally cannot be configured to optimize the measurement of all of these. Invariably, some choices must be made as to which are most important to the issues under investigation.

The second-harmonic response of metal surfaces is generally significantly greater than the second-harmonic response of molecules. A Q-switched, doubled

Nd:YAG laser at 532 nm, with a pulse duration of 8 ns and only 5 mJ per pulse, focused to a spot size of 0.16 cm^2 on a Ni surface, will give 10^4 SH photons per pulse. For comparison, the same field intensity incident on a monolayer of molecules such as pyridine would result in approximately 1 SH photon per pulse.

With metal surfaces and nanosecond laser pulse durations, pump laser intensities higher than that noted above run the risk of inducing desorption of adsorbed layers or of damaging the metal surface. For nonabsorbing, nonmetallic substrates, it is possible to use significantly higher pump laser intensities. For metal surfaces, there is an advantage to pump lasers having short (picosecond) pulse durations in that higher peak intensities can be tolerated before laser-induced thermal desorption or surface damage becomes a limitation.

SH experiments using a cw mode-locked Nd:YAG laser with [6.30] and without [6.31] Q-switching have also been reported. The fundamental light must be tightly focussed in order to achieve requisite energy densities. A cw laser may have an advantage over low repetition-rate pulsed lasers in duty cycle, or average power level. For time resolved measurements, time resolutions ranging from 30 to 100 ms might be expected [6.2]. With pulsed lasers, time-resolved pump-SH probe studies with sub-nanosecond time resolution have been reported [6.32–36].

Tunable lasers allow measurement of the frequency dependence of SHG. For example, the SH intensity can be enhanced when ω or 2ω (or both) is near resonance with an electronic transition in the substrate or an optical transition of an adsorbate. A recent study [6.30] exploited an enhancement of several orders of magnitude to demonstrate that SH light from a dye monolayer is detectable using only a continuous wave 20 mW diode laser as the excitation source. A discussion of results on resonance effects in SHG is given below.

The relative phase of the SH field can be measured by interfering the sample SH field with that from a quartz plate in the beam path at a variable distance following the sample [6.37, 38]. A pressure scanning interferometric method was first introduced by *Chang* et al. [6.39]. The absolute phase of the SH from the interface can be measured if the phase of the reference material susceptibility is known [6.40, 41]. Several interferometric SH methods have been used to measure the relative [6.42] or absolute [6.43] magnitude and the sign [6.43] of the second order susceptibility of the sample.

As for all quantitative surface methods, careful surface preparation is critical to the SHG technique. Sputtering and heating or annealing followed by Auger analysis of impurities are standard methods in the SH studies of metals in UHV. Low energy electron diffraction (LEED) is used to check surface order. Thin metal films are prepared by vapor deposition under conditions of UHV cleanliness.

6.1.4 Experimental Results

(a) SHG from Simple Metals. Quantitative measurements of the second-harmonic response of simple metals provide a useful test of available theories.

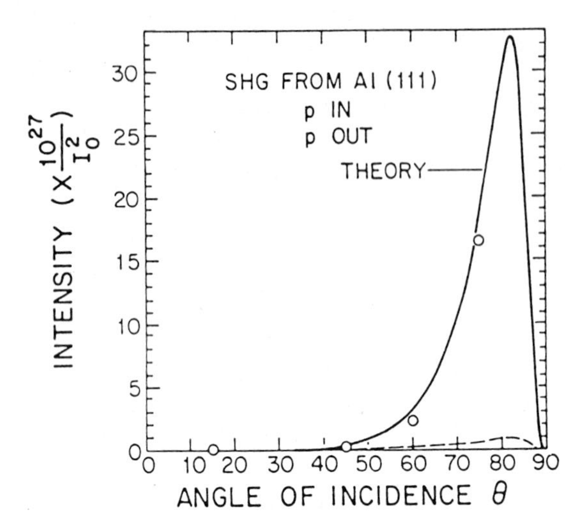

Fig. 6.4. Second-harmonic conversion efficiency ($cm^2 \cdot s/erg$) from Al{111} for p-polarized fundamental and p-polarized second harmonic vs the angle of incidence with respect to the crystal normal. The experimental intensity scale has an associated +/ − 20% error not displayed. The solid line is the $a = -36 - 9i$ time dependent local-density result. From [6.28]

Unfortunately, only a few such measurements have been made under sufficiently controlled conditions that the results are reliable [6.28, 44]. In a recent paper, *Plummer* and coworkers made quantitative measurements of the SH response of Al{111}, Al{110} and polycrystalline Al surfaces in UHV to 1.06 μ radiation [6.28]. They measured the dependence of the SH signal on polarization and on angle of incidence.

An example of their results is shown in Fig. 6.4. Here, the measured second-harmonic conversion efficiency as a function of angle of incidence is compared to calculations by *Weber* and *Schaich* [6.45]. The calculation used time-dependent, local-density theory and a Lang-Kohn surface charge-density profile. In these experiments, the excitation was p-polarized, and only the p-polarized SH signal was detected. Under these conditions, the dominant response is due to the perpendicular component of the susceptibility. The results show that the magnitude of the SH response is close to that predicted by the calculations, and significantly higher than that predicted by hydrodynamic calculations with a step-function profile for the surface charge-density [6.46]. The hydrodynamic calculations give a value for the perpendicular susceptibility, $a(\omega)$, of $-\frac{2}{9}$. The local-density calculations give $a(0) = -30$, and $a(\omega = 1.17\,\text{eV}) = -39 - 9i$. This variation corresponds to more than a factor of 10^4 in the SH intensity. These results verify the theoretical prediction that, at least at low frequencies, the second-harmonic response is sensitive to the local charge-density at the surface.

Interestingly, with s-polarization, they observe a significant azimuthal anisotropy relative to the crystal axis in the SH signal. They suggest that this anisotropy is an indication of significant bulk and/or surface band-structure effects. These effects, not included in any of the electron-gas or jellium models, will be discussed below.

(b) SHG from Metals with Adsorbed Layers. One of the most striking early observations was that the intensity of SHG from a Ag surface is dramatically reduced by small quantities of adsorbates [6.13]. For simple metals, the theories both of *Rudnick* and *Stern* and of *Weber* and *Leibsch* predict that the SH intensity is a function of the free-electron density at the surface. Furthermore, surface currents normal to the surface play a particularly important role. It is just this quantity that is likely to be most influenced by the formation of surface chemical bonds.

Studies of adsorption onto clean, single crystal metal surfaces in ultra-high vacuum conditions should provide a test of these issues. Adsorbates that localize the free-electron density at the surface by the formation of covalent bonds should reduce the SHG intensity. Thus, adsorbates that form strong chemisorption bonds, such as O, H, and S are expected to reduce the nonlinear response of the metal, while contributing little to the total hyperpolarizability. Conversely, adsorbates with low ionization potentials or low work functions that donate charge to the surface, such as alkali metals, should increase the nonlinear response.

Shen and coworkers were the first to report UHV studies of the effects of adsorbates on SHG. They measured the dependence of the SHG intensity from a Rh{111} surface on adsorption of CO, and O_2 [6.47], and of Na, K, and Cs [6.48]. Since then, there have been a number of similar studies, including, CO [6.49,50], H_2, [6.49,51] and O_2 [6.49] on Ni{111}, CO and H_2 on Pt{111} [6.49], pyridine and benzene on Rh{111} [6.52], NH_3 on Re{0001} [6.53], and O_2 [6.54] and alkali metals [6.55] on Ag.

A representative example of the results obtained from these studies, taken from our own work [6.49], is given in Fig. 6.5. The intensity of the second-harmonic signal, normalized to the clean surface signal, is shown as function of

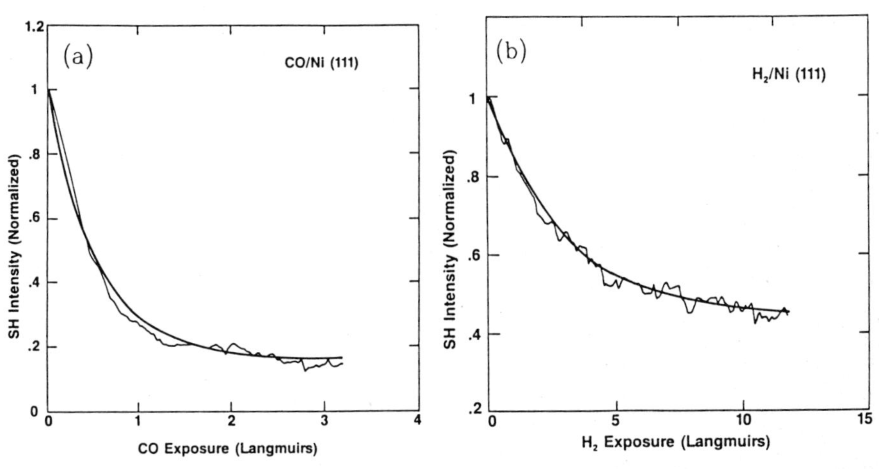

Fig. 6.5a,b. Normalized SH signal intensity from a Ni{111} surface as a function of exposure to CO (a) and hydrogen (b). The signals are normalized to the SH intensity from the clean surface. The excitation wavelength is 532 nm. The units of exposure are in Langmuirs, 1 Langmuir = 1×10^6 Torr s. The solid lines are fits to a Langmuir adsorption kinetic model (see text). After [6.49]

exposure of the surface (at 100 K) to hydrogen and to CO. It can be seen that there is a monotonic decrease of intensity to a limiting value, which occurs at approximately 1 monolayer coverage. Qualitatively, these data fit the expectation that the SH intensity will decrease with formation of bonds that localize surface-charge. The fact that nearly 90 % of the bare Ni{111} SH signal is extinguished upon adsorption of a monolayer of CO demonstrates the surface specificity of the measurement. It is unlikely that a monolayer of CO significantly affects the bulk charge-density. Most of the SH signal must come from the surface region.

The exposures were made at a fixed pressure of 1×10^{-8} Torr so that the abscissa in Fig. 6.5 corresponds to time, with 1 Langmuir being 10^{-6} Torr s. Thus, when there is a direct relationship between coverage and the change in the SH signal, these data provide direct information on the kinetics of adsorption.

The assumption is generally made in modeling the dependence of the SHG intensity on adsorption that the last two terms in (6.4) can be combined into a single term, and that this term varies in direct proportion to the surface coverage [6.47, 49]. Thus, the surface nonlinear susceptibility is written as:

$$\chi^{(2)}_{\mathrm{s,eff}} = A + B(\vartheta/\vartheta_{\mathrm{satn}}) , \tag{6.7}$$

where $A = \chi^{(2)}_{\mathrm{ss}}$, $B = (\chi^{(2)}_{\mathrm{A}} + \chi^{(2)}_{\mathrm{I}})$, ϑ is the fractional coverage of adsorbate relative to metal surface atoms, and $\vartheta_{\mathrm{satn}}$ is the saturation value of ϑ. The Langmuir model dictates that the rate of change of adsorbate coverage with time varies as

$$d\vartheta/dt = \kappa p(1 - \vartheta/\vartheta_{\mathrm{satn}}) , \tag{6.8}$$

where p is the adsorbate pressure, and κ is related to the initial sticking coefficient, S_0, by $\kappa = (S_0/N_{\mathrm{s}})(RT/2\pi M_{\mathrm{g}})^{1/2} N_{\mathrm{g}}$, where N_{s} is the density of surface adsorption sites, M_{g} is the molecular weight of the gas, and N_{g} is the gas density. Integration of (6.8) yields

$$\vartheta(t) = \vartheta_{\mathrm{satn}} \left[1 - \exp(-\kappa pt/\vartheta_{\mathrm{satn}})\right] . \tag{6.9}$$

The intensity of the second-harmonic signal as a function of coverage, $I_{2\omega}(\vartheta)$, is proportional to the square of the nonlinear susceptibility, $|\chi^{(2)}_{\mathrm{s,eff}}|^2$. The relative intensity, i.e. $I_{2\omega}(\vartheta)$ divided by the SH intensity for the clean surface, $I_{2\omega}(\vartheta = 0)$, has the following form for Langmuir adsorption:

$$\frac{I_{2\omega}(\vartheta)}{I_{2\omega}(\vartheta = 0)} \propto \left|1 + \frac{B}{A} - \frac{B}{A}\exp(-\kappa pt/\vartheta_{\mathrm{satn}})\right|^2 . \tag{6.10}$$

The solid line in Fig. 6.5 indicates the fit of the data to Langmuirian adsorption model. The quality of the fit suggests that there is indeed a direct proportionality between coverage and the change in the SH signal, and that the kinetics of adsorption closely resemble those predicted by a Langmuir model. For CO adsorption on Ni{111}, the fitting parameters obtained are, $B/A = 1.1 \exp(\mathrm{i}(163°))$, and $S_0 = 1.0 \pm 0.05$. For hydrogen adsorption, the values are, $B/A = 0.31 \exp(\mathrm{i}(177°))$, and $S_0 = 0.15 \pm 0.05$. The initial sticking coefficients obtained for both gases are in good agreement with independent measurements of these quantities [see 6.49 and references therein].

Other workers have similarly found that the SH signal decreases monotonically with adsorption and that they can fit their data to a Langmuir model. Included are studies of the adsorption of CO on Rh{111} [6.47], CO adsorption and desorption on Cu{100} [6.56], and oxygen on Ag{110} [6.54]. In addition, it is possible to measure equilibrium surface coverages as a function of gas pressure or solute concentration. A recent example is the measurement of the adsorption isotherms of NH_3 on Re{0001} as a function of surface temperature and NH_3 pressure [6.53].

Perhaps a more interesting result is that not all adsorption processes exhibit Langmuirian kinetics. In the case of adsorption of CO on Rh{111}, an abrupt change in slope of SH intensity versus coverage at $\frac{1}{3}$ monolayer was observed [6.47]. This corresponds to a well known transition from CO binding at top sites to binding at both top and bridge sites, indicating that SHG is sensitive to the details of the binding site. For pyridine adsorption on Ag{110}, a complex coverage dependence of the SH signal was interpreted to be indicative of a two site adsorption process that included a phase transition between two orientations of pyridine on the surface [6.54].

Indeed, even the data for CO adsorption on Ni{111} shown in Fig. 6.5 does not fit a Langmuir model nearly as well as does the data for hydrogen adsorption. Much better agreement with the CO adsorption data is obtained with a more generalized model, one in which the coverage dependence of the sticking coefficient is not constrained to a Langmuir form. The more general model suggests that CO actually adsorbs via an extrinsic precursor-state that allows the sticking coefficient to remain close to unity up to high coverages [6.49]. In the same work, even more pronounced evidence for precursor-state mediated adsorption was observed for the adsorption of oxygen on Ni{111}.

The assumption that the change in SH intensity can be related to a specific coverage can be complicated by a number of effects. As noted above, a change in the binding site, in adsorbate orientation, or in long-range order (phase), can lead to a change in the SH response. Any contributions from interband resonances, surface/adsorbate charge-transfer resonances, or resonances within the adlayer can also complicate the picture. At least two of these complications are observed in the adsorption of CO on Pt{111} [6.49]. The dependence of the SHG signal on CO coverage at two excitation wavelengths is shown in Fig. 6.6. There is a qualitative difference in the behavior at the two excitation wavelengths. The behavior for 1.06 μm excitation is attributed to a surface-state or interband resonance in Pt [6.49]. Ni does not have states in the same region and no such wavelength dependence is observed for CO adsorption on Ni{111}.

The coverage dependence of the Pt{111} SH signal at 532 nm is not believed to be influenced by the resonance. These results have a striking correlation with the change in work function with adsorption of CO [6.49]. At a coverage of 0.33 monolayers, there is a change in the binding site from atop to bridge sites. This coverage is marked by the appearance of a $(\sqrt{3} \times \sqrt{3})\,R30°$ LEED pattern. It

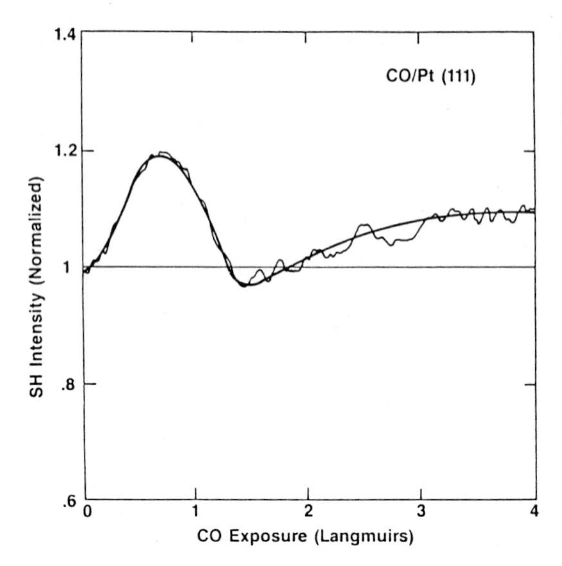

Fig. 6.6. Normalized SH signal intensity from a Pt{111} surface as a function of exposure to CO. The signal is normalized to the SH intensity from the clean surface. From [6.49]

occurs at an exposure of 0.9 Langmuirs in Fig. 6.6. The molecular orbital most involved in bonding to top sites is the filled 5σ orbital of CO. The bonding involves a partial transfer of charge from the 5σ orbital to the surface. With this, the SHG signal increases and the work function decreases. Bonding to bridge sites, on the other hand, involves back donation from the surface to the $2\pi^*$ antibonding orbitals of CO. As the coverage at bridge sites increases, more charge is removed from the surface. With this, the SHG signal decreases and the work function increases. The SHG signal also mirrors the change in the work function at high coverages (> 1.5 Langmuirs), where a compression structure occurs. A correspondence between ΔSHG and $\Delta\phi$ might be expected if the surface nonlinear susceptibility is dominated by surface free-electrons. These results demonstrate clearly that the SH response is a function of binding site.

A number of studies have shown that adsorption of electron-donating species leads to a dramatic increase in the SH intensity. Adsorption of Na onto single-crystal Ge in UHV was found to increase the SH intensity by roughly a factor of 10 [6.57]. A factor of 7 increase in the SH signal from Re{0001} is observed on adsorption of a monolayer of NH_3 [6.53]. *Tom* et al. observed increases of more than 10^2 in the SH intensity from 1.06 μm excitation of Rh{111} upon adsorption of Na, K, or Cs [6.48]. This increase is too large to be due to electron donation effects alone. Furthermore, different results are obtained for 532 nm excitation. The dispersion, or wavelength dependence, of the enhancement strongly suggests that the increase is due to a resonance effect. Tom et al. attributed the increase at low coverages to a resonance between the second-harmonic wave at 2ω and coverage dependent optical transitions from occupied states below the Fermi level to higher, unoccupied states associated with the alkali adatoms. At coverages above 0.5 monolayers, the adlayer becomes more metallic and the enhancement does not continue to increase so dramatically.

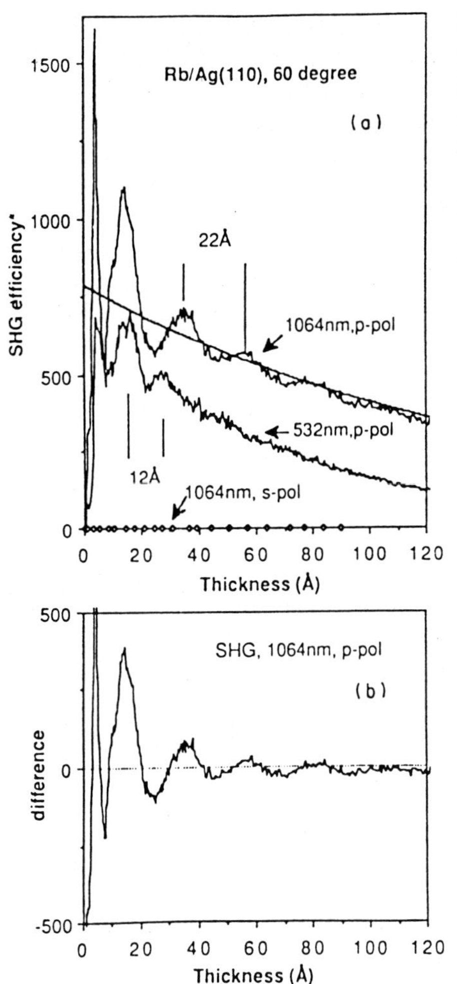

Fig. 6.7. (**a**) SH signal as a function of Rb film thickness for two different wavelengths for *p*-polarized light. The SH signal is normalized to the Ag{110} signal at 1064 nm, while the efficiency is the SH energy flux divided by the incident energy flux squared. The *o*'s are the 1064 nm data for *s*-polarized light. The angle of incidence is 60°. (**b**) The oscillatory part of the 1064 nm SH signal after the smoothly varying background shown by the dashed line in (a) has been subtracted. From [6.55]

Similarly dramatic enhancements of the SH signal from Ag{110} upon adsorption of alkali metals were observed by *Plummer*, *Dai*, and coworkers [6.55]. In this interesting study, an oscillatory behavior in the intensity was observed as a function of overlayer film thickness (Fig. 6.7). With 1064 nm excitation, there is an increase in the SH intensity of a factor of about 1600 with respect to the Ag surface for roughly one monolayer of Rb. With 532 nm excitation, the increase is about a factor of 10. The increase is presumed to be caused by intra-atomic resonances, either within the broadened alkali valence orbitals or between a valence orbital and the next unoccupied level.

There is a general decline in the SH intensity as the Rb film thickness is increased. Superimposed on this decrease, is a short wavelength oscillation. The wavelengths are 22 Å and 12 Å for 1064 and 532 nm excitation, respectively.

These oscillations are the result of longitudinal waves set up in the Rb film by the incident transverse wave. The oscillations can come from a three-wave mixing process where E'_L and E'_T mix at the buried Ag surface to produce a signal at 2ω. The wavelength of the transverse wave is far too long to result in any interference over the distances involved here. The possibility of an interference between the nonlinear longitudinal electric fields at the two interfaces was not ruled out, however. The authors showed that the oscillations could be understood in terms of the critical branches of the excitation spectrum for a free-electron metal with the density of Rb. These results show that SHG can reveal important details of the structure and electronic properties of thin metal films.

(c) Measurements of Surface Diffusion and Reaction Kinetics. In many cases, the change in the SH response of a metal is proportional to the density of surface chemical bonds [6.2, 47, 49]. Resonance effects can complicate this, but a simple proportionality may hold even in the presence of a resonance. In this event, the SH intensity can be used to monitor surface processes in real time.

In one particularly interesting example, *Shen* and coworkers demonstrated that it is possible to measure surface diffusion of adsorbates by diffraction of SHG from a monolayer grating [6.58]. Results for the diffusion of CO on Ni{111} are shown in Fig. 6.8. Two pulsed laser beams, intersecting at the surface so as to produce an interference pattern, are used to burn a grating in the CO overlayer by laser-induced thermal desorption. A separate SHG laser beam is then diffracted from this grating. The grating relaxes to a uniform coverage by surface diffusion, and the rate of this relaxation is monitored by the decrease in the diffracted SHG intensity.

The time dependence of the first-order diffracted SHG signal at several surface temperatures is shown in Fig. 6.8b. The relaxation is analyzed by a simple Fick's law diffusion model with a coverage-independent diffusion constant, $D = D_0 \exp(-E_{\mathrm{diff}}/RT)$. With the characteristic, macroscopic diffusion length given by the grating spacing, $10\,\mu$m in this case, a least-squares fit of the data gives $E_{\mathrm{diff}} = 6.9 \pm 0.1$ kcal/mol, and $\ln(D_0) = -11.36 \pm 2$. This is in reasonable agreement with previous measurements of the surface diffusion of CO using a hole burning technique, which also is based on laser-induced thermal desorption [6.59].

This technique, although it requires careful control of the various laser excitation processes, has some significant advantages. Since the grating spacing is in the μm range, it is possible to measure the macroscopic diffusion of adsorbed species with binding energies of up to roughly 25 kcal/mol. As with hole burning, this technique is capable of varying the orientation of the concentration gradients so that anisotropy of surface diffusion can be measured, and it is applicable to insulator and semiconductor surfaces as well as metal surfaces.

In another application, we have found that it is possible to measure the kinetics of some surface chemical reactions using time-resolved SHG. One of the required conditions for this is that the second-order susceptibility for the surface plus reactants be different than that of the surface plus products. One example is

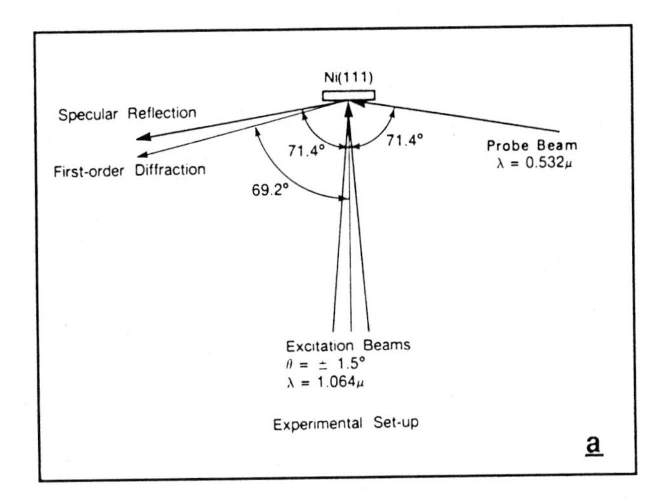

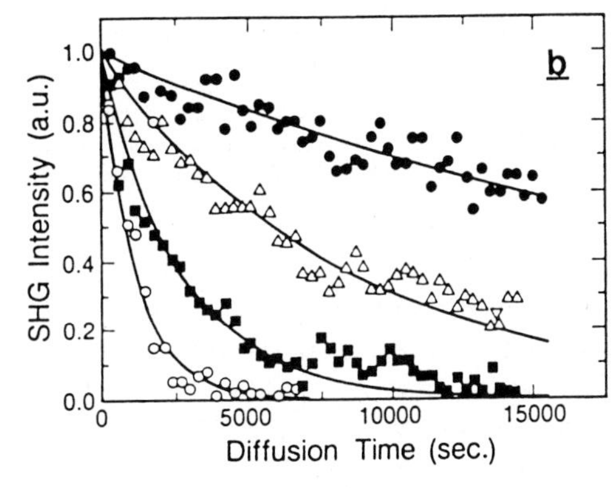

Fig. 6.8a,b. The experimental setup for the surface diffusion measurement. The Ni{111} sample was placed in an ultrahigh vacuum chamber. The two excitation beams overlap and set up an interference pattern on the sample surface covered by CO, generating a CO monolayer grating via thermal desorption. The probe beam monitors the grating via specular and diffracted second-harmonic generation from the grating. From [6.58]

shown in Fig. 6.9. Here, we show the SHG signal from a Ni{111} surface during the adsorption of methanol and, subsequently, the decomposition of CH_3O (methoxy) to CO and 3 hydrogen atoms. At $t = -400$ s in Fig. 6.9, 1×10^{14} methanol molecules per cm^2 (0.12 monolayer) are dosed onto the surface. The surface temperature during this stage of the experiment is 100 K. At this temperature, methanol is stable on the surface. At $t \approx -100$ s, the surface temperature, indicated by the dotted line and the right-hand axis of Fig. 6.9, is increased to 195 K. This is sufficient to decompose the methanol to methoxy and a surface hydrogen atom, but not sufficient to decompose methoxy [6.60]. For this first reaction step, apparently, the second-order susceptibility for the surface plus reactant is approximately equal to that of the surface plus products, and it is not possible to use SHG to follow the kinetics of this reaction. The reason the SHG signal does not change with this reaction, despite a significant change in the surface bonding, is associated with a resonance enhancement of the SHG signal for methanol. This is discussed below.

The decomposition of methoxy is initiated, at $t = 0$, by rapidly heating the surface to 274 K. It should be pointed out that the rate determining step in

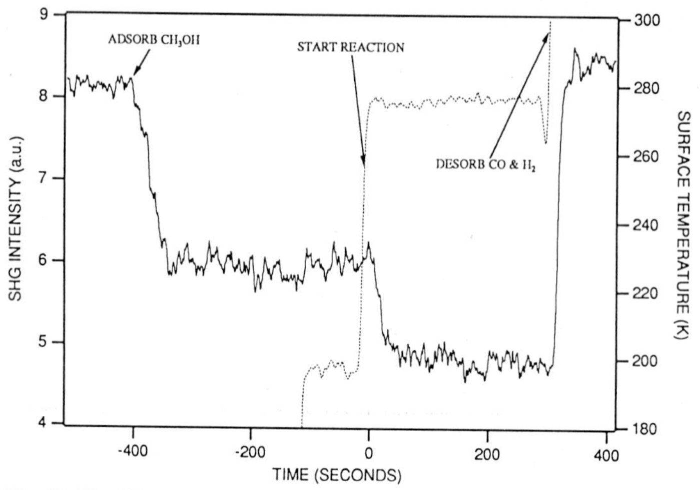

Fig. 6.9. The SH intensity (*solid line*), in arbitrary units, from a Ni{111} surface as a function of time while the following events occurred. (The surface temperature is indicated by the dotted line and the right hand axis.) At $t = -400$ s, 0.012 ML of methanol is adsorbed at a surface temperature of 100 K. At $t = -100$ s, the surface temperature is warmed to 190 K, and the methanol decomposes to methoxy and surface hydrogen. At $t = 0$ s, the surface is heated to 280 K, and methoxy decomposes to CO and surface hydrogen. Finally, at $t = 320$ s, the surface is heated to 600 K, desorbing both CO and hydrogen

the decomposition of methoxy is the dissociation of the first C–H bond. The rate for the dissociation of the subsequent C–H bonds is significantly faster. (In fact, formaldahyde decomposes on Ni at temperatures below 150 K.) As methoxy decomposes, the SHG signal decreases to a new asymptotic value. The asymptotic SHG intensity is the same as that observed in an independent experiment in which CO and H are coadsorbed at coverages equivalent to that produced by complete decomposition of the methoxy, i.e. 1×10^{14} CO molecules and 4×10^{14} H atoms per cm^2. The time required to reach the final SHG intensity is determined by the reaction time. After the reaction has reached completion, the surface is heated to 500 K. This desorbs all of the CO and hydrogen from the surface. Notice that the SHG signal in Fig. 6.9 returns to the intensity originally observed for the clean surface.

An expanded plot of the time response of the SHG intensity during the decomposition of methoxy is shown in Fig. 6.10. The solid line is a fit of the data to the following, simple kinetic model. The second-order susceptibility is assumed to be given by,

$$\chi^{(2)}_{\mathrm{s,eff}} = A + B_{\mathrm{CH_3O}}\vartheta_{\mathrm{CH_3O}}(t) + B_{\mathrm{CO}}\vartheta_{\mathrm{CO}}(t) + B_{\mathrm{H}}\vartheta_{\mathrm{H}}(t)\ , \tag{6.11}$$

where $A = \chi^{(2)}_{\mathrm{ss}}$, and the constants, B_{i}, include both the susceptibility of the adsorbate and the contribution from adsorbate/surface interaction. For example, $B_{\mathrm{CH_3O}} = (\chi^{(2)}_{\mathrm{CH_3O}} + \chi^{(2)}_{\mathrm{I_{CH_3O}}})$. The various B values can be obtained from separate measurements for the individual species. The surface coverages, $\vartheta_{\mathrm{i}}(t)$, are time dependent. Initially, both ϑ_{CO} and ϑ_{H} are zero.

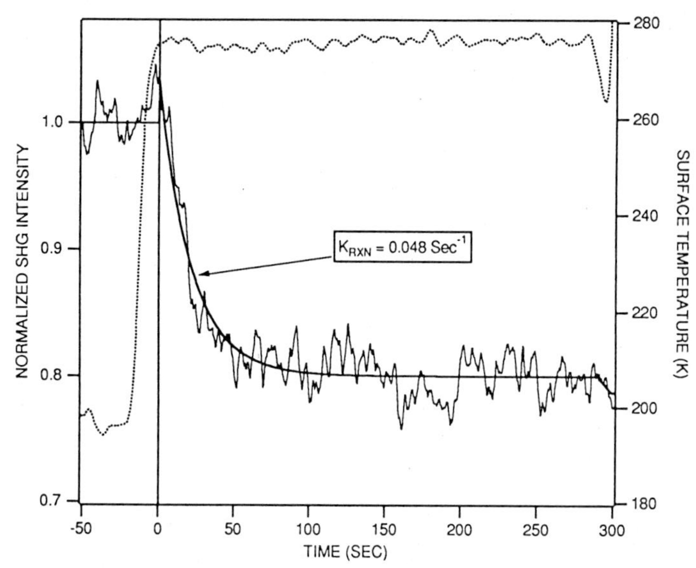

Fig. 6.10. The normalized SH intensity (*solid line*) from a Ni{111} surface with 0.012 ML of methoxy and 0.012 ML of hydrogen atoms. The surface temperature is indicated by the dotted line and the right hand axis. At time = 0 s, the surface is heated from 190 K to 276 K, inducing the decomposition of methoxy into CO and 3 H atoms. The SH intensity is normalized to the SH intensity before the reaction. The solid line is a fit to a reaction kinetic model described in the text. The rate constant, K_{RXN}, obtained from the fit is indicated

The reaction is assumed to be unimolecular with,

$$\frac{-d\vartheta_{CH_3O}(t)}{dt} = \frac{d\vartheta_{CO}(t)}{dt} = \frac{1}{3}\frac{d\vartheta_{H}(t)}{dt} = k_{RXN}\vartheta_{CH_3O} \,. \tag{6.12}$$

The factor of $\frac{1}{3}$ in the rate of production of H atoms comes from the stoichiometry of the reaction. The data in Fig. 6.10 are fitted using (6.11) and (6.12). The parameter, B_{CH_3O}, is obtained from the SHG intensity prior to the temperature jump at $t = 0$. Values for B_{CO} and B_H from previous measurements are given above in the section on SHG from adsorbed layers. The sum of these components, with the stoichiometry appropriate to this reaction, gives a value for the SH intensity that agrees with the asymptotic intensity shown in Fig. 6.10. Thus, k_{RXN} is the only adjustable parameter in fitting the time dependence shown in Fig. 6.10. The solid line in Fig. 6.10 shows the fit to the data for k_{RXN} = $0.048\,s^{-1}$.

Additional experiments were carried out for temperature jumps to reaction temperatures ranging from 230 to 280 K. A "vant't Hoff" plot of the temperature dependence of k_{RXN}, for both CH_3O and CD_3O, is shown in Fig. 6.11. A significant kinetic isotope effect is observed. This supports the fact that the time dependence of the transition is determined by the reaction kinetics. The activation energies and pre-exponential factors obtained from the data in Fig. 6.11 are: $Ea_{CH_3O} = 17.0 \pm 0.2$ K cal/mol, $\nu_{CH_3O} = 2 \times 10^{12}$, $Ea_{CD_3O} = 17.8 \pm 0.2$ K cal/mol, and $\nu_{CD_3O} = 6 \times 10^{12}$.

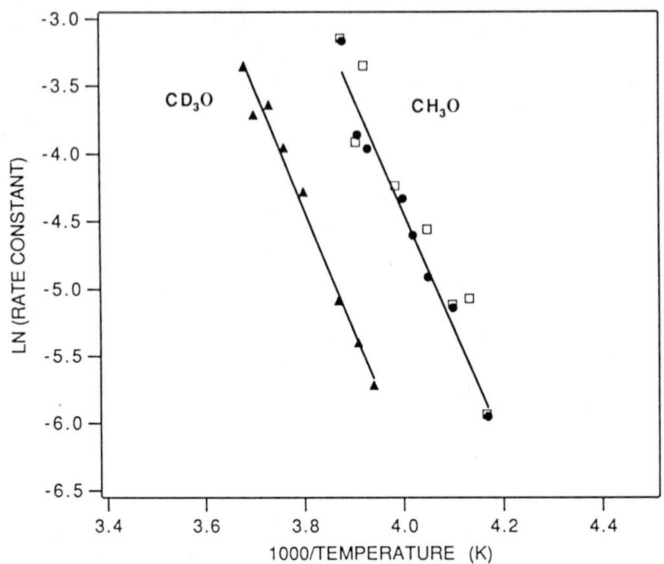

Fig. 6.11. Temperature dependence of the rate constants for the decomposition of CD_3O (*triangles*) and CH_3O (*circles*). The rate constants are obtained from the kinetic analysis described in the text. The closed symbols are for reaction on a Ni{111} surface, the open symbols are for reaction on a Ni{100} surface

A similar set of experiments and analysis was carried out on Ni{100}. Interestingly, the SHG signal from Ni{100} increases upon dissociation of methoxy, rather than decreases as it does on Ni{111}. For Ni{100}, the susceptibility of the surface plus reactants is *less* than that of the surface plus products. The fractional decrease in the SHG signal caused by adsorption of hydrogen and CO is similar on the two surfaces. The fractional decrease caused by methoxy, however, is greater on Ni{100} than it is on Ni{111}. Hence the SHG intensity increases when methoxy dissociates. Importantly, we find that the reaction kinetics on Ni{100} are indistinguishable to those observed on Ni{111}. The results for Ni{100} are indicated by the open symbols in Fig. 6.11. The values obtained for the kinetic parameters for methoxy decomposition on Ni{100} are: $Ea_{CH_3O} = 16.6 \pm 0.4$ K cal/mol, and $\nu_{CH_3O} = 1 \times 10^{12}$.

These values are in good agreement with values for the decomposition of methoxy on Ni{110}, obtained using a time-resolved, high-resolution electron-energy loss (HREELS) technique [6.61]. This reaction is a good example of a surface reaction that is insensitive to surface structure.

At a reaction temperature of 255 K, we find on both Ni{100} and Ni{111}, that $k_{CH_3O}/k_{CD_3O} \approx 6.0$. According to Transition State Theory, the relative rates of reaction of isotopically labelled species are determined predominately by the difference in zero-point energies for the vibrational mode in each isotope associated with the critical nuclear motion for reaction. The general application of this theory to the interpretation of kinetic isotope effects is described by *Melander* and *Saunders* [6.62]; recent applications to specific surface reactions are given by *Gates* et al. [6.63], *Richter* and *Ho* [6.64], and *Hall* et al. [6.65]. In the current case, methoxy bonds to the Ni surface through an oxygen $2p$ orbital, with the O–C bond tilted roughly 35° from the surface normal (see below, and [6.66]). Although, undoubtedly, some of the bending or rocking motions that bring the methoxy hydrogens closer to the surface are important for reaction, the magnitude of the kinetic isotope effect observed here suggests that the critical reaction coordinate is a higher energy, C–H stretching mode. A more detailed account of this analysis and the application of time-resolved SHG to the measurement of surface reaction kinetics will be published at a later date.

(d) Resonance Effects. Most of the above are examples of nonlinear responses that can be described in terms of a jellium metal and the charge localization associated with surface bonding. Some brief mention, in the discussion of alkali metal adsorbates, was made about enhancement of the SH intensity due to resonant interaction with an electronic transition associated with a surface state. Electronic transitions, close in energy to either the incident wave at ω or to the outgoing SH wave at 2ω, can significantly affect the SH response. Among the transitions that need to be considered for a metal surface with adsorbed overlayers are metal interband transitions, transitions involving surface states, transitions between molecular levels in the adsorbed molecules, and surface-adsorbate charge transfer processes.

Resonance enhancement of SHG was first demonstrated by *Heinz* et al. in studies of a fused silica surface covered with either rhodamine 110 or with rhodamine $6G$ [6.67]. They observed an enhancement in the SHG intensity, as a function of the photon energy of the incident laser field, corresponding to the $S0 \rightarrow S2$ transition for each adsorbed species. Since the adsorbates have different resonance energies, and since the differences were reflected in the wavelength dependence of the SH signal, these results demonstrate that SHG can provide molecularly specific information, in some cases.

The above is a simple example, in the sense that there is no contribution from any electronic structure in the substrate. On metal surfaces, this is not the case. Many more states are involved and it is possible to have more complex interference effects. In their studies of alkali metals on Rh$\{111\}$, *Tom* et al. found that the signal intensity had a complicated dependence on coverage [6.48]. As a function of coverage, changes could occur in: the position of the Fermi level with respect to the alkali derived states, a change in the substrate nonlinear response, and a change in the nature of the alkali bonding. *Dick* [6.68] has explored interference effects in systems having a nonlinear susceptibility consisting of several components. For a resonance contribution which is similar in magnitude to other components, both constructive and destructive interferences are predicted.

Both *Tom* et al. [6.48], in their studies of alkali adsorbates, and *Grubb* et al. [6.49], in their studies of CO on Pt$\{111\}$, found that the SH response to 1064 nm excitation as a function of coverage was qualitatively different than the response to 532 nm excitation. For the latter, while the SHG data for 532 nm excitation displayed a striking correspondence to the change in work function upon adsorption, the data at 1064 nm excitation did not. No such wavelength dependence was observed for CO on Ni{111}, leading the authors to postulate a resonance contribution from an interband transition or surface state specific to platinum. *Mate* et al. also reported differences in the nonlinear response to these two wavelengths in studies of adsorption of benzene and pyridine on Rh{111} [6.62]. They correlate the decrease in the SH intensity with coverage for both molecules with the intensity of an energy loss at roughly 4 eV in the EELS spectrum.

There have been only a few studies of the wavelength dependence of SHG from clean metal surfaces. *Giesen* et al., observed a maximum in the SHG intensity from Ag{111} at a SH photon energy of 3.85 eV [6.69]. The enhancement of the SHG signal at this energy was attributed to a resonance with an optical transition from a surface state into an image potential state located 0.7 eV below the vacuum level. The SH resonance energy agrees well with the energy for this transition measured in the same work by two-photon photoemission. They also found that the SHG intensity increases as the photon energy approaches the surface plasmon energy at 3.6 eV.

Hicks et al. also reported a maximum in the SHG intensity from Ag{111} at $\sim$ 3.9 eV [6.70]. They found that the SH intensity at this energy was dependent on the surface temperature, and postulated that this dependence was due to the temperature dependent population of the surface states. Using the SH intensity as an indicator of surface temperature, the authors were able to measure the time dependence of the surface temperature-jump induced by pulsed-laser excitation that preceded the SH generation pulses [6.70].

In our work on Ni{111} and Ni{100} surfaces, we find no dependence on wavelength for the change in the SHG intensity upon adsorption of electron withdrawing adsorbates such as CO, H, O, and CH_3O. However, the change in the SHG intensity induced by adsorption of electron donating adsorbates, such as CH_3OH, H_2O, NH_3, and C_2H_4, has a pronounced dependence on both wavelength and coverage. For example, in Fig. 6.12, the SH intensity is shown as function of methanol coverage for various incident-photon energies. A different coverage dependence is observed for each photon energy. For 2.33 eV excitation (532 nm), there is a monotonic decrease in the SH intensity for coverages up to 1 monolayer. The qualitative behavior is much the same as that for adsorption of CO or hydrogen, although the magnitude of the decrease of the SH intensity is significant greater for methanol. However, as the photon energy is decreased, the coverage dependence changes. At photon energies < 2.2 eV, the SH intensity first increases as the methanol coverage increases, and then, above roughly 0.4 monolayers, decreases with increasing coverage. The maximum SH intensity is observed at a coverage of 0.4 monolayers and an excitation energy of 1.68 eV, where it is about a factor of 2.5 greater than the intensity from the clean surface. As the excitation energy is decreased from 1.68 to 1.35 eV, the peak SH intensity decreases. For clarity, the data at these lower photon energies are not included in Fig. 6.12. However, the SH response at 1.17 eV (1064 nm) is given in the upper

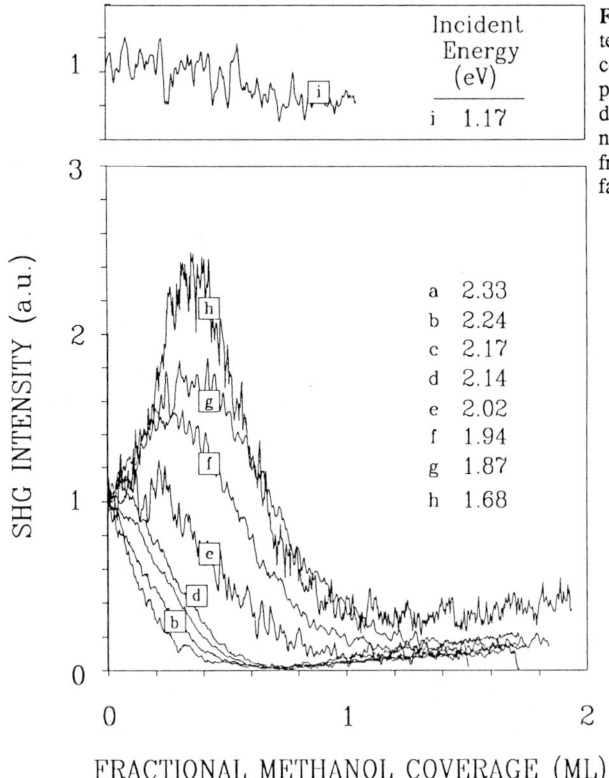

Fig. 6.12. Normalized SH intensity as a function of methanol coverage, at several different photon energies for the incident field. The SH signal is normalized to the SH intensity from the clean surface. The surface temperature is 100 K

panel. There is very little change in the SH intensity with coverage at this energy, suggesting a compensation from two or more offsetting effects.

The relative SH intensity at fixed methanol coverage as a function of incident-photon energy can be extracted from the data in Fig. 6.12. This is shown in Fig. 6.13, for several coverages ranging from 0.1 to 0.4 methanol per surface Ni atom. The solid lines in Fig. 6.13 are calculated from a model which we now describe.

Dick [6.68] has shown that the interference amongst components of the second-order susceptibility can result in a nonlinear response that has complex wavelength dependence. The dispersion of the data in Fig. 6.13 is certainly suggestive of an interference. To model the data, a simple phenomenological extension of (6.4) is taken for the effective total susceptibility:

$$\chi_{s,\mathrm{eff}} = \chi_{ss} + \chi_{nr} + \chi_{res} \,, \tag{6.13}$$

where, as before χ_{ss} is the nonlinear response of the clean surface, χ_{NR} is the nonresonant contribution associated with the surface free-electrons, and χ_{res} is a resonant contribution. The nonresonant term includes both χ_A for methanol, and χ_I for the perturbation of χ_{ss} by methanol bonding to the surface. Assum-

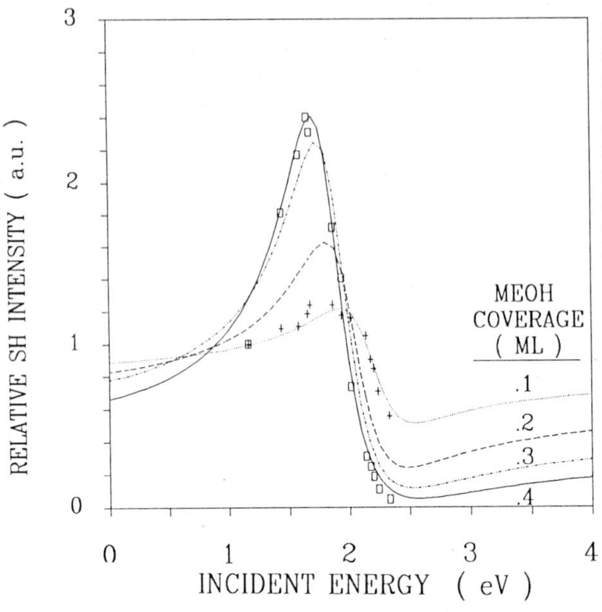

Fig. 6.13. Second-harmonic intensity (relative to the intensity from the clean surface) as a function of the photon energy of the incident field, at several different coverages of methanol. These data are taken from the data in Fig. 6.12. The solid lines are a fit to a resonance model described in the text

ing a single resonance, which has a Lorentzian dependence on wavelength, the nonlinear susceptibility, normalized to the clean surface value, can be written as:

$$\chi_{\text{normalized}} = 1 + \left(\frac{B}{A}\right) \mathrm{e}^{-\mathrm{i}\phi} \left(\frac{\vartheta}{\vartheta_{\text{satn}}}\right) + \frac{(C/A)\mathrm{e}^{-\mathrm{i}\psi}}{\hbar(\omega - \omega_{\text{res}} + \mathrm{i}\Gamma_{\text{res}})} \left(\frac{\vartheta}{\vartheta_{\text{satn}}}\right) . \quad (6.14)$$

The parameters, A, B, ϑ and ϑ_{satn} are defined as in (6.7) the parameter, C, is the magnitude of the resonant component, ϕ and ψ, respectively, are the phases of the nonresonant and resonant components relative to the phase of the clean surface response, ω is the frequency of the driving field, ω_{res} is the frequency of the resonance, and Γ_{res} is the characteristic width of the resonance.

This model for the susceptibility can account for the observed dependence on both wavelength and coverage. The curves in Fig. 6.13 are calculated from a least squares fit, using (6.3) and (6.14) to the SH intensity data as a function of wavelength at 20 separate coverages. Curves for only 4 coverages, and data for only two coverages, are shown. The remainder are omitted from the figure for clarity. The fitting was done with ϕ fixed at 180° (approximately the value found for CO and H), and Γ fixed at 0.34 eV FWHM. The remaining parameters were allowed to vary in order to minimize the deviation from the data. The values for the parameters obtained from this procedure are, B/A = 1.0, C/A = 3.5, and $\psi \approx 0°$. The values obtained for ω_{res} as a function of coverage are plotted in Fig. 6.14. A satisfactory fit to the data was obtained with these parameters. A somewhat improved fit was obtained when C/A or Γ is allowed to vary with coverage. We are currently investigating whether these variations are significant within the accuracy of the available data.

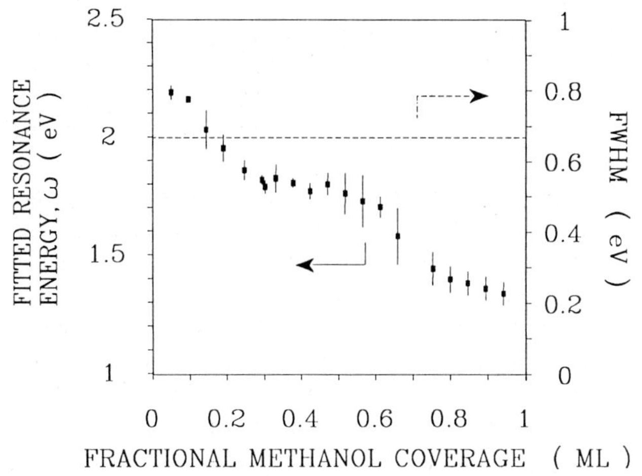

Fig. 6.14. Fitted resonance energy (*symbols*) as a function of methanol coverage (see text). Also shownis the FWHM of the resonance (dotted line and right hand axis), which was taken to be independent of coverage for the calculations here

Both constructive and destructive interference, with respect to the clean surface SH intensity, is observed in the lineshapes shown in Fig. 6.13. It is only when the magnitude of the resonant contribution is much greater than the nonresonant contribution, or when the two components are in quadrature, that the SH lineshape reflects the Lorentzian lineshape of the resonance. For methanol on Ni{100}, apparently, the resonant and nonresonant components are nearly out of phase. The coverage dependence appears to come largely from a shift in the fitted resonance energy, varying from 2.3 eV at low coverages to ~ 1.4 eV at 1 monolayer.

It is important to note that (6.14) is not unique in its ability to fit the data. Other expressions may provide an equally satisfactory description of the dispersion and its dependence on coverage. In addition, there can be more than one physical origin for this phenomenological expression. For example, there may be a resonance associated with the metal surface that is affected in direct proportion to the adsorbate coverage, or the resonance may originate in the adlayer or between surface states and adlayer states, again varying directly with coverage. Thus, fitting the data to this form is more a test of the plausibility of one of the more simple phenomenological expressions for the susceptibility than it is a rigorous test of physical reality. Nevertheless, the model provides some hints as to the properties a resonance must have to account for the data.

This resonance behavior is somewhat surprising because there are no electronic transitions in methanol, or interband transitions in Ni{111} that are close in energy to the incident photon energy, or to the SH photon energy. There is, however, an image-potential state in Ni{111} at the right energy to be nearly resonant with the SH photon. An electronic transition has been observed at 4.69 eV using two-photon photo-emission [6.71]. The transition is from a surface state, located 0.25 eV below the Fermi level [6.72], to the image-potential state, located

0.8 eV below the vacuum level [6.71]. The SH photon energy, over the range studied here, is slightly below this transition energy. We believe this transition is responsible for the wavelength dependence of the change in the SH intensity induced by electron donating adsorbates, and that it is an important factor with electron withdrawing adsorbates as well.

The transition energy between the surface state and the image-potential state is expected to be influenced by both charge donating and charge accepting adsorbates. For methanol, there is a relatively weak interaction with the surface through charge donation from the oxygen lone-pair electrons. There is little perturbation of the electronic energy levels in methanol and only a slight change in the electron density at the surface. There is, however, a significant reduction in the work function of the surface due to the dipole field in the adlayer. The work function decreases monotonically with increasing methanol coverage. At, $\theta_{CH_3OH} = 1$ monolayer, the change in the work function, $\Delta\phi$, is -1.6 eV [6.72]. As the work function decreases, the surface state and the Fermi level move closer in energy to the vacuum level. To a first approximation, the transition energy will be given by, $\Delta E_{res} \approx (4.69\,\mathrm{eV} - \Delta\phi)$. For a linear relationship between $\Delta\phi$ and θ_{CH_3OH}, this becomes, $\Delta E_{res} \approx (4.69\,\mathrm{eV} - (\theta_{CH_3OH} \times 1.6\,\mathrm{eV}))$.

This change in the resonance energy with work function (coverage) can account for all of the effects observed. We find that the fitted resonance energies, plotted in Fig. 6.14, are equal to $\Delta E_{res}/2$, within experimental error. (The fitted resonance energies are equal to $\Delta E_{res}/2$ because the resonance term in (6.14) is for a resonance at frequency ω. For a resonance at frequency 2ω, the fitted resonance energies would equal ΔE_{res}.) For the clean surface, the SH photon energy is nearly at the peak of the resonance at an excitation energy of 2.33 eV (532 nm). As methanol is added, ΔE_{res} shifts to lower energies, away from the SH energy. This is consistent with the monotonic decrease in the SH intensity with methanol coverage for this excitation energy. At lower excitation energies, the SH photon energy is at lower energy than the peak energy of the resonance. As methanol is added, ΔE_{res} shifts initially into resonance with the SH photon energy, and then past it. Thus, there is first an increase in the SH intensity, and then a decrease.

The lack of a wavelength dependence of the relative SH intensity for electron withdrawing adsorbates, such as CO, oxygen and hydrogen, is not inconsistent with the above model. All of these adsorbates increase, rather than decrease, the work function. Thus, for these species, the effect of adsorption is to shift ΔE_{res} to higher energies than all of the SH photon energies generated in our experiments. In addition, photoemission measurements show that these species reduce electron density in the states near the Fermi level so that the free-electron component of the nonlinear response also is decreased by the formation of these surface bonds. On both counts, the relative SH intensity will decrease with adsorption, at all excitation frequencies in this study.

Currently, it is difficult to quantify the relative contribution of each of the terms in (6.14). The wavelength dependence indicates that the resonant contribution is important. However, the lineshape indicates that the nonresonant component is also significant. To determine the relative importance of these two components, it is necessary to know, among other attributes, the *absolute* SH intensity as a function of wavelength. At present, we have only a measure of the SH intensity *relative* to the signal from the clean surface.

(e) Rotational Anisotropy. Jellium models of the second-order polarizibility predict that the SH response from a cubic metal is independent of crystal face or orientation of the polarization in the plane of the crystal surface. It is clear that a free-electron model is not an adequate description of a semiconductor, and many studies have demonstrated that there is significant rotational anisotropy on semiconductor surfaces [6.73–78]. More surprisingly, a number of recent studies have shown that there is rotational anisotropy in the SH response of cubic metal surfaces [6.1, 51, 79, 80]. This is an important characteristic of SHG because it can provide direct information on the structural symmetry of the surface. Furthermore, such measurements can be made under reaction conditions and in real time.

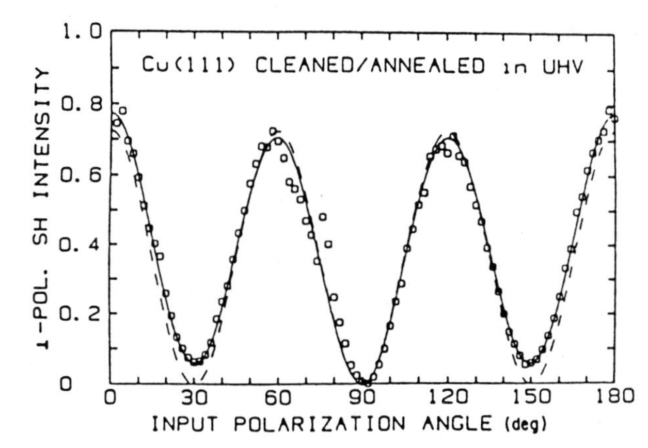

Fig. 6.15. Second-harmonic intensity from a Cu{111} surface as a function of the input polarization angle, where the SH field is measured with polarization perpendicular to the input polarization. Data taken with 0.25° angle of incidence (*open circles*). Theory assuming exactly normal incidence (*dashed line*). Theory accounting for nonzero angle of incidence (*solid line*). From [6.79]

In Fig. 6.15, we reproduce a figure from *Tom* and *Aumiller* [6.79] which shows the rotational anisotropy of a Cu(111) surface in UHV. In this case, the input polarization angle is varied and the SH signal polarized perpendicular to the input field is measured. The angle of incidence is 0.25° from the surface normal. In the limit of exactly normal incidence, this measurement is equivalent to measuring the p-polarized SH signal generated from s-polarized excitation and rotating the sample about its surface normal, i.e., measuring the dependence of the SH signal on the azimuthal angle.

Macroscopic theories for the rotational anisotropy have been developed by *Tom* et al. [6.73, 79], and by *Sipe* et al. [6.81]. The surface nonlinear susceptibility, $\chi_s^{(2)}$, and the bulk anisotropic susceptibility, ζ, can both contribute to the rotational anisotropy. There are four independent tensor elements for a $\{111\}$ surface with $3m$ symmetry. Three are isotropic and the fourth, $\chi_{\xi\xi\xi}$, is anisotropic. Expressions have been derived for the total reflected s- and p-polarized SH fields from {111}, {100}, and {110} crystals under s- and p-polarized excitation [6.73, 81]. They are of the general form:

$$I(2\omega) = |A + Bf(\psi)|^2 . \quad (6.15)$$

The constant A represents a linear combination of the isotropic components, the constant B is proportional to the anisotropic susceptibilities $\chi_{\xi\xi\xi}$ and ζ. These constants depend on the dielectric constants and on the excitation geometry. The azimuthal angle, ψ, is usually taken to be the angle between the plane of incidence and a mirror plane or the projection of a crystal axis direction onto the surface. The function, $f(\psi)$, is a linear combination of $\cos(m\pi)$ and $\sin(m\psi)$, and reflects the 2 mm, 3 m and 4 mm symmetry of the {110}, {111}, and {100} surfaces, respectively.

The polarization dependence, illustrated in Fig. 6.15, exhibits 3m symmetry. The topmost atomic layer of Cu{111} has 6m symmetry. Inclusion of the next atomic layer reduces the symmetry to 3m. Thus, the observed response originates from at least the first two layers. The surface electric-dipole contribution to the nonlinear susceptibility is thought to arise from the first 10 Å [6.79]. The bulk contributions extend over the penetration depth of the field, approximately 100 Å. Thus, the reduction to 3m symmetry seems reasonable.

Tom and *Aumiller* also investigated the effects of oxygen adsorption on the SH anisotropy from Cu{111} [6.79]. An exposure of 2000 L reduced the SH intensity to 9 % of the intensity observed for the clean surface, but the SH rotational anisotropy, and presumably the surface order, persisted. Even after exposure to air, no further change in the SH signal was observed. This suggests that the structure of the Cu layers underlying the surface oxide is not substantively altered. This illustrates the ability of SHG to provide information on the structure as well as on the electronic properties of buried interfaces.

The effect of an adsorbate on the SH rotational anisotropy was also investigated in several articles by *Anderson* and *Hamilton* [6.50, 51]. They studied the adsorption and desorption of CO and hydrogen on Ni{111}. Again, the magnitude of the SH intensity declines with coverage, but a threefold anisotropy is retained even to saturation. Interestingly, at saturation coverages, long range order in the adlayer is disrupted, but the rotational anisotropy remains. This indicates that SHG is sensitive to the local site symmetry, rather than longer range effects.

The SH rotational anisotropy also has been studied from metal films [6.82, 83], and from electrochemical electrode surfaces in solution [6.84, 85]. In an interesting series of articles, *Richmond* and coworkers have pioneered the application of SHG to the study of surface structure and symmetry at the electrode/electrolyte interface. For Ag{111} electrodes, they observed a 6-fold symmetric pattern which agreed with the macroscopic theory outlined above. These results demonstrated that an ordered surface is present in the electrolyte solution [6.84]. Evidence for surface restructuring was found as the potential was biased within the limits of the ideally polarizable region [6.85]. The bias-potential induced restructuring was found to be independent of the electrolyte and prior electrode treatment. Similar studies of a Cu{111} electrode revealed a three-fold symmetric pattern that did not change in form or in magnitude as a function of bias potential [6.86]. The differences between Ag and Cu were attributed to differences in the relative contributions of anisotropic surface and bulk components of the susceptibility. Modifications of the surface induced by the deposition of thallium overlayers were also studied in this work. Finally, Richmond and coworkers demonstrated the ability to perform time-resolved SH measurements to obtain intriguing new information on the kinetics of adlayer deposition and stripping [6.1].

6.1.5 SHG Summary

There have been significant advances in the theory and application of SHG over the past 6 years. A number of examples have been discussed to show that a rich variety of information can be obtained. The technique can be applied to many problems in surface and interfacial sciences. The specificity to the interface, sensitivity, and time response make it a valuable complement to existing methods. There are several new directions and challenges for the future. For metal surfaces especially, advances in the microscopic theory are needed in order to derive full benefit from the information obtainable. With improved models for the nonlinear response of transition metals, SHG could provide new information on the nature of the electrons involved in the formation of surface chemical bonds. It is likely that the emphasis for experimental efforts in the near future will turn to the study of surfaces in reactive environments (CVD, ion etching, thin film growth), higher pressures (catalytic reactivity, sensors), and buried interfaces (electrochemistry, polymer coatings, corrosion).

6.2 Sum-Frequency Generation

6.2.1 Introduction and Background

Following on the pioneering work of *Shen* and coworkers [6.87–90], a considerable interest has developed in the use of sum-frequency generation (SFG) for the measurement of surface electronic states and electronic and vibrational states of molecules adsorbed at an interface. Among the various possible sum- and difference-frequency combinations, IR-visible sum-frequency generation has commanded the most interest. For this review, we will focus exclusively on IR-visible sum-frequency generation.

Infrared-visible sum-frequency generation has several advantages for applications in surface spectroscopy. Like second-harmonic-generation, SFG is inherently sensitive only to the interfacial region between centrosymmetric media. Like IR and Raman spectroscopy, IR-visible SFG provides information on molecular vibrational modes, and thereby, on molecular interactions and orientations. This molecular specificity overcomes the chief limitation of second-harmonic-generation. Finally, SFG has the potential to provide sub-picosecond time resolution. The capability to detect specific molecular motions at buried interfaces and to provide short time-scale resolution are the chief advantages of SFG over conventional IR and Raman techniques.

Most of the initial SFG work was done in air, on glass or water surfaces. *Zhu* et al. reported the first vibrational SFG spectrum, measuring the vibrational spectrum in the vicinity of 10 μm of a monolayer of cumarin 504 adsorbed on glass [6.87]. They proposed that the relatively strong SFG signals obtained from this adlayer is the result of an additional enhancement due to a resonance of the SFG frequency with the frequency of a visible transition in cumarin. They found that the difference-frequency signal, the frequency of which is not near the frequency of the visible transition, is much weaker. *Hunt* et al. measured the C–H stretch modes around 3μm of methanol and pentadecanoic acid adsorbed on glass and at an air/water interface [6.88]. By measuring the polarization dependence of the SF response from pentadecanoic acid in water, they determined that the molecules in the liquid phase have the alkane chains extended and oriented nearly normal to the surface [6.89].

Superfine et al. reported the first spectra of molecules adsorbed on metal and semiconductor surfaces. They measured the spectra of octadecyltrichlorosilane on silicon and aluminum films and ethylidyne (-C–CH_3) on Rh$\{111\}$ [6.90]. They obtained spectra from the Al film with visible pulse energies just below the surface damage threshold. A similar limitation was encountered with the Rh$\{111\}$ surface. *Harris* and *Levinos* measured the spectra of n-octadecyl thiol on a gold film, of cadmium stearate on a silver film, and of stearic acid on germanium, again in the 3μm region of the spectrum [6.91]. All of these spectra were taken with the samples in air and the effects of unknown surface impurities and adlayer morphology are difficult to assess.

6.2.2 Theory

The process of SFG at an interface or surface has been discussed in a number of earlier studies [6.89, 91]. Briefly, IR-visible SFG is a second-order, nonlinear optical process in which an infrared (ω_{IR}) and visible (ω_{vis}) wave couple via the nonlinear susceptibility, generating a third wave (ω_{sum}) whose frequency is the sum of the frequencies of the two input fields. The nonlinear polarization that arises from the two wave excitation can be written as [6.3, 89, 90]

$$\boldsymbol{P}^{(2)}\left(\omega_{\mathrm{sum}} = \omega_{\mathrm{IR}} + \omega_{\mathrm{vis}}\right) = \chi^{(2)} : \boldsymbol{E}\left(\omega_{\mathrm{IR}}\right)\boldsymbol{E}\left(\omega_{\mathrm{vis}}\right) , \qquad (6.16)$$

where $\chi^{(2)}$ is the total nonlinear surface susceptibility, $\boldsymbol{P}^{(2)}(\omega_{\mathrm{sum}})$ is the polarization induced at the sum frequency, and the $\boldsymbol{E}(\omega_{\mathrm{i}})$ are the local field magnitudes at the IR and visible frequencies respectively. One may model the surface susceptibility in lowest order as the sum of resonant and nonresonant terms, where, in this case, the resonant term is associated with a vibrational mode of the adsorbate, and the nonresonant term is associated with the substrate plus adsorbate.

$$\chi^{(2)} = \chi^{(2)}_{\text{res}} + \chi^{(2)}_{\text{nr}} , \tag{6.17}$$

where $\chi^{(2)}_{\text{res}}$ and $\chi^{(2)}_{\text{nr}}$ are resonant and nonresonant components of the susceptibility, respectively. The SFG intensity arises from this local nonlinear polarization and is proportional to the product of the local visible and IR field intensities with $|\chi^{(2)}|^2$.

The *resonant* part of the SF susceptibility (derived from second-order perturbation theory) has a Lorentzian form: [6.2, 89, 90]

$$\left(\chi^{(2)}_{\text{res}}\right)_\nu = \frac{N A_\nu M_\nu \Delta\varrho}{\hbar\left(\omega_{\text{IR}} - \omega_\nu + \mathrm{i}\Gamma_\nu\right.} , \tag{6.18}$$

where N is the surface density of molecules, A_ν is the infrared transition moment and M_ν is a term that is proportional to the Raman transition moment of the vibrational mode ν, ω_ν is the vibration frequency of this mode, Γ_ν the damping width, and $\Delta\varrho$ the population difference between the ground and excited states. The resonance contribution becomes significant as the infrared laser at ω_{IR} is tuned through a vibrational transition provided the mode is both Raman and infrared active, i.e., both A_ν and M_ν must be nonzero. The SFG spectrum thus provides information similar to that obtained from conventional vibrational spectroscopy but the signal originates predominantly from the surface. It should be noted that $\chi^{(2)}_{\text{res}}$ can have a slowly varying dependence on the visible frequency, through the M_ν term [6.2].

The *nonresonant* part of the SF susceptibility can have several components, in principle having a contribution from the adsorbate, a contribution from the surface, and a cross term resulting from the adsorbate/surface interaction, much like the SH susceptibility. Earlier studies have shown that for glass and water surfaces, the adsorbate vibrational resonance terms dominate the SFG spectrum, resulting in lineshapes similar to those obtained in IR absorption [6.87–89]. For metal and semiconductor surfaces, however, the magnitude of the nonresonant contribution from the surface can be on the same order as the magnitude of the resonant contribution. For a Lorentzian resonant term, the relative phase will vary rapidly as the frequency of the IR field is tuned through the vibrational resonance. In addition, a relative phase difference can exist between the nonresonant and resonant polarizations. Since the signal is proportional to the square of the complex sum of the resonant and nonresonant terms, phase differences are important and can lead to different lineshapes in SFG than those observed in a linear absorption process.

6.2.3 Experimental Methods

Many of the experimental considerations discussed in the SHG experimental methods section also pertain to SFG. However, SFG involves the coupling of two optical fields, and the experimental requirements for SFG tend to be more complex. The generation of two excitation fields usually necessitates the use of separate laser sources as well as separate optical trains. The k-vector matching condition for the photon fields causes the propagation direction of the SF beam to change as the frequency of the tunable excitation beam is changed. This necessitates more complex signal-collection optics. Also, since two excitation fields are employed, care must be taken into provide temporal, as well as spatial, overlap of the beams at the surface.

A schematic of a generalized SFG experimental arrangement, for reflection geometry, is shown in Fig. 6.16. Although, an SF field can be generated utilizing an internal reflection geometry for the pump beams [6.1], this configuration is generally not applicable to metal substrates thicker than $\sim$ 100 Å and will not be discussed here. There has been one report in which ω_1 and ω_2 were both

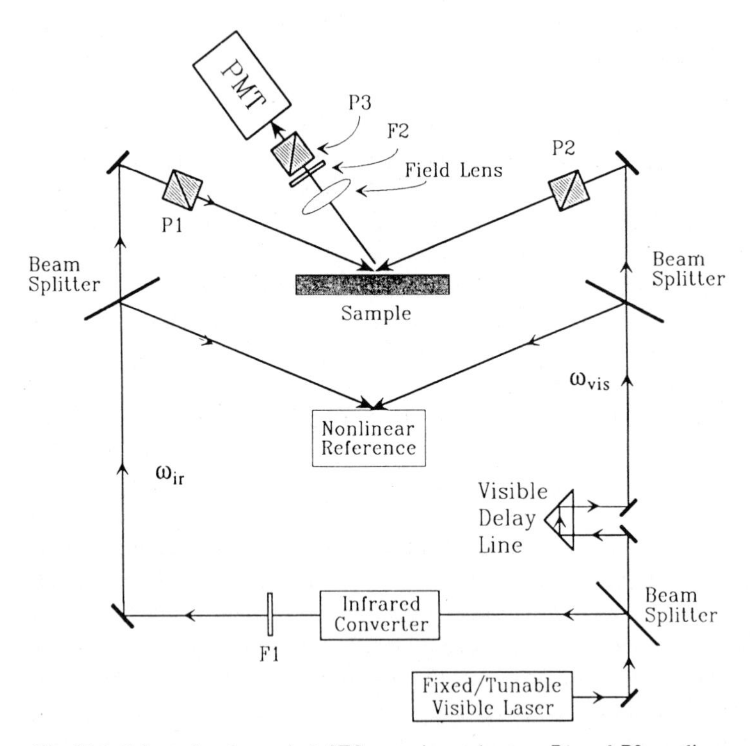

Fig. 6.16. Schematic of a typical SFG experimental setup. P1 and P2 are linear polarizers for the input infrared and visible pump fields, respectively. The delay line in the visible input provides the proper temporal overlap of the visible and infrared intensities. The generated SFG signal is focused onto a photomultiplier tube (PMT) via a field lens. Scattered visible light is filtered from the signal by F2, and its polarization is analyzed with P3. See the text for a discussion of the laser sources and the nonlinear reference

visible laser beams [6.92]. Typically, however, one fixed-frequency visible laser and one tunable IR laser are employed. In Fig. 6.16, linearly polarized light of frequencies ω_{IR} and ω_{vis} are shown, incident in a counterpropagating geometry on the sample surface; these fields generate a reflected field at the frequency $\omega_{sum} = \omega_{IR} + \omega_{vis}$. Temporal overlap of the incident beams is achieved by placing a variable delay line in the optical path of the visible beam. The propagation direction of the reflected SFG signal is determined by matching of the k-vectors of the incident IR and visible waves at the sample surface. The incident angles of the pump beams are generally selected so that the SFG signal will *not* have the same propagation direction as that of the specularly reflected visible beam. Thus, the SFG signal may be separated from the reflected visible beam by spatial filtering. The detection optics consist of a field lens, bandpass filters and/or a monochromator (F2), and a polarization analyzer for the SF beam (P3). The propagation direction of the SF beam will change when ω_{IR} (or ω_{vis}) is changed. The field lens can compensate for small angle changes by imaging the signal from the sample surface onto the photon detector. The combination of bandpass filter and/or monochromator provides rejection of visible light from scattered laser light and from other sources.

After the last optical stage, the SF photons are detected with a photomultiplier tube (PMT), the output of which is monitored by a gated integrator. As with SHG, the SF signal from the sample generally is normalized to an SF signal from a nonlinear reference to improve the signal-to-noise ratio.

The generation of a detectable SF signal requires high intensities for both the visible and IR input beams. In most investigations, Q-switched, doubled Nd:YAG laser radiation (532 nm) provides visible input for SFG at the sample surface as well as the pump for the IR converter. Both picosecond and nanosecond lasers have been used as primary excitation sources. A dye laser may be used to provide a tunable visible source. For studies of metal surfaces, the incident visible power density on the sample surface must be chosen so as to avoid laser desorption of adsorbed layers as well as ablation of the metal surface. The advantages of using picosecond pulses rather than nanosecond pulses apply to SFG as well as SHG.

To date, two methods have been employed in the generation of high IR pulsed laser power for surface SFG. The first involves the stimulated Raman conversion of tunable visible radiation from a dye laser (560 to 800 nm) into tunable near infrared radiation (2 to 5 μm). Examples include stimulated vibrational Raman scattering in hydrogen gas [6.93], and stimulated electronic-state Raman scattering in cesium vapor [6.91]. The former has been used to generate 1–10 mJ per pulse with a pulse duration of 7 nanoseconds. The latter has been used to generate 10–30 μJ per pulse with a pulse duration of 3 picoseconds. Bandwidths of 0.1 cm^{-1} are typically achieved in both processes.

Another well-developed conversion procedure involves a nonlinear, three-wave coupling process known as optical parametric conversion [6.88, 94]. In this process, a fixed frequency pump laser field (usually 1.06 μ) is converted in a nonlinear media such KDP or $LiNbO_3$ into two longer wavelength fields, and a field at the difference frequency is generated. A requirement of this process is that the generated fields conserve both the energy and k-vector of the incident field. Because the k-vectors are functions of the refractive index of the material, which in turn is a function of the angle between the propagation direction of these fields in the crystal and its optical axis, it is possible to tune the output frequency by simply varying the angle of propagation of the pump field through the crystal. In most commonly used nonlinear crystals, IR tunability between 1.5 and 4.0 μm with a bandwidth of approximately 4 to 6 cm^{-1} is achieved [6.94]. An advantage of this technique over Raman conversion is in its ability to generate high power picosecond laser pulses throughout this wavelength region.

As a final comment, the sample surface must be clean and well characterized. The procedures mentioned in SHG section for sample preparation also apply here.

6.2.4 Experimental Results

In Fig. 6.17, we reproduce the SFG spectrum of methanol on a glass surface, from *Hunt* et al. [6.90]. The authors used a visible beam of 1 mJ/pulse at 532 nm, and an IR beam (from an optical parametric amplifier) of 0.2 mJ/pulse in the range of 3.2 to 3.6 μm. The pulse width of both lasers was $\approx$ 15 picoseconds. The polarizations of the beams were not specified. Prior to the adsorption of methanol, the SFG spectrum from the glass surface was "very weak" and showed little structure. A saturated layer of methanol was obtained by rinsing the glass once with liquid methanol. Additional rinsing did not increase the SFG signal, but the density of methanol on the surface was not determined by any alternative methods. The strong peak and its shoulder in Fig. 6.17 was identified, by analogy with the spectrum for liquid methanol, as the asymmetric C–H stretch and a Fermi resonance (or a second component of the asymmetric stretch), respectively. The authors expect the symmetric C–H stretch to appear at around 2840 cm^{-1}, but

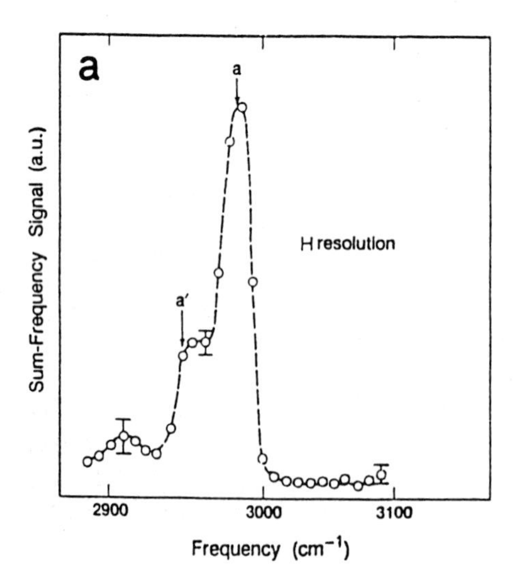

Fig. 6.17. The SFG spectrum of methanol on glass. The open circles are the experimental points. The dotted line is a guide to the eye. Arrows a and a' indicate the CH_3 resonances of methanol in the liquid phase. From [6.88]

did not attempt to detect it. The small peak at $2910\,cm^{-1}$ was not identified. The lineshapes of the SFG resonances are quite broad, but they resemble the lineshapes observed in IR absorption in the liquid phase. An SFG spectrum of methanol on a Ni{100} surface will be presented below for comparison.

We have measured the SFG spectra of a number of adsorbates, including CO and methanol, on a Ni{100} surface in UHV. The nonlinear polarizability of the metal surface is not insignificant in comparison to the vibrational resonances and it is important to understand this in some detail if, for example, the SFG lineshapes are to be understood. The ability to control the surface cleanliness, and the coverage and composition of the adlayer, is essential in this.

The experimental setup and sample preparation for these studies have been discussed previously [6.95]. The excitation beams consisted of a visible beam of 2 mJ/pulse at 532 nm, and an IR beam (generated by stimulated Raman scattering from H_2 gas [6.93]) of 2 to 6 mJ/pulse in the range from 3.2 to 5.4 μm. Both laser beams had a pulse width of roughly 7 nanoseconds. The SFG signal from the Ni surface was normalized to the SFG signal from a ZnSe sample so as to minimize shot-to-shot fluctuations in the laser amplitude. All scans were recorded in $0.5\,cm^{-1}$ increments of the infrared frequency, each increment being an average of 20 laser pulses. Peak SFG intensities correspond to approximately 100 photons per pulse.

In Fig. 6.18, SFG spectra for CO on Ni{100} as a function of coverage are shown. An SFG signal is obtained from the clean Ni{100} surface prior to the

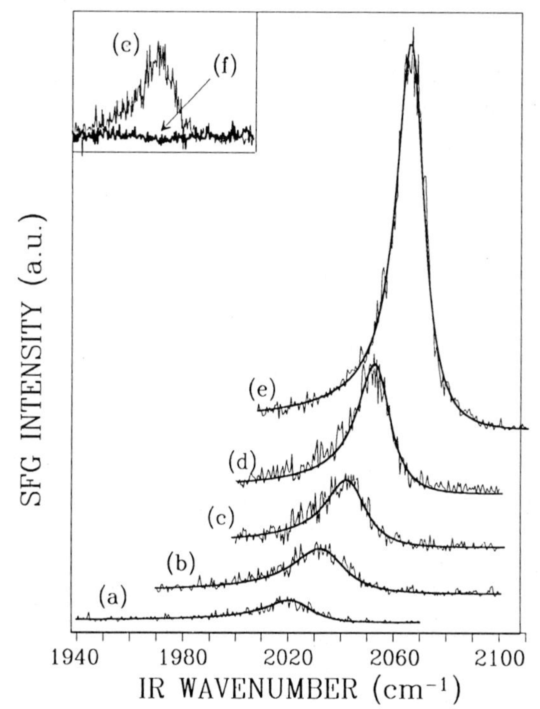

Fig. 6.18. The SFG spectrum of CO on Ni{100}, for various coverages of CO. (*a*) 0.09 ML CO; (*b*) 0.17 ML CO; (*c*) 0.25 ML CO; (*d*) 0.33 ML CO; (*e*) 0.48 ML CO. Solid line through the data is a least squares fit to the data using model described in the text. Inset compares the SFG signal from the clean Ni{100}, (*f*), surface with curve (*c*)

adsorption of CO, as shown in the inset of Fig. 6.18. This signal is roughly independent of the input IR frequency from 1850 to 2100 cm^{-1}. Upon adsorption of CO at 90 K, a single resonance feature is observed in the SFG spectrum. The frequency of the resonance is characteristic of the stretching vibration of CO bound to an on-top site on a nickel atom (a linear bond). No resonance corresponding to bridge-bonded CO is observed for coverages up to 0.48 monolayers (ML). As the coverage increases from 0.09 to 0.48 ML, the resonance shifts from 2023 cm^{-1} to 2067 cm^{-1}. At the same time, the linewidth decreases from 21.1 cm^{-1} to 12.4 cm^{-1}.

Previous studies of CO on Ni{100} using linear reflectance and IR emission spectroscopy have reported similar changes in the resonance frequency over this coverage range [6.96–98]. In the earlier works, the frequency shift was attributed to the change with coverage of the dynamic dipole coupling among neighboring CO molecules and their image dipoles. This is the dominant effect here because there is no change in the adsorbate bonding site with coverage, and changes in the population of the $2\pi^*$ anti-bonding orbital are small [6.99, 100]. Larger effects are observed for bridge bonded CO where binding site and chemically-induced changes in the charge-density in the bonding orbitals can be significant [6.101].

A relatively broad linewidth at low coverage, and a narrowing of the linewidth at high coverages are observed both in the SFG spectra reported here and in previous linear spectroscopy studies [6.95]. The linewidth is determined by homogeneous and inhomogeneous broadening. Homogeneous broadening is due to both vibrational-phase disruption (dephasing) and vibrational energy relaxation. Inhomogeneous broadening typically is due to a distribution of local molecular environments [6.102–104]. Relaxation and dephasing processes for the CO stretching are unlikely to result in more than approximately a 6 to 8 cm^{-1} linewidth at liquid nitrogen temperatures [6.105, 106]. The remaining width is likely due to inhomogeneous broadening. LEED studies have shown that, in fact, the CO overlayer remains disordered until reaching 0.5 ML [6.107]. Under these conditions, a distribution of intermolecular spacings between CO molecules exists on the surface, resulting in a distribution of adsorbate-adsorbate interactions in the overlayer. This distribution leads to a resonance linewidth greater than the homogeneous linewidths. The distribution becomes more nearly uniform as the long-range order sets in at high coverages, resulting in a decrease in the linewidth.

In a number of earlier studies an asymmetry was observed in the C–O resonance lineshape at low surface coverages [6.96–98]. Generally, in linear spectroscopies, an asymmetric lineshape suggests a degree of inhomogeneity. An asymmetry is also observed in the SFG spectra in Fig. 6.18; however, the asymmetry in the resonance lineshape is in part due to interference effects. Inherently, these phase effects are not present in a linear optical process.

Our approach to modelling the SFG lineshape is similar to that used in earlier works [6.88, 91]. For equations (6.17) and (6.18), $\chi^{(2)}_{\mathrm{res}}$ is assumed to be a single Lorentzian line, and $\chi^{(2)}_{\mathrm{nr}}$ is assumed to be a complex, wavelength independent term. In this simplified case, the SFG lineshape is determined by the magnitudes of the resonant and nonresonant terms and by the relative phase between them. A least squares fitting routine is used to fit the spectra in Fig. 6.18 to a Lorentzian resonance of adjustable position, width, and amplitude, plus a nonresonant term of adjustable amplitude and phase. The solid lines in Fig. 6.18 indicate the fit.

A more realistic model for the lineshape would include inhomogeneous broadening in the resonant term. It would also allow for frequency dependence of the nonresonant term in order to satisfy causality (Kramers-Kronig relations). However, with the signal to noise of the current experiments, there is little to be gained by a more sophisticated model. The fitting routine statistics showed that the probability of finding a better fit to the data was near zero. At the lower coverages for which we know that inhomogeneous broadening occurs, that probability did show a tendency to increase, suggesting that a single Lorentzian line may be inadequate to describe the lineshape. However, it should be emphasized that the asymmetry observed in these lines has its principal source in the interference effects that strongly depend on the relative phase of the resonant to nonresonant generated polarizations.

The physical origins of this optical phase difference are not presently understood. We note that the phase differences obtained from the modeling remain roughly constant at –1.13 radians for coverages up to 0.48 ML, where the phase shift becomes 0.76 radians. Since overlayer ordering occurs only at these higher coverages on the Ni{100} surface, the change may result from adsorbate-adsorbate interactions which alter the phase with which the overlayer responds to the driving fields.

We have also taken SFG spectra during the adsorption and reaction of methanol on the Ni{100} surface. The reactions are identical to those indicated in Sect. 6.1.4 (c) for methanol on Ni{111}. In Fig. 6.19a, the SFG spectra obtained from the clean Ni surface in the C–H stretching region, from 2775 to 3000 cm^{-1}, is shown. There is a significant, nonresonant, SFG response originating from the nonlinear polarizability of the metal electrons. It should be possible to relate the SFG signal intensity to the SHG signal from the clean surface since both arise from the same second-order susceptibility. However, that relationship can be determined only if $\chi^{(2)}$ is independent of frequency or if its full frequency dependence is known. Although $\chi^{(2)}_{nr}$ seems to be nearly constant over the range of a typical infrared scan, about 250 cm^{-1}, we find significant differences in the background SFG intensity when changing dyes and going over larger spectral regions, e.g., from 3000 cm^{-1} to 2000 cm^{-1}. Currently, the uncertainties in normalizing signals obtained using different dyes is too large to determine the wavelength dependence over a sufficiently wide range to relate the SFG and SHG intensities.

In Fig. 6.19b, the SFG spectrum is shown after adsorbing 0.12 monolayers (ML) of methanol onto the surface at 90 K. Two distinct peaks are observed at 2803 and 2909 cm^{-1} with linewidths of approximately 30 cm^{-1}. These peaks are assigned as the CH_3 symmetric and asymmetric stretches respectively. The frequencies of these peaks are very close to the frequencies observed in investigations of methanol adsorption on Ni{111} and Ni{110} using high-resolution electron energy-loss spectroscopy (HREELS) [6.108] and on Cu{100} using IR

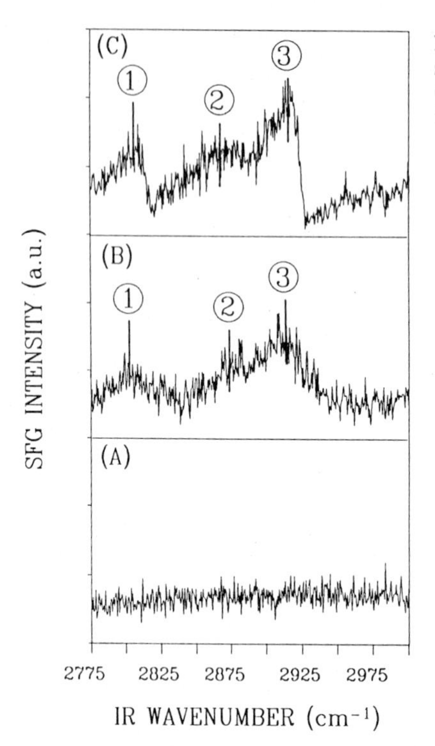

Fig. 6.19. The SFG spectra from Ni{100} in the C–H stretching region. (*a*) Clean Ni{100} surface at 90 K; (*b*) with 0.12 monolayers of CH_3OH; (*c*) after heating to 200 K for 1 minute. From [6.95]

absorption-reflection spectroscopy (IRAS) [6.109, 110]. We also observe a shoulder in the SFG spectrum at 2880 cm^{-1}. This shoulder is probably due to a Fermi resonance between the C–H symmetric stretch and the first overtone of the symmetric bend of the methyl group. It cannot be resolved in HREELS but has been observed in IR absorption studies of methoxy on Cu{100} [6.110] and of the methyl group modes of n-alkyl carboxylic acids [6.111].

The SFG spectrum in the carbon monoxide stretching region, from 1850 to 2100 cm^{-1}, showed no resonance features for adsorbed methanol. The nonresonant SFG response in this region of the spectrum is approximately 20 % lower than that observed for the clean surface.

These spectra are re-examined after the surface is heated to 200 K for 1 minute. This temperature is sufficient to break the O–H bond in methanol and produce surface bound CH_3O and hydrogen atoms. The stability of CH_3O at this temperature is supported by the SFG spectrum, shown in Fig. 6.19c, which retains that all three C–H stretch features. Importantly, however, there is a significant change in the lineshape of these features. The change in the C–H lineshapes suggests that there is a change in the interference between the resonant and nonresonant components of $\chi^{(2)}$ when the O–H bond is broken. In order to compare the SFG spectra to more conventional vibrational spectra, the resonant and nonresonant components of $\chi^{(2)}$ must be deconvoluted. The deconvolution discussed below gives the values for the linecenter, linewidth and relative intensities of the C–H features of methoxy that are shown in Table 6.1. A shift to higher frequen-

Table 6.1. Parameters from the nonlinear least-squares fit for the resonant features of methanol and methoxy

Methanol				
mode	R_ν [a.u.]	ω_ν	[cm^{-1}]	Γ [cm^{-1}]
1	6.6 ± 0.9	2803.8	$\pm$ 1.9	28.4 ± 7
2	6.8 ± 1.4	2876.8	$\pm$ 2.0	28.2 ± 10
3	14.0 ± 0.9	2909.5	$\pm$ 1.4	31.8 ± 5
Nonresonant Background	28.0 ± 0.7		—	—
		Relative Phase = $-91 \pm 8°$		
Methoxy				
mode	R_ν [a.u.]	ω_ν	[a.u.]	Γ [cm^{-1}]
1	14.0 ± 1.0	2813.1	$\pm$ 0.4	9.4 ± 1
2	5.7 ± 1.4	2885.6	$\pm$ 1.8	23.8 ± 9
3	35.0 ± 1.8	2919.5	$\pm$ 0.5	15.6 ± 1
Nonresonant Background	27.0 ± 0.6		—	—
		Relative Phase = $171 \pm 3°$		

cies of about $10\,\mathrm{cm}^{-1}$ and a significant narrowing of the linewidths occurs upon reaction. The intensity of the nonresonant response in the CO stretch region of the spectrum is the same as it was prior to the decomposition of the adsorbed methanol.

Finally, the surface is heated to 260 K for 1 minute. This is sufficient to decompose the methoxy to CO and H, and eliminate all three C–H resonance features from the SFG spectra (not shown). A nonresonant SFG response in the C–H stretch region is observed with an intensity roughly 30 % lower than the clean surface value. Interestingly, no carbon monoxide stretch is observed until the surface H atoms are desorbed by heating to 350 K. Previous IRAS investigations of methoxy decomposition on nickel substrates found that the stretch frequency for CO on Ni{100} with coadsorbed hydrogen is located at $1800\,\mathrm{cm}^{-1}$ [6.112]. Unfortunately, with the current experimental configuration, scans below $1850\,\mathrm{cm}^{-1}$ are not possible because of attenuation of the IR laser beam due to atmospheric water absorption bands. After desorbing the hydrogen, however, an easily detectable resonance is observed at $2024\,\mathrm{cm}^{-1}$ with a width of approximately $25\,\mathrm{cm}^{-1}$.

In Fig. 6.20a, the results of a nonlinear least squares fit to the methoxy spectrum is shown. The fit uses three Lorentzian resonances of adjustable position, width and amplitude plus a nonresonant background of adjustable amplitude and phase. It can be seen that this simple model enables quite a good representation of the data. The parameters obtained from this fit are listed in Table 6.1. In Fig. 6.20b, the magnitude of the imaginary part of $\chi^{(2)}_{\mathrm{res}}$ obtained from our fitting

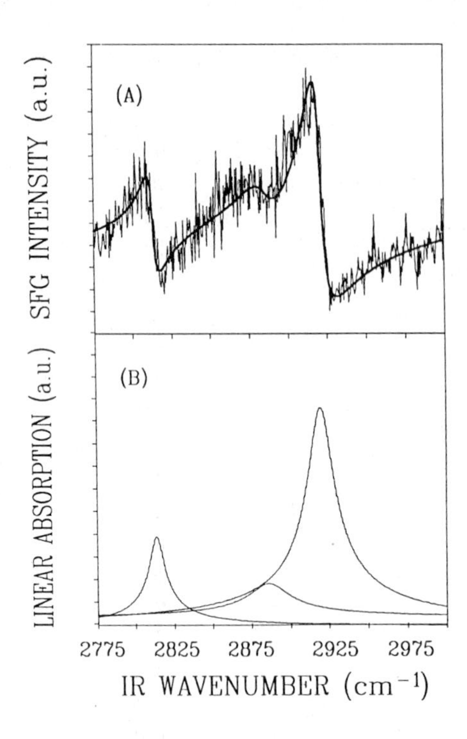

Fig. 6.20. (*a*) Comparison of the SFG spectrum of CH_3O to a nonlinear least squares fit (*dark solid line*) using model for lineshapes described in the text. (*b*) Components of the simulated linear absorption spectrum calculated from the fitting parameters obtained in the least squares fit. From [6.66]

is shown in order to enable a comparison of our results to that which would be obtained from a measurement of IR absorbance.

The same model is used to fit the methanol spectrum. The fitting parameters are given in Table 6.1. As can be seen, the major change from methanol to methoxy is the relative phase of the resonant to the nonresonant term. In methanol it is approximately $-90°$, while in methoxy it is $171°$. For methanol, the phase quadrature results in a vanishingly small interference between the resonant and nonresonant terms, producing symmetric, nearly Lorentzian features that ride on top of a constant nonresonant background.

Some limitations of this simple model should be noted. A number of complications arise when the resonances cannot be described by single Lorentzian lines. The imaginary part of the linear susceptibility and the magnitude squared of the second order susceptibility are identically equal only in one specific case, when the adsorbate resonant lineshape can be described by a single Lorentzian and the background nonresonant term is vanishingly small. We have examined the effect of other lineshapes analytically and found that if the absorption line is not Lorentzian, but, e.g., a convolution of a Lorentzian with some other distribution function, the SFG lineshape will be broadened relative to that found in absorption measurements. Furthermore, if the distribution function is skewed, as often is the case, then the resonant peak also will be shifted in frequency. These effects taken together with the nonresonant background interference can make it difficult to determine the extent to which inhomogeneous broadening or phase shifts are responsible for asymmetries. Additional details concerning these effects will be presented elsewhere.

An estimate of the orientation of the methoxy group with respect to the surface normal can be made from the relative magnitudes of the symmetric and antisymmetric C–H resonances, given in Table 6.1. SFG like reflectance IR, is insensitive to vibrational modes oriented parallel to a metal surface [6.114]. This is due to the fact that the local field intensity parallel to the surface of a metal is nearly zero because the reflected beam is 180° out of phase with respect to the incident beam. Thus, only those vibrational modes with a sizable component of the dipole oriented perpendicular to the surface of a metal can be detected. If the O–C bond in methanol is normal to the surface, we would expect to detect only the symmetric mode because the dynamic dipole of the symmetric stretch is parallel to the O–C bond, while that of the asymmetric methyl C–H stretch is orthogonal to the O–C bond. On the contrary, as shown in Fig. 6.20b, the asymmetric stretch feature is more intense than the symmetric stretch feature. Assuming the relative oscillator strengths of these modes are similar to those in the gas phase [6.115], and adjusting for the parallel and perpendicular local field intensities calculated from the optical properties of Ni, the observed intensity ratio indicates that methanol is bound to the surface with the O–C bond tilted $25 \pm 10°$ from the surface normal. This is consistent with the bonding angle expected if methanol is bound to the surface via the oxygen lone pair orbital. Similarly, it is found that methoxy is bound with the O–C bond tilted $35 \pm 10°$ from the surface normal. This is consistent with the angle expected for bonding through the oxygen $2p$ orbital initially associated with the alcohol hydrogen. A similar result has been obtained for the orientation of methoxy on Cu{100} on the basis of reflectance IR measurements, and the same gas phase oscillator strengths used in our analysis [6.109, 110].

An additional constraint can be placed on the tilt angle of adsorbed methanol from the fact that we are unable to detect the O–H stretch at coverages below 1 monolayer. The O–H stretch of methanol adsorbed on Ru and Pd has been measured by HREELS to be at 3190 and 3345 $\mathrm{cm^{-1}}$ respectively [6.113]. SFG measurements of the 3000 to 3500 $\mathrm{cm^{-1}}$ region reveal a nonresonant signal of intensity comparable to that observed in the C–H region, but no resonance contribution can be detected. For a 25° tilt angle for the Ni–O–C bond, and an H–O–C bond angle close to the gas phase value of 108° [6.115], the O–H bond would be roughly 8° away from being parallel to the surface. From the gas phase oscillator strength of the O–H stretch, and the signal to noise of our SFG experiments, we estimate that we would not be able to detect the O–H stretch if it is within 10° of being parallel to the surface. Thus, our inability to detect the O–H stretch suggests that the tilt angle of the methyl group ranges from 8 to 28°. This is consistent with the 25° estimated from the relative intensities of the C–H modes.

When picosecond laser pulses are used to generate the sum-frequency signal, it is possible to use a repetitive pump/probe technique to measure the response of individual vibrational modes with picosecond time resolution. Vibrational level lifetimes provide information on the coupling of adsorbate nuclear motion to the

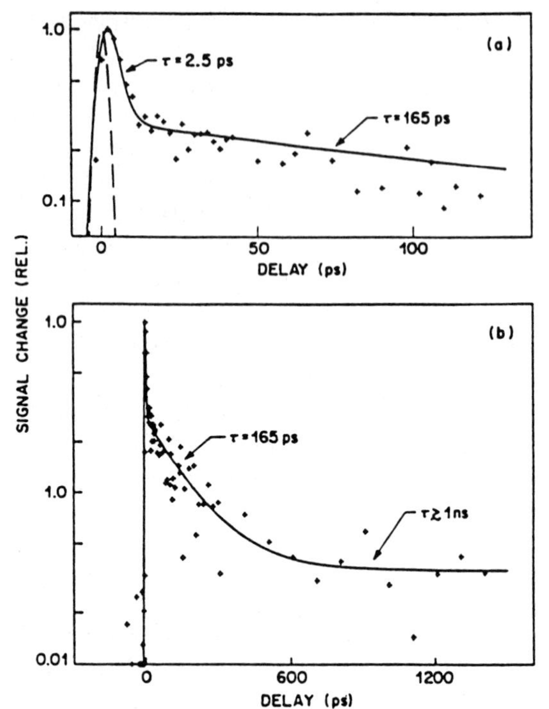

Fig. 6.21. Transient signal change for pump and probe at $2878\,\mathrm{cm}^{-1}$ on two time scales. The negative of the change in the square root of the sum signal is plotted, normalized to the maximum at +2 ps. (*Points*) experimental data. (*Solid line*) biexponential fit (including a long time offset) convoluted with pump and probe pulses of 3 ps duration. From [6.116]

substrate. This information is important for understanding energy transfer, energy accommodation, and sticking in gas/solid collisions, and in understanding the role that these couplings play in surface chemical reactions. Presently, information on the rates and mechanisms of vibrational energy coupling to surfaces comes from theory or is inferred from lineshapes of vibrational spectra. As discussed above, lifetimes of vibrational-levels can seldom be obtained from the lineshape because phase disruption and inhomogeneous ensemble effects generally broaden the line. Thus, time-resolved measurements of the relaxation of individual homogeneous lines are required.

Recently, *Harris* and *Levinos* reported time-resolved measurements of the lifetime of C–H stretch modes of methyl groups of Cd stearate on an evaporated Ag film [6.116]. Their results are reproduced in Fig. 6.21. In these experiments, a monolayer of Cd stearate is formed by Langmuir-Blodgett deposition with the $(CH_2)_{16}CH_3$ chains in an all-trans configuration and oriented nearly normal to the surface. The methyl group symmetric C–H stretch vibration is excited by an intense, picosecond infrared laser pulse that is in resonance with the transition from the ground state to first excited vibrational state of this mode. The maximum excited state population is $\sim 15\,\%$. The subsequent vibrational relaxation is probed by a time-delayed infrared-visible pulse pair that generates a SF signal. The SF intensity is a measure of the population of the ground state if the vibrational mode is sufficiently anharmonic that the excited state is not in resonance with the probe pulse-pair [see (6.18)].

Because the methyl groups are close packed and are so far from the Ag surface (20 Å), the authors attribute the observed relaxation kinetics to intramolecular energy transfer in the adlayer. The multi-lifetime decay shown in Fig. 6.21 indicates that the relaxation processes are complex. The relaxation data are fit to a biexponential decay with a long time offset. The early time decay is fit by a 2.5 picosecond lifetime. For delay times between 15 and 300 ps, the lifetime is 165 ps. The long time offset, shown in Fig. 6.21b, is only 4 % of the initial transient. Interestingly, the authors state that the transient response of the other two C–H modes is different, both with a weak saturation (2–3 %) and a fast recovery (< 3 ps). No microscopic model to account for these results is yet available. These results, and others like them [6.106], will undoubtedly be important in providing information with which improvements in theory can be made.

6.2.5 SFG Summary

Many of the opportunities and applications discussed in the section on SHG also apply to SFG. The results discussed here demonstrate that infrared-visible SFG can provide information specific to a given adsorbed molecule even for adlayers containing a mixture of species. This is a significant advantage over SHG. In addition, SFG can measure specific vibrational modes, providing information on the orientation of functional groups as well as overall molecular orientations. The time-resolution inherent in SFG will provide new opportunities to study the kinetics and dynamics of a number of surface processes.

The cost of these advantages is that both the interpretation of the data and the experimental apparatus are more complex. For metal surfaces, the interference between the substrates and adsorbate second-order polarizations can lead to complex lineshapes. A better theoretical foundation for understanding of the second-order response of transition metals would help in this regard. Also, absolute SF intensities and phases should be measured rather than relative values.

Some additional development in laser systems is probably required before SFG can become a routine technique. Presently available IR lasers cover only limited portions of the IR spectrum and scans over more than $500\,\text{cm}^{-1}$ are tedious. Advances expected in new, tunable laser systems, such as those based on Ti/Sapphire, may provide some help. Other areas for additional research include the extension of the tuning range to wavelengths longer than $5\,\mu$m, and the generation of tunable IR pulses less than 1 picosecond in duration.

In spite of these complications, SFG offers some important advantages in investigations of surface structure, electronic properties, surface bonding, and the kinetics of surface processes. The expectation is that, in the next several years, a number of important advances will be made.

References

6.1 G.L. Richmond, J.M. Robinson, V.L. Shannon: Prog. Surf. Sci. **28**, 1 (1988)
6.2 Y.R. Shen: In "Chemistry and Structure at Interfaces: New Laser and Optical Techniques", ed. by R.B. Hall, A.B. Ellis (VCH, Deerfield Beach, Florida 1986), p. 151;
Y.R. Shen: "Principles of Nonlinear Optics" (Wiley, New York 1984)
6.3 N. Bloembergen, P.S. Pershan: Phys. Rev. **128**, 606 (1962)
6.4 J. Ducuing, N. Bloembergen: Phys. Rev. Lett. **10**, 474 (1963)
6.5 F. Brown, R.E. Parks, A.M. Sleeper: Phys. Rev. Lett. **14**, 1029 (1965);
F. Brown, R.E. Parks: Phys. Rev. Lett. **16**, 507 (1966)
6.6 S.S. Jha: Phys. Rev. **140**, A2020 (1965)
6.7 N. Bloembergen, Y.R. Shen: Phys. Rev. **141**, 298 (1966)
6.8 C.H. Lett, R.K. Chang, N. Bloembergen: Phys. Rev. Lett. **18**, 167 (1967)
6.9 H. Sonnenberg, H. Heffner: J. Opt. Soc. Am. **58**, 209 (1968)
6.10 C.C. Wang, A.N. Duminiski: Phys. Rev. Lett. **20**, 688 (1968)
6.11 C.C. Wang: Phys. Rev. **178**, 1457 (1969)
6.12 N. Bloembergen, R.K. Chang, S.S. Jha, C.H. Lee: Phys. Rev. **174**, 813 (1968); **178**, 1528E (1969)
6.13 F. Brown, M. Matsuoka: Phys. Rev. **185**, 985 (1969)
6.14 J.M. Chen, J.R. Bower, C.S. Wang, C.H. Lee: Opt. Commun. **9**, 132 (1973)
6.15 C.K. Chen, T.F. Heinz, D. Ricard, Y.R. Shen: Phys. Rev. Lett. **46**, 1010 (1981)
6.16 C.K. Chen, T.F. Heinz, D. Ricard, Y.R. Shen: Phys. Rev. Lett. **48**, 478 (1982)
6.17 D. Guidotti, T.A. Driscoll, H.J. Gerriberr: Solid State Commun. **46**, 337 (1983)
6.18 H.W.K. Tom, G.D. Aumiller: Phys. Rev. B **33**, 189 (1986)
6.19 P. Guyot-Sionnest, W. Chen, Y.R. Shen: Phys. Rev. B **33** (1986)
6.20 J. Rudnick, E.A. Stern: Phys. Rev. B **4**, 4274 (1971)
6.21 Guyot-Sionnest et al. have shown using macroscopic formalisms that structural asymmetry and field discontinuity at an interface contribute separately to the optical nonlinearity of the interface [6.19]. If the refractive indices of the bulk media are not matched, there is a field variation that produces a non-local response involving not only an electric dipole contribution, but all the multipole contributions. Structural asymmetry contributes a local nonlinear polarization, represented by a electric-dipole allowed surface nonlinear susceptibility. Structural asymmetry contributes roughly as much as the field discontinuity. This was shown in a series of investigations in which the SH response of a surface in air was compared to that of the surface under an index-matched fluid [6.22]
6.22 P. Guyot-Sionnest, Y.R. Shen: unpublished results
6.23 J.E. Sipe, V.C.Y. So, M. Fukui, G.I. Stegeman: Phys. Rev. B **21**, 4389 (1980)
6.24 M. Corvi, W.L. Schaich: Phys. Rev. B **33**, 3688 (1986)
6.25 M. Weber, A. Liebsch: Phys. Rev. B **35**, 7411 (1987)
6.26 M. Weber, A. Liebsch: Phys. Rev. B **37**, 1019 (1988)
6.27 A. Chizmeshya, E. Zaremba: Phys. Rev. B **37**, 2805 (1988)
6.28 R. Murphy, M. Yeganch, K.J. Song, E.W. Plummer: Phys. Rev. Lett. **63**, 318 (1989)
6.29 J.R. Bower: Phys. Rev. B **14**, 2427 (1976)
6.30 G.T. Boyd, Y.R. Shen, T.W. Hänsch: Opt. Lett. **11**, 97 (1986)
6.31 J. Miragliotta, T.E. Furtak: Phys. Rev. B **37**, 1028
6.32 C.V. Shank, R. Yen, C. Hirlimann: Phys. Rev. Lett. **51**, 900 (1983)
6.33 T.F. Heinz, G. Arjavalingam, M.T. Loy, J.H. Glownia: In Proc. of the XIV Int. Quantum Electronics Conference, June 9–13, 1986, San Francisco, CA. Paper THII 1
6.34 G. Arjavalingam, T.F. Heinz, J.H. Glownia: In *Ultrafast Phenomena V*, ed. by G.R. Fleming, A.E. Siegman (Springer, Berlin, Heidelberg 1986), p. 370
6.35 H.W.K. Tom: In *Advances in Laser Science*, 11. AIP Conference Proceedings Series No. 180, ed. by W.C. Stwalley, M. Lapp, G.A. Kenney-Wallace, AIP, NY (1987)
6.36 H.W.K. Tom, G.D. Aumiller, C.H. Brito-Cruz: Phys. Rev. Lett. **60**, 1438 (1988)
6.37 T.F. Heinz: PhD. Dissertation, University of California, Berkeley, 1984
6.38 H.W.K. Tom: PhD. Dissertation, University of California, Berkeley, 1984
6.39 R.K. Chang, J. Ducuing, N. Bloembergen: Phys. Rev. Lett. **15**, 6 (1965)
6.40 T.F. Heinz, K. Kemnitz, K. Bhattacharyya, J.M. Hicks, G. Pinto, K.B. Eisenthal: In the Proc. of the XV Int. Conf. on Quantum Electronics, Baltimore, MD, 1987, Paper W664

6.41 K. Kemnitz, K. Bhattacharyya, J.M. Hicks, G.R. Pinto, K.B. Eisenthal, T.F. Heinz: Chem. Phys. Lett. **131**, 285 (1986)
6.42 P.D. Maker, R.W. Terhune, M. Nisenoff, C.M. Sauvage: Phys. Rev. Lett. **8**, 21 (1962)
6.43 J.J. Wynne, N. Bloembergen: Phys. Rev. **188**, 1211 (1969)
6.44 J.C. Quail, H.C. Simon: Phys. Rev. B **31**, 4900 (1985)
6.45 A. Liebsch, W.L. Schaich: Phys. Rev. B (to be published)
6.46 W.L. Schaich, A. Liebsch: Phys. Rev. B **37**, 6187 (1988)
6.47 H.W.K. Tom, C.M. Mate, X.D. Zhu, J.E. Crowell, T.F. Heinz, G.A. Somorjai, Y.R. Shen: Phys. Rev. Lett. **52**, 348 (1984)
6.48 H.W.K. Tom, C.M. Mate, X.D. Zhu, J.E. Crowell, Y.R. Shen, G.A. Somorjai: Surf. Sci. **172**, 466 (1986)
6.49 S.G. Grubb, A.M. DeSantolo, R.B. Hall: J. Phys. Chem. **92**, 1419
6.50 R.J.M. Anderson, J.C. Hamilton: Chem. Phys. Lett. **151**, 455 (1988)
6.51 R.J.M. Anderson, J.C. Hamilton: Bull. of the Am. Phys. Soc. **33**, 387 (1988)
6.52 C.M. Mate, G.A. Somorjai, H.W.K. Tom, X.D. Zhu, Y.R. Shen: J. Chem. Phys. **88**, 441 (1988)
6.53 Z. Rosenzweig, M. Asscher: Surf. Sci., to be published
6.54 D. Heskett, K.J. Song, A. Burns, E.W. Plummer, H.L. Dai: J. Chem. Phys. **85**, 7490 (1986); D. Heskett, L.E. Urbach, K.J. Song, E.W. Plummer, H.L. Dai: Surf. Sci. **197**, 225 (1988)
6.55 K.J. Song, D. Heskett, H.L. Dai, A. Liebsch, E.W. Plummer: Phys. Rev. Lett. **61**, 1380 (1988)
6.56 X.D. Zhu, Y.R. Shen, R. Carr: Surf. Sci. **163**, 114 (1985)
6.57 J.M. Chen, J.R. Bower, C.S. Wang, C.H. Lee: Opt. Commun. **2**, 132 (1973)
6.58 X.D. Zhu, Th. Rasing, Y.R. Shen: Phys. Rev. Lett. **25**, 2883 (1988)
6.59 J.N. Russell, R.B. Hall: to be published
6.60 R.B. Hall, A.M. Desantolo: Surf. Sci. **137**, 421 (1984); R.B. Hall, A.M. Desantolo, S.J. Bares: Surf. Sci. **161**, L533 (1985)
6.61 Lee J. Richter, W. Ho: J. Chem. Phys. **83**, 2869 (1986)
6.62 L. Melander, W.H. Saunders, Jr.: *Reaction Rates of Isotopic Molecules* (Wiley, New York 1980); P.J. Robinson, K.A. Holbrook: *Unimolecular Reactions* (Wiley, New York 1972)
6.63 S.M. Gates, J.N. Russell, Jr., J.T. Yates, Jr.: Surf. Sci. **146**, 199 (1984)
6.64 Lee J. Richter, W. Ho: J. Vac. Sci. Technol. **A3**, 1549 (1985)
6.65 R.B. Hall, S.J. Bares, A.M. DeSantolo, F. Zaera: J. Vac. Sci. Technol. **A4**, 1493 (1986)
6.66 J. Miragliotta, R.S. Polizzotti, P. Rabinowitz, S.D. Cameron, R.B. Hall: Chem. Phys. **143**, 123 (1990)
6.67 T. Heinz, H.W.K. Tom, Y.R. Shen: Phys. Rev. Lett. **48**, 478 (1982)
6.68 B. Dick, R.M. Hochstrasser: J. Chem. Phys. **78**, 3398 (1983)
6.69 K. Geisen, F. Hage, H.J. Reiss, W. Steinmann, R. Haight, R. Beigang, R. Dreyfus, Ph. Avouris, F.J. Himpsel: Physica Scripta **35**, 578 (1987)
6.70 J.M. Hicks, E. Urbach, E.W. Plummer, Hai-Lung Dai: Phys. Rev. Lett. **61**, 2588 (1988)
6.71 K. Giesen, F. Hage, F.J. Himpsel, H.J. Riess, W. Steinmann: Phys. Rev. B **33**, 5241 (1986)
6.72 F.J. Himpsel, D.E. Eastman: Phys. Rev. Lett. **41**, 507 (1978)
6.73 D. Guidotti, T.A. Driscoll, H.J. Gerritsen: Solid State Comm. **46**, 337 (1983)
6.74 H.W.K. Tom, T.F. Heinz, Y.R. Shen: Phys. Rev. Lett. **51**, 1983 (1983)
6.75 J.A. Litwin, J.E. Sipe, H.M. van Driel: Phys. Rev. B **31**, 5543 (1985)
6.76 T.F. Heinz, M.M.T. Loy, W.A. Thompson: Phys. Rev. Lett. **54**, 63 (1985)
6.77 C.V. Shank, R. Yen, C. Hirlimann: Phys. Rev. Lett. **51**, 900 (1983)
6.78 H.W.K. Tom, G.D. Aumiller, C.H. Brito-Cruz: Phys. Rev. Lett. **60**, 1438 (1988)
6.79 H.W.K. Tom, G.D. Aumiller: Phys. Rev. B **33**, 8818 (1988)
6.80 V.L. Shannon, D.A. Koos, G.L. Richmond: J. Phys. Chem. **91**, 5548 (1987)
6.81 J.E. Sipe, D.J. Moss, H.M. van Driel: Phys. Rev. B **35**, 1129 (1987)
6.82 J.F. McGilp, Y. Yeh: Solid State Commun. **59**, 91 (1986)
6.83 J.F. McGilp: J. Vac. Sci. Tech. A **5**, 1442 (1987)
6.84 V.L. Shannon, D.A. Koos, G.L. Richmond: Appl. Opt. **26**, 3579 (1987)
6.85 V.L. Shannon, D.A. Koos, G.L. Richmond: J. Phys. Chem. **91**, 5548 (1987)
6.86 V.L. Shannon, D.A. Koos, S.A. Kellar, G.L. Richmond: J. Phys. Chem. (1989)
6.87 X.D. Zhu, Hajo Suhr, Y.R. Shen: Phys. Rev. B **35**, 3047 (1987)
6.88 J.H. Hunt, P. Guyot-Sionnest, Y.R. Shen: Chem. Phys. Lett. **133**, 189 (1987)
6.89 P. Guyot-Sionnest, J.H. Hunt, Y.R. Shen: Phys. Rev. Lett. **59**, 1597 (1987)

6.90 R. Superfine, P. Guyot-Sionnest, J.H. Hunt, C.T. Kao, Y.R. Shen: Surf. Sci. **200**, L445 (1988)
6.91 A.L. Harris, C.E.D. Chidsey, N.J. Levinos, D.N. Lioacono: Chem. Phys. Lett. **141**, 350 (1987)
6.92 D.C. Nguyen, R.E. Muenchausen, R.A. Keller, N.S. Nogar: Opt. Commun. **60**, 111 (1986)
6.93 P. Rabinowitz, B.N. Perry, N. Levinos: J. Quantum Electron. **22**, 797 (1986)
6.94 H. Graener, A. Laubereau: Appl. Phys. **B29**, 213 (1982)
6.95 J. Miragliotta, R.S. Polizzotti, P. Rabinowitz, S.D. Cameron, R.B. Hall: Chem. Phys., to be published
6.96 R. Klauer, W. Spiess, A.M. Bradshaw: J. Elect. Spec. and Rel. Phen. **38**, 187 (1986)
6.97 J.D. Fedyak, M.J. Dignam: In *ACS Symp. Series No.137*, ed. by A.T. Bell, M.L. Hair (Am. Chem. Soc. 1980), Chap. 5
6.98 R.G. Tobin, S. Chiang, P.A. Thieland, P.L. Richards: Surf. Sci. **140**, 393 (1984)
6.99 R.G. Shigeishi, D.A. King: Surf. Sci. **58**, 379 (1976)
6.100 A. Ortega, F.M. Hoffmann, A.M. Bradshaw: Surf. Sci. **119**, 79 (1982)
6.101 M. Trenary, K.J. Uram, F. Bozso, J.T. Yates, Jr.: Surf. Sci. **146**, 269 (1984)
6.102 J.W. Gadzuk, A.C. Luntz: Surf. Sci. **144**, 429 (1984)
6.103 R.G. Tobin: Surf. Sci. **183**, 226 (1987)
6.104 R.M. Shelby, C.B. Harris, P.A. Cornelius: J. Chem. Phys. **70**, 34 (1979)
6.105 F.M. Hoffmann, B.N.J. Persson: Phys. Rev. **B34**, 4354 (1986)
6.106 J.D. Beckerle, M.P. Cassassa, R.R. Cavanagh, E.J. Heilweil, J.S. Stephenson: Phys. Rev. Lett. **64**, 2090 (1990)
6.107 J.C. Tracy: J. Chem. Phys. **56**, 2736 (1972)
6.108 J.E. Demuth, H. Ibach: Chem. Phys. Lett. **60**, 395 (1979);
L.J. Richter, W. Ho: J. Chem. Phys. **83**, 2569 (1985)
6.109 R. Ryberg: Chem. Phys. Lett. **83**, 423 (1981)
6.110 R. Ryberg: Phys. Rev. B **31**, 2545 (1985)
6.111 I.R. Hill, I.W. Levin: J. Chem. Phys. **70**, 842 (1979)
6.112 F.L. Baudais, A.J. Borschke, J.D. Fedyk, M.J. Dignam: Surf. Sci. **100**, 210 (1980)
6.113 J. Hrbek, R.A. dePaola, F.M. Hoffmann: J. Chem. Phys. **81**, 2818 (1984);
K. Christmann, J.E. Demuth: J. Chem. Phys. **76**, 6318 (1982)
6.114 R.B. Greenler: J. Chem. Phys. **44**, 310 (1966)
6.115 A. Serrallach, R. Meyer, Hs.H. Gunthard: J. Molec. Spectrosc. **52**, 94 (1974)
6.116 A.L. Harris, N.J. Levinos: J. Chem. Phys. **90**, 3878 (1989)

7. Surface Physics and Chemistry in High Electric Fields

H.J. Kreuzer

Department of Physics, Dalhousie University Halifax, N.S. B3H 3J5, Canada and Fritz-Haber-Institut der Max-Planck-Gesellschaft, Faradayweg 6–8, D-1000 Berlin 33, Germany

To specify the term "high electric fields" in the title of this chapter, we mention typical static field strengths encountered in a variety of situations. To start with, we note that the maximum field strength that can be maintained between two conductors in air is limited to less than about 10^4 V/cm above which dielectric breakthrough leads to the formation of an ionized plasma. In semiconductors, fields of the order of 10^6 V/cm can be maintained, whereas fields within the double layer at the electrolyte-electrode interface can reach 10^7 V/cm. Around localized charges in zeolite cavities, electric fields of the order of 10^8 V/cm = 1 V/Å have been estimated on the basis of Coulomb's law

$$F = \frac{3.4}{r^2} \frac{q}{e} \text{ [V/Å]} \tag{7.1}$$

at a distance r, measured in Ångstroms, away from a charge of magnitude q/e, measured in units of the elementary charge e. Fields of this order can also be established within 10^3 Å of a metal tip with a tip radius of less than 10^3 Å, provided dielectric breakthrough is avoided by working in ultrahigh vacuum. The upper limit of electric field strength that can be maintained over macroscopic distances is dictated by the onset of field emission and field evaporation, and is of the order of 6 V/Å.

Electric field effects on matter can be classified, rather arbitrarily, into two categories: (i) in low fields, i.e., below roughly 10^{-1} V/Å, atoms, molecules and condensed matter only get polarized; we will call such effects physical. (ii) In fields larger than typically 10^{-1} V/Å, chemical effects come into play in addition in that the electronic orbitals get distorted to such a degree as to affect the chemical characteristics of an atom or molecule, for example, by establishing new bonding orbitals. In this way, molecules, unstable in field free situations, may be stabilized by a strong electric field. Also, new pathways in chemical reactions, and in particular in heterogeneous catalysis, may be established.

In this chapter we will be mainly concerned with the physics and chemistry at metal tips in high static electric fields as they occur in the field ion microscope, although the theory can be carried over to other high field situations. We first deal with the local variation of electric fields within a few Ångstroms of a metal surface. Next, in Sect. 7.2 we will review the theory of dispersion and polarization forces in high electric fields. Section 7.3 then reviews field-induced chemisorption with the aim of explaining field adsorption, in particular of helium, and in Sect. 7.4, we will apply this theory to understand field evaporation. In Sect. 7.5 we will set up a master equation to study the kinetics in high electric fields concentrating on thermal field desorption. The last section will then be devoted to give a brief overview of field-induced chemistry.

7.1 Electric Fields at Metal Surfaces

Breaking the crystal symmetry at the surface of a solid leads to a relaxation of the lattice and a rearrangement of the electron distribution resulting in a dipole layer across which the electrical potential rises from the Fermi level inside the metal to the vacuum level outside, the difference being the electron work function. If the metal is bounded by inequivalent surfaces, then small excess charges of surface density $\sigma(r)$ build up on each surface giving rise to stray fields, extending over microscopic distances, obtained from Gauss' law to be

$$(\boldsymbol{F}_2 - \boldsymbol{F}_1) \cdot \boldsymbol{n} = 4\pi\sigma \ , \tag{7.2}$$

where $\boldsymbol{n}$ is the surface normal.

To establish an electric field over macroscopic distances we add additional charges onto the surface of the metal. As this is usually done by connecting the metal to a battery, an equal but opposite charge is induced on a counterelectrode assumed to be far away. Let us assume that our metal specimen is a tip, which we can approximate around the apex as a paraboloid. The field in the vicinity of the apex is then given by

$$F = -\frac{4\pi\sigma_0}{\beta\sqrt{\alpha^2+\beta^2}} \ , \tag{7.3}$$

where the parabolic coordinates are given as $\alpha^2 = r + z$ and $\beta^2 = r - z$ in terms of cartesian coordinates. Thus, if we establish a voltage of 10 kV over 10 cm then just above a tip of curvature of 10^2 Å, the field is of the order of Volts per Ångstrom. We note that it is now possible to make tips with a single atom at the apex, i.e., with a curvature of one Ångstrom!

The surface of a solid is, of course, not mathematically flat. To see to what extent electric fields are modified by roughness, we consider a spherical boss of radius R on top of a flat surface, see Fig. 7.1. The electric potential is then given by

$$\Phi = -4\pi\sigma_0 z(1 - R^3/r^3) \ . \tag{7.4}$$

Locally the excess charge redistributes itself into

$$\sigma = \sigma_0(1 - R^3/r^3) \tag{7.5}$$

on the plane, and

$$\sigma = 3\sigma_0 z/R \tag{7.6}$$

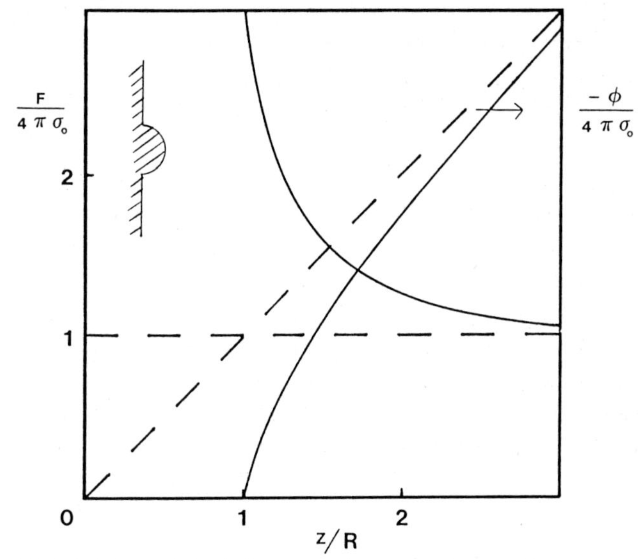

Fig. 7.1. Electric field (*left scale*) and potential (*right scale*) as a function of distance at a metal surface with a spherical boss of radius R (see insert) far from the boss (*dashed lines*) and along the apex of the boss (*solid lines*)

on the sphere. We note that at the apex of the boss the electric potential is zero and the field is three times its value at infinity. For $z = 2R$ above the apex, potential and field values agree to within 20% of their values laterally far from the boss.

So far, we have described electric fields by using classical electromagnetic theory which assumes that the surface of a metal is a mathematical plane with excess charges and a dipole layer at which the normal component of the electric field drops discontinuously to zero, at least for a perfect conductor. On real surfaces, however, the electron distribution and also electric fields vary smoothly over distances of a few Ångstroms. A simple model [7.1–3] that bears out these features is the jellium model of a metal in which we assume that the ionic lattice can be smoothed into a uniform positive charge density n_+ that drops to zero abruptly half a lattice constant above the topmost layer of ion cores. It is given in terms of the Wigner-Seitz radius as

$$n_+ = \frac{3}{4\pi}\left(\frac{r_s}{a_0}\right)^3 , \tag{7.7}$$

where r_s is given in units of the Bohr radius a_0. Within the framework of density functional theory, the exact ground state electron density is then given as the selfconsistent solution of the equations;

$$n(z) = \frac{m}{\pi\hbar^2}\sum_\nu (\varepsilon_F - \varepsilon_\nu)\Theta(\varepsilon_F - \varepsilon_\nu)|\psi_\nu(z)|^2 , \tag{7.8}$$

$$\left(-\frac{\hbar^2}{2m}\frac{d^2}{dz^2} + v_{\text{eff}}[n;z] - \varepsilon_\nu\right)\psi_\nu(z) = 0 \,. \tag{7.9}$$

In the local density approximation, the effective potential

$$v_{\text{eff}}[n;z] = \varphi(z) + \mu_{xc}(n(z)) \,, \tag{7.10}$$

is the sum of the electrostatic potential, determined from Poisson's equation

$$\frac{d^2}{dz^2}\varphi(z) = 4\pi e[n_+ - n(z)] \,, \tag{7.11}$$

and the exchange and correlation energy, μ_{xc}, for which various approximations such as Wigner's are employed [7.1].

In Fig. 7.2, we present the results of a density functional calculation for a jellium surface. In panel (a) we show the selfconsistent electron distribution in the absence of an external field with the local deviation from charge neutrality,

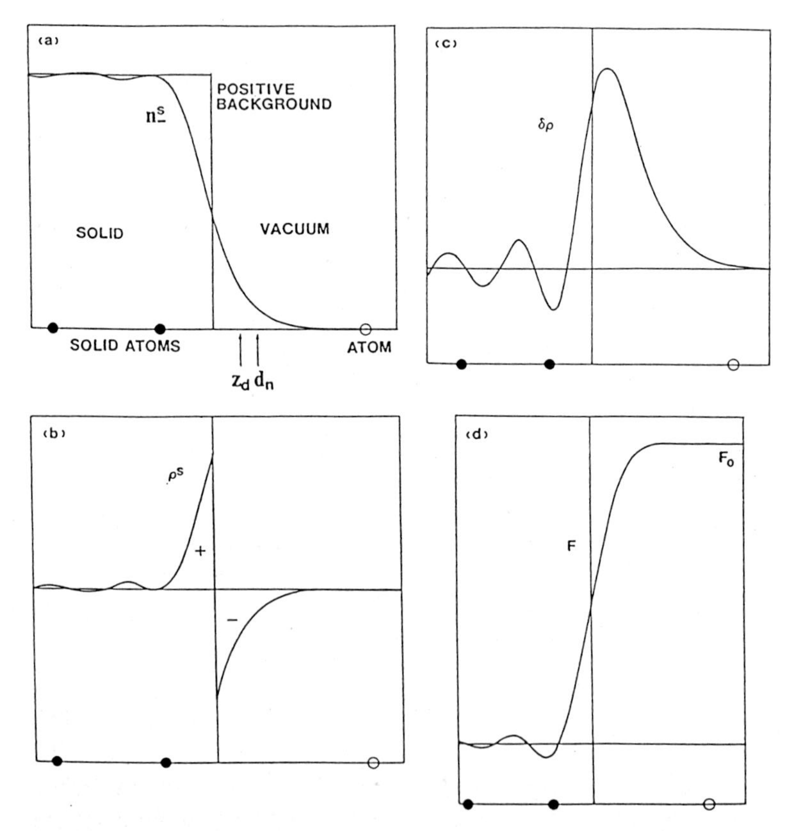

Fig. 7.2a–d. A schematic view of **(a)** the charge distribution at a metal surface without a field ($F_0 = 0$), the positive jellium background is indicated, Z_d and d_n are the dynamic and static image plane positions; **(b)** the surface dipole layer $\varrho^s = n_+^s - n_-^s$ for $F_0 = 0$, constructed from (a); **(c)** the field-induced surface charge $\delta\varrho$; and **(d)** the applied electric field F at a metal surface; from [7.18]

i.e., the dipole layer, given in panel (b). In panel (c) we have added some excess charge, $\delta\varrho$, that gives rise to the external field in panel (d). We note that the field decays smoothly into the metal with appreciable strength left at the position of the top most ion layer. This can be viewed as partial penetration of the field into the metal, or as incomplete expulsion of the field from the metal; the words make little difference to the physics. To compare these quantum mechanical calculations with classical results from Maxwell's theory, we note that the plane at which boundary conditions are imposed on the classical fields, i.e., the discontinuous drop of the normal component of the electric field to zero, is given by the center of gravity of the excess charge $\delta\varrho$, i.e., roughly the point where the field has dropped to half its value at infinity. For future reference we note here that this plane does not remain constant but moves towards the ion cores as the asymptotic field strength increases, due to the fact that the electrons are pushed into the metal, increasing the field penetration and the Friedel oscillations.

7.2 Dispersion and Polarization Forces

Atoms and molecules approaching the surface of a solid experience attractive dispersion forces due to mutually induced fluctuating multipole moments which, in the dipole approximation, leads to the van der Waals interaction with potential energy

$$V_{\mathrm{vdW}}(z) = -C_3/(z - z_{\mathrm{d}})^3 \ , \tag{7.12}$$

where z is the distance between the atom and the surface of the solid. A macroscopic theory based on electromagnetic flucutations has been presented by *Dzyaloshinskii* et al. [7.4] who also summarize and extend earlier work. They find for the van der Waals constant

$$C_3 = \frac{\hbar}{4\pi} \int_0^\infty \frac{\varepsilon(\mathrm{i}\omega) - 1}{\varepsilon(\mathrm{i}\omega) + 1} \alpha(\mathrm{i}\omega)\, d\omega \ , \tag{7.13}$$

where α is the polarizability of the adsorbing atom or molecule, evaluated at imaginary frequency and

$$\varepsilon(\mathrm{i}\omega) = 1 + \frac{2}{\pi} \int_0^\infty x \frac{\varepsilon''(x)}{x^2 + \omega^2}\, dx \ , \tag{7.14}$$

is the dielectric function of the solid, again at imaginary frequency, which can be evaluated from the above Kramers-Kronig relation involving the imaginary part of the dielectric function, $\varepsilon''(x)$, at real frequencies. More recently, *Zaremba* and *Kohn* [7.5], *Holmberg* and *Apell* [7.6], and Liebsch [7.7] have developed a density functional theory of the van der Waals interaction, adopting a jellium model for the metal, and inferring the position of the dynamical image plane z_{d}, from second order perturbation theory. Values for C_3 have been recently collected and analysed by *Bruch* [7.8].

If the dispersion force is the only attractive one, as is the case for rare gases and inert molecules such as CH_4, we say that such atoms and molecules get physisorbed, in contrast to chemisorbed particles that develop bonding orbitals as they approach the surface of a solid to within 1–2 Å. Concentrating on physisorption, we note that to construct the full adsorption potential, a repulsive part due to electron overlap must be added to (7.12). One then finds that for helium on transition metals, for instance, this potential has a minimum some 3.5 Å above the topmost layer of ion cores in the metal, with a depth of typically between 50 and 100 K or 5–10 meV.

It has been known since the invention of the field ion microscope that the imaging gas, usually helium or neon, adsorbs on the tip in electric fields of the order of the best image field strength, i.e., several volts per Ångstroms, at standard operating temperatures around 80 K [7.9–11]. In particular, field-adsorbed helium on a tungsten tip is bound by about 250 meV in fields of the order of 4–5 V/Å. We will treat field-adsorption as our first example of high electric field effects. Two mechanisms have been proposed to explain field adsorption: (i) polarization (dipole-dipole) forces [7.12–13] and (ii) field-induced chemisorption [7.14–15]. In this section we will review dispersion and polarization forces between atoms and solids in high electric fields.

The first, and most widely used model of field adsorption [7.12–13] assumes that the adsorbing atom, otherwise weakly physisorbed, only interacts with the atom of the solid on top of which it gets adsorbed. Next, one assumes that the external field $\boldsymbol{F}_0$ penetrates fully into the metal beyond the surface atom. In the light of our earlier discussion of electric fields at metal surfaces, this is not quite justified, but rather one should assume that the field at the adsorbed atom position is more or less equal to its value at infinity, $F_0^{(a)} \approx F_0$, but that at the metal atom it is reduced to $F_0^{(m)} < F_0$. It is then argued that the total field at the adsorbed atom is given by

$$F_a = F_0^{(a)} + \frac{2p_m}{d^3} , \tag{7.15}$$

where

$$p_m = \alpha_m F_m \tag{7.16}$$

is the induced dipole moment of the metal atom in a field

$$F_m = F_0^{(m)} + \frac{2p_a}{d^3} , \tag{7.17}$$

with

$$p_a = \alpha_0 F_a . \tag{7.18}$$

The energy gain by bringing the adatom from infinity to a distance d above the metal atom is then given by

$$Q = \frac{1}{2}\alpha_0 \left[\frac{F_0^{(a)}/F_0 + 2\alpha_m d^{-3} F_0^{(m)}/F_0)^2}{(1 - 4\alpha_m \alpha_0 d^{-6})^2 - 1} \right] F_0^2 , \tag{7.19}$$

which reduces for $F_0^{(a)} \approx F_0$ and neglecting the denominator in the first term to

$$Q \approx \frac{2\alpha_0\alpha_m}{d^3} F_0 F_0^{(m)} . \tag{7.20}$$

This expression with $F_0^{(m)}$ replaced by F_0 has been used to fit data by adjusting the polarizability α_m of the surface atom and by setting d equal to the sum of the van der Waals radii of the adatom and the surface atom. However, we should note that the expression (7.15) for the field at the adatom position is rather dubious as it accounts for the electric field effect twice. This can be seen if we recall from the previous section that the modification of the surface charge, i.e., local excess charge and dipole, give rise to the external field, they are not additional effects.

To account for the local field enhancement at a kink site atom or above an isolated metal atom on top of a plane, one might use (7.4–6) as a model which gives

$$Q = \tfrac{1}{2}\alpha_0 F_0^2[(1+2(R/d)^3)^2 - 1] \approx 2\alpha_o(R/d)^3 F_0^2 . \tag{7.21}$$

For He $\alpha_0 = 1.38$ a.u., so that in a field of $F_0 = 5\,\mathrm{V/Å} = 5 \times 0.019$ a.u. $Q \approx 100$ meV for $d = 2R$ and $Q \approx 30$ meV for $d = 3R$. This suggests that one should take the local field enhancement seriously, and also that one should worry about higher gradient corrections. This problem has been attacked by *Forbes* and *Wafi* [7.17] at the phenomenological level in setting up an array model in which the surface of a metal is represented by a layer of localized dipoles and charges. Their estimates for Q amount to 50–60 meV for He on W.

We now want to go beyond phenomenological models and try to understand dispersion and polarization forces in high electric fields from first principles. *Watanabe* et al. [7.18] have developed the theory by considering a model in which the metal is represented by a jellium for which selfconsistent electric field calculations exist, as reviewed in the previous section. We should stress here that the concepts of dispersion and polarization forces are only applicable at distances large compared to atomic dimensions, e.g., the length scale of the induced dipoles. Local crystalline structure should therefore not be too important. This is also corroborated by the fact that, at least on perfect surfaces, the field variation a few Ångstroms above the surface is negligible as shown by *Inglesfield* [7.19] for Al{001}. The jellium model, therefore, does not address the additional field enhancement at kink sites.

A theory of dispersion and polarization forces between atoms and solids in high electric fields should start from a Hamiltonian

$$H = H_s + H_a + H_{sa} . \tag{7.22}$$

Here H_a is the hamiltonian for an isolated atom and H_s is that for a solid filling a half space and having a surface charge $\sigma = F_0/4\pi$ with F_0 being the asymptotic value of the external electric field. H_s also includes a solid with the opposite surface charge a large distance away from the surface to preserve neutrality for the system as a whole. The interaction between the atom and the solid is then given by

$$H_{\rm sa} = \int d\boldsymbol{r} \int d\boldsymbol{x}\, \frac{\hat{\varrho}^{\rm s}(\boldsymbol{r})\hat{\varrho}^{\rm a}(\boldsymbol{x})}{|\boldsymbol{r}-\boldsymbol{x}|}\,, \tag{7.23}$$

with

$$\hat{\varrho}^{\rm s,a}(\boldsymbol{r}) = n_+^{\rm s,a}(\boldsymbol{r}) - \hat{n}_-^{\rm s,a}(\boldsymbol{r})\,, \tag{7.24}$$

where $n_+^{\rm s,a}$ are the ion densities and $\hat{n}_-^{\rm s,a}$ are the electron density operators for the solids and the atom, respectively. We use atomic units throughout this chapter. Obviously, in this generality, $\hat{\varrho}^{\rm s}(\boldsymbol{r})$ contains all the surface features like dipole layers, excess surface charge, and their selfconsistent modifications in an electric field. However, the unperturbed atomic Hamiltonian $H_{\rm a}$ does not contain any field effects, which only come about through the interaction with the solid, $H_{\rm sa}$. This is an important point because inclusion of field effects in $H_{\rm a}$ amounts to accounting for the field twice and will lead to serious errors.

Assuming that the unperturbed atom far from the solid is neutral, spherically symmetric and does not carry a permanent dipole moment, one can calculate the interaction energy between the atom and the solid in an electric field in perturbation theory in $H_{\rm sa}$ up to fourth order. One notes immediately that the first and third order energy shift vanish for neutral, spherically symmetric atoms without a permanent dipole moment. Following the perturbation scheme of *Zaremba* and *Kohn* [7.5], one readily obtains the van der Waals or dispersion energy (7.12) in second order with C_3 given by (7.13). The position of the dynamic or van der Waals image plane, $z_{\rm d}$, is determined by

$$z_{\rm d}(F_0) = \frac{1}{4\pi C_3}\int_0^\infty du\, \frac{\varepsilon({\rm i}u)-1}{\varepsilon({\rm i}u)+1}\, \alpha({\rm i}u)\, d_{\rm IP}({\rm i}u, F_0)\,, \tag{7.25}$$

and

$$d_{\rm IP}({\rm i}u, F_0) = \frac{\varepsilon({\rm i}u) d_n({\rm i}u, F_0) + d_p(F_0)}{\varepsilon({\rm i}u)+1}\,. \tag{7.26}$$

The length scales

$$\begin{aligned} d_{\rm n} &= \int_{-\infty}^{\infty} dz\, \frac{z\,\delta\varrho(z)}{\sigma}\,,\\ d_{\rm p} &= \frac{1}{n_\infty}\int_{-\infty}^{\infty} \varrho^{\rm s}(z)dz = \frac{\sigma}{e n_\infty}\,, \end{aligned} \tag{7.27}$$

are related to the center of gravity of the surface charge and the effective width of the field induced excess charge, respectively. Here $\delta\varrho(z)$ is the induced charge density, n_∞ is the positive ion density deep inside the solid, and $\sigma = F_0/4\pi$ is the induced surface charge density. The subscripts stand for normal (n) and parallel (p) to the surface.

To carry out the integration in (7.25) analytically we adopt a single Lorentzian for the polarization function of the atom

$$\alpha(\mathrm{i}u) = \frac{\alpha_0 \bar{u}^2}{\bar{u}^2 + u^2} , \tag{7.28}$$

where

$$\bar{u} = \frac{4}{3} \frac{C_{\mathrm{aa}}}{\alpha_0^2} , \tag{7.29}$$

with C_{aa} the dipole dispersion coefficient and α_0 the static polarizability. Taking the solid to be a metal, we approximate its dielectric function by

$$\varepsilon(\mathrm{i}u) = 1 + \frac{\omega_{\mathrm{p}}^2}{u^2} , \tag{7.30}$$

$$\frac{\varepsilon(\mathrm{i}u)}{\varepsilon(\mathrm{i}u)+1} d_{\mathrm{n}}(\mathrm{i}u, F_0) = \frac{\omega_{\mathrm{sp}}^2}{\omega_{\mathrm{sp}}^2 + u^2} d_{\mathrm{n}}(0, F_0) . \tag{7.31}$$

Here ω_{p} and $\omega_{\mathrm{sp}} = \omega_{\mathrm{p}}/\sqrt{2}$ are the bulk and surface plasma frequencies, respecitvely. We, thus, get for the van der Waals constant (7.13)

$$C_3 = \frac{1}{8} \frac{\alpha_0 \bar{u} \omega_{\mathrm{sp}}}{\bar{u} + \omega_{\mathrm{sp}}} , \tag{7.32}$$

and for the position of the dynamic image plane (7.25)

$$z_{\mathrm{d}}(F_0) = \frac{\bar{u} + 2\omega_{\mathrm{sp}}}{2(\bar{u} + \omega_{\mathrm{sp}})} \left[d_{\mathrm{n}}(0, F_0) + \frac{\bar{u} F_0}{8\pi n_\infty (\bar{u} + 2\omega_{\mathrm{sp}})} \right] . \tag{7.33}$$

Numerical data for $d_{\mathrm{n}}(0, F_0)$ are available from Table 1 of Ref. [7.2]. In addition, the second order perturbation also produces the (free space) polarization energy

$$E_{\mathrm{pol}}^0 = -\tfrac{1}{2} \alpha_0 F_0^2 . \tag{7.34}$$

The fourth order energy shift can be written formally as

$$E^{(4)} = - \sum_{r,s,t} \frac{\langle 0|H_{\mathrm{sa}}|r\rangle \langle r|H_{\mathrm{sa}}|s\rangle \langle s|H_{\mathrm{sa}}|t\rangle \langle t|H_{\mathrm{sa}}|0\langle}{(\varepsilon_{\mathrm{r}} - \varepsilon_0)(\varepsilon_{\mathrm{s}} - \varepsilon_0)(\varepsilon_{\mathrm{t}} - \varepsilon_0)}
- E^{(2)} \sum_t \frac{\langle 0|H_{\mathrm{sa}}|t\rangle \langle t|H_{\mathrm{sa}}|0\rangle}{(\varepsilon_{\mathrm{t}} - \varepsilon_0)^2} . \tag{7.35}$$

From the first term, we obtain three contributions: The first is independent of the electric field and constitutes a higher order correction to the van der Waals interaction. The second is a higher order (free space) hyper-polarization energy which is independent of the distance. The third is identified as the field-induced dipole-dipole or polarization energy; it can be written, after a series of approximations, as [7.18]

$$E_{\mathrm{pol}} = -\frac{1}{32} \frac{\alpha_0^2 F_0^2}{[z - d_{\mathrm{n}}(0, F_0)]^3} . \tag{7.36}$$

It has been argued that the polarization energy (7.36) should have a factor 1/8.

This is the case for a permanent dipole interacting with its image in the metal. For an induced dipole a further reduction by 1/4 pertains leading to overall factor 1/32 in (7.36). This concludes the derivation of the dispersion and polarization energies between an atom and a solid.

This calculation shows that an electric field has two effects on the interaction between an atom and a solid: (i) It modifies the van der Waals, or dispersion, interaction and (ii) it induces a dipole-dipole, or polarization, interaction. Both have been calculated in lowest, i.e., fourth order perturbation theory. We proceed now to estimate their magnitudes.

The modification of the dispersion energy (7.12) is due to the field dependence of the position of the dynamic image plane, $z_d(F_0)$, as given by (7.33). It essentially follows the field dependence of $d_n(0, F_0)$ and decreases from values around 1.5 Å above the topmost lattice plane in zero field to 1.3 Å in fields of the order of 5 V/Å. There is little dependence on the properties of the adsorbing atom or the metal. A further decrease in the strength of the van der Waals interaction must be expected from higher order corrections because spontaneous charge fluctuations around a polarized atom are less likely than around an unpolarized one.

For an estimate of the polarization energy (7.36) we note that the static image plane position $d_n(0, F_0)$ changes typically from 1.7 Å to 1.4 Å (measured from the topmost lattice plane, not from the jellium edge) as the field is increased from zero to 5 V/Å. This shift is negligible at the equilibrium position for adsorption, $z_{eq} \approx 3.5$ Å. For Helium, with $\alpha_0 = 1.38$ a.u., the polarization energy (7.36) contributes about 1 meV to the binding in the highest field. The situation is not much different for the other rare gases.

To obtain the dependence of the total adsorption energy on F_0, and, thus, to determine the equilibrium position z_{eq} of the adsorbed atom, one has to add a repulsive energy to (7.12) and (7.36) which is usually assumed to be proportional to the electron density $n(z, F_0)$ at the position of the adsorbed atom. To avoid an (artificial) singularity at the position of the image plane, one also introduces a heuristic [7.21] damping factor $D(z)$ multiplying the dispersion and polarization energies. The sum of (damped) dispersion and polarization energies and the repulsive energy then yields the binding energy as a function of field. One finds that an increase in binding energy for He from about 5–10 meV in zero field to 20–40 meV in fields of the order 5 V/Å is mainly due to the shift in the image plane as a result of the electrons being pushed into the metal. However, we should point out that, although the dispersion and polarization energies are calculated rather accurately, the assumptions of adding a repulsive energy and including a damping factor are rather ad hoc. Yet the main conclusion stands, namely that (i) the polarization energy is rather small, and (ii) the modification of the dispersion energy due to the shift in the image plane does not account for the observed field-induced binding energy of He. What is still noteworthy is the trend that the equilibrium position moves towards the solid with increasing field. This indeed aids in the formation of chemisorption bonds that result ultimately in field-adsorption [7.14–16].

The calculation by *Watanabe* et al. [7.18] of dispersion and polarization forces in electric fields is rigorous in the context of the adopted model, namely, a flat structureless metal surface. There is no doubt, however, that the electric field at a kink site or around a single metal atom is different from the field along an extended crystal plane. A first step towards taking such local field variations into account would be the calculation of dispersion and polarization forces, following the approach by Watanabe et al. [7.18], for an atom on top of a boss, i.e., for the geometry of Fig. 7.1.

7.3 Field-Induced Chemisorption

Electric fields of the order of volts per Ångstroms are comparable to those experienced by valence electrons in atoms and molecules. One should, therefore, expect that in external fields of that magnitude a redistribution of the valence electrons in the coupled adsorbate-solid system takes place which affects both the orbitals of the surface bond as well as internal bonds in an adsorbed molecule. Whether this redistribution leads to enhanced or reduced binding depends on whether bonding or antibonding orbitals are more strongly affected. We will refer to this phenomenon as field-induced chemisorption. Very surprisingly, it is important even for the most inert atom, namely helium. We recall that in zero field, chemisorption is rather unimportant for the lighter rare gases because for them, physisorption is dominant, arising from mutually induced fluctuating dipole-dipole interactions of the van der Waals type. We note, however, that even for argon some weak chemisorption effects, i.e., formation of bonding orbitals, take place. Returning to helium, we note that in fields of the order of 5 V/Å, polarization induces the occupation of excited states at the level of a few percent. Thus, even helium cannot, in such fields, be regarded as a closed shell atom, with the consequence that it forms weak covalent bonds as it approaches a metal surface.

An early microscopic calculation of field-induced chemisorption by *Kahn* and *Ying* [7.22] was based on the local density approximation of the density functional formalism; they calculated the potential energy curves, and thus the activation energy, for alkali atoms on W, the latter being treated within the jellium model. *Kingham* [7.23] has presented some preliminary results for the field evaporation of W obtained within a tight binding cluster model with field effects and charge transfer treated in an ad hoc manner. In our work [7.14–16] on field-induced chemisorption, we have used the atom superposition and electron delocalization molecular orbital method (ASED-MO) [7.24]. Based on a charge partitioning model, one writes the total energy E of a cluster of interacting atoms as a sum of a repulsive term E_r that accounts for the Coulomb interaction of isolated atoms with each other and a remainder E_{npf} that entails the (non perfectly following) rearrangement within the atoms in the presence of each other. The latter is calculated from a Hamiltonian:

$$H = \sum_{i\alpha} H_{ii}^{\alpha\alpha} |\varphi_i^\alpha\rangle\langle\varphi_i^\alpha| + \sum_{ij\alpha\beta} H_{ij}^{\alpha\beta} |\varphi_i^\alpha\rangle\langle\varphi_j^\beta| , \qquad (7.37)$$

where, in the spirit of an extended Hückel scheme, one puts the diagonal elements

$$H_{ij}^{\alpha\beta} = E_i^{\alpha}\delta_{ij}\delta_{\alpha\beta} , \tag{7.38}$$

equal to the negative of the ionization energy of level i on atom α, taken from experiment. The remaining off-diagonal elements are a modification of the extended Hückel formula

$$H_{ij}^{\alpha\beta} = \kappa(H_{ii}^{\alpha\alpha} + H_{jj}^{\beta\beta})S_{ij}^{\alpha\beta}\exp(-aR_{\alpha\beta}) , \tag{7.39}$$

with

$$S_{ij}^{\alpha\beta} = \langle\varphi_i^{\alpha}|\varphi_j^{\beta}\rangle \tag{7.40}$$

being the overlap integral between the i−th atomic orbital on atom α and the j−th orbital on atom β, the latter being a distance $R_{\alpha\beta}$ away. Fitting bond strengths and lengths to first row diatomics, one determines the parameters in (7.39) to be $\kappa = 1.125$ and $a = 0.13\,\text{Å}^{-1}$.

To include the effect of a strong electric field within the ASED-MO model, we must add terms

$$H_{ij}^{\alpha\beta}(F) = \int \varphi_i^{\alpha}(\boldsymbol{r})^* V_{\mathrm{F}}\varphi_j^{\beta}(\boldsymbol{r})d\boldsymbol{r} , \tag{7.41}$$

where the field potential is given by

$$V_{\mathrm{F}}(\boldsymbol{r}) = e\int_{-\infty}^{\boldsymbol{r}} \boldsymbol{F}(\boldsymbol{r}')\cdot d\boldsymbol{r}' . \tag{7.42}$$

Additionally, one must, of course, also add the field energy of the nuclei

$$-\sum_{\alpha}\frac{Z_{\alpha}}{e}V_{\mathrm{F}}(\boldsymbol{R}_{\alpha}) , \tag{7.43}$$

where Z_{α} is the charge of the α-th nucleus.

In (7.42) we write

$$V_{\mathrm{F}}(\boldsymbol{r}) = V_{\mathrm{F}}(\boldsymbol{R}_{\alpha}) + V_{\mathrm{F}}(\boldsymbol{r}) - V_{\mathrm{F}}(\boldsymbol{R}_{\alpha}) \tag{7.44}$$

and get

$$H_{ij}^{\alpha\beta}(F) = V_{\mathrm{F}}(\boldsymbol{R}_{\alpha})S_{ij}^{\alpha\beta} + \int\varphi_i^{\alpha}(\boldsymbol{r})^*[V_{\mathrm{F}}(\boldsymbol{r}) - V_{\mathrm{F}}(\boldsymbol{R}_{\alpha})]\varphi_j^{\beta}(\boldsymbol{r})d\boldsymbol{r} . \tag{7.45}$$

The diagonal parts of the first term in (7.45) lead to the raising of the energy levels in (7.38) by the field energy, i.e.,

$$H_{ij}^{\alpha\beta} = (E_i^{\alpha} + V_F(\boldsymbol{R}_{\alpha}))\delta_{ij}\delta_{\alpha\beta} . \tag{7.46}$$

The off-diagonal parts of the first term in (7.45) are obviously proportional to the overlap integrals between atomic orbitals on different atoms. To evaluate the second term in (7.45) for $\alpha = \beta$, we note that over the extent of the atom α

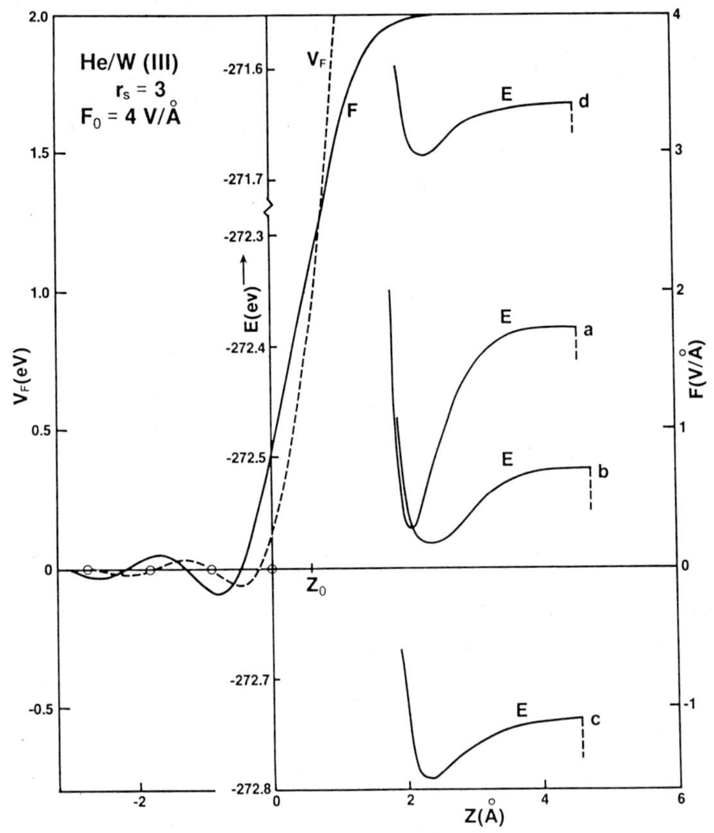

Fig. 7.3. Binding energy curves, E(ev), for He on a W_4 cluster in an asymptotic field $F_0 = 4$ V/Å in various approximations: (*a*) shifting the energy levels in (7.38) and (7.39) by $V_F(z_\alpha)$, (*b*) including also the linear polarization of He, (*c*) adding the linear polarization on the topmost W, and (*d*) including all $z-$ and z^2 matrix elements. The field F and the associated field energy V_F are taken from *Gies* and *Gerhardts* [7.2] for $r_s = 3.0$; from [7.33]

the electric field varies little. We can, therefore, employ a Taylor expansion (we restrict ourselves to the case where the electric field depends on z only) to obtain

$$V_F(z) - V_F(z_\alpha) = (z - z_\alpha)eF(z_\alpha) + \tfrac{1}{2}(z - z_\alpha)^2 eF'(z_\alpha) + \dots , \tag{7.47}$$

so that the overlap integrals reduce to matrix elements of powers of z.

As a first example, we show in Fig. 7.3 the potential energy of He adsorbed on the apex of a tetrahedral W_4 cluster in a field of 4 V/Å as a function of the He position above the topmost lattice plane, the W atoms being held fixed at their equilibrium lattice sites. The relevant electronic energies and Slater exponents for isolated He and W atoms are given in Table 1 of Ref. 7.15. In curve (a), we have neglected all polarization effects, i.e., we only shift the energies E_i^α according to (7.46) for the adsorbing He and also for the topmost W atom by the local field energy. Including the linear polarization term for the He atom,

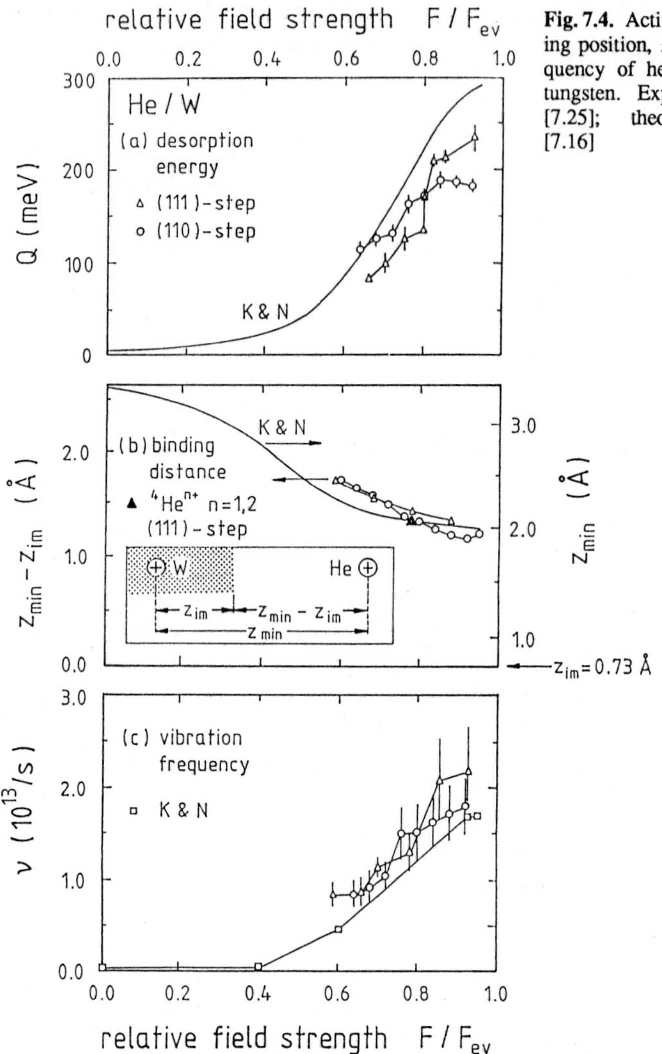

Fig. 7.4. Activation energy, Q, binding position, $z_{\min}$, and vibrational frequency of helium field adsorbed on tungsten. Experimental points from [7.25]; theoretical curves from [7.16]

curve (b), we find that the activation energy gets reduced and that the adsorption potential gets wider. Including the linear polarization term for the topmost W atom as well, curve (c), reduces the activation energy some more. Finally, in curve (d) we include the z- and z^2-matrix elements for He and for the topmost W.

The complete field dependence of the activation energy, the position of the minimum in the adsorption potential and the vibrational frequency of the He atom in the surface potential is depicted in Fig. 7.4 together with experimental result [7.16, 25]. To facilitate a direct comparison, we have normalized the field strength by the evaporation field strength of W as measured and calculated, respectively. The overall agreement is rather encouraging. In particular, the results prove the point that field-induced chemisorption accounts for most of the binding of He in strong electric fields. An interesting bonus emerging from Fig. 7.4 results from the fact that experimentally, one can only determine the position of the adsorbed He atom relative to the position of the image plane, the latter being the reference plane from which the acceleration of He^+ starts. With the theory referring the position of the adsorbed He to the position of the topmost W atoms in the metal, we can shift the experimental points onto the theoretical curve and, thus, get an estimate of the position of the image plane with respect to the metal ion cores. One finds that the image plane lies (roughly) 0.7 Å above the topmost lattice plane, i.e., about half an interlattice distance outside the metal!

Exciting as these results may be, one must be aware of a number of caveats: (i) The ASED-MO method is semi-empirical and lacks selfconsistency. The latter, however, is crucial when large charge transfers occur as happens at and beyond the apex of the ground state energy curves in Fig. 7.3. Thus, with He having 2 electrons in the $1s$ level, once the latter is raised in the electric field above the Fermi level of the metal, 2 electrons cross over into the metal. What should rather happen is that 1 electron crosses over with the energy level of the second electron dropping to $-54.4\,\mathrm{eV}$. (ii) With finite clusters of metal atoms, image charge effects are not accounted for. This should not be too serious an error as we have seen in the previous section that such effects account for only a fraction of the total field-induced binding energy. (iii) The electric field was taken from jellium calculations for a flat surface. Thus, local enhancements at kink sites or above single metal atoms on top of a plane are not included. Again, the estimate around (7.21) suggests corrections of the order of a few tens meV. In all, the results obtained for field-induced chemisorption with the ASED-MO method are interesting enough to work on an ab initio calculation, which takes chemisorption and field effects into account selfsonsistently.

7.4 Field Evaporation

Field evaporation is the removal of lattice atoms as singly or multiply charged positive ions from a metal in a strong electric field F of the order of several V/Å, as it occurs at field ion tips [7.26, 27]. The term field desorption is usually reserved for the process of removing field-adsorbed atoms or molecules from a field ion tip [7.28–30]. Field evaporation and field desorption are crucial in cleaning and preparing field ion tips. They are thermally activated processes; as such, their rate constants can be parametrized according to Frenkel-Arrhenius as

$$r_{\mathrm{d}} = \nu(T, F)\,\mathrm{e}^{-Q(F)/k_{\mathrm{B}}T} \,. \tag{7.48}$$

Here $Q(F)$ is the field dependent height of the activation barrier to be overcome by the desorbing particle. The prefactor $\nu(T, F)$ can also be field dependent and is usually weakly temperature dependent as well. The minimum field strength beyond which, at low temperatures, the metal tip evaporates is termed the evaporation field strength; it varies from 2.5 V/Å for Ti to 6.1 V/Å for W with a typical experimental error margin of 10–15% [7.30]. *Ernst* has measured $Q(F)$ and $\nu(F)$ for Rh [7.31], and *Kellogg* [7.32] has presented data for W in the field range 4.7 to 5.9 V/Å.

Two phenomenological models have been used in the past to calculate the activation energy $Q(F)$. In the "image-force" model [7.27], field evaporation is envisaged as the activation of an ion of charge n over an activation barrier that results from the superposition of the field potential $-neFz$ (assumign a constant electric field), and the image potential of the ion $-n^3e^2/4z$. One gets

$$Q(F) = \Lambda + \sum_n I_n - n\Phi - \sqrt{n^3e^3F} + \tfrac{1}{2}(\alpha_a - \alpha_i)F^2 \,, \qquad (7.49)$$

where Λ is the field-free sublimation (cohesive) energy of the metal, I_n is the n-th ionization potential of the desorbing ion, Φ is the work function of the surface, F is the applied electric field, and α_a and α_i are the polarizability of the surface and the desorbing atom, respectively.

In the "charge-exchange" model [7.28] one assumes that ionization and desorption occur at the crossover point z_c between the atomic and ionic potential energy curves. One finds

$$Q(F) = \Lambda + \sum_n I_n - n\Phi - \frac{(ne)^2}{4z_c} - neFz_c - \Gamma + \frac{1}{2}(\alpha_0 - \alpha_i)F^2 \,, \qquad (7.50)$$

where Γ is the halfwidth of the ionic level broadened by interaction with the atomic curve. It seems to be the consensus in the literature [7.31–32] that the charge-exchange model is the more realistic, albeit less useful, due to the fact that the atomic and ionic potential energy curves are not known. The image-force model on the other hand, has been used to predict evaporation field strengths (dropping the polarization terms) agreeing very well with existing data, in some cases at the one percent level [7.28–30]. This agreement is rather surprising because (i) experimental values are only accurate to within 10–15% [and Kellogg's field dependence [7.32] of Q disagrees with both (7.49) and (7.50)] and (ii) some of the assumptions underlying (7.49) are highly oversimplified. We are referring in particular to the assumption that the electric field maintains its constant value beyond the position of the field desorbing lattice atom.

Field evaporation is a dramatic demonstration of the limitation of classical concepts in solid state physics. Maxwell's theory says that the electric field drops to zero at the image plane, i.e., just outside the metal. Thus the electric field has, classically, no effect on the ion cores of the metal, and thus, classically, field evaporation is not possible. On the other hand, we have seen in Sect. 7.1 that field expulsion from the metal is not complete and that the field strength at the topmost ion cores in a metal can be substantial, and in particular strong enough to cause field evaporation.

Experimental results suggest that field evaporation of metal atoms occurs most likely at steps, kinks and edges or for small clusters of atoms on larger planes. Theoretically, one should be able to calculate the electric field, the electron density, and the geometry of the ion cores for such configurations selfconsistently.

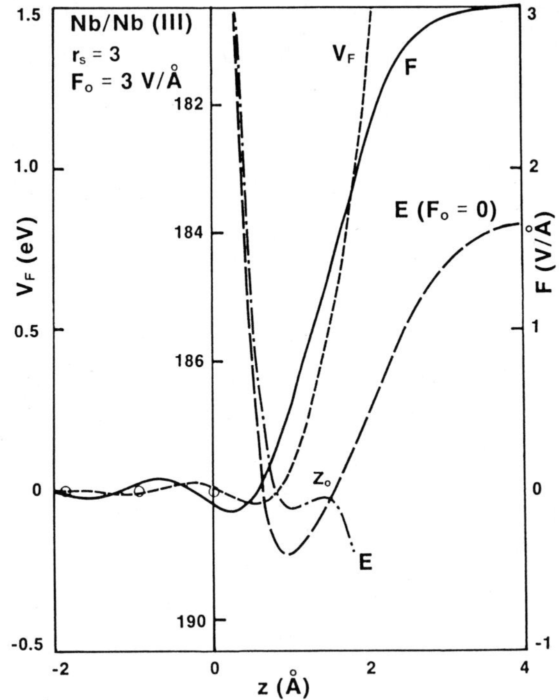

Fig. 7.5. Binding energy curves for Nb on Nb_4 in F_0 = 0 and 3 V/Å; from [7.33]

So far, this has not been done. However, *Kreuzer* and *Nath* [7.33] have used their ASED-MO cluster method taking the electric field again from jellium calculations, i.e., foregoing the selfconsistency requirement. In Fig. 7.5 is plotted the energy of a Nb atom at a distance z from the bottom of a tetrahedral Nb_4 cluster with the electric field applied at the apex as indicated, for a Wigner-Seitz radius $r_s = 3.07$. We define the activation energy $Q(F)$ for evaporation as the potential difference between the bottom of the potential well and the height of the activation barrier. This is plotted in Fig. 7.6, together with several other metals, as $Q(F)/Q(F=0)$ as a function of F/F_{ev} where the (low temperature) evaporation field strength F_{ev} is that field at which the activation barrier dissappears. Also plotted are Kellogg's data [7.32] for $Q(F)$ for W showing excellent agreement with the theory.

For W and Nb, an increasing electric field leads to a continuous draining of electrons from the bonding orbitals. This leads to the observed monotonic decrease in the activation energy. For Fe, one finds initially a small rearrangement between bonding and antibonding orbitals so that the activation energy remains rather constant for small fields. The scaling behavior of the activation energy curves for W and Nb in Fig. 7.6 is astounding. Attempting an explanation, we recall that the binding energy curves for metals and for bimetallic interfaces have been found to be universal [7.34] if one scales the energy with the cohesive (adhesive) energy and the distance with the Thomas-Fermi screening length

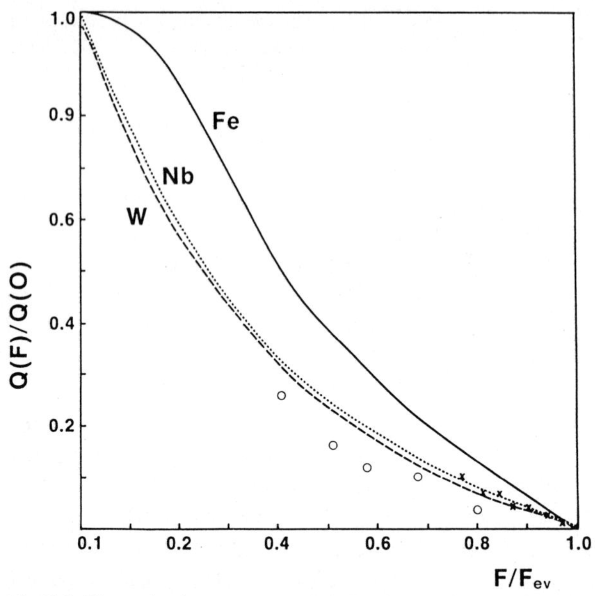

Fig. 7.6. The activation energy, $Q(F)$, for field evaporation, normalized with the cohesive energy, $Q(0) = V_c$, as a function of asymptotic field strength, F, normalized with the evaporation field strength, F_{ev}, from [7.33]. Crosses are data for W from [7.32]; circles are data for Rh from [7.31]

$\lambda = (9\pi/4)^{1/3} r_s^{1/2}/3$. To model the binding energy curves in the presence of an electric field, e.g., those in Fig. 7.5, we could argue that as long as the field-induced charge transfer to the topmost metal atom is minimal, it is essentially moving in the field-free binding energy curve out to the point where its highest occupied electron level crosses the Fermi level leading to ionization. From thereon, it is accelerated by the electric field. The retarding attractive forces, for which the image hump model takes the image force [7.28, 29], are taken care of by keeping the (field-free) cohesive energy curve, thus, indeed, neglecting modification due to the electric field. In this simple model one would then write the binding energy in a field as

$$\frac{V_F(z^*)}{V_c} = v(z^*) - \frac{1}{2} f z^* , \tag{7.51}$$

where

$$z^* = \frac{z}{\lambda} , \quad f = \frac{F}{F_{ev}} , \quad F_{ev} = \frac{3V_c}{2ne(9\pi/4)^{1/3}\sqrt{r_s}} , \tag{7.52}$$

$v(z^*)$ is the universal cohesive energy curve [7.34] which we parametrize for simplicity by a Morse potential

$$v(z^*) = e^{-2(z^* - z_0^*)} - 2e^{-(z^* - z_0^*)} . \tag{7.53}$$

The activation energy for field evaporation is then given by the energy difference between the minimum and the maximum of the potential energy curve (7.51); it is

$$\frac{Q(f)}{Q(f=0)} = \sqrt{1-f} + \frac{1}{2} f \ln\left[\frac{1-\sqrt{1-f}}{1+\sqrt{1-f}}\right] . \tag{7.54}$$

This simple expression follows the curves for W and Nb in Fig. 7.6 remarkably well. A less straightforward behavior, e.g., that for Fe, which results from complicated electronic rearrangement, cannot be reproduced by this simple model. Absolute values of $Q(F = 0) = V_c$ and F_{ev}, calculated from the full theory or predicted from the simple scaling law (7.52) and (7.54), agree well with experimental data [7.33].

7.5 Thermal Field Desorption

Thermal field desorption is the removal of field-adsorbed species from the surface of a field ion tip which can be achieved by raising the temperature of the metal tip. Being an activated process, one could argue that its rate constant should follow an Arrhenius parametrization (7.49) where E_d is the activation energy, obtained from adiabatic energy curves, and the prefactor ν reflects the ionization probability of the adsorbed molecule. This simple-minded approach, however, ignores a number of interesting questions, e.g., whether the desorbing species emerge as ions or neutrals, whether postionization occurs, and what the energy distribution of the desorbing species is. To answer such questions, one has to set up a kinetic theory that accounts for energy and charge transfer by formulating the appropriate master equation for the problem and calculating all transition probabilities from first principles. This theory was formulated recently [7.35, 36] and will be reviewed briefly.

We start our discussion of field desorption by considering a helium atom adsorbed on a field ion tip. As temperature is raised, it will eventually get the chance to escape from its binding potential. If its escape from the surface is slow enough, it will get ionized at the hump of the ground state energy curve and reach the detector as an ion. However, if ionization, i.e., tunneling of an electron from the adatom to the metal, is too slow, the adparticle will escape as a neutral atom with a kinetic energy that has no relation to the ground state energy curve. Indeed, a neutral atom in a field past a critical distance no longer corresponds to the ground state of the system but to some excited state.

From this short discussion it should be obvious that adiabatic states, for which the nuclei of all atoms are held fixed, are not the most intuitive basis to set up a kinetic theory of field desorption and field ionization. Rather a new basis must be constructed in which the motion of the gas species is explicitly taken into account; they are known as diabatic states. For field desorption of helium, two diabatic states are relevant, namely those for a neutral atom and for a singly charged ion. They are obtained from the adiabatic potential energy curves, $V_i(\boldsymbol{R})$, by an unitary transformation which reads explicitly [7.35]:

$$W_{00}(\boldsymbol{R}) = \cos^2\Theta(\boldsymbol{R})V_1(\boldsymbol{R}) + \sin^2\Theta(\boldsymbol{R})V_0(\boldsymbol{R}) , \tag{7.55}$$

$$W_{++}(\boldsymbol{R}) = \cos^2\Theta(\boldsymbol{R})V_0(\boldsymbol{R}) + \sin^2\Theta(\boldsymbol{R})V_1(\boldsymbol{R}) \,, \tag{7.56}$$

$$W_{0+}(\boldsymbol{R}) = \tfrac{1}{2} \sin 2\Theta(\boldsymbol{R})[V_1(\boldsymbol{R}) - V_0(\boldsymbol{R})] \,. \tag{7.57}$$

Here W_{00} is the potential curve for a neutral atom, and W_{++} that of an ion. The off-diagonal terms $W_{0+} = W_{+0}$ couple these states together and are responsible for ionization or neutralization. The transformation angle Θ was calculated by assuming that the many-electron wavefunctions of the cluster are given as Slater determinants with the constituent single electron wavefunctions obtained from the ASED-MO programme (including electric field effects). The relevant excited state is selected as belonging to the same symmetry representation as the ground state, e.g., if we model the gas-metal system by a tetrahedral cluster of metal atoms with helium adsorbed on its apex, it has C_{2v} symmetry. With He in its $1s$ ground state, the total ground state wave function Φ_g, belongs to the A_1 representation. Because W_{0+} preserves this symmetry, the relevant excited state has a wave function, Φ_e, with this symmetry. Because the atom or ion will desorb along the steepest field gradient, i.e., perpendicular to the surface, we can neglect any lateral variation, so that, by assumption, $\Theta = \Theta(z)$ depends on the distance, z, from the metal only. In Fig. 7.7, we present adiabatic and diabatic potential energy curves for several field strengths. Such curves have been drawn qualitatively to serve as a basis of Müller's image hump model, Gomer's charge exchange model, and for discussions of charge hopping and charge draining mechanisms [7.37]. According to the discussion above, the difference between the adiabatic potential energy curves reaches a minimum at the apex and is of the order of the interaction energy between the highest occupied and the lowest unoccupied orbitals. With increasing field, the apex will move towards the metal surface, resulting in an increase in the interaction energy and, thus, in an increase in the energy difference between the adiabatic energy curves. We note that Θ varies rapidly from 0° to 90° over a very short distance, i.e., less than 0.1 Å around the apex of the adiabatic ground state energy curve, indicative of the narrowness of the ionization zone.

To calculate the kinetics of field desorption and field ionization, a master equation has been derived which reads for our present problem [7.38] as

$$\begin{aligned} \frac{dn_{i\nu}}{dt} &= \sum_{\mu} [R_i(\nu,\mu)n_{i\mu} - R_i(\mu,\nu)n_{i\nu}] \\ &\quad + \sum_{\mu j} [T_{ij}(\nu,\mu)n_{j\mu} - T_{ji}(\mu,\nu)n_{i\nu}] \,. \end{aligned} \tag{7.58}$$

Here $i = 0, +$ refers to the neutral and ionic state of helium. The symbols ν and μ label the states in the diabatic potentials, $W_{00}(z)$ and $W_{++}(z)$, among which phonon-induced transitions take place with rates $R_i(\nu,\mu)$. The tunneling rates connecting neutral and ionic states are given by

$$T_{0+}(\mu,\nu) = 2\pi/\hbar \left| \int d\boldsymbol{R}\, \eta^*_{0\mu}(\boldsymbol{R}) W_{0+}(\boldsymbol{R})\, \eta_{+\nu}(\boldsymbol{R}) \right|^2 \Delta(E_{+\nu} - E_{0\mu}, \Gamma_{0\mu}), \tag{7.59}$$

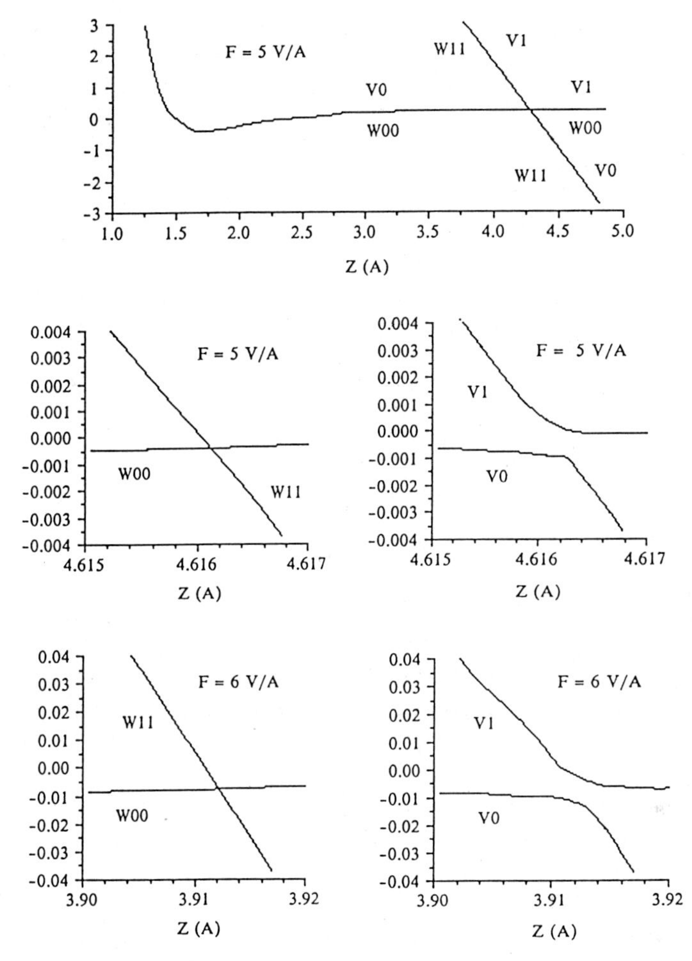

Fig. 7.7. Adiabatic (V) and diabatic (W) energy curves for He on W in a field of 5 V/Å (upper set and enlarged version in center) and in 6 V/Å (lower set); from [7.36]

with

$$\Delta(\varepsilon, \Gamma) = \pi^{-1} \frac{\Gamma/2}{\varepsilon^2 + \Gamma^2/4} . \tag{7.60}$$

In (7.59), $\Gamma_{0\mu}$ is the half width of the (discrete) level μ in W_{00} due to phonon transitions and is given as

$$\Gamma_{0\mu} = \hbar \sum_{\nu} R_0(\nu, \mu) . \tag{7.61}$$

The ion yield can be calculated from (7.58) and is given approximately by

$$Y_{\text{ion}} = \sum_{\nu,\mu} T_{0+}(\mu, \nu) \exp(-E_{0\mu}/k_B T) / \sum_{\mu} \exp(-E_{0\mu}/k_B T) . \tag{7.62}$$

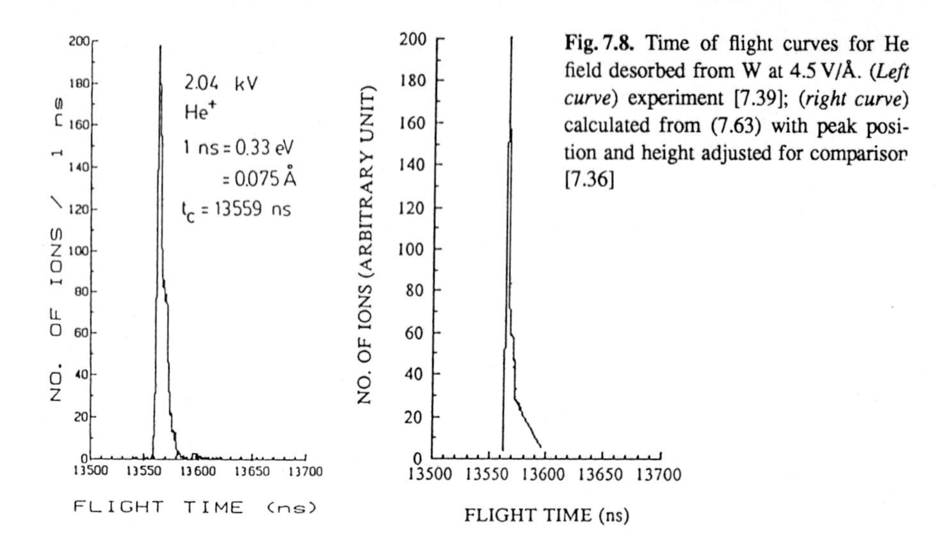

Fig. 7.8. Time of flight curves for He field desorbed from W at 4.5 V/Å. (*Left curve*) experiment [7.39]; (*right curve*) calculated from (7.63) with peak position and height adjusted for comparison [7.36]

In Fig. 7.8, we show the ion yield as a time of flight curve adjusted in position and height to allow direct comparison with experimental data by *Tsong* [7.38]. We note that the theoretical curve has an energy width of about 0.6 eV as compared to about 1 eV in the experiment. Considering a number of theoretical approximations and also that the experiment yields an upper limit, these two estimates compare rather favorably.

Thermal field desorption being an activated process, one can evaluate the yield (7.62) according to an Arrhenius parametrization (7.49). One finds a dramatic increase in the prefactor as a function of field, more or less in an exponential fashion. We recall that in ordinary, i.e., field free, thermal desorption, the effective prefactor is a product of a sticking coefficient and an attempt frequency to desorb. In thermal field desorption, the role of the sticking coefficient is replaced by the ionization probability at the apex of the adiabatic ground state energy curve. It varies from zero in $F = 0$ to a saturation limit in high fields, as borne out by the calculations. For He, this strong field dependence of the prefactor has an interesting consequence. We note that for He the activation energy for field desorption is roughly equal to the depth of the diabatic curve for the neutral species. Thus thermal desorption of neutral and ionic He in a field have the same desorption energy, implying that the ratio of ions to neutrals is proportional to the ratio of the prefactors. Since the prefactor for the desorption of neutral He is not a strong function of field strength, one predicts that the ratio of ion to neutral yield is an exponential function of field strength. In particular, well below the best image voltage, thermal desorption will only yield neutral He. To test this idea, an experiment should start from a well defined, constant coverage of He that is totally removed by a fast temperature rise so that the total number of desorbed species remains constant. If the detector only registers ions, one should see the exponential increase in ion yield directly.

The master equation (7.58) can also be used as a basis to study field ionization, i.e., the process by which a particle impinging from the gas phase gets ionized within the ionization zone. This process is complicated by the fact that the gas particles do not arrive with thermal velocities but are accelerated in the inhomogeneous field in front of the metal tip. It is crucial to understand this process in detail in order to develop a concise theory of the image formation in the field ion microscope.

7.6 Field-Induced Chemistry

Electric fields of sufficient strength must influence thermodynamic equilibrium between chemically reacting species and also the reaction pathways and the kinetics. To understand the electric field effect on the equilibrium properties, one argues [7.40] that the dependence of the equilibrium constant, K, of a reactive system on the electric field strength, F, is controlled by a van't Hoff equation

$$\left(\frac{\partial \ln \mathrm{n} K}{\partial F}\right)_{\mathrm{P,T}} = \frac{\Delta M}{RT}, \tag{7.63}$$

where P and T indicate constant pressure and temperature, respectively, and ΔM is a partial molar energy related to the change in electric moment in the reaction, i.e.,

$$\Delta M = \Delta p F + \tfrac{1}{2}\Delta\alpha F^2 + \dots . \tag{7.64}$$

Here Δp is the difference in the permanent dipole moments of the products and the reactants, and $\Delta\alpha$ the change in their polarizabilities. In order to achieve values for ΔM comparable with typical reaction enthalpies or volumes, one needs fields in excess of 0.1 V/Å.

Block and coworkers have developed a field pulse technique in the field ion microscope that allows the investigation of the field effect on chemical reactions; a detailed account has been given by *Block* [7.41]. Systems that have been studied by this technique are the formation of metal subcarbonyls, the polymerization of acetone, reaction of sulphur on metal surfaces, decomposition of methanol on metal surfaces, hydride formation on semiconductors, NO reactions on metals and many more.

A micrcoscopic theory can elucidate the changes in reaction pathways induced by high electric fields. As a precursor to such field-induced chemistry we will now look at the effect of high electric fields on the vibrational frequency of an adsorbed diatomic molecule. As an example we review a study of N_2 on a Fe{111} surface using the ASED-MO cluster model [7.42]. It was found that upon adsorption in zero field, the frequency of the N_2 stretch vibration gets reduced mainly due to occupation of N_2 antibonding orbitals. As the electric field is increased from zero to 0.5 V/Å, these orbitals are drained leading to a stiffening of the N–N stretch vibration. At somewhat higher fields, a dramatic

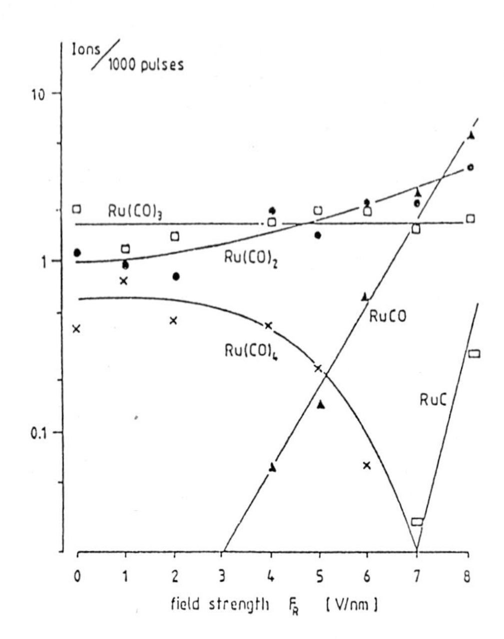

Fig. 7.9. Experimental intensities of $Ru(CO)_n$ as a function of steady field strength; from [7.44]

decrease occurs in the stretch vibration due to draining of N–N bonding orbitals into the metal which for fields of the order of 1 V/Å eventually leads to field dissociation. It is obvious that in the presence of other reactants, the electric field would influence the reaction mechanism drastically. With reference to the van't Hoff equation (7.64), the electric field, by modifying bonding and antibonding oribitals, affects the dipole moment and the polarizability of the adsorbed species, and thus drastically alters the equilibrium constant of the reaction.

As a second example of field-induced chemistry, we look at the formation of Ru-subcarbonyls in high electric fields. The interaction of CO with a Ru field emitter tip has been studied using pulse field desorption mass spectrometry [7.43, 44]. It was found that for fields less than about 0.3 V/Å, mostly $Ru(CO)_n^{2+}$ with $n = 2, 3, 4$ was produced. For larger fields, $Ru(CO)_4^{2+}$ disappeared and RuCO was produced, see Fig. 9. To understand this, *Wang* and *Kreuzer* [7.45] have used the ASED-MO cluster model to calculate the binding energy of Ru subcarbonyls; the results are shown in Fig. 7.10. The formation of subcarbonyls is envisaged as a sequential process that starts with a CO molecule attaching itself to a Ru atom at a kink step site. The higher carbonyls are then formed by successive addition of CO. It emerges from the calculations that with increasing field strength, more electronic charge is transferred from the CO to the metal with the net charge on the O atom changing from negative to positive. This leads, for example, for the $Ru(CO)_2$ species to a decrease in the C-O bonding angle, which in turn makes more room for sequential adsorption of additional CO up to the point where, at high fields, the repulsion between CO molecules on the same Ru becomes too strong. These findings are in qualitative agreement with experiment. It is clear that more calculations of this nature are needed to

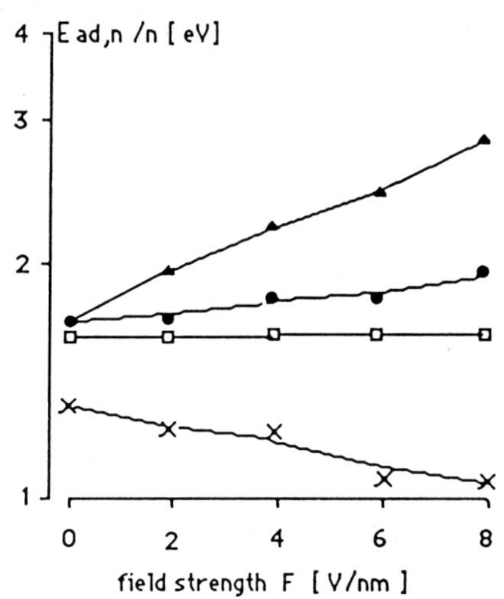

Fig. 7.10. Adsorption energies of n(CO) to form $Ru(CO)_n$ on a Ru cluster; from [7.45]

understand the stabilization and destabilization of adsorbed molecules in high electric fields. Moreover, using the kinetic theory of the last section, one should also calculate reaction rates. This, together with thermodynamic considerations will then allow us to understand field-induced chemistry quantitatively.

In closing, we would like to point out a link to heterogeneous catalysis. It is well known that adsorbed alkali atoms act as promoters to many catalytic reactions. In the context of field-induced chemistry this can be readily understood if we observe that alkalis adsorb as ions which in turn generate fields of the order of Volts per Ångstrom. As an example, beyond the formation of subcarbonyls, we have studied the formation of N_2O from NO, and found that on top or next to a K^+ ion on a metal surface, N_2O becomes much more stable than in the gas phase, basically due to increased charge transfer, in agreement with experimental results [7.46]. It is obvious that a halogen ion would act as a poison for the same reaction.

7.7 Concluding Remarks

The field ion microscope, invented more than 50 years ago, has opened up a new avenue to study the properties of matter in high electric fields. In recent years, a theory has emerged that looks at the new physics and chemistry from a microscopic point of view. We now understand field adsorption as field-induced chemisorption, a mechanism that affects even the lightest rare gases. In this field, a lot of work still remains to be done to understand local field enhancements and ultimately image formation in the field ion microscope.

Even more exciting is the emerging field of field-induced chemistry. Because fields of the order of Volts per Ångstrom affect the valence electrons of atoms, new molecular species are stabilized in the high fields thus opening up new reaction pathways in heterogeneous catalysis. Most work so far has been concentrated on static electric fields. However, many new phenomena are also to be expected in alternating fields as the work on photon-induced field desorption suggests [7.41].

References

7.1 N.D. Lang, W. Kohn: Phys. Rev. B **1**, 4555 (1970); **3**, 1215 (1966); **7**, 3541 (1973)
7.2 P. Gies, R.R. Gerhardts: Phys. Rev. B **33**, 982 (1986)
7.3 F. Schreier, F. Rebentrost: J. Phys. C **20**, 2609 (1987)
7.4 I.E. Dzyaloshinskii, E.M. Lifshitz, L.P. Pitaevskii: Adv. Phys. **10**, 165 (1961)
7.5 E. Zaremba, W. Kohn: Phys. Rev. B **13**, 2270 (1976)
7.6 C. Holmberg, P. Apell: Phys. Rev. B **30**, 5721 (1984)
7.7 A. Liebsch: Phys. Rev. B **35**, 9030 (1987)
7.8 L.W. Bruch: Surf. Sci. **125**, 194 (1983)
7.9 E.W. Müller: Quart. Rev., Chem. Soc. **23**, 177 (1969)
7.10 E.W. Müller, S.B. MacLane, J.A. Panitz: Surf. Sci. **17**, 430 (1969)
7.11 E.W. Müller: Naturwissenschaften **57**, 222 (1970)
7.12 T.T. Tsong, E.W. Müller: Phys. Rev. Lett. **25**, 911 (1970)
7.13 T.T. Tsong, E.W. Müller: J. Chem. Phys. **55**, 2884 (1971)
7.14 D. Tomanek, H.J. Kreuzer, J.H. Block: Surf. Sci. **157**, L315 (1985)
7.15 K. Nath, H.J. Kreuzer, A.B. Anderson: Surf. Sci. **176**, 261 (1986)
7.16 N. Ernst, W. Drachsel, Y. Li, J.H. Block, H.J. Kreuzer: Phys. Rev. Lett. **57**, 2686 (1986)
7.17 R.G. Forbes, M.K. Wafi: Surf. Sci. **93**, 192 (1982)
7.18 K. Watanabe, S.H. Payne, H.J. Kreuzer: Surf. Sci. **202**, 521 (1988)
7.19 J.E. Inglesfield: Surf. Sci. **188**, L701 (1987)
7.20 Y. Takada, W. Kohn: Phys. Rev. B **37**, 826 (1988)
7.21 K.T. Tang, J.P. Toennies: J. Chem. Phys. **80**, 3726 (1984)
7.22 L.M. Kahn, S.C. Ying: Solid State Commun. **16**, 799 (1975); Surf. Sci. **59**, 333 (1976)
7.23 D.R. Kingham: In *Proceedings of the 29th International Field Emission Symposium*, ed. by H.-O. Andren, H. Norden (Almquist and Wiksell, Stockholm 1983)
7.24 A.B. Anderson, R.G. Parr: J. Chem. Phys. **53**, 3375 (1970); A.B. Anderson: J. Chem. Phys. **60**, 2477 (1974); **62**, 1187 (1975); **63**, 4430 (1975)
7.25 N. Ernst: Surf. Sci. **200**, 275 (1988), ibid **219**, 1 (1989)
7.26 E.W. Müller: Naturwiss. **29**, 533 (1941)
7.27 E.W. Müller: Phys. Rev. **102**, 618 (1956)
7.28 E.W. Müller, T.T. Tsong: *Field-Ion Microscopy, Principles and Applications*. (Elsevier, Amsterdam 1969)
7.29 E.W. Müller, T.T. Tsong: Progr. Surf. Sci. **4**, 1 (1973)
7.30 T.T. Tsong: Surf. Sci. **70**, 211 (1978)
7.31 N. Ernst: Surf. Sci. **87**, 469 (1979)
7.32 G.L. Kellogg: Phys. Rev. B **29**, 4304 (1984)
7.33 H.J. Kreuzer, K. Nath: Surf. Sci. **183**, 591 (1987)
7.34 J.H. Rose, J. Ferrante, J.R. Smith: Phys. Rev. Lett. **47**, 675 (1981)
7.35 H.J. Kreuzer, K. Watanabe: J. de Physique **49**, C6–7 (1988)
7.36 H.J. Kreuzer, L.C. Wang: J. de Physique **50**, C8–9 (1989)
7.37 R.G. Forbes: Surf. Sci. **102**, 255 (1981)
7.38 M. Tsukada, Z.W. Gortel: Phys. Rev. B **38**, 3892 (1988)
7.39 T.T. Tsong: Surf. Sci. **177**, 593 (1986)
7.40 K. Bergmann, M. Eigen, L. de Maeyer: Ber. Bunsenges. Phys. Chem. **67**, 819 (1963)
7.41 J.H. Block: In *Chemistry and Physics of Solid Surfaces IV*, ed. by R. Vanselow, R. Howe, Springer Ser. Chem. Phys. Vol. 10 (Springer, Berlin, Heidelberg 1982)
7.42 D. Tomanek, H.J. Kreuzer, J.H. Block: Surf. Sci. **157**, L315 (1985)
7.43 N. Kruse: Surf. Sci. **178**, 820 (1986)
7.44 N. Kruse, G. Abend, J.H. Block, E. Gillet, M. Gillet: J. de Phys. **47**, C7–87
7.45 L.C. Wang, H.J. Kreuzer: J. de Phys. **50**, C8–53 (1989)
7.46 N. Kruse, J.H. Block, G. Abend: J. Chem. Phys. **88**, 1307 (1988)

8. Chaos in Surface Dynamics

J.W. Gadzuk

Center for Atomic, Molecular, and Optical Physics,
National Institute of Standards and Technology, Gaithersburg, MD 20899, USA

An undeniable reality of current thinking in many areas of intellectual inquiry, both at the "arm-chair physics" as well as serious-research level, is an awareness of the possibility of "chaotic behavior" in one's favorite system. The reasons for such an awareness are many-fold due in part to: i) the easy availability of the entire spectrum of computers (from simple PCs to super computers) being utilized in state-of-the art numerical experiments characteristic of the field; ii) real cross-disciplinary studies in which the eccentricities of nonlinear systems form common links between previously unrelated areas [8.1–7]; iii) the appearance of many excellent review articles directed at the "non-involved but interested physical scientist layperson" [8.8–14], of which the one by Jensen is particularly lucid [8.13]; iv) the timely publication of the justifiably best selling book "Chaos" by James Gleick [8.15] which has gone a long way towards making the concept of chaos a household word, but in a scientifically and/or mathematically enlightened way. Still, one may wonder if and how this explosion of colorful pictures [8.16] and frequently humorous (e.g., "The Chaotic Behavior of the Leaky Faucet" [8.17] or the exposition of the "Ding-a-Ling Model" [8.18]) but legitimate numerical studies impacts on ones own field of work. One asks, "Is it necessary for me to learn what is going on in chaos and is there anything that I would do or think differently if I was to avail myself of modern (chaotic) thinking?" [8.19] It is the intent of this chapter to provide some guidance to those physicists and chemists involved in surface studies such that they might better answer these questions for themselves. To implement this goal, a number of possibly relevant aspects of nonlinear studies will be brought out featuring some of the "paradigms of chaos". Then a few of the already existing examples of surface physics and chemistry research, which have drawn in one way or another upon developments from the chaos world will be aired [8.20–30]. In the process, questions pertaining to whether chaos is necessary or not and where is it all going, will be touched upon [8.19].

8.1 Concepts in Chaos

For present purposes we will consider the issue of "chaotic behavior" almost entirely within the context of the dynamics or time evolution of a few-degrees-of-freedom, deterministic and conservative (rather than dissipative) classical system which often provides an adequate description of both intramolecular dynamics and also kinetics for multi-component, chemically reactive systems. All sorts of new issues arise when one considers the possibility and/or relevance of chaos to quantum systems [8.31-35]. Since these go far beyond the present intended scope, they will mainly be ignored here. (Parenthetically note though that many of the conflicts between the concept of classical and quantum chaos appear in the following way: Since nonlinearity is an essential ingredient of chaotic time evolution; and since the state-vector of a quantum system time evolves according to $i\hbar\partial\psi(t)/\partial t = H\psi(t)$, the time-dependent Schrödinger equation, which is *linear* in $\psi(t)$; therefore, "truly chaotic" time evolution in quantum systems is impossible, at least in the form known in classical systems.)

Operationally, theoretical studies into chaos are mainly in the form of numerical experiments. One settles upon a set of evolution equations which contain the relevant elements of the dynamical system under consideration (e.g., Hamilton's equations for a mechanical system, a set of nonlinearly coupled rate equations for reaction kinetics, etc.). The system under study is then allowed to (time) evolve numerically in accord with some set of initial conditions which have been selected by the experimenter. An outcome, perhaps as a function of time in the long time limit after decay of "uninteresting" transients, is noted and the exercise is repeated, maybe with another relevant set of initial conditions or with a change of system parameters. Suggestive of chaotic dynamics is a sequence of outcomes that look erratic or unpredictable or better still, that show a sensitive dependence upon neighboring initial conditions. Such behavior is exemplified by the breakdown of periodic or quasi-periodic behavior in an oscillating system, the example which will be most heavily treated here.

It is soon apparent that the active participation of the researcher in "pattern-recognition" plays a crucial role in chaos studies. With this in mind, it is important to draw distinctions between *complicated* versus *chaotic* dynamics. Not all time evolution that "looks" complex and complicated is chaotic, and conversely, not all chaotic time evolution "looks" complicated, at least when viewed for short times. From these somewhat provocative pronouncements, it should be guessed that there is a more rigorous meaning to the term *chaotic behavior* than simply how something looks. This shall now be expanded upon.

8.2 Examples

Schuster [8.5] has assembled a nice compendium of various realizations of the transition from regular or periodic to chaotic dynamics as shown in Fig. 8.1. Not only are the physical systems and their mathematical representation displayed in the left and center columns, but also the respective quantitative indications of the dynamics are depicted on the right. In each of these examples, the behaviour of the physical system is described in terms of a system of (at least 3) ordinary differential equations (ODE) which must then be solved numerically.

We note without much further comment, that for conservative systems at least a "3-component flow" through the relevant state-space is required for chaotic motion. This makes intuitive sense in the following way. Since chaos is intimately connected with loss of retrievable information or predictability, a physical system must be "big enough" to lose information, a criterion which is not fulfilled for a 2-component conservative flow. For instance, if energy or population or whatever is initially described by a displacement of variable #1, call it $x_1 = x_1(t)$ then as time evolves any loss or further displacement in $x_1(t)$ is compensated by a gain or reactive displacement in $x_2(t)$ which can be inverted to give $t = f(x_2)$ and then $x_1 = x_1(f(x_2))$. Thus, since a one-to-one correspondence exists between a displacement in x_1 and in x_2, no loss of information (or gain of entropy) results from such evolution and, in spite of the possibility that both $x_1(t)$ and $x_2(t)$ may show *complicated* looking behavior, this motion is *not chaotic*. On the other hand, for a 3-component flow in which there is nonlinear mixing or coupling between the components, a loss or displacement in one quantity can lead to an infinity of combinations of displacements in the other 2 components and it is this possibility that is ultimately responsible for information loss and thus chaotic evolution.

The conservative system of two nonlinearly coupled oscillators will be the workhorse throughout this chapter. With regard to the criteria just stated, this system has 2 degrees of freedom in the sense of Hamiltonian mechanics. However, each degree of freedom is characterized by canonically conjugate variables, p_i and q_i, momentum and displacement. The time evolution of the system characterized by a Hamiltonian function $H = H(p_1, q_1, p_2, q_2; t)$ is thus described in terms of trajectories within a 4-dimensional phase space.

For the conservative systems of interest here, H does not contain an explicit time dependence and as a result, is equal to the total energy [i.e., $E = H(p_1, q_1, p_2, q_2)$]. Thus, the time evolution of the 4-dimensional phase space points is restricted to an energetically allowed $3d$ "surface" within the $4d$ space and as such, can be usefully represented in terms of a less-restricted flow in a 3-dimensional space which is nice for visual presentation. Depending upon the strength of the nonlinearities, this flow may or may not be chaotic. This is *the important issue* with regard to the applicability of statistical mechanics analysis of such systems. Fundamental to statistical mechanics is the assumption of ergodicity; that is throughout the time history of the system, all energetically allowed points in phase space are equally likely to be visited/occupied [8.36, 37]. This condition is fulfilled for simple Hamiltonian systems when the dynamics is rigorously chaotic but not necessarily when it is just complicated [8.38]. It is this reason amongst others which justifies the effort in developing a more rigorous science of chaos.

8.2.1 The Direct Signal

As the first example, consider the driven and damped pendulum shown in the top row of Fig. 8.1. Although this is a dissipative rather than conservative system, it

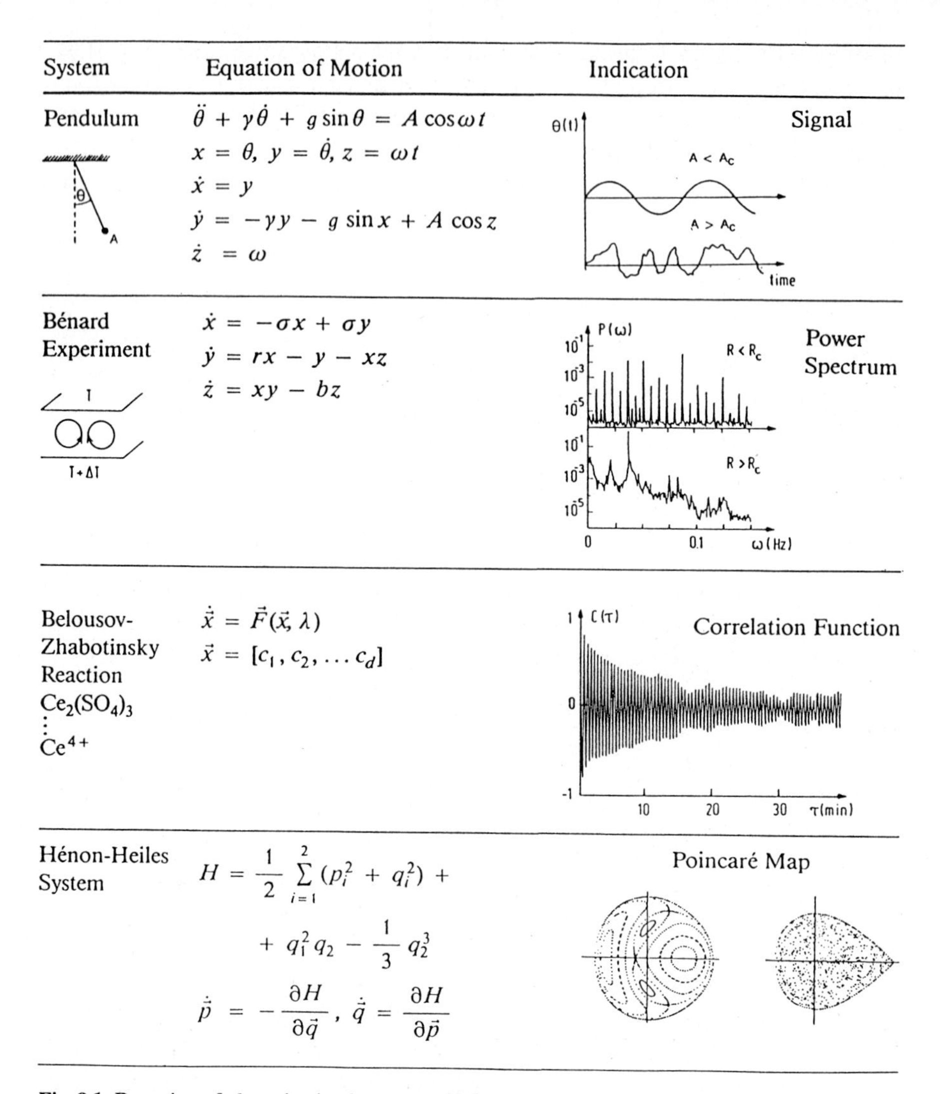

Fig. 8.1. Detection of chaos in simple systems [8.5]

still serves to illustrate the intended point. Computationally, it is best to convert the inhomogeneous 2nd-order differential equation of motion into a system of 3 1st order ODE, as has been done in Fig. 8.1. It is then straightforward to numerically solve this system for a prescribed set of parameters: γ the damping strength and A or ω the driving force amplitude and frequency respectively and for given initial conditions. Typical traces of the direct signal, angular displacement as a function of time, are shown on the right for two choices of force amplitude. The upper curve shows the regular periodic motion expected for small amplitude ($\sin\theta \approx \theta$) driven harmonic oscillations which occur when $A < A_c$, where A_c is some critical force amplitude, above which the angular displacement as a function of time is irregular, perhaps chaotic, as in the lower θ vs time curve. Use of the direct signal as a signature of chaotic dynamics (i.e., chaos occurs when the signal "looks" irregular, hence chaotic) is the weakest and least precise of the numerical criteria considered here.

8.2.2 Power Spectrum

The transition to chaotic dynamics in complex systems which, for small values of control parameters, are inherently periodic or multiply periodic is easily followed by application of Fourier analysis to the relevant signal or variable of interest. If $\xi_i = \xi_i(t)$ gives the time dependence of the signal, then the power spectrum of this signal, defined as

$$P_i(\omega) = \lim_{T\to\infty} \frac{1}{T}\left| \int_0^T dt\, e^{i\omega t}\xi_i(t)\right|^2 , \tag{8.1}$$

will consist of a set of discrete lines at the characteristic frequencies of the system when the dynamics is regular. However, as chaos sets in, say by a denumerable sequence of period doubling bifurcations, $\xi_i(t)$ acquires Fourier components at all frequencies, and the power spectrum converts to an irregular series of broadened lines on top of a broad band continuous background.

For an illustration of this, consider the Rayleigh-Bénard experiment sketched in the second row of Fig. 8.1 which demonstrates the transition between convective and turbulent heat transfer as chaos is established. The physical system is one in which a fluid characterized by a viscosity η, expansion coefficient α, thermal diffusivity D_{T}, and mean density ϱ_0, fills the space between two heatbaths maintained at temperature T and $T + \Delta T$ and separated by a distance L. When the so-called Rayleigh number $R_{\mathrm{a}} \equiv \varrho_0\, g\alpha L^3 \Delta T/\eta D_{\mathrm{T}}$ is less than some critical value, heat transfer between the plates is convective since the competing forces of hot fluid rising and cold falling, viscosity, and gravitation (g = gravitational acceleration) balance to produce an oscillatory velocity field within the fluid. The reduced system of equations shown here follow from the Navier-Stokes, Fourier heat conduction, and continuity equations. The power spectrum of the fluid velocity in the periodic regime where $R_{\mathrm{a}} < R_{\mathrm{c}}$ is shown at the top of $P(\omega)$ vs ω as a series of well defined lines. In contrast, when $R_{\mathrm{a}} > R_{\mathrm{c}}$ and irregular motion or chaos is established, the power spectrum takes on the broadband character shown at the bottom of this plot. In this example, although pattern recognition is required from the observer, the distinctions between the signatures of quasi-periodic and chaotic dynamics are much less ambiguous than in the pendulum example using the direct signal. We will make heavy use of power spectra analysis throughout this chapter.

8.2.3 Autocorrelation Functions

Yet another quantitative measure of chaotic dynamics is the form of the autocorrelation function of the variable being monitored, that is

$$C_i(\tau) = \lim_{T\to\infty} \frac{\int_0^T dt\, \xi_i(t)\xi_i^*(t-\tau)}{\int_0^T dt|\xi_i(t)|^2} . \tag{8.2}$$

In the case of purely periodic (with frequency ω_0) time evolution, $\xi_i(t)$ remains correlated with its past and $C_i(\tau) = \cos\omega_0\tau$. In contrast, as a system approaches chaotic time evolution, the dynamic variable loses memory of its origin and thus the autocorrelation function shows a decay which tends to exponential.

An example which benefits from autocorrelation function analysis is the transition between oscillatory and equilibrium chemical reaction kinetics [8.39], exemplified by the much-documented [8.1–7] Belousov-Zhabotinsky reaction (often considered as a chemical clock) which has been schematized in the third row of Fig. 8.1. In this reaction, a complicated nonlinear set of kinetic equations are proposed to characterize the time evolution of the reaction for which the concentrations of the various reactants and products are the dynamic variables. Under certain choices of control parameters labeled as the set (λ) (i.e., temperature, pressure, reactor geometry, incoming fluxes of reactants, etc.) the concentrations of the constituents may show regular oscillatory behavior (frequently observed as a dramatic change of color of the reaction "soup"), hence the identification as a chemical clock. With other choices of (λ), the oscillatory reaction might die out in favor of a chaotic steady state, noted by a decay of the autocorrelation of a species concentration, as suggested in the "Indication" column. We will later return to the topic of oscillatory reactions in the context of a simpler CO oxidation reaction on a Pt surface studied by the group of *Ertl* [8.29].

8.2.4 Poincaré Map/Surface-of-Section

This final example of "quantitative chaos" will be illustrated in terms of a Hamiltonian model which, in one form or another, provides the basis for most of the hard content in Sect. 8.4. We have already mentioned that the time evolution of a 2-degree-of-freedom (DoF), conservative mechanical system can be visualized in terms of trajectories throughout a 3-dimensional sub-space of the 4-dimensional phase space, the reduction from 4 to 3 dimensions made possible by energy conservation. (More generally, for an N degree-of-freedom mechanical system, the reduction is from a $2N$ dimension phase space to a $2N-1$ dimensional subspace or 1 less dimension per conserved quantity.) A three-body system can be cast into the form of two coupled oscillators showing both bound and free motion (plus the irrelevant total system center-of-mass motion). This type of model can be useful in studying periodic to chaotic transitions in not only astronomical-scale planetary motion within the solar system but also astronomically small-scale intramolecular motion such as that of vibrational properties of molecules adsorbed upon surfaces. The procedure is the same in both cases. Construct a Hamiltonian as in the bottom row of Fig. 8.1 using the potential energy functions characterizing the interacting system. The time evolution of the canonical variables is determined by Hamilton's equations which provide a set of ODEs to be solved numerically. If H contains cross terms between the degrees of freedom which are higher than quadratic, then the resulting coupled equations of motion are nonlinear which allows for the possibility of chaotic motion.

Long ago, it was recognized that if the complicated phase space trajectories were followed, and a record was made each time they passed through a specified plane (the $p_2 - q_2$ plane where $q_1 = 0$ for example) in only one direction ($p_1 < 0$ perhaps), then under conditions in which regular quasi-periodic motion prevailed, the locus of points cutting the plane, called the Poincaré Surface-of-Section, would form closed orderly figures, as in the Poincaré Map on the left in Fig. 8.1. In contrast, under conditions in which the motion through phase space is chaotic, the surface-of-section appears as an area-filling random distribution of points as depicted on the right. Further, mixed regimes can exist in which the relative area of closed curves versus random-hits is used as a quantitative index of the degree of chaos. Illustrative phase space trajectories are shown in Fig. 8.2 depicting the making of surface-of-sections for various types of motion.

The model proposed and analyzed by *Hénon and Heiles* [8.40] is one of the simplest imaginable extensions beyond coupled harmonic oscillators, namely, inclusion of the cubic interaction terms shown in the bottom row of Fig. 8.1. Still, the inherent complexity and richness of this apparently simple model, with respect to the whole spectrum of subtleties of nonlinear dynamics, has conspired to make this one of the ultimate paradigms in the study of chaos in small conservative Hamiltonian systems.

8.2.5 Summary

The dynamical systems shown in Fig. 8.1 and the methods used to quantify their time evolutions have provided a good overview of many of the aspects of

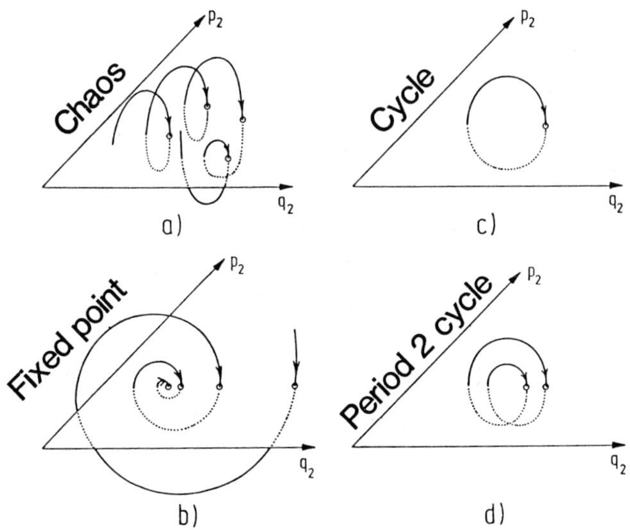

Fig. 8.2a–d. Qualitatively different phase space trajectories which can be distinguished by their Poincaré surface-of-sections: (**a**) chaotic, (**b**) approach to fixed point; (**c**) period-one cycle; (**d**) period-two cycle [8.5]

nonlinear dynamics/chaos research that might find some enlightening applications in the world of surface dynamics. A few general messages should be "taken home" from this section.

To begin with, we all have the intuitive feeling that there is some intrinsic feature about chaotic systems that makes reliable, longtime predictions impossible. (Why else do you frequently not have your umbrella on days when, in spite of the TV weather persons advice, rain has substituted for the promised sun?). In fact, those nonlinear systems which do show chaotic behavior, are extremely sensitive to the choice of initial conditions in a numerical simulation. The nonlinearities, which are a necessary but not sufficient condition for chaos, over the course of time amplify the initially small differences in neighboring trajectories. While at first, close-by trajectories may stick close together, they eventually go their own way and separate exponentially if chaotic, as quantified in terms of Lyapunov exponents [8.1–7]. This divergence appears as a decay of not only the cross-correlation function of trajectories starting from similar (but not identical) initial conditions, but also of the auto-correlation function of the (calculated) trajectory with itself. Because of numerical noise and round off error inherent in the finite truncated number system of modern computers, after a large but finite number of iterations in a calculation, the uncertainties due to the calculation itself will be large enough to generate the exponential divergences of chaotic systems, hence the inescapable decay of a calculated auto-correlation function. As a result, we must accept real life consequences such as limited expectations for long term, accurate weather forecasts.

The four examples shown in Fig. 8.1 have displayed aspects of dynamic systems which facilitate meaningful identification of chaotic motion, namely:

i) Time dependence of direct signal "looks chaotic". This is a very weak criterion.
ii) Broadband noisy power spectrum rather than discrete frequencies of multiply-periodic system.
iii) Decay of correlation functions, due to separating trajectories.
iv) Poincaré surface of section is erratically space filling.

As mentioned early on, in spite of whatever numerical sophistication, the wisdom and trained eye of the researcher is still one of the ultimate pieces of equipment in this field.

Finally, some cautionary messages before plunging in. Most of what is currently being done in this field works very well and provides informative graphics output when the numerical inquiry is confined to a 3-dimensional flow such as the 2-degree-of freedom conservative Hamiltonian systems. For larger systems, the numbers can be calculated, but not easily displayed in a useful way. Since pattern recognition is such a crucial part of this work and this requires suggestive graphics displays, the preferred strategy for using the modern methodologies from nonlinear dynamics is to somehow first map your particular problem onto an equivalent problem for a 2-DoF system, coupled in an unrestricted manner. This being done, one is then able to tap into the vast array of numerics and graphics of the chaos world, hopefully in ways which provide new scientific insights. We have already found this to be a very effective way to proceed, both in the context of surface vibrational spectroscopy [8.24–27] and in molecular collision phenomena at surfaces [8.27, 41–45] where the separation into a relevant internal molecular mode and a center-of-mass translational or reaction coordinate have been made.

As the final word on this point take Pierre Hohenberg's quote [8.46]: "The miracle is that in systems that are interesting, you can still understand behavior in detail by a model with a small number of degrees of freedom."

8.3 Period Doubling

Having accepted the common view that chaos is tied to the demise of periodic motion, one might wonder how this demise comes about. Rather remarkably, there is a large class of phenomena which show a very orderly and systematic approach to the state of disorder called chaos. We have in mind the situation in which chaos is attained through a sequence of period doubling bifurcations as the strength of the nonlinearities in the system is increased. Obviously, the longer is the extent or period of the system motion with respect to any other relevant time scale, the less significant is the fact that such a period even exists. A few examples of the manifestations of the period doubling route to chaos will now be put forth.

8.3.1 Driven Oscillator

A one-dimensional nonlinear oscillator (such as the pendulum considered in Fig. 8.1 and Sect. 8.2.1) is a nice vehicle for displaying various aspects of the approach to chaos because such systems are rich in possibilities and their behavior is clearly and unambiguously presented in a 2-dimensional phase diagram. For specificity consider the so-called Duffing oscillator whose equation of motion is

$$\ddot{x} + \gamma\dot{x} + x^3 = A\cos\omega t \ .$$

A combination of parameter values for the damping γ, driving amplitude A, and frequency ω have been used to obtain the sequence of phase plots $\dot{x}$ vs x and the displacement vs time shown in Fig. 8.3. Although the period doubling shown in Figs. 8.3a–c is much more easily discerned in the phase portraits on the left than in the displacement curves on the right, both presentations contain the same information. Finally, in Fig. 8.3d, an example in the chaotic extreme is shown.

8.3.2 Logistic Map

Perhaps the most well known demonstration of nonlinearity leading to a period-doubling progression to chaos has been with the iterated logistic map. Its pop-

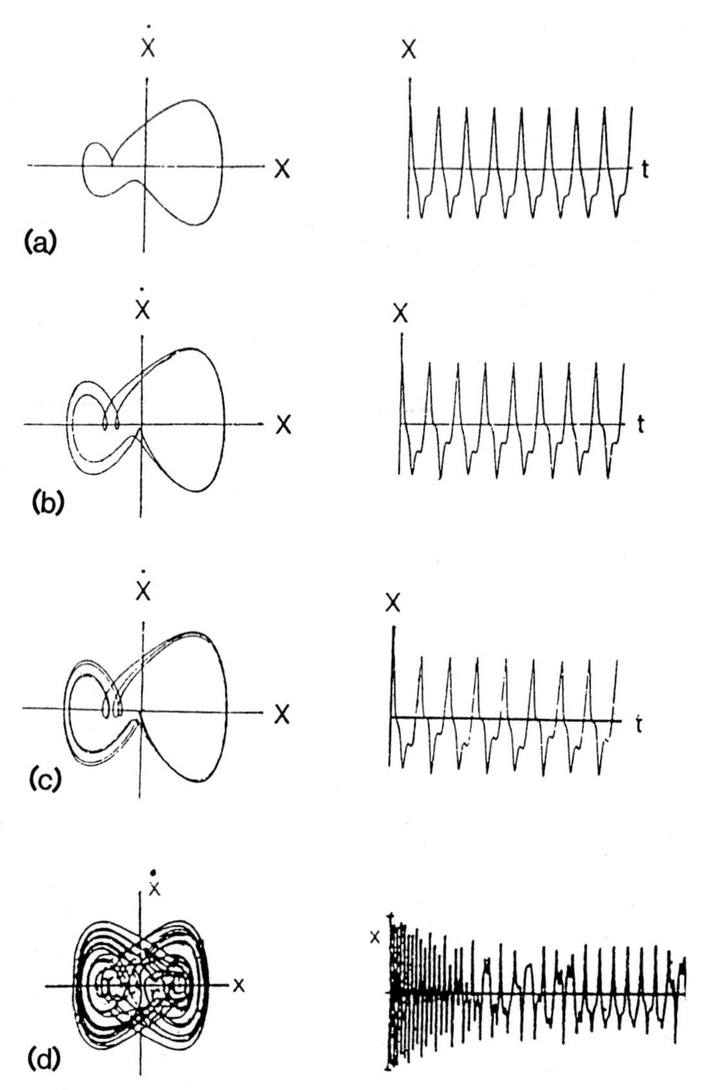

Fig. 8.3a–d. Behavior of Duffing oscillator shown as a trajectory in phase space (*left*) and as a displacement against time (*right*), for parameter variations showing period doubling. In (**a**) a stable cycle occurs in each period of the oscillation. In (**b**) the stable cycle repeats after every two periods and in (**c**) the period doubles to every fourth cycle. This development continues until there are innumerable stable points and chaos ensues, as shown in (**d**) [8.14]

ularity is undoubtedly due to some combination of its simplicity, profundity, and universality. Applications have appeared in almost every human endeavor in which some sort of quantification is possible. The mathematical expression of a general mapping is that the $n+1$'th iterative value of some dynamic variable is related to its value at the n'th iteration via $x_{n+1} = F(x_n)$ and for the logistic map, this is given as

$$x_{n+1} = \lambda x_n(1 - x_n) \,. \tag{8.3}$$

The usual example given to show the relevance of this map to something in the real world is the population growth model in which the population of a species at the beginning of the n + 1'th season (or whatever sets a natural sampling time iteration) is proportional not only to x_n, the population at the beginning of the n'th iteration, but also to perhaps the amount of food available per creature which, for a limited supply, must decrease with increasing number of hungry mouths, here proportional to $(1 - x_n)$. The parameter λ is a measure of the importance of the nonlinear term in (8.3) with respect to the simple linear growth. Such a map could just as well be used to categorize some sort of an oscillatory surface chemical reaction in which the reaction probability at the $n+1$'th iteration depends upon the number of reactants on the surface (x_n) multiplied by the sticking probability for more reactants which is proportional to the fraction of unoccupied sites, $(1 - x_n)$. This will be discussed a bit more in the next section.

Since the numerical implications and geometrical interpretation of the logistic map have been given in almost every book or general article on chaos [8.1–15], brevity will be attempted here. The bifurcation diagram shown in Fig. 8.4 shows

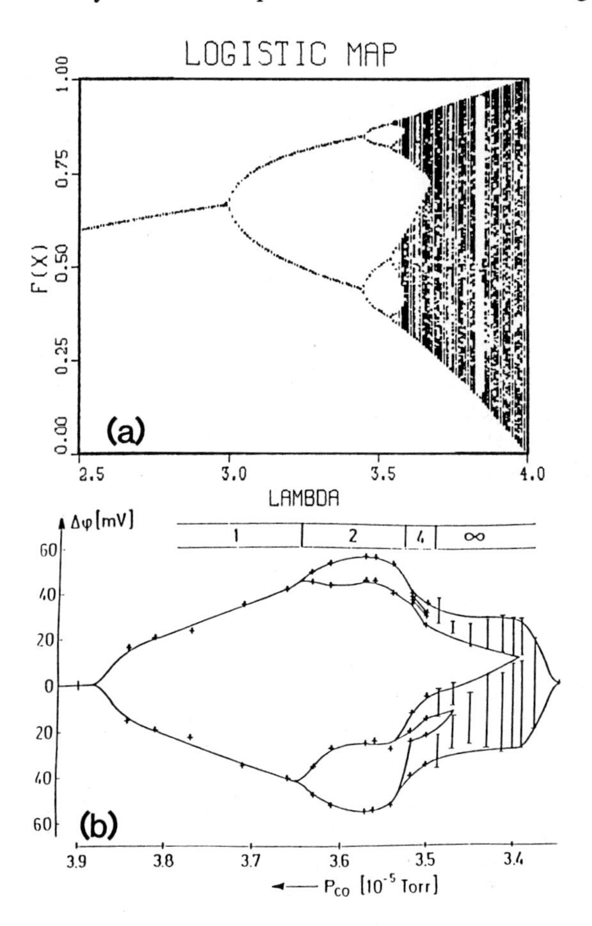

Fig. 8.4. (**a**) Bifurcation diagram for the iterative logistic map (8.3). (**b**) Bifurcation diagram for the approach to chaos of the oscillatory CO oxidation reaction on a Pt{110} surface [8.29]

what amounts to the steady state values of x, as a function of the nonlinearity parameter λ. For $\lambda < 3$, a single steady state value exists. Precisely at $\lambda = 3$, a bifurcation occurs, two steady state values obtain and the subsequent periodicity has been doubled. Further increase of λ leads to yet another bifurcation. This period doubling continues and for $\lambda > 3.57$ almost continuous bands of steady state values follow, signifying the demise of periodicity, hence chaos.

8.3.3 Oscillatory Surface Reactions

The period doubling route to chaos has been studied by *Eiswirth*, *Ertl*, and coworkers [8.29] in the context of the oscillatory CO oxidation reaction on Pt{110} surfaces. Under properly chosen ambient conditions (e.g., temperature, background, CO and O_2 pressure), they have shown that the (oxidation) reaction rate exhibits sustained temporal oscillations (with a period of approximately 2 s) due to the interplay of the various steps in the Langmuir-Hinshelwood mechanism of recombinative desorption of dissociatively adsorbed O_2 and adsorbed CO molecules. Further, they have demonstrated that small changes in P_{CO}, the background pressure of CO, can lead to dramatic and profound changes in the time dependence of this oscillatory reaction. Rather remarkably, they have observed that as P_{CO} (and hence the surface CO replenishment rate) is decreased in small steps from a starting pressure above 3.5×10^{-5} Torr, the oscillatory character of the reaction rate (monitored by work function changes, as detailed in their papers) goes through a number of period-doubling bifurcations, as it proceeds towards a completely erratic time structure suggestive of a periodic-to-chaotic transition. They were able to construct the bifurcation diagram shown in Fig. 8.4b, in which P_{CO} plays the role of the analogous control parameter λ in the logistic map shown in Fig. 8.4a. Other presentations of their data in terms of power spectra, correlation functions and Lyapunov exponents were consistent with their suggestion that in fact a period doubling sequence was pushing the surface reaction towards a chaotic kinetics. Undoubtedly, future studies and analysis of this sort will provide signficant new insights into surface kinetics that were never imagined prior to the formal introduction of nonlinear kinetics and chaos into surface thinking.

8.4 Hamiltonian Systems

At the close of Sect. 8.2, some encouragement was given in support of the strategy of mapping one's favorite problem onto that of a 2-DoF conservative Hamiltonian system, if this could be done in a credible way, since numerical and visual techniques are well developed for treating this class of problems. To a certain limited extent, this is being done in the areas of surface molecular vibrational spectroscopy [8.24–27, 47], scattering [8.41, 48], and the union of the two [8.27].

Here, a flavor of some of the issues being addressed, particularly as they pertain to interests from the chaos world, will be offered.

8.4.1 Vibrational Spectroscopy

Simulated vibrational spectroscopy, will be carried out using the spectral analysis technique developed by *Marcus* and coworkers [8.49, 50]. In this method, the vibrational lineshape associated with excitation of a displacement coordinate $q(t)$, is given by (8.1), the power spectrum of $q(t)$, where $q(t)$ is the trajectory followed by the coordinate subject to prescribed initial conditions characterizing the excitation/absorption processes. The time evolution is obtained numerically from Hamilton's equations ($\dot{q}_i = \partial H/\partial p_i$, $\dot{p}_i = -\partial H/\partial q_i$) using a classical Hamiltonian that contains the model potential energy surface (PES) characterizing the physical phenomenon under investigation. Numerical studies on model systems involving nonlinearly coupled degrees of freedom (e.g., high-frequency intramolecular stretch and low-frequency hindered translation) frequently show that, below a (system- and parameter-dependent) critical energy, the dynamics of the nonlinear system displays quasi-periodic behavior. A rapid transition to chaotic dynamics occurs at a fairly well-defined energy [8.40]. This energy is often associated with the condition for accessing portions of the PES in which a hidden symmetry is destroyed [8.51]. For instance, the onset of chaos on the Hénon-Heiles PES is due to the symmetry breaking of the potential caused by the three dissociative necks or exit channels, a feature not present on the related Toda PES (which coincidentally displays regular dynamics at all energies) [8.2, 27]. A number of possible examples of such behavior in surface spectroscopy and dynamics will now be put forth.

a) Fermi Resonances. As a first example, consider the scaled Hamiltonian

$$H = \tfrac{1}{2}\left(p_x^2 + p_y^2\right) + \tfrac{1}{2}\left(x^2 + 4y^2 - 8yx^2\right) . \tag{8.4}$$

The quadratic terms make up the Hamiltonian for two, uncoupled harmonic oscillators where $\omega_y = 2\omega_x$, which choice clearly displays the role of Fermi resonances [8.52], and also simulates the interplay between a high-frequency stretch mode and a low-frequency mode associated with the bonding of a diatomic molecule to a surface. The cubic term destroys the rotational symmetry of the potential, providing forces which "steer" the trajectory from x to y motion. For instance, the trajectory (y vs x) obtained from the set of Hamilton's equations with (8.4) is shown in Fig. 8.5a, superimposed on constant energy contours for the PES implicit in (8.4). The initial conditions were set at $y_{\text{in}} = 0.1$, $x_{\text{in}} = 0.01$, $p_x = p_y = 0$ to simulate a spectroscopic event in which the "high frequency mode" was excited. For a purely harmonic system, the trajectories would be Lissajous figures confined to the very narrow strip defined by the caustics in which $|y| \leq 0.1$ and $|x| \leq 0.01$. The time evolution $y(T)$ and $x(T)$ vs T ($T = \pi$ corresponds to one y cycle) are shown in Figs. 8.5b and c where the relation between loss of

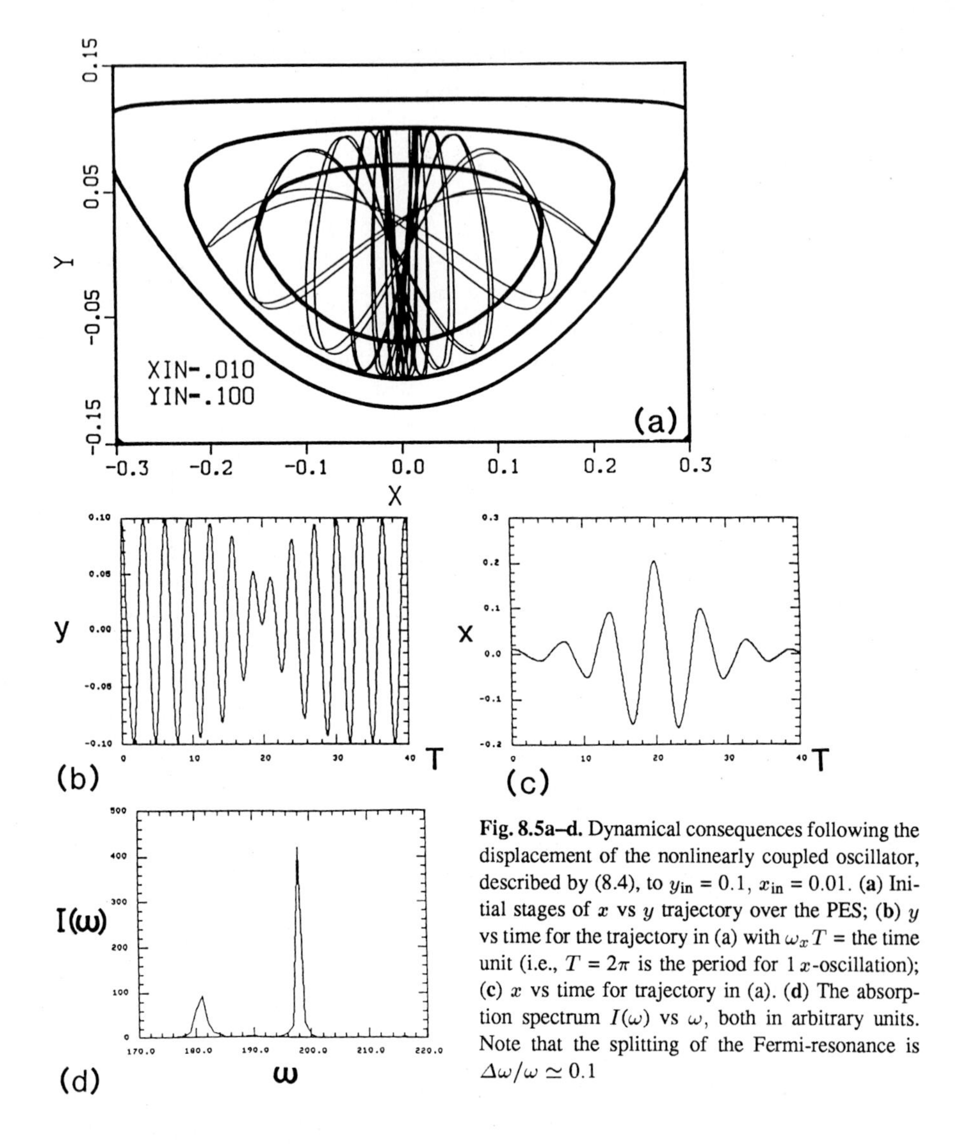

Fig. 8.5a–d. Dynamical consequences following the displacement of the nonlinearly coupled oscillator, described by (8.4), to $y_{\text{in}} = 0.1$, $x_{\text{in}} = 0.01$. (**a**) Initial stages of x vs y trajectory over the PES; (**b**) y vs time for the trajectory in (a) with $\omega_x T$ = the time unit (i.e., $T = 2\pi$ is the period for 1 x-oscillation); (**c**) x vs time for trajectory in (a). (**d**) The absorption spectrum $I(\omega)$ vs ω, both in arbitrary units. Note that the splitting of the Fermi-resonance is $\Delta\omega/\omega \simeq 0.1$

y and x energy is apparent. Finally, the absorption spectrum appropriate to this trajectory, obtained from (8.1) and (8.4), is shown in Fig. 8.5d over a limited range of frequency. This type of feature, a discrete line narrowly split ($\Delta\omega/\omega \simeq 0.1$ in Fig. 8.5d) into slightly broadened lines (unresolved discrete levels) of various intensities by anharmonic coupling is an example of a Fermi resonance between the originally excited y oscillator and some overtone or combination excited state of the rest of the system (doubly excited x oscillator in this case).

It is enlightening to look for systematic changes in the vibrational spectrum as the strength of the anharmonic coupling, which scales linearly with the initial value of y (for fixed x), is increased. On the left in Fig. 8.6, are shown $x(T)$ vs T, in which $y_{\text{in}} = 0.025$, 0.05, and 0.075, respectively. The increasing anharmonicity

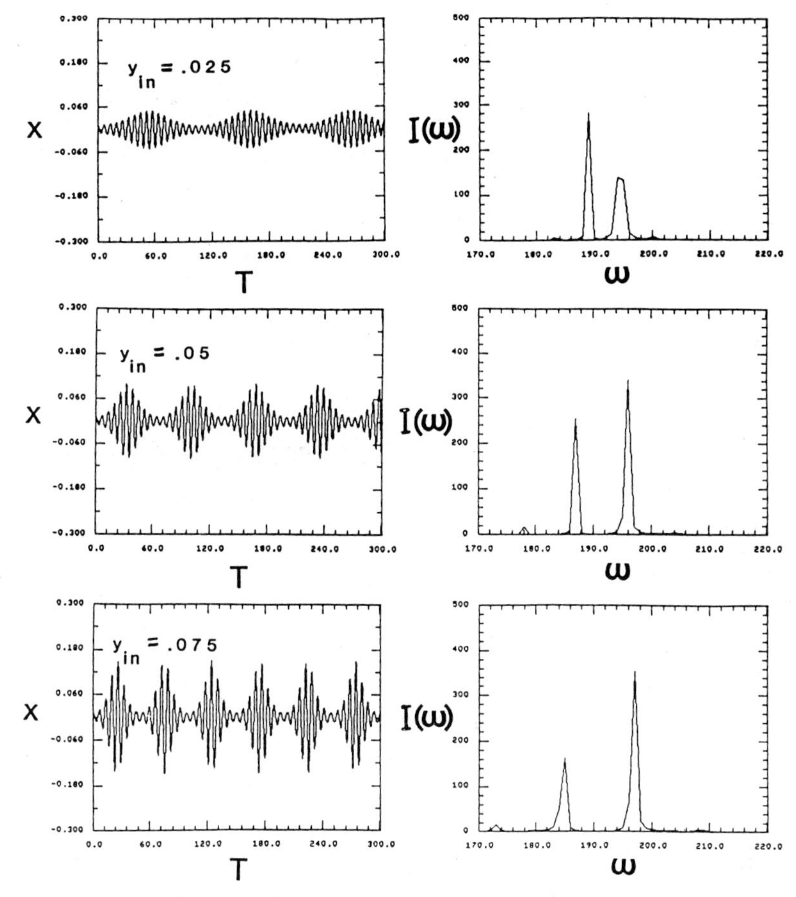

Fig. 8.6. x vs T (*left*) and $I(\omega)$ vs ω (*right*) for x_{in} = 0.01, y_{in} = 0.025, 0.05, and 0.075 (*top to bottom*)

increases the recurrence frequency, but is not sufficient to destroy the quasi-periodic motion. The enhanced amplitude is due to the larger amount of energy available as y_{in} is raised. The companion spectra are shown on the right. In these cases, the splitting scales linearly with y_{in} and thus anharmonicity but otherwise the spectra remain fairly similar.

Interesting behavior sets in for $y_{\text{in}} \geq 0.1$. Calculated spectra are shown on the left in Fig. 8.7 for y_{in} = 0.11, 0.12, and 0.13 (note scale change for 0.13). With increasing anharmonicity, the dominant line rapidly broadens, and intensity appears as a spread out background. This is a manifestation of the transition from quasi-periodic to chaotic dynamics as it would appear to the vibrational spectroscopist. In the right column of Fig. 8.7 are shown x and y vs T for y_{in} = 0.13 where, although some regularity still survives, the envelope structure is indeed quite different from those in Fig. 8.6. The bottom line in this particular example is that for coupled oscillators with rational frequency ratios, mild anharmonic

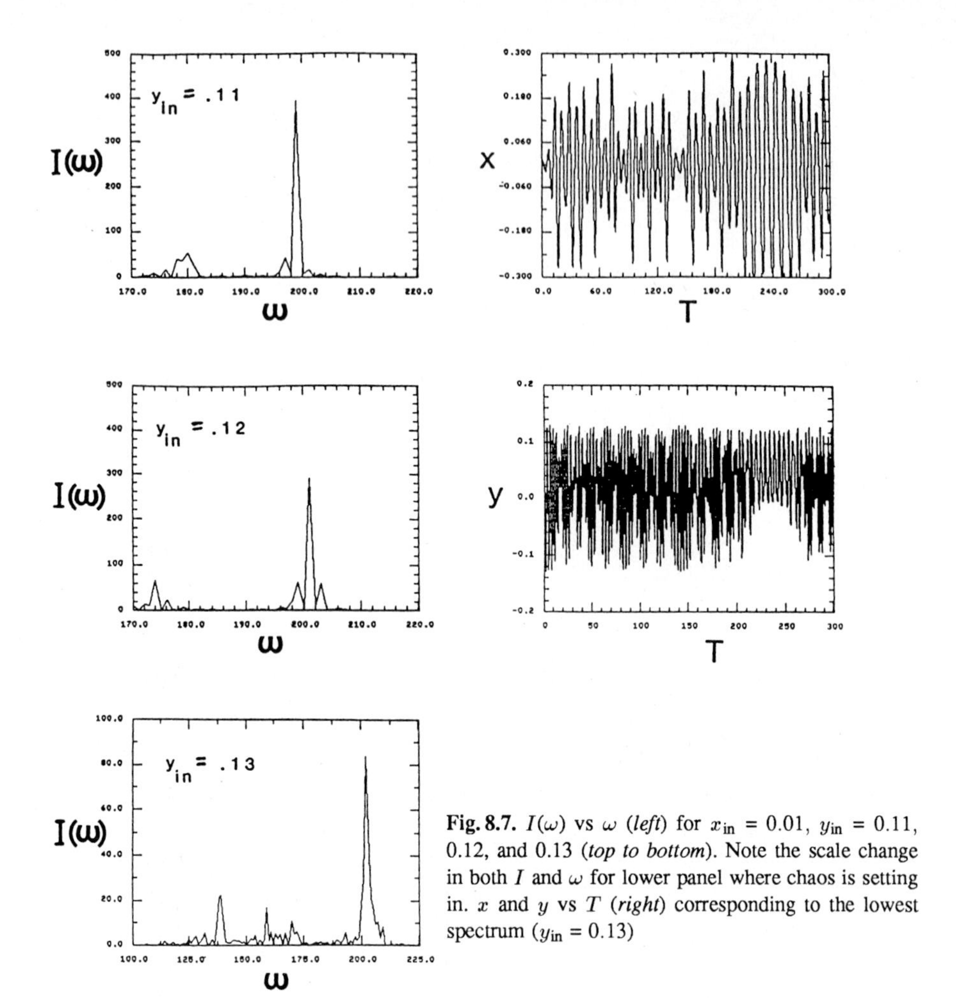

Fig. 8.7. $I(\omega)$ vs ω (*left*) for $x_{\text{in}} = 0.01$, $y_{\text{in}} = 0.11$, 0.12, and 0.13 (*top to bottom*). Note the scale change in both I and ω for lower panel where chaos is setting in. x and y vs T (*right*) corresponding to the lowest spectrum ($y_{\text{in}} = 0.13$)

coupling results in well-defined, sharp Fermi-resonance structure which degrades as the anharmonic coupling is increased to the point where chaotic time evolution sets in.

b) Toda vs Hénon-Heiles Surface Resonances. There is a vast literature on two closely related models from nonlinear dynamics, the Toda and Hénon-Heiles systems which have been used [8.27] to model the nonlinearly coupled molecule-surface system, often considered as a resonance [8.41–45, 48]. The Hamiltonian for the Toda system is

$$H_{\text{Toda}} = \tfrac{1}{2}\left(p_x^2 + p_y^2\right) + \tfrac{1}{24}\left(\exp(2y + 2\sqrt{3}x) + \exp(2y - 2\sqrt{3}x) + \exp(-4y)\right) - \tfrac{1}{8}\,. \quad (8.5)$$

The potential curves implied by (8.5) and shown in Fig. 8.8a, vary smoothly outward from the origin in both x and y, displaying a threefold symmetry. If (8.5)

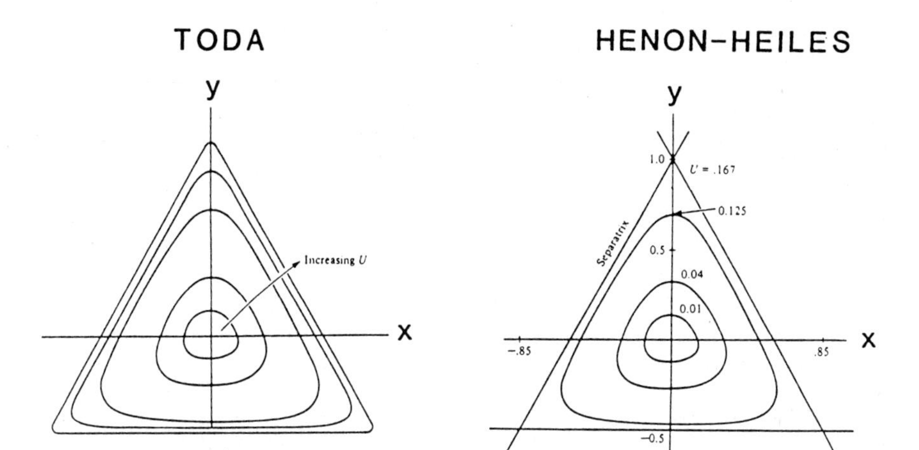

Fig. 8.8. (**a**) Potential well for the Toda Hamiltonian showing lines of constant potential U. (**b**) Potential well for the Hénon-Heiles Hamiltonian showing lines of constant potential U, for closed equipotentials ($U \leq \frac{1}{6}$) only

is expanded to cubic terms in x and y, we obtain the Hénon-Heiles Hamiltonian

$$H_{\rm HH} = \tfrac{1}{2}\left(p_x^2 + p_y^2\right) + \tfrac{1}{2}\left(x^2 + y^2 + 2x^2y - \tfrac{2}{3}y^3\right) , \tag{8.6}$$

whose PES is shown in Fig. 8.8b. This is quite similar to the Toda PES except for the fact that $H_{\rm HH}$ shows a separatrix and thus unbound motion for total energies in excess of $\varepsilon_{\rm tot} = \frac{1}{6}$. From a dynamics point of view, these related systems are interesting because motion governed by $H_{\rm Toda}$ is quasi-periodic for all energies, whereas complete quasi-periodic dynamics with $H_{\rm HH}$ occurs only for $\varepsilon_{\rm tot} < 0.1$. The onset of chaos begins at this point and increases as $\varepsilon_{\rm tot}$ is raised, in spite of the fact that the (classical) trajectories remain confined if $\varepsilon_{\rm tot} < \frac{1}{6}$. High resolution vibrational spectroscopy simulations have been performed on molecule-surface systems in which the potential energy function of the intramolecular stretch anharmonically coupled to the vibrational mode associated with the rigid molecule-surface bond has been taken to be either of the Toda or Hénon-Heiles form. (Although not essential for this expository exercise, one could rationalize the three-fold symmetry in terms of bonding at a bridge site.) As with the Fermi resonance example, lineshapes were obtained as power spectra of trajectories on the Toda and Hénon-Heiles PES, treating the initial y displacement parametrically, and some results are shown in Fig. 8.9. The Toda lineshape, Fig. 8.9a, remains sharp for all initial conditions (and thus total energy). This is a characteristic of the quasi-periodic motion on the Toda PES. In contrast, the Hénon-Heiles lineshape shows an evolution as the initial displacement and hence total energy increases. For small displacements in which $\varepsilon_{\rm tot} < 0.1$, the lineshape is sharp as shown in Fig. 8.9b. For larger displacements in which $\varepsilon_{\rm tot}$ exceeds the chaotic threshold, the lineshape broadens and spreads throughout the spectral

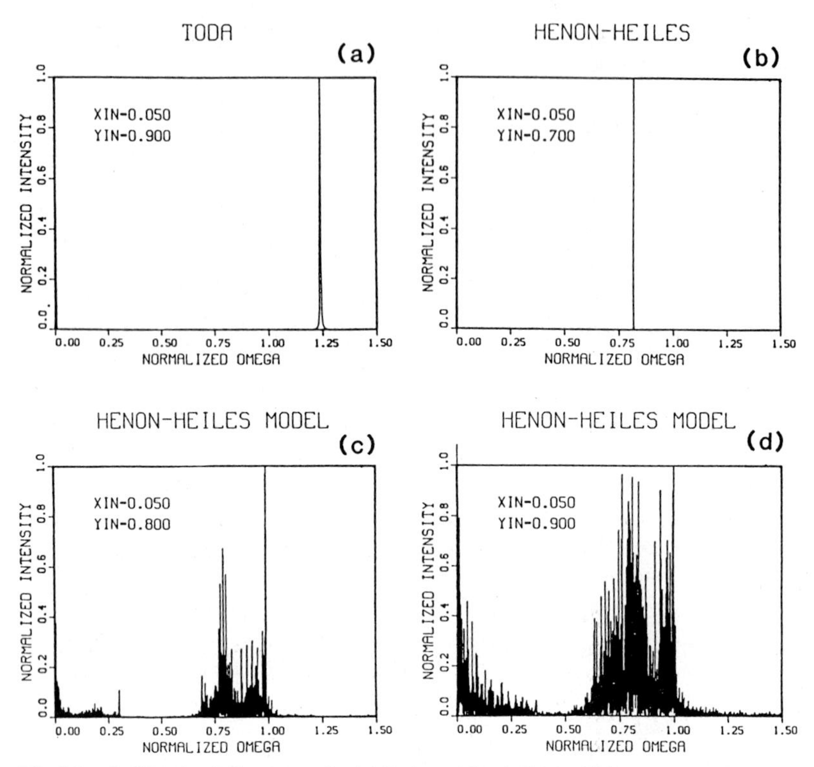

Fig. 8.9a–d. Vibrational lineshapes for (a) Toda and (b–d) Hénon-Heiles systems. The Toda lineshape (**a**) remains sharp for all initial displacements whereas the originally sharp (**b**) Hénon-Heiles lineshape speads out (**c–d**) for initial displacements corresponding to energies above the chaotic threshold

range, as seen in Figs. 8.9c and d. It thus seems reasonable that vibrational spectroscopy could exploit these characteristic signatures of quasi-periodic vs chaotic dynamics in order to demonstrate the relevance (or lack of it) of chaotic behavior to surface vibrational spectroscopy and "reaction dynamics".

c) Multiple Sites. Yet another type of PES topology that can provide a breakdown in quasi-periodic motion is one with multiply connected minima along one of the orthogonal coordinates. While motion within any one of the wells might be perfectly periodic, if the trajectory is incident upon the saddle point separating wells, and if sufficient energy has been directed into this "reaction coordinate" to overcome the activation barrier, then passage into the neighboring well can occur. This breaks the temporal periodicity of the initial motion within the internal degree of freedom (the intramolecular mode). While realization of such motion is well known to control isomerization dynamics [8.53], there are also some intriguing hints that site migration could occur in surface vibrational spectroscopy [8.25, 54]. This has been stimulated by *Hayden* and *Bradshaw*'s observation of a temperature-dependent intramolecular doublet (1810 and 1850 cm^{-1}) in the IR absorption spectrum of CO adsorbed on Pt{111} and on Cu{111}, possibly due to rapid site conversion between the two-fold bridge and neighboring three-fold sites. This observation has motivated studies on the 3-site model;

$$H = \frac{1}{2}\left(p_x^2 + p_y^2\right) + \frac{1}{2}\left(y^2 + \frac{1}{4}x^2 f(x) - 2x^2 y f(x)\right) , \tag{8.7}$$

with $f(x) \equiv 1 - 2(x/x_0)^2 + (x/x_0)^4$ and $x = \pm x_0$ the positions of the exterior minima with respect to the central one at the origin. The basic new features expected in the dynamics and absorption lineshapes associated with the three-site

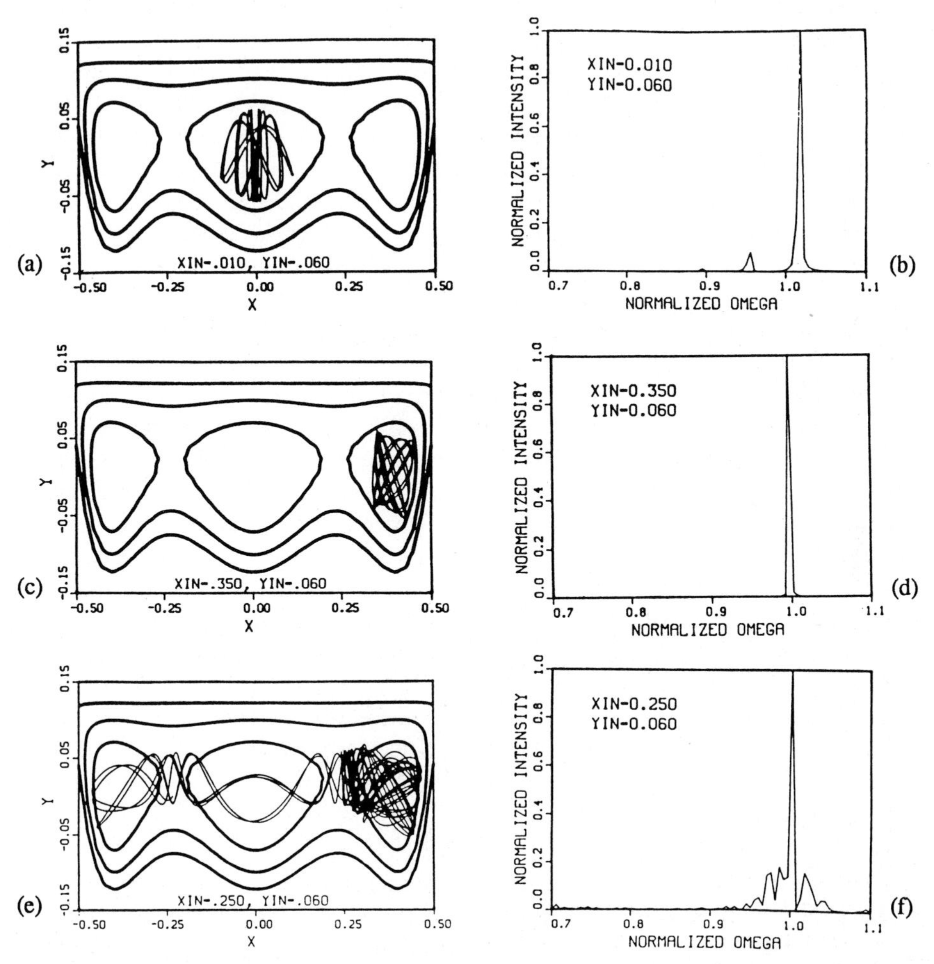

Fig. 8.10a–f. Trajectories and spectra for three-site PES of (8.7) with $y_{in} = 0.06$ in all cases but with different values of x_{in}, chosen to illustrate qualitatively different behaviors. (**a**) Motion localized to central well, similar to that in Fig. 8.5a. (**b**) Remains of Fermi resonance in spectrum. (**c**) Motion localized in side well shows Lissajous trajectory characteristic of uncoupled harmonic oscillators. (**d**) Single-peaked spectrum characteristic of pure harmonic motion of y coordinate. (**e**) Delocalized trajectory over all sites. (**f**) Spectrum showing sidebands and broadening characteristic of the breakdown in perfect periodic motion

PESs are illustrated in Fig. 8.10. Here the trajectories determined by the Hamiltonian of (8.7) and its concomitant lineshapes [from (8.1)] are shown for initial conditions $y_{\mathrm{in}} = 0.06$ and various x_{in}, in a manner similar to the single-site spectra of Figs. 8.6 and 8.7. In the upper example ($x_{\mathrm{in}} = 0.01$), the motion is initiated within the central site with insufficient total energy to pass over the saddle point; hence the trajectory (and thus the vibrationally excited molecule) remains localized on the site where it was excited. The basic shape of the trajectory is similar to that of Fig. 8.5a, and the spectrum in Fig. 8.10b shows remnants of the Fermi resonance, as in Fig. 8.6. The middle set, Figs. 8.10c and d, corresponds to excitation and subsequent localization in the right-hand well. For this particular choice of initial conditions ($y_{\mathrm{in}} = 0.06$ and $x_{\mathrm{in}} = 0.35$), the trajectory appears to be a Lissajous figure confined by distorted but well-defined caustics corresponding to two uncoupled harmonic oscillators with a rational frequency ratio [8.55]. The *single*, narrow-peaked excitation spectrum (no Fermi resonance) bears this out. Finally, in Figs. 8.10e and f, the initial conditions $y_{\mathrm{in}} = 0.06$ and $x_{\mathrm{in}} = 0.25$) are such that the system has enough energy to overcome the activation barrier and migrate from site to site. Although the early stages of the trajectory ($0 \leqq T \leqq 75$ corresponding to ~$12y$ oscillations) in Fig. 8.10e show almost complete free motion along the x axis, after a few passes among the three sites, enough energy in the x oscillator has been directed into y oscillation through anharmonic forces, that the oscillator becomes trapped in the right-hand well, where it will stay until the trajectory once again become focused into the narrow bottleneck for site migration [8.53]. The companion spectrum in Fig. 8.10f is quite different from those of Figs. 8.10b and d for localized oscillators. Considerable intensity has been redistributed into sidebands around the main line, which would appear as a broadened absorption line if proper averaging over initial conditions were carried out. This final example suggests the intriguing possibility that the actual act of performing vibrational spectroscopy on an initially localized molecule, that is, depositing a quantum of energy into a single oscillator, could by itself initiate the site migration that manifests itself as a broadened spectral line. In this sense, the phenomenon is not unrelated to bond-selective laser chemistry [8.56], an area of pursuit whose major goal is to defeat the uncontrollable curse of randomness and ergodicity resulting from chaotic intramolecular dynamics!

8.4.2 Scattering

A certain class of molecule-surface collisions [8.41–45, 48] are characterized by the formation of some temporary resonance state in which the intramolecular geometry of the molecule differs from that of the electronic-ground-state molecule far removed from the target surface. Surface collisions involving such resonances exhibit unusual dynamic behavior resulting from greatly enhanced redistribution of translational energy into the vibrational degrees of freedom of the molecule. Such enhancement should have important implications in vibrational state-to-state [8.41] and dissociative scattering [8.42], activated dissociative adsorption [8.43], selectivity in fragmentation distributions of scattered polyatomic molecules [8.44], and vibrationally assisted sticking [8.45, 57] for instance. Since the dynamics in both scattering and vibrational spectroscopy involve nuclear motion over the same multi-dimensional potential energy surface (PES), albeit continuum states for molecular center-of-mass translational motion in scattering as opposed to bound states in vibrational spectroscopy, it seems reasonable to suppose that vibrational spectroscopy has something to tell the dynamicist and vice versa [8.27]. This is certainly the case for modeling in terms of classical trajectories. As with spectroscopy, it has proven useful to consider a 2 DoF model for a molecule-surface collision in which one coordinate (ϱ) describes the intramolecular vibrations and the other (z) represents the center-of-mass motion of the molecule with respect to the surface. Model PES have been constructed which are similar to the PES considered in Sect. 8.4.2, but modified by the attachment of both an entrance channel in the z direction representing a molecule incident upon the surface and an exit channel along ϱ, say for a dissociatively adsorbed molecule [8.58].

A molecular beam scattering event is run in the following way. The molecule is placed in the entrance channel far from the surface with a prescribed initial translational and vibrational energy and with momentum directed towards the resonance. Either the molecule is directly scattered back out of the entrance channel or it enters the resonance. If the trajectory enters the resonance then it will bounce around until it finds the energetically allowed window for departure, either into the initial entrance channel, in which case it is likely to be vibrationally excited, or into the dissociative exit channel. There is a two-fold role played by the resonance PES. First, it merely refocuses the trajectory, allowing it to search out the saddle points [8.59]. More interesting is the role played by nonlinear mixing within the resonance PES. This gives rise to serious energy scrambling of the initial partitioning of energy, in effect causing any subsequent outcomes to depend more on the total energy available than on how it was initially divided. Furthermore, if the dynamics within the resonance is quasi-periodic, then memory of the initial conditions will be retained, thus permitting some degree of selectivity in outcomes [8.44, 60]. On the other hand, if the dynamics is chaotic, then the resulting ergodic behavior will lead to a statistical distribution of outcomes, for given total energy. Issues of this sort define one of the most exciting forefront areas of research in which reaction dynamics [8.61], collision theory [8.62], and surface dynamics [8.63] are all coming together under a common umbrella of chaos.

Acknowledgement. Portions of this work were facilitated by a grant (# RG. 85/0231) from the NATO Scientific Affairs Division.

References

8.1 S. Jorna (ed.): *Topics in Nonlinear Dynamics*, AIP Conf. Proc. #46 (AIP; NY 1978)
8.2 A.J. Lichtenberg, M.A. Lieberman: *Regular and Stochastic Motion* (Springer, New York, Heidelberg, Berlin 1983)

8.3 P. Bergé, Y. Pomeau, C. Vidal: *Order Within Chaos* (Wiley, New York 1984)
8.4 J.M.T. Thompson, H.B. Stewart: *Nonlinear Dynamics and Chaos* (Wiley, Chichester 1986)
8.5 H.G. Schuster: *Deterministic Chaos*, 2nd ed. (VCH, Weinheim 1988)
8.6 S. Lundqvist, N.H. March, M.P. Tosi (eds.): *Order and Chaos in Nonlinear Physical Systems* (Plenum, New York 1988)
8.7 N.G. Cooper (ed.): *From Cardinals to Chaos: Reflections on the Life and Legacy of Stanislaw Ulam* (Cambridge University Press, Cambridge 1989)
8.8 R.M. May: Nature **261**, 459 (1976)
8.9 M.J. Feigenbaum: Los Alamos Science **1**, 4 (1980)
8.10 L.P. Kadanoff: Physics Today (Dec. 1983)
8.11 J.P. Crutchfield, J.D. Farmer, N.H. Packard, R.S. Shaw: Sci. American **255**, 46 (1986)
8.12 C. Grebogi, E. Ott, J.A. Yorke: Science **238**, 632 (1987)
8.13 R.V. Jensen: Amer. Sci. **75**, 168 (1987)
8.14 R.G. Harrison: Contemp. Phys. **29**, 341 (1988)
8.15 J. Gleick: *Chaos: Making a New Science* (Viking, New York 1987)
8.16 H.-O. Peitgen, P.H. Richter: *The Beauty of Fractals* (Springer, Berlin, Heidelberg 1986)
8.17 P. Martien, S.C. Pope, P.L. Scott, R.S. Shaw: Phys. Letters **110A**, 399 (1985)
8.18 G. Casati, J. Ford, F. Vivaldi, W.M. Visscher: Phys. Rev. Letters **52**, 1861 (1984)
8.19 R. Pool: Science **245**, 26 (1989)
8.20 M.V. Berry: in Ref. [8.1], p. 16
8.21 Z. Penzar, M. Sunjić: Fizika **13**, 79 (1981)
8.22 R.V. Jensen: Phys. Rev. A **30**, 386 (1984)
8.23 U. Landman, R.H. Rast: In *Dynamics of Gas-Surface Interactions*, ed. by B. Pullman, J. Jortner (Reidel, Paris 1984)
8.24 J.W. Gadzuk: J. Elec. Spect. **38**, 233 (1986)
8.25 J.W. Gadzuk: J. Opt. Soc. Am. B **4**, 201 (1987)
8.26 T.S. Jones, S. Holloway, J.W. Gadzuk: Surface Sci. **184**, L421 (1987)
8.27 J.W. Gadzuk: J. Elec. Spect. **45**, 371 (1987)
8.28 M.M. Millonas, R.V. Jensen: Surface Sci. **179**, L33 (1987)
8.29 M. Eiswirth, K. Krischer, G. Ertl: Surface Sci. **202**, 565 (1988)
8.30 V.P. Zhdanov: Surface Sci. **214**, 289 (1989)
8.31 G. Casati (ed.): *Chaotic Behavior in Quantum Systems* (Plenum, New York 1985)
8.32 T.H. Seligman, H. Nishioka (eds.): *Quantum Chaos and Statistical Nuclear Physics* (Springer, Berlin, Heidelberg 1986)
8.33 P. Brumer, M. Shapiro: Adv. Chem. Phys. **70**, 365 (1988)
8.34 J. Manz: J. Chem. Phys. **91**, 2190 (1989)
8.35 R. Pool: Science **243**, 893 (1989)
8.36 I.E. Farquhar: *Ergodic Theory in Statistical Mechanics* (Interscience Wiley 1964)
8.37 E.B. Stechel, E.J. Heller: Ann. Rev. Phys. Chem. **35**, 563 (1984)
8.38 P.W. Milonni, J.R. Ackerhalt, M.E. Goggin: Adv. Chem. Phys. **73**, 867 (1989)
8.39 E.V. Mielczarek, J.S. Turner, D. Leiter, L. Davis: Am. J. Phys. **51**, 32 (1983)
8.40 M. Hénon, C. Heiles: Astron. J. **69**, 73 (1964)
8.41 J.W. Gadzuk: J. Chem. Phys. **79**, 6341 (1983); J.W. Gadzuk, S. Holloway: Phys. Scripta **32**, 413 (1985); J.W. Gadzuk: J. Chem. Phys. **86**, 5196 (1987)
8.42 J.W. Gadzuk: J. Chem. Phys. **84**, 3502 (1986)
8.43 J.W. Gadzuk, S. Holloway: Chem. Phys. Letters **114**, 314 (1985); S. Holloway: J. Vac. Sci. Tech. **5**, 476 (1987)
8.44 J.W. Gadzuk: Chem. Phys. Letters **136**, 402 (1987)
8.45 J.W. Gadzuk, J.K. Nørskov: J. Chem. Phys. **81**, 2828 (1984)
8.46 P. Hohenberg: in Ref. [8.15], p. 208
8.47 H. Ueba: Prog. Surf. Sci. **22** (3), 181 (1986); J.E. Adams: J. Chem. Phys. **84**, 3584 (1986); L.J. Richter, T.A. Germer, J.P. Sethna, W.Ho: Phys. Rev. B **38**, 10403 (1988–II); J.T. Muckerman, T. Uzer: J. Chem. Phys. **90**, 1968 (1989); I. Benjamin, W.P. Reinhardt: J. Chem. Phys. **90**, 7535 (1989); A.A. Stuchebrukhov: Chem. Phys. Letters **158**, 274 (1989)
8.48 R.B. Gerber: Chem. Rev. **87**, 29 (1987); J.W. Gadzuk: Ann. Rev. Phys. Chem. **39**, 395 (1988); A.E. De Pristo, A. Kara: Adv. Chem. Phys. (in press)
8.49 D.W. Noid, M.L. Koszykowski, R.A. Marcus: J. Chem. Phys. **67**, 404 (1977); Ann. Rev. Phys. Chem. **32**, 267 (1981)
8.50 R.S. Dumont, P. Brumer: J. Chem. Phys. **88**, 1481 (1988)
8.51 A. Ankiewicz, C. Pask: Phys. Letters **101A**, 239 (1984)

8.52 E.J. Heller, E.B. Stechel, M.J. Davis: J. Chem. Phys. **73**, 4720 (1980); E.J. Heller: Acc. Chem. Res. **14**, 368 (1981)
8.53 N. DeLeon, B.J. Berne: J. Chem. Phys. **75**, 3495 (1981); ibid **77**, 283 (1982)
8.54 B.E. Hayden, A.M. Bradshaw: Surface Sci. **125**, 787 (1983); B.E. Hayden, K. Kretzschmar, A.M. Bradshaw: Surface Sci. **155**, 553 (1985)
8.55 S. Holloway, J.W. Gadzuk: J. Chem. Phys. **82**, 5283 (1985)
8.56 A.D. Bandrauk (ed.): *Atomic and Molecular Processes With Short Intense Laser Pulses* (Plenum, New York 1988); D.J. Tannor, S.A. Rice: Adv. Chem. Phys. **70**, 441 (1988); A.H. Zewail: Science **242**, 1645 (1988)
8.57 S. Holloway, A. Hodgson, D. Halstead: Chem. Phys. Letters **147**, 425 (1988)
8.58 J.C. Polanyi: Acc. Chem. Res. **5**, 161 (1972): P.J. Kuntz: In *Dynamics of Molecular Collisons B*, ed. by W.H. Miller (Plenum, New York 1976) , p. 53; G.C. Schatz: Rev. Mod. Phys. **61**, 669 (1989)
8.59 R.L. Sundberg, E.J. Heller: J. Chem. Phys. **80**, 3680 (1984)
8.60 D.J. Tannor, S.A. Rice: J. Chem. Phys. **83**, 5013 (1985); J.W. Gadzuk: Surface Sci. **184**, 483 (1987)
8.61 D.C. Clary (ed.): *The Theory of Chemical Reaction Dynamics* (Reidel, Dordrecht 1986)
8.62 B. Eckhardt: Physica D **33**, 89 (1988); S. Bleher, E. Ott, C. Grebogi: Phys. Rev. Letters **63**, 919 (1989)
8.63 V. Balasubramanian, N. Sathyamurthy, J.W. Gadzuk: Surface Sci. **221**, L741 (1989)

9. Ten Years of Low Energy Positron Diffraction

K.F. Canter[1], *C.B. Duke*[2,4], *and A.P. Mills*[3]

[1] Department of Physics, Brandeis University, Waltham, MA 02254, USA
[2] Pacific Northwest Laboratory* P.O. Box 999, Richland, WA 99352, USA
[3] AT & T Bell Laboratories, Murray Hill, NJ 07974, USA
[4] Current Address: Xerox Webster Research Center, Webster, NY 14580,USA

It was not until five years after the discovery of low-energy electron diffraction (LEED) [9.1] that the positron was first observed [9.2]. By the time it became possible to form controllable, albeit very weak, low-energy positron beams in the 1970s, LEED had become an ubiquitous and powerful tool in surface structure determination. Thus, when the first observation of low energy positron diffraction (LEPD) was made in 1979 [9.3] one could well question whether it was too late for LEPD to be of any practical value in surface physics. As this review will chronicle, during the past ten years, LEPD has evolved from being a novelty to becoming a valuable technique for surface structure determination. The differences between LEPD and LEED are nontrivial; they not only raise questions of fundamental interest but also suggest that in some cases LEPD ultimately may prove superior to LEED for quantitative surface structure determination.

The fact that the positron is the antimatter partner of the electron means that the positron has the following properties: 1) it has the same mass as the electron, 2) it has the same magnitude of charge, but opposite sign, as the electron, 3) it is quantum mechanically distinguishable from the electron, and 4) it can undergo annihilation with an electron, i.e., the masses of the annihilating positron-electron pair are converted to electromagnetic energy in the form of γ-rays. Although the last property is perhaps the most spectacular, it is surprisingly of no significance in LEPD. This is because the long range nature of the positron-electron coulomb interaction gives rise to a scattering cross section which is 5–6 orders of magnitude larger than the cross section for annihilation, which requires positron-electron encounters within a range comparable to the Compton wavelength [9.4]. The first property results in the most trivial consequence, i.e., the diffracted beam angles for LEPD and LEED are the same for the same incident angle and energy. Properties 2 and 3 are responsible for important differences between LEPD and LEED which are discussed in Sect. 9.2. One consequence of the positron having a positive charge that is only of a small consequence to LEPD theoretically, but has enormous consequences experimentally, is the fact that the positron has a negative work function in many metals, notably Cu, Ni, and W. Irradiating such metals or "moderators" with fast positrons, usually obtained from β^+-emitting isotopes such as ^{58}Co or ^{22}Na, can result in highly monoenergetic low energy ~ 1 eV positron sources [9.5].

* The Pacific Northwest Laboratory is operated by Battelle Memorial Institute for the U.S. Department of Energy under Contract DE-AC06-76RLO 1830.

This review will discuss the technology of using negative work function positron emission to produce low energy positron beams of sufficient brightness for high precision LEPD studies, and other applications. The use of the term "LEPD" is principally restricted to structure determination experiments, although there are other diffraction processes of fundamental interest which have been observed, mainly at energies below $\sim$ 30 eV. Such energies fall below the energy range that is covered in conventional structure determinations with LEED. Positron diffraction at these "very low" energies are briefly reviewed in Sect. 9.5. The best context in which to discuss LEPD is within the well established discipline of LEED. Following a brief review of LEED, LEPD studies are discussed mainly with respect to the differences between LEPD and LEED.

9.1 Low-Energy Electron Diffraction

Low-energy electron diffraction (LEED) was discovered in 1927 by *Davisson* and *Germer* [9.1] who recognized its sensitivity to the atomic structure of surfaces from which the electrons were reflected. However, this sensitivity was not exploited for quantitative surface structure determinations until the mid 1970s [9.6], due to the lack of an adequate theory of the electron-solid interactions for electrons in the "low" ($1\,\mathrm{eV} \lesssim E \lesssim 500\,\mathrm{eV}$) energy range. Such a theory was constructed during the period 1968–75 by exploiting the important role of inelastic electron scattering processes in formulating a multiple-scattering theory of elastic low-energy electron diffraction [9.7–10]. Thus, the birth of LEED as a modern method of surface structure determination followed its discovery by over four decades. This development was complete by 1979 when the initial LEPD experiments were performed, and provided the foundation for the exceptionally rapid rise of LEPD from an interesting new phenomenon to a technique for quantitative surface structure analysis.

Concurrent with the development of the theory of LEED, computer programs were written to analyze experimental LEED intensities. Specifically, programs were constructed to evaluate the crystal potential of multilayer periodic arrays of ion cores parallel to a given crystal surface (with or without adsorbed overlayers), to reduce the individual ion core potentials to spherical form, to calculate the atomic phase shifts of the resulting spherical potentials, to modify these phase shifts to account for the atomic vibrations of the atomic scatterers, and to use the resulting modified phase shifts as inputs in a multiple scattering calculation of the LEED intensities for a given surface atomic geometry. Three groups constructed such libraries of programs in the early 1970s: *Duke* and co-workers [9.7, 11, 12], *Marcus* et al. [9.13, 14], and *Pendry* and co-workers [9.8]. Each of these libraries has been expanded and extended over the years by a wide variety of workers, and additional ones have also been assembled [9.15]. Measured LEPD intensities have been analyzed in detail by workers using the program libraries of *Duke* et al. [9.16–18] and of *Marcus* et al. [9.19, 20]. A detailed description of this process (for Cu{001} and Cu{111}) is given by *Weiss* et al. [9.16].

The utilization of these computer program libraries for surface structure determination has evolved rapidly since the mid 1970s. The promise of LEED for the determination of surface atomic geometries had been recognized for decades prior to that time, but could not be brought to fruition because of the lack of an adequate theoretical model of the low-energy electron diffraction process. An excellent synopsis of the situation just prior to the rapid development of LEED theory during 1968–75 is given by *Lander* [9.21], one of the major pioneers in the field. He noted the promise of LEED for surface structural chemistry, summarized the experimental situation and the ingredients of an adequate theory of LEED, and observed that at the time (1965), "LEED faces a critical period of maturation in which, through the development of more accurate scattering formulations, it should become a precise means of structure analysis" [9.21]. His words proved harbingers of the future, for within a decade this goal was well on its way to being achieved. By 1975, rudimentary quantitative structure analyses were being reported for clean metal surfaces and ordered adsorbed overlayer thereon [9.6], although semiconductor surface structure analyses remained in the future [9.22]. The first of these was reported in 1976 [9.23], and by 1978 they were becoming routine [9.24–26]. The field of quantitative semiconductor surface structure determination via LEED intensity analysis has continued to evolve, with its development being traceable through a sequence of major reviews [9.24, 27, 28]. The LEED intensity analysis structure determination methodology has been tested intensively over the years, especially in the case of GaAs{110} [9.29]. The results of careful analyses have been confirmed repeatedly by independent workers using other experimental techniques [9.28–30]. Therefore, once reliable, reproducible experimental LEED intensity data are obtained, a successful structure determination usually can be anticipated with confidence.

9.2 The First LEPD Experiments (1979)

The first observation of LEPD was carried out with a slow positron beam whose characteristics were on the borderline of making a credible LEPD demonstration possible [9.3]. To allow for the low beam flux and brightness, the incident beam angular divergence and diameter were maximized and, at the same time, kept within limits such that some semblance of typical LEED data could be obtained. Figure 9.1 shows both the diffractometer and the beam characteristics (at 400 eV incident energy) used in the first LEPD demonstration. To attain the emittance shown, it was necessary to aperture away 80% to 90% of the slow positrons emitted from a parallel vane W moderator [9.31]. In order to attain a 1mm-degree emittance at 40 eV, which is typical of LEED, it would have been necessary to aperture the $4 \times 10^4\, e^+/s$ incident beam down by a factor of 10^4, which would have made any LEPD observations virtually impossible. The fact that one can now produce positron beams having LEED-like emittances, without at the same time incurring drastic reductions in beam flux, is due to the real-

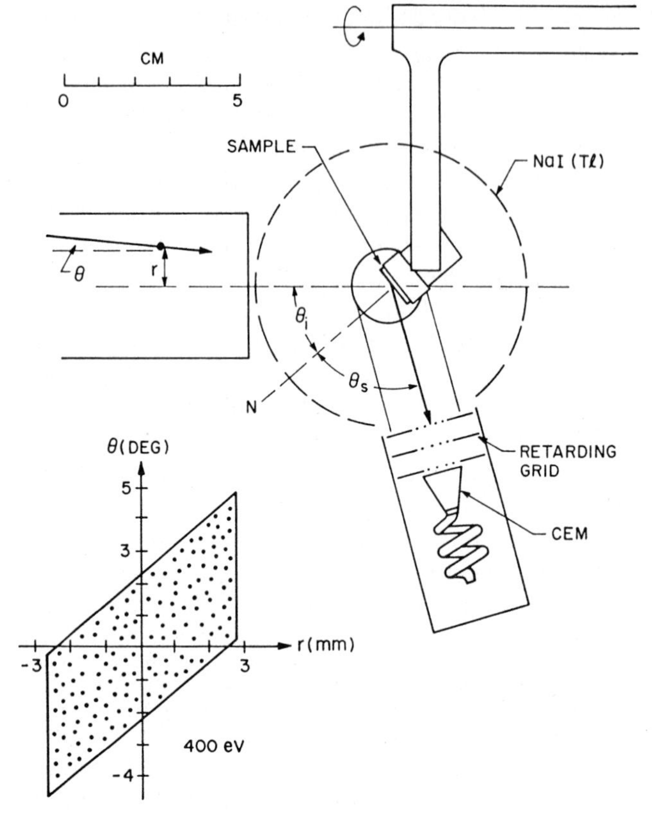

Fig. 9.1. Diffractometer for first generation LEPD measurements. The channel electron multiplier (CEM) detected scattered positrons and the NaI(T*l*) annihilation gamma-ray detector monitored the incident flux. Motion of the sample out of the scattering plane enabled sputtering and Auger analysis of the surface. The lower left insert indicates the angular and spatial spread of the incident beam at 400 eV [9.3]

ization of the "brightness enhancement" concept [9.32] which is discussed in Sect. 9.3. In spite of the poor brightness in the first LEPD work, it was not only possible to determine the magnitude of diffracted beam intensities (an important question because of incident beam flux limitations) but to also gain insight into the differences between LEED and LEPD.

9.2.1 Comparison of LEPD and LEED Cu Results

The first observed LEPD diffraction "spots", obtained from Cu{111}, are shown in Fig. 9.2. The angular width of the peaks was exaggerated by the angular spread of the incident beam as well as the large acceptance angle of the scattered positron detector. The intensity of the (00) specular beam as a function of incident positron energy is shown in Fig. 9.3. Also shown is the corresponding I-V profile obtained in the same system using electrons. The electron beam was produced

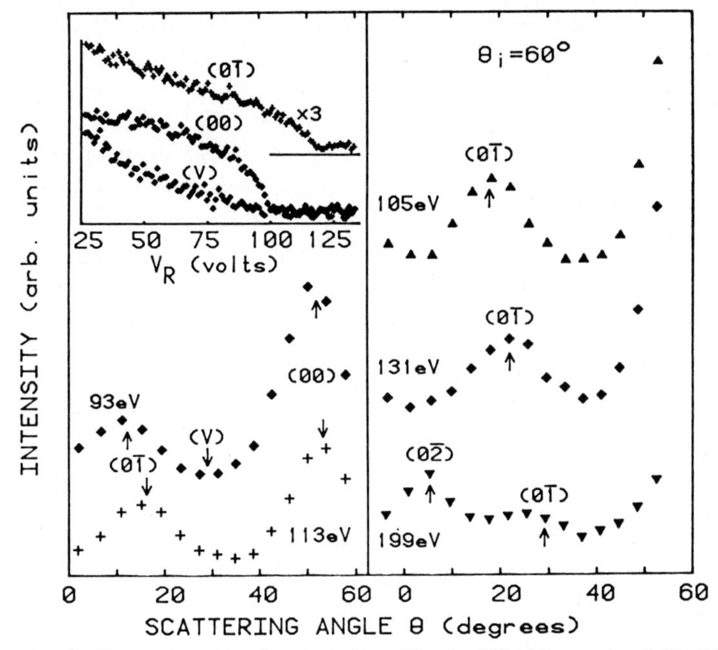

Fig. 9.2. Elastically reflected positron intensities for 53° (*left panel*) and 60° (*right panel*) incident angle. The upper left insert shows the retarding field energy spectra for positrons observed in the (00) and (0T) diffracted beams as well as the region (V) between diffraction peats. V_R is the voltage on the retarding grid in front of the CEM [9.3]

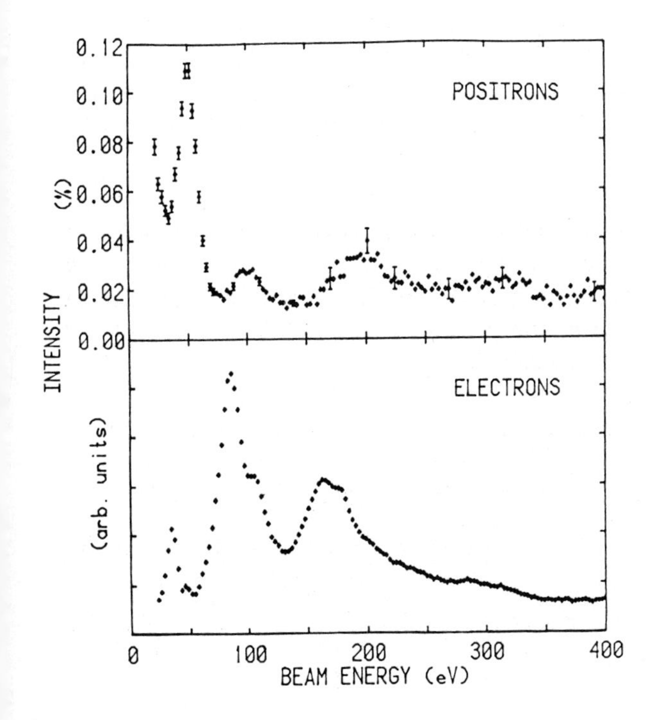

Fig. 9.3. Specular reflection intensities from Cu{111} for positrons and electrons incident at $\theta_i = 30°$ [9.3, 19]

by reversing the polarity of the positron beam optics and transporting secondary electrons produced at the moderator due to β and γ irradiation. Although the angular distribution I(θ) at the same incident energy were of course the same for electrons and positrons, after correcting for differences in the incident angular distributions [9.33], the I-V profiles as shown in Fig. 9.3 differed dramatically.

The large differences in LEPD and LEED I-V profiles led to the speculation that requiring a surface structure model to result in agreement between theoretical and experimental I-V profiles for LEPD and LEED data obtained from the same sample would better constrain the choice of surface model parameters than requiring a good fit for LEPD or LEED alone. Binary systems, for example, were in particular expected to benefit from direct LEPD/LEED comparisons since the relative "contrast" to the A and B atoms would be different for positrons and electrons [9.16]. Questions such as how to define a reliability factor that was a measure of simultaneous agreement between theory and experiment for LEPD and LEED I-V profiles were not proposed at the time. As will be seen later in this review, devising a LEPD/LEED combined reliability factor is complicated by the possibility that LEPD and LEED data might not warrant equal weighting when trying to make a structural determination.

9.2.2 Predicted Advantages of LEPD over LEED

As previously mentioned, the large differences between LEPD and LEED I-V profiles alone suggested that it would be valuable to pursue LEPD as a complement to LEED. However, there are differences between positrons and electrons that prompted predictions that LEPD could ultimately prove to be superior to LEED as a structural determination tool. Although most of the predicted advantages of LEPD over LEED were originally discussed in the context of structural studies of metal surfaces, many of these considerations are relevant to non-metals as well.

a) *Elastic Mean Free Path.* The electron elastic mean free path in a metal reflects the density of occupied electron states in the metal as well as the availability of unoccupied states for the scattered electron. However, there are no excluded final states for the positron since there is only one positron at a time in the sample [9.34]. This results (for a free electron gas model) in a 30% reduction in the elastic mean free path for positrons as compared to the elastic mean free path of electrons at 50 eV, for example [9.35]. Above 150 eV, the differences in inelastic scattering for positrons and electrons become negligible. The end result is that for energies below 150 eV one would expect LEPD to exhibit more sensitivity than LEED to the structural parameters of the outermost layers of a metal.

b) *Inner Potential.* Accurate knowledge of the real part of the inner potential V_0 as used in LEED structural determinations is difficult to obtain and thus often requires treating V_0 as an adjustable non-structural parameter [9.36]. For a zero-energy incident electron, we have $V_0 = -\Phi^- - E_f$, where Φ^- is the electron work function and E_f is the Fermi energy relative to the muffin-tin zero of the crystal. Although Φ^- can be measured experimentally, E_f must be calculated. *Read* and *Lowy* pointed out that in addition to E_f being energy-dependent, its zero-energy value as determined from calculation may have an uncertainty of $\pm$ 2 eV for heavy metals [9.37]. In the case of positrons, it should be possible to determine V_0 more accurately since $E_f = 0$ and Φ^+, the positron work function, can be measured directly for negative positron work function metals.

c) *Relativistic Effects*. The difference in the sign of the charge is also responsible for differences in the relativistic effects for the scattering of electrons and positrons. In particular, Coulomb repulsion between the positron and the atomic nucleus greatly reduces the spin-orbit effects observed for LEED from high-Z materials [9.38]. This in turn means that it is not necessary, in most cases, to spin-average the theoretical diffracted beam intensity profiles. Besides the resulting savings of a factor of two in computing time, the insensitivity of LEPD to positron spin orientation is fortuitous since beam polarization is difficult to avoid when using β^+ emitting isotopes [9.39]. Coulomb repulsion of the positron by the nucleus also prevents positrons from gaining relativistic velocities as do electrons in close encounters with the nucleus. Consequently LEPD intensities for high-Z materials can be calculated with the non-relativistic Schrödinger equation.

d) *Exchange-Correlation*. In addition to the Coulomb interaction between an incident electron and the ion core, the effects of exchange-correlation with the ion core electrons also play a large role in the scattering process. Although exchange is automatically eliminated in LEPD, the difficult correlation interaction, or equivalently, polarization in excess of the Hartree term, is still present. Since the Coulomb repulsion of the positron by the ion inhibits the positron from correlating with the core electrons [9.40,41], one might expect that the normal exchange-correlation interaction used in LEED would be less significant in LEPD. A dramatic demonstration of how unimportant it is to have to deal with the correlation interaction for LEPD is shown in Fig. 9.4. The I-V pro-

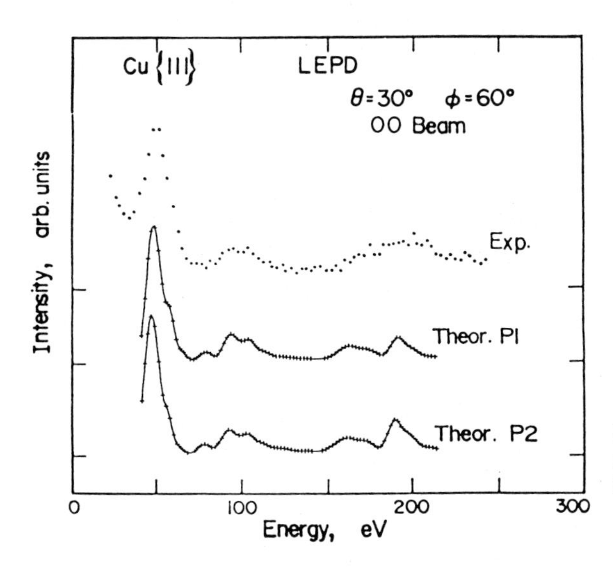

Fig. 9.4. Experimental and theoretical specular reflection LEPD intensities for $\theta_i = 30°$ [9.19]

file "Theor. P1" is a theoretical calculation for the (0,0) specular reflection of positrons from Cu{111} which employed the Hedin-Lundquist correlation term in the scattering potential, as a reasonably accurate way of including the effects of positron-core electron correlation. In "Theor. P2" the same exchange term needed to treat electron scattering was included in the calculation. Since this term is approximately four times the correlation term in P1, a comparison of P1 and P2 can be regarded as a demonstration of the consequences of introducing a 400% "error" in how correlation is treated in a LEPD calculation. As can be seen, the difference between P1 and P2 is very small. In LEED, a variation in the exchange-correlation term by one-tenth as much as this would produce a much more discernible change in the calculated profiles [9.42]. It should also be noted that Fig. 9.4 uses the actual computer generated I-V profiles, instead of the hand-traced curves that appeared in the original version of this figure by *Jona* et al. [9.19]. Subsequent calculations on the Cu{111} data and additional Cu{100} data showed that it was sufficient to drop the correlation interaction altogether in order to get good agreement between theoretical and experimental LEPD I-V profiles [9.16].

The insensitivity of LEPD to positron-core-electron correlation was originally predicted by *Read* and *Lowy* [9.37]. The particular case that Read and Lowy pointed out as one in which the lack of correlation would put LEPD at an advantage over LEED was the determination of the first interlayer contraction δ_1 of W{001} and Mo{001}. A comprehensive study of several LEED determinations of the W{001}-(1×1) surface was carried out by *Read* and *Russel* [9.42]. They concluded that the uncertainty in treating exchange-correlation contributed to half the uncertainty in determining δ_1. The remaining half of the uncertainty in δ_1 was attributed to experimental errors. Although it was not reasonable to expect that LEPD could do any better than LEED in eliminating this source of uncertainty in δ_1, a direct comparison of LEPD and LEED on the same sample was expected to greatly reduce the effects of most of the sources of experimental error. In order to ultimately realize the promise of LEPD, however, it was first necessary to improve the brightness of slow positron sources.

9.3 Brightness Enhancement

In order to effectively use the fast positrons that irradiate a slow positron moderator, the moderator has to be fairly large, i.e., at least 10 mm in diameter. Until recently, this large emitting area greatly restricted how well the positrons could be collimated or focussed down to a small spot, due to Liouville's Theorem as applied to focussing particles with conservative forces [9.43]. The thermalization of fast positrons in a solid and reemission into the vacuum, however, can be regarded as a dissipative mechanism not subject to Liouville's Theorem. This was the basis of the brightness enhancement concept: accelerate to a few keV slow positrons emitted from the large primary moderator and focus down to a spot size, limited by Liouville's Theorem, one-tenth to one-hundredth the diameter of the primary moderator [9.32]. The reemitted positrons will then have a higher brightness-per-volt since they are emitted from a smaller area, but at low ($\sim 1\,\text{eV}$) energies, and with relatively small flux loss in the process. Not only is brightness enhancement crucial for modern LEPD, but it enables other applications such as the positron annihilation microbe [9.44] and brightness enhanced positron reemission microscopy [9.45].

In practice, there are two ways of obtaining brightness enhancement: using transmission or reflection modes. In the transmission mode, positrons are focussed onto the backside of a thin (≈ 1000 Å) negative positron work function, single crystal, self supporting film, with Cu, Ni, and W being the best candidates. Typically, $\approx 20\%$ of the positrons will diffuse to the other side of the film and emerge as slow positrons [9.46, 47]. In the reflection mode, positrons are focussed onto a thick single crystal, again Cu, Ni, or W. For incident penetration depths less than the thermal positron diffusion length, $\approx 40\%$ of the incident positrons can be expected to reemerge from the same face of the crystal [9.5]. The original difficulty with the reflection mode concept was that electrostatic lenses capable of accelerating and focussing positrons down to a sufficiently small spot (i.e., one hundredth the original beam diameter) at the remoderator surface also produce electric fields which interfere with subsequent extraction of the reemitted slow positrons from the same surface.

Although use of the transmission mode offers a simple solution to the electron optics of brightness enhancement, it presents a more difficult materials science task, i.e., producing self supporting, thin single crystal films [9.48]. The first practical solution of how to employ the reflection mode geometry for brightness enhancement was based on the realization that 1) reducing the beam by a factor of only 10, with a weak lens in the reflection mode, eliminated the problem of the focussing interfering with the extraction of the remoderated positrons and 2) doing this twice yielded the desired reduction of the original beam diameter by a factor of 100 with nearly the same efficiency anticipated for the one stage transmission mode [9.49]. The principle of two stage weak focussing reflection mode brightness enhancement is illustrated in Fig. 9.5.

The first actual demonstration of brightness enhancement was made by *Frieze* et al. and used a variation of the two stage reflection approach that employed stronger focussing lenses with gridded terminations in order to avoid interference with the remoderated positron extraction optics [9.50]. The fact that *Frieze* et al. only achieved a factor of 50 in brightness enhancement was primarily due to a large loss in brightness when trying to transport their primary beam to the first remoderator rather than the use of gridded lenses. The high brightness beam at Brandeis (Fig. 9.6) achieves a factor 500 in brightness enhancement by using the weak focussing method shown in Fig. 9.5 and well characterized transport optics [9.51].

The first application of the BNL brightness enhanced beam was made by *Mayer* et al. [9.20] who carried out a reexamination of the early Brandeis

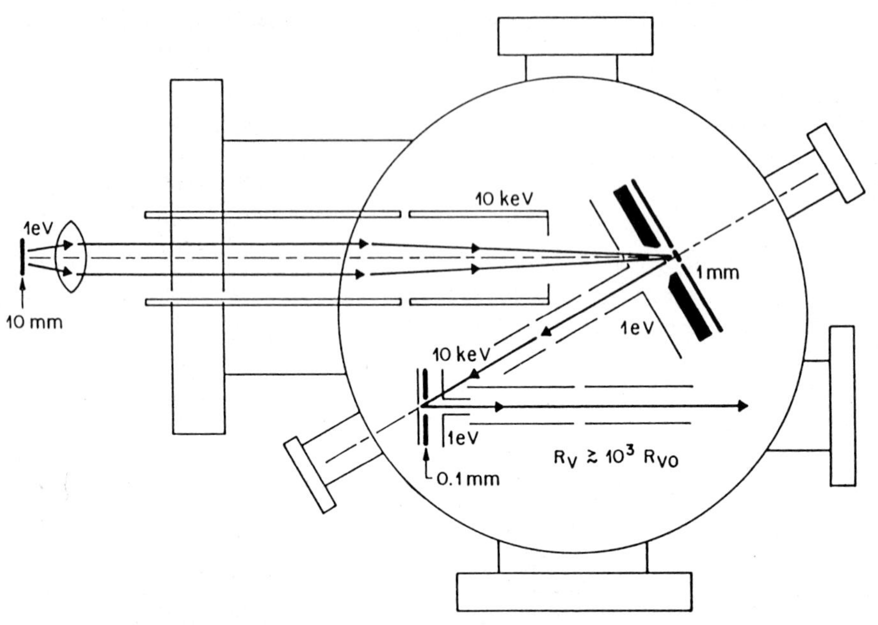

Fig. 9.5. Schematic illustration of the two-stage, weak-focussing, reflection mode brightness enhancement scheme. The 10 keV positrons are reemitted from the 1mm and 0.1mm W{110} crystals with ~ 40% probability [9.49]

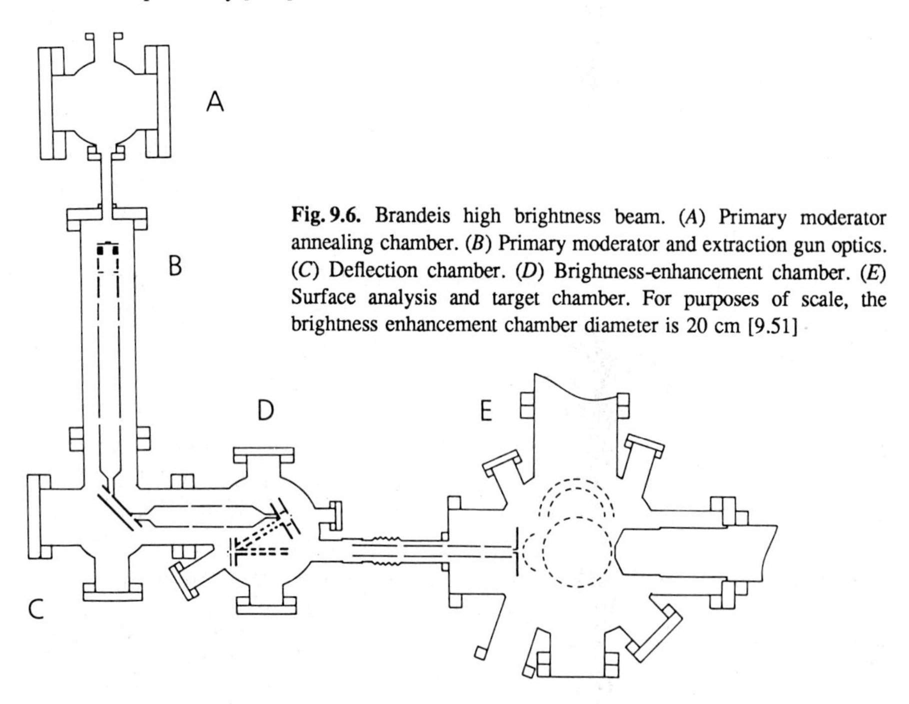

Fig. 9.6. Brandeis high brightness beam. (*A*) Primary moderator annealing chamber. (*B*) Primary moderator and extraction gun optics. (*C*) Deflection chamber. (*D*) Brightness-enhancement chamber. (*E*) Surface analysis and target chamber. For purposes of scale, the brightness enhancement chamber diameter is 20 cm [9.51]

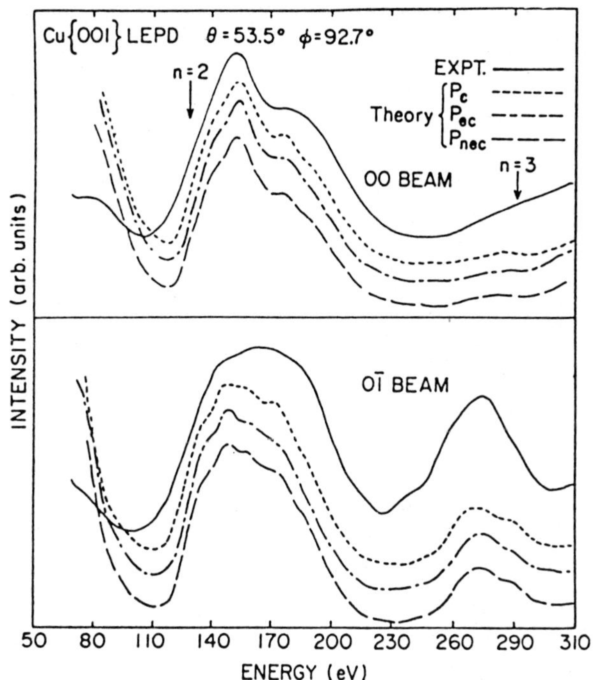

Fig. 9.7. Experimental and calculated normalized LEPD I-V curves for (00) and (0$\bar{1}$) beams. Arbitrary units used in vertical scale. Experimental and calculated curves are shifted to aid visual comparison. Arrows in 00 curve denote n = 2,3 Bragg maxima. Theory curves with $\frac{1}{6}$ exchange (P_c), full exchange (P_{ec}), and no exchange (P_{nec}) [9.20]

Cu{100} LEPD measurements obtained without brightness enhancement [9.3]. Surprisingly, the BNL results (Fig. 9.7) showed only a small improvement over the early Brandeis data, with regard to fine features in the I-V profiles. However, this was not due to insufficient brightness enhancement, but mainly to the BNL measurements being restricted to rather large angles of incidence ($\geq$ 50°). The peaks in the positron I-V spectra, due to increased inelastic scattering and consequent sampling of fewer sub-surface layers when the incident particle enters the sample far-off normal incidence, are broadened. This reduces the benefit that would be anticipated from having a better collimated incident beam. With the same BNL system *Mayer* et al. subsequently produced a dramatic demonstration of this difference between LEPD and LEED at near glancing angles of incidence [9.52]. Figure 9.8 shows that the complex structure in the LEED I-V profiles is absent in LEPD at the same large ($\geq$ 75°) angles of incidence. *Mayer* et al. attribute the striking difference between the LEPD and LEED profiles to a novel role played by the differences in the real part of the inner potential V_0 for positrons and electrons. They reasoned that the repulsive nature of $V_0 \sim +1$ eV for positrons would result in the positron not penetrating as far below the surface as would be the case for electrons where $V_0 \sim -8$ eV. Such differences would not be evident at small angles of incident where the perpendicular energy, which determines the number of layers sampled, is typically $\gtrsim$ 40 eV as opposed to $\gtrsim$ 10 eV in the case of $\gtrsim$80° angles of incidence.

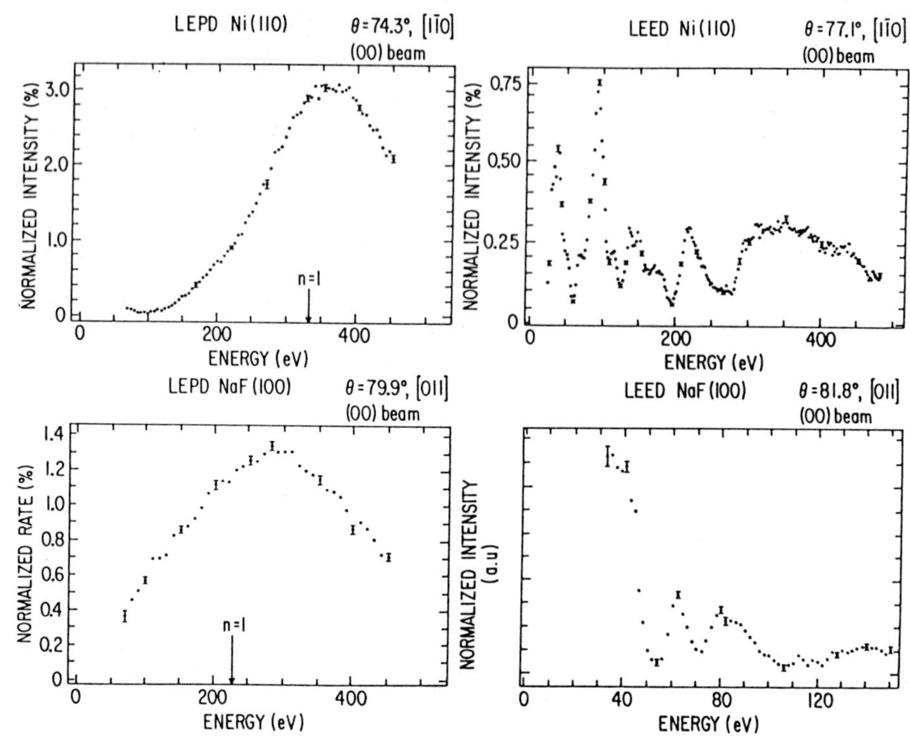

Fig. 9.8. Comparison of normalized specular LEPD and LEED spectra from NaF{100} and Ni{110}. The arrow below $n = 1$ denotes the first-order Bragg peak position [9.52]

While large angle incidence LEPD studies might not be of any direct value in structural determinations, they could, however, increase our understanding of the role played by the surface barrier in diffraction. Although a large number of fundamental experimental studies of the differences between positrons and electrons can be continued to be carried out with modest brightness beams, practical structural determinations with LEPD requires a high brightness beam such as the Brandeis beam described here.

9.4 Surface Structure Determinations with Modern LEPD: CdSe$\{10\bar{1}0\}$ and CdSe$\{11\bar{2}0\}$

The possible advantages of LEPD over LEED were originally anticipated mainly for metals. However, the first modern LEPD study was carried out on a compound semiconductor CdSe. This material not only enabled a direct comparison between LEPD and LEED for a binary system [9.18], but also offered the opportunity to make a structural determination of a previously unexamined surface CdSe$\{11\bar{2}0\}$ [9.17].

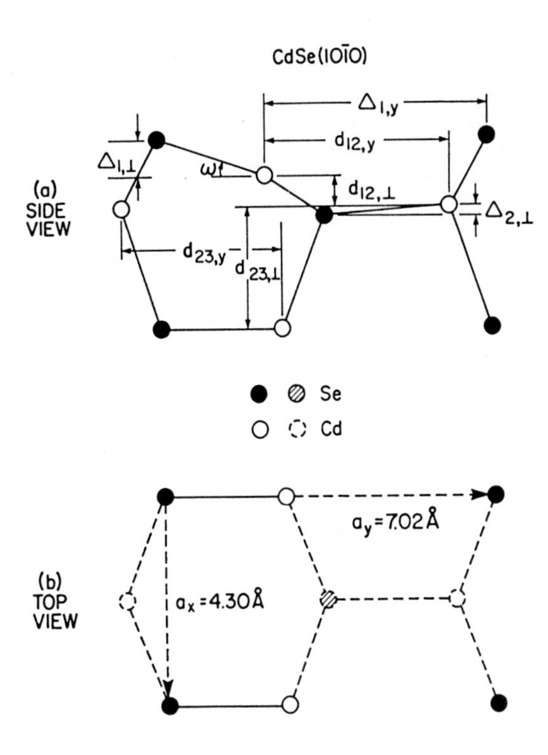

Fig. 9.9. Schematic indication of the structural variables for CdSe{10$\bar{1}$0} [9.17]

9.4.1 Relaxation Models

Wurtzite-structure compound semiconductors exhibit two cleavage faces, the {10$\bar{1}$0} and {11$\bar{2}$0} surfaces, both of which are electrically neutral in that the surface atomic plane contains equal numbers of anions (Se) and cations (Cd). Early LEED intensity data on these surfaces for ZnO and CdS were reported by *Mark* et al. [9.22]. Complete LEED intensity analysis structure determinations have been reported for the {10$\bar{1}$0} surfaces of ZnO [9.53, 54] and CdSe [9.55]. Originally thought to exhibit truncated-bulk structures [9.22], both the {10$\bar{1}$0} and {11$\bar{2}$0} surfaces are now known to exhibit large nearly-bond-length-conserving relaxations with the anion displaced outward and the cation inward [9.17, 18, 53–56]. Schematic representations which illustrate the independent structural variables associated with these relaxations are shown in Figs. 9.9 and 9.10 for the {10$\bar{1}$0} and {11$\bar{2}$0} surfaces, respectively.

9.4.2 I-V Profile Measurements

Typically, a laboratory positron beam will have a primary flux of $10^5 - 10^6$ e^+/s, depending on the source strength and the primary moderator efficiency. After two stages of remoderation, one can anticipate being able to carry out a LEPD study with 10^4–10^5 e^+/s, or $< 10^{-14}$A incident current. Most of the final

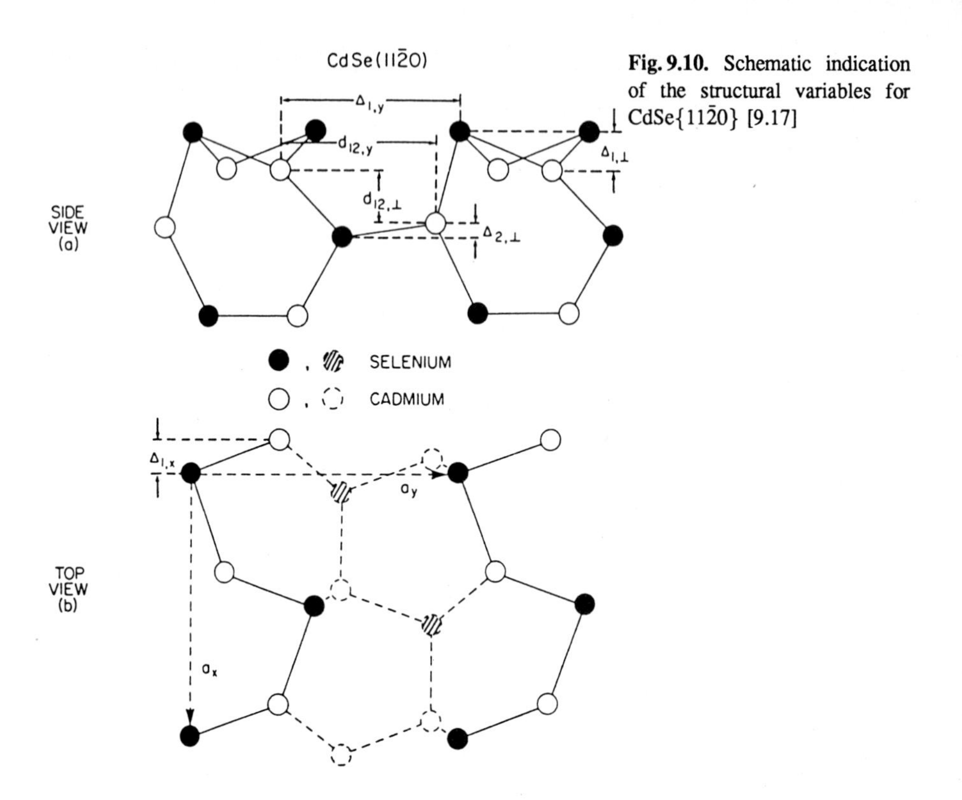

Fig. 9.10. Schematic indication of the structural variables for CdSe{11$\bar{2}$0} [9.17]

data in the CdSe LEPD measurements at Brandeis were obtained with the ^{58}Co primary positron source having decayed to $\approx$ 100 mCi, with a corresponding brightness enhanced beam of 5×10^3 e^+/s. Such a low incident flux required the use of "digital LEED" techniques, wherein the diffracted particles are individually counted [9.57]. The diffractometer used at Brandeis is shown schematically in Fig. 9.11. The experimental apparatus included an in situ sample cleaver, and an electrostatic mirror mounted to the sample manipulator. By swinging the mirror to face the beam, the incident flux, mirrored back into the detector, could routinely be measured. Measuring the incident beam current with the same detector as that used for the diffracted intensity measurements was a valuable aid in obtaining accurate absolute diffraction intensities.

Essentially, all of the diffracted beams are recorded simultaneously in the digital LEED method. Thus, the CdSe LEPD I-V profiles could be obtained by scanning over the incident energy range many times and then appropriately combining the recorded 2D spot patterns to form a final cumulative spot pattern histogram at each individual incident energy. With off-line computer analysis the intensity of each diffracted beam spot versus incident energy could then be determined. The resulting I-V profiles of four of the fourteen diffracted beams observed are shown in Fig. 9.12a. The cumulative run time at each incident energy value was 4000 s. The CdSe samples, after cleaving, were periodically cleaned by

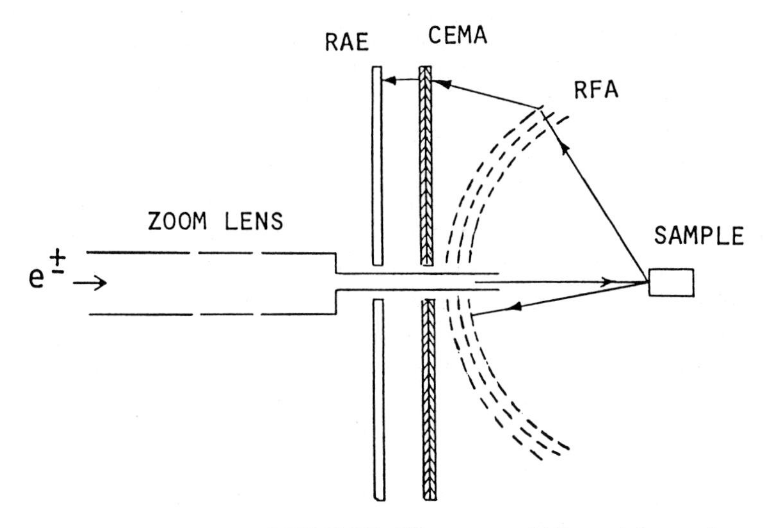

Fig. 9.11. Brandeis digital LEED/LEPD diffractometer. *RAE* – resistive anode encoder. *CEMA* – channel electron multiplier array. *RFA* – retarding field analyzer (supressor) [9.17, 62]

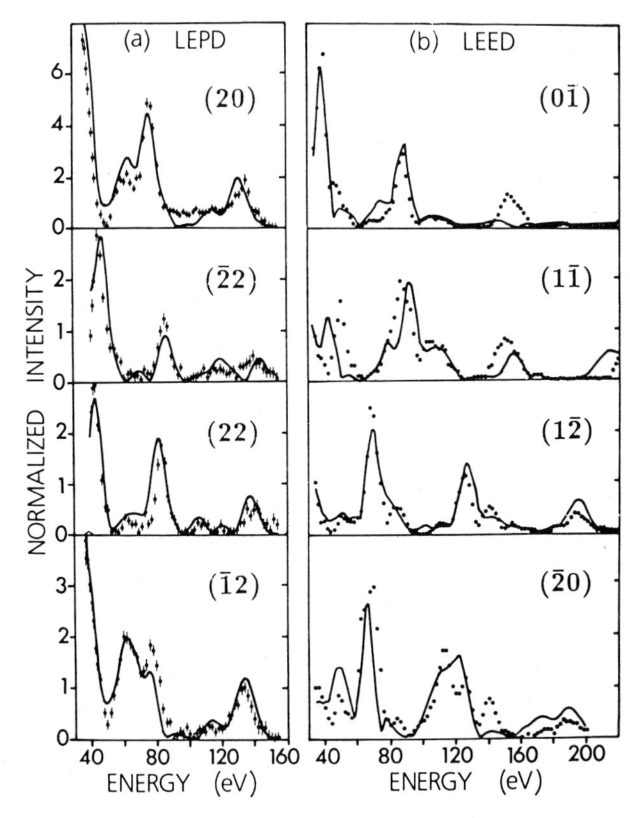

Fig. 9.12a,b. Data (*points*) and theory (*lines*) for four beams diffracted from the CdSe{11$\bar{2}$0} surface, utilizing the best-fit structures of **(a)** LEPD and **(b)** LEED. Intensities are given in absolute reflectivity multiplied by 10^4 for (a) and in arbitrary units for (b) [9.18]

heating and diffracted intensities were monitored to insure that the total $\sim$100 hr run period for a complete energy scan corresponded to clean surface conditions in the mid-10^{-11} Torr sample chamber.

The Brandeis LEPD system could also be used to obtain LEED data by reversing the polarity of the optics following the final remoderator and using an electron filament in the vicinity of the last remoderator stage. The purpose of the LEED measurements was not only to provide an in situ comparison of LEPD and LEED, but to also provide an important "calibration" of the Brandeis LEPD/LEED diffractometer by comparing the Brandeis LEED experiment with a completely independent set of CdSe $\{10\bar{1}0\}$ and $\{11\bar{2}0\}$ LEED measurements obtained using a spot photometer diffractometer at Princeton University. The Princeton $\{10\bar{1}0\}$ results [9.55] had been published prior to the Brandeis experiments while the $\{11\bar{2}0\}$ work proceeded essentially in parallel with the Brandeis measurements. The agreement between the Brandeis and Princeton LEED I-V profiles was very good, especially considering that the data were taken with different CdSe samples, produced by the same manufacturer, and with different cleavers and diffraction systems. However, the comparison between theory and experimental I-V profiles, i.e., $I_{th}(V)$ and $I_{ex}(V)$, respectively, was found to be slightly better for the Princeton data. Thus the four "best" beams from the Brandeis $\{11\bar{2}0\}$ LEPD data and the Princeton $\{11\bar{2}0\}$ LEED data, shown in Figs. 9.12a and 9.12b, respectively, were chosen as an illustration of the relative agreement between $I_{th}(V)$ and $I_{ex}(V)$ for LEPD versus LEED. Also shown are the theoretical profiles $I_{th}(V)$ obtained using the CdSe structural determinations as discussed below.

9.4.3 Structure Analysis

Surface atomic geometries are determined by comparing calculated and measured LEPD (and LEED) intensities for a systematic series of hypothesized structures. For each structure, the comparison between the calculated and measured intensities is rendered quantitative by using the X-ray (R_x) and integrated intensity (R_I) R factors as described in the report [9.55] of the LEED analysis of CdSe$\{10\bar{1}0\}$. R_x is essentially a measure of the square deviation between $I_{th}(V)$ and $I_{ex}(V)$, suitably averaged over all of the incident energy values and for all of the diffracted beams observed. Specifically, R_x is evaluated from Eqs. (3), (8), (13), (14) and (16) of *Zannazi* and *Jona* [9.58] and R_I as described by *Duke* et al. [9.59]. In general, R_x is calculated by varying V_0 to yield the minimum value for a given structural model. In the vicinity of the best-fit structures, this procedure leads to $V_0 = -2 \pm 1\,\mathrm{eV}$ (LEPD) and $V_0 = -13 \pm 1\,\mathrm{eV}$ (LEED) for both CdSe$\{10\bar{1}0\}$ and CdSe$\{11\bar{2}0\}$.

For each surface, the structure analysis is performed via comparison of the measured and calculated intensities, using the R-factor method noted above, for a systematic variation of the independent structural variables. For CdSe$\{10\bar{1}0\}$, the variables are defined in Fig. 9.9. The vector shear between the Cd and Se in each layer, $\boldsymbol{\Delta}_i$, has two independent components: one normal to the surface, $\Delta_{i,\perp}$, and one along the y axis, $\Delta_{i,y}$. The symmetry of the measured intensities requires that $\Delta_{i,x}$ equal its value in the bulk, i.e., $\Delta_{i,x} = 0$. It is convenient to initiate the structure search by linking the values of $\Delta_{i,\perp}$, $\Delta_{1,y}$, $d_{12,\perp}$ and $d_{12,y}$ in such a fashion that all the bond lengths remain constant as the surface species are displaced from their bulk positions ("bond-length-conserving" rotations). In this case, the angle ω between the plane of the uppermost dimers of Cd and Se and that of the truncated bulk surface is utilized as an independent structural variable in lieu of $\Delta_{1,\perp}$. For such a rotation $\Delta_{1,\perp}$ and $\Delta_{1,y}$ are related by

$$\Delta_{1,\perp} = d \sin \omega \ , \tag{9.1}$$

$$\Delta_{1,y} = a_y - d \cos \omega \ , \tag{9.2}$$

in which $d = 0.375 a_y = 2.63$ Å is the top-layer Cd-Se bond length. The condition of constant bond length also generates changes in $d_{12,\perp}$ and $d_{12,y}$ with changes in ω. Variations in ω for bond-length-conserving rotations are utilized to fix $\Delta_{1,\perp}$ and subsequently $\Delta_{1,y}$ is varied independently in order to determine the shear vector $\boldsymbol{\Delta}_1$ characteristic of the uppermost layer of Cd and Se species. The third independent structural variable is taken to be the uppermost layer spacing $d_{12,\perp}$. Finally, the shear vector in the second layer $\boldsymbol{\Delta}_2$ is defined by its perpendicular component $\Delta_{2,\perp}$ alone because the LEED and the LEPD analyses are not sufficiently accurate to determine ω_2 or $\omega_{2,y}$ separately.

The structure analysis for CdSe$\{11\bar{2}0\}$ was performed analogously to that for CdSe$\{10\bar{1}0\}$ with a search over bond-length-conserving relaxations being followed by refinements in $d_{12,\perp}$ and $\Delta_{2,\perp}$. The bond-length-conserving relaxation is more complicated in this case because the atoms in the top layer pucker causing it to become nonplanar. The angular variable in this case is that between the normal to each Cd–Se–Cd or Se–Cd–Se triplet and that of the unrelaxed surface. If $\hat{n}$ designates the former and $\hat{z}$ the latter, the polar angle ω is given by

$$\hat{n} \cdot \hat{z} = \cos \omega \ . \tag{9.3}$$

This leads to a complicated dependence between $\Delta_{1,\perp}$ and ω to replace the simple expression $\Delta_{1,\perp} = d \sin \omega$ for the $(10\bar{1}0)$ surface. It is of the general form

$$\Delta_{1,\perp} = f(\Delta_{1,x}, \Delta_{1,y}) \tan \omega \ , \tag{9.4}$$

in which variations of $\Delta_{1,\perp}, \Delta_{1,x}, \Delta_{1,y}$ are constrained to keep all the bond lengths constant. These constraints can be satisfied, however, for $0 \lesssim \omega \lesssim 45°$ corresponding to a range in $\Delta_{1,\perp}$ from 0 to 1.02 Å.

The values of the X-ray R factor R_x, which measures the descrepancies in the shapes of the current-voltage profiles for the various beams, are shown in Fig. 9.13 as functions of the top-layer bond-length-conserving rotation angles, ω, for both LEPD and LEED from CdSe$\{10\bar{1}0\}$ [panel (a)] and CdSe$\{10\bar{2}0\}$ [panel (b)] [9.18].

It is evident from Figure 9.13 that both surfaces are reconstructed with $15° \lesssim \omega \lesssim 22°$ for CdSe$\{10\bar{1}0\}$ and $27° \lesssim \omega \lesssim 34°$ for CdSe$\{11\bar{2}0\}$. Error estimates

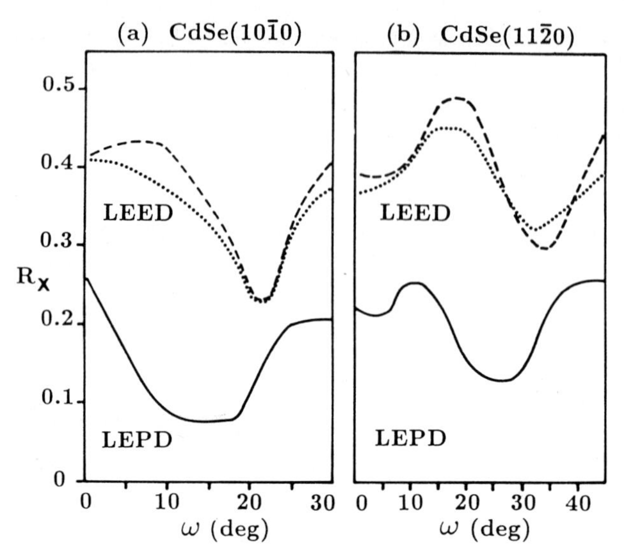

Fig. 9.13a,b. X-ray R-factor R_x as a function of the bond-rotation angle ω for **(a)** the CdSe$\{10\bar{1}0\}$ surface and **(b)** the CdSe$\{11\bar{2}0\}$ surface. (*Solid lines*) LEPD. (*Broken lines*) Princeton LEED. (*Dotted lines*) Brandeis LEED [9.18]

for the LEED analysis have been studied in detail for zincblende $\{110\}$ surfaces [9.60] with the result that $\Delta R_x = 0.04$ is a significant discriminator between different structures. For CdSe$\{10\bar{1}0\}$ this criterion yields $\omega = 15 \pm 5°$ from LEPD and $\omega = 21 \pm 4°$ from LEED. These results are compatible both with each other (in the range $17° \lesssim \omega \lesssim 20°$) and with the predicted [9.53] value of $\omega = 18°$. For CdSe$\{11\bar{2}0\}$ the projected ranges are $\omega = 27 \pm 7°$ (LEPD) and $\omega = 34 \pm 6°$ (LEED), leading to overlap between the two for $28° \lesssim \omega \lesssim 34°$ and compatibility with the predicted [9.61] value of $\omega = 32°$.

Table 9.1. Anion-cation perpendicular shear $\Delta_{1,\perp}$, first interlayer spacing $d_{12,\perp}$ and bond-rotation angle ω for the best-fit structures for LEED and LEPD from CdSe cleavage faces, and their associated R-factors. Uncertainties in ω are indicated in the text. For $d_{12,\perp}$ and $\Delta_{1,\perp}$ these uncertainties are approximately ±0.1Å

CdSe(10$\bar{1}$0)	ω(°)	$\Delta_{1,\perp}$(Å)	$d_{12,\perp}$(Å)	R_x	R_I
Bulk	0	0	1.24	–	–
LEPD	15	0.68	0.65	0.08	0.05
Theory[(a)]	17.7	0.78	0.64	–	–
LEED[(b)]	21.5	0.96	0.45	0.19	0.05
LEED[(c)]	21.5	0.96	0.45	0.21	0.03
CdSe(11$\bar{2}$0)					
Bulk	0	0	2.15	–	–
LEPD	27	0.61	1.62	0.12	0.04
Theory[(a)]	32	0.71	1.52	–	–
LEED[(b)]	33	0.73	1.51	0.32	0.22
LEED[(c)]	34	0.76	1.47	0.29	0.12

(a) Ref. 9.61
(b) Brandeis data
(c) Princeton data

Additional small refinements in both structures result upon optimization of the remaining independent structural variables [9.17, 18, 62] and are predicted by a total-energy minimization calculation [9.61]. Although they improve the R values of the best fit, they lie within the $\Delta R_x = 0.04$ ranges of uncertainty. The final optimal parameters and associated R factor values are given in Table 9.1.

9.4.4 Differences Between LEED and LEPD Structural Determinations

Although the differences in the LEPD and LEED structural determinations might be viewed as being relatively minor, the difference in the degree of agreement between $I_{ex}(V)$ and $I_{th}(V)$ for LEPD and LEED is significant. Visual inspection of Fig. 9.12 suggests that $I_{th}(V)$ agrees with $I_{ex}(V)$ for LEPD better than for LEED. A more quantitative measure is provided by the best fit structure R_x factors in Table 9.1. The R_x values are significantly better for LEPD than for LEED. Determining the reason(s) why theory is able to account for LEPD intensity profiles more accurately than it can for LEED is as important as the structural determinations themselves. In the CdSe LEPD/LEED comparison by *Horsky* et al. [9.18], two possible reasons were proposed as being responsible for the improved agreement between $I_{th}(V)$ and $I_{ex}(V)$ for LEPD while being consistent with the fact that LEED theory agrees much better with experiment for metals than it does for semiconductors.

The optimum positron elastic mean free paths for R_x minimization were found to be nearly half those of electrons in the CdSe study. Thus, complex relaxations propagating below the surface (which were not incorporated in the diffraction computations) could make the single bilayer relaxation picture more amenable to LEPD than to LEED, if such relaxations are indeed present. The other possibility suggested was that positrons might be less sensitive than electrons to the non-spherical nature of the valence electron spatial distribution within the CdSe bonds. This insensitivity could be due to the absence of exchange in LEPD and/or due to the ion core potential effects recently proposed by *Duke* and *Lessor* [9.63].

The detailed analysis of the ion-core scattering factors for CdSe by Duke and Lessor supports the argument that the generally better description of the LEPD data is attributed mainly to the utilization of more accurate ion-core scattering factors relative to the case of LEED. These scattering factors are essentially identical for Cd and Se, so the atomic geometry exerts a more significant influence on the calculated LEPD intensities. Specifically, positrons experience a repulsive ion-core potential which augments the angular momentum barrier in repelling them from the core region and is responsible for the lack of positron-core electron correlation discussed in Sect. 9.2.2d. Moreover, the total partial-wave potential is observed to be insensitive to the atomic number (and hence electronic change density) of the scatterers. Thus, Cd and Se exhibit nearly identical positron scattering phase shifts and cross sections, the latter of which are slowly varying with energy and are close to the geometrical size of the ion-core. It was

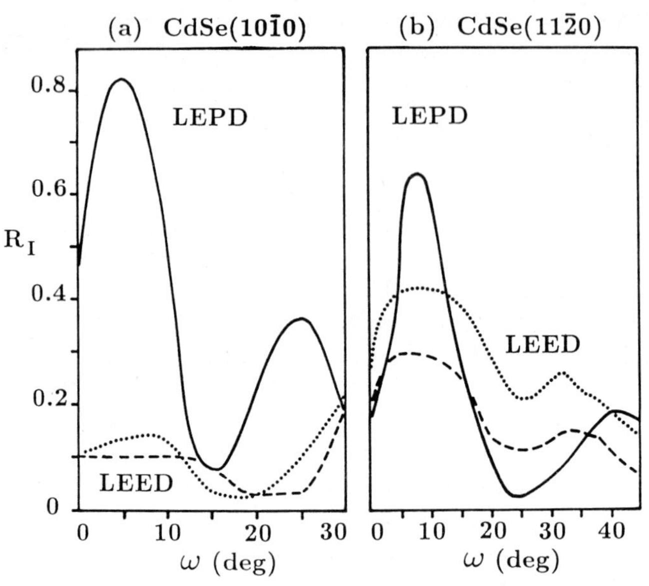

Fig. 9.14a,b. Integrated beam R-factor R_I as a function of ω for CdSe. Same key as for Fig. 9.13 [9.18]

this similarity of the scattering of positrons from Cd and Se that was proposed by *Horsky* et al. as an explanation for the LEPD R_I values having the enhanced sensitivity to ω shown in Fig. 9.14. Electrons, on the other hand, exhibit attractive ion core potentials which cancel the effect of the angular-momentum barrier leading to attractive partial-wave potentials at intermediate distances for angular momentum quantum numbers $L < 3$. Exemplary calculations of positron and electron partial-wave potentials for Cd and Se carried out by Duke and Lessor are shown in Fig. 9.15. For electrons these potentials generate scattering resonances, the energies of which depend on the atomic number, as illustrated in Fig. 9.16. As evident from this figure, these resonances for Cd and Se in turn lead to significant variations in the scattering factors for atoms in different rows of the periodic table, Cd and Se in particular. In addition, Duke and Lessor find that the energies of these electron-ion core scattering resonances depend upon the model used for the exchange-correlation potentials, (e.g., *Hara* [9.64, 65] as opposed to *Kohn-Sham* [9.64, 66]), thereby rendering the electron atomic scattering factors more sensitive to the details of the model than the positron scattering factors. Since the calculated LEED R factors are known to depend significantly on the model potential [9.6–9, 64, 67], it is most likely that the improved R factors in LEPD result from a better description of the positron-ion-core scattering.

Other factors also influence the differences between LEED and LEPD, as discussed in Sect. 9.2.2. For example, the nearly zero real part of the inner potential reduces the uncertainties inherent in the boundary conditions at the solid vacuum interface [9.68, 36]. Also, at energies below 50 eV the differences in the energy-loss processes between electrons and positrons can exert an impor-

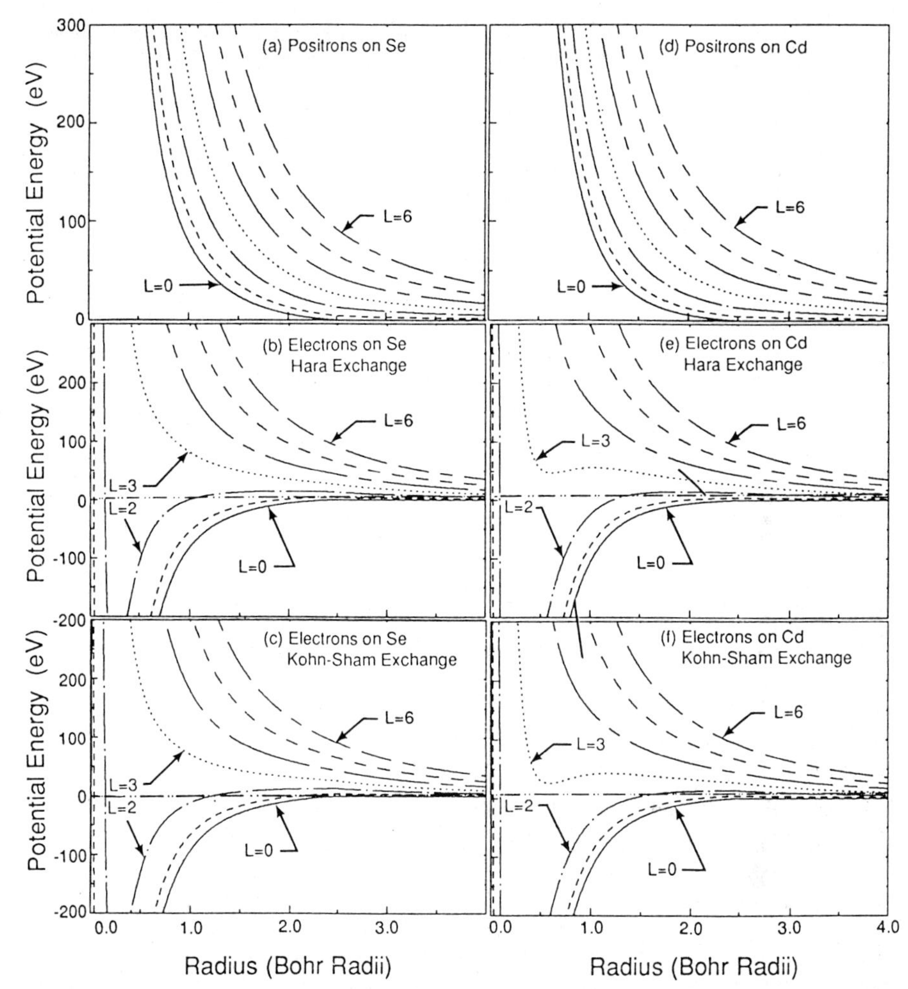

Fig. 9.15a–f. Calculated partial wave-potentials $V_L(r)$ for electrons and positrons from muffin-tin ion cores of CdSe. (**a**) The positron potential of Se. (**b**) The electron potential of Se using Hara exchange and E=100 eV. (**c**) The electron potential of Se using Kohn-Sham exchange. (**d**) The positron potential of Cd. (**e**) The electron potential of Cd using Hara exchange and E = 100 eV. (**f**) The electron potential of Cd using Kohn-Sham exchange. The angular momentum quantum number is designated by L in the figure [9.63]

tant effect on the diffracted intensities [9.7–9]. For the energy range used in the surface-structure analysis ($40\,\text{eV} \lesssim E \lesssim 200\,\text{eV}$), however, the improved description of positron relative to electron ion-core scattering is proposed by Duke and Lessor to be the most important factor in the improved theoretical description of LEPD relative to LEED intensities.

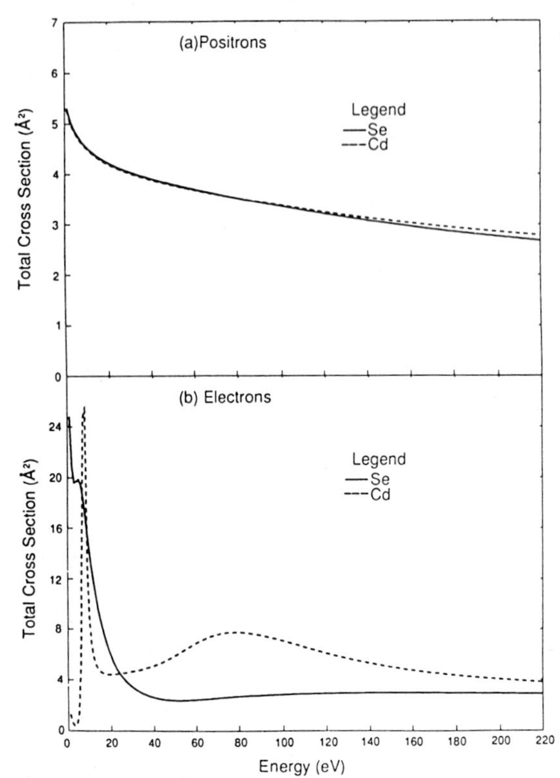

Fig. 9.16a,b. Total Cd and Se scattering cross sections for (**a**) positrons and (**b**) electrons with Hara exchange [9.63]

9.5 Positron Diffraction at Very Low Energy

While it is not often possible to detect the lowest order Bragg peak using LEED, the positron inner potential is only of the order of few eV, and, thus, it is possible to explore diffraction effects at low energies using externally implanted positrons. In addition, a small inner potential makes the existence of surface resonances more likely.

9.5.1 Threshold Effects in LEPD from NaF and LiF

Just below the threshold energy for the emergence of a new diffracted beam, one might have the conditions for populating a traveling-wave surface state [9.69]. After many years of study, there is general agreement that various grazing emergence effects observed for electron scattering from metallic surfaces are caused by multiple scattering interference of waves that are not necessarily confined to the surface [9.70]. Since positrons have only a small or even negative affinity for many solids [9.5, 71], true surface resonances might be expected to be a common occurence. In a recent experiment, *Horsky* et al. [9.72] measured the energy dependence of the intensity of a positron beam specularly reflected from

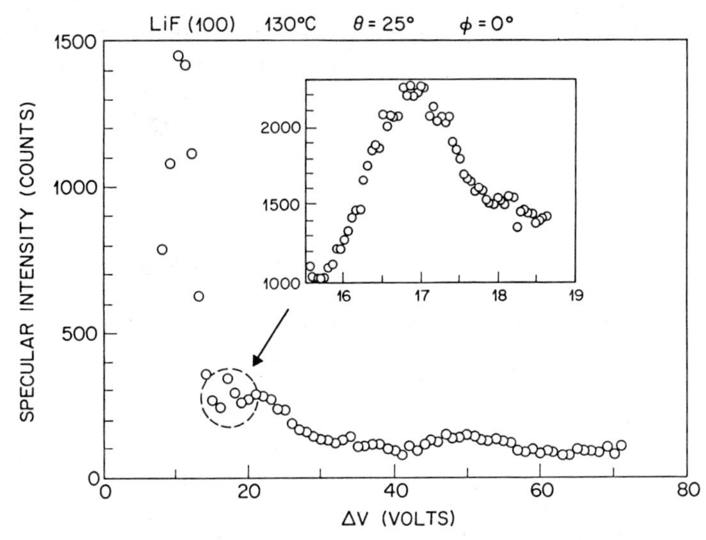

Fig. 9.17. Intensity of the specular (00) beam for positrons scattered from LiF{100} at a 25° angle of incidence. δV is the bias voltage between the W{110} positron moderator and the sample. The inset shows a detail of the narrow peak. The data have not been normalized or corrected for background [9.72]

the {100} surfaces of NaF and LiF. A narrow peak was observed near the $(\bar{1}\bar{1})$ threshold as shown in Fig. 9.17. The location of the peak energy decreases with increasing angle of incidence in qualitative agreement with what one would expect for a beam threshold effect. However, the dispersion shown in Fig. 9.18

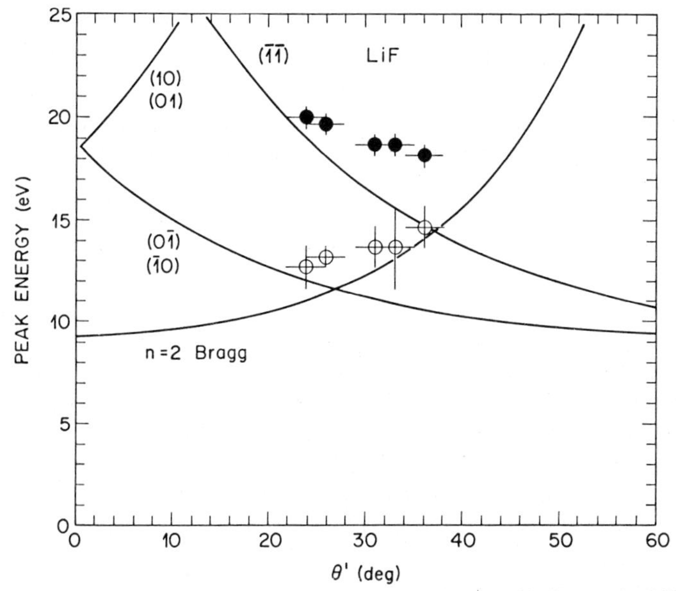

Fig. 9.18. Energy of the narrow peak vs corrected angle of incidence for LiF [9.72]

appears to depart from the expected energy versus angle trajectory parallel to the kinematic threshold. There is also no evidence for the expected Rydberg series of fine structure peaks that should accompany a true surface resonance. Further work will be required to map the dispersion with improved resolution and to identify with certainty the origin of the narrow peak.

9.5.2 Lowest Order LEPD Bragg Peak and the Darwin Top Hat

Since the bottom of the positron band in a solid is typically within a few eV of the vacuum level, the lowest order Bragg peak will be visible in the specularly reflected beam. In Fig. 9.19 the $n = 1$ peak for positrons back-reflected from three faces of single crystal aluminum can be seen [9.73]. The peak positions are shifted to lower energy partly because of the electronic contribution to the effective mass [9.74], and the width of the peaks is principally due to the positron energy gap in the periodic solid, with some contribution due to inelastic effects.

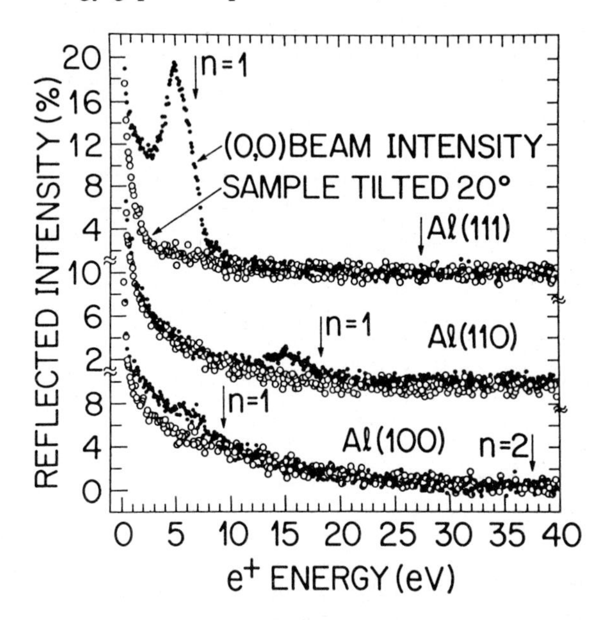

Fig. 9.19. Fractions of positrons back-reflected from Al samples. The arrows show the positions of Bragg peaks calculated assuming a weak potential and positron effective mass equal to m_e [9.73]

Positron Bragg reflection from the rare gas solids exhibits with textbook simplicity the lowest energy gap for a particle in a solid. The nearly perfect reflection of positrons over a range of energies below the threshold for the inelastic scattering results in a Bragg peak profile having the classic shape of the "Darwin top hat" [9.75]. The width and position of the peak allow us to determine the positron band gap and inner potential with precision. Figure 9.20 shows the positron reflection probability versus positron energy for the $\{111\}$ surfaces of four rare gas solids [9.76]. The arrows indicate the inner potentials deduced under the assumption that the electronic contributions to the positron effective

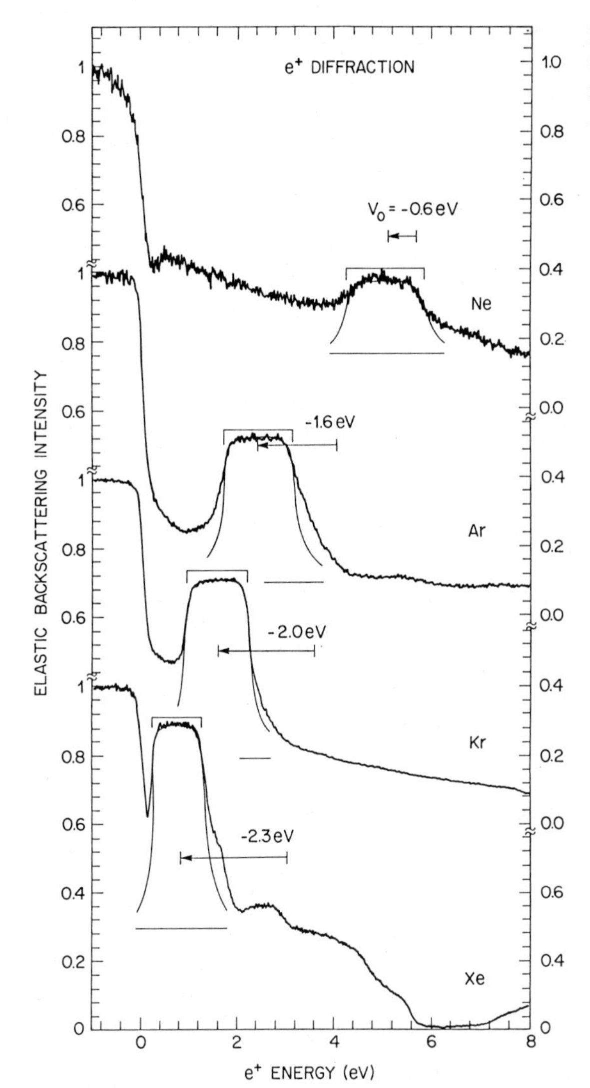

Fig. 9.20. Positron specular reflection probability vs positron energy for the {111} surfaces of solid Ne, Ar, Kr, and Xe

mass are negligible. The departure of the top hats from their ideal shape can be used to estimate that the positron mean free path is a few hundred angstroms at the Bragg energies. Ionic crystals [9.77] also exhibit the Darwin top hat effect, but the shape of the peak (Fig. 9.21) implies that the inelastic potential is much larger than for the rare gas solids, possibly due to the strong coupling to the optical phonons.

A final example of the use of very low energy positron diffraction is in the determination of the inner potential of positrons in graphite [9.78]. Plotting the positron Bragg peak energies versus the square of the index number gives an estimate for the inner potential $V_0 = -1.4\,\mathrm{eV}$. The data plotted on a momentum scale that includes the shift due to the inner potential (Fig. 9.22) shows the flat-

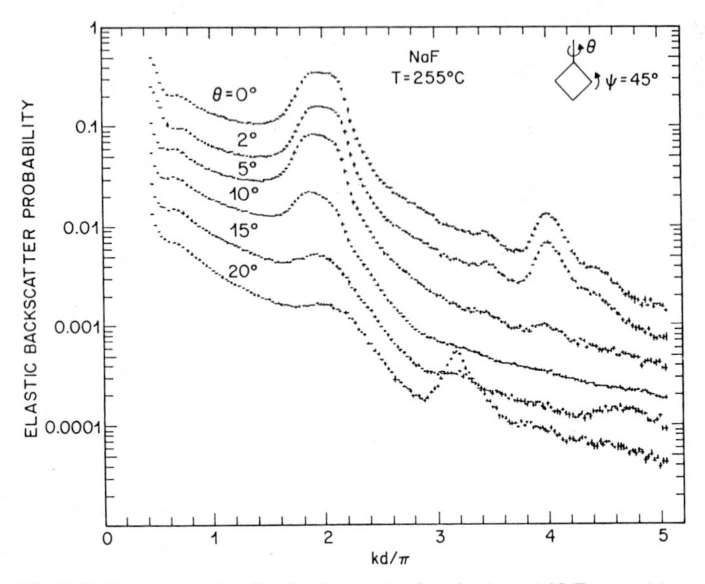

Fig. 9.21. Positron back reflection intensities for air-cleaved NaF rotated by various amounts about a $\{110\}$ axis. The curves are separated for display by powers of two. k, the incident wavevector, is corrected for an inner potential $V_0 = -0.5\,\text{eV}$ and $d = 4.6\,\text{Å}$ [9.77]

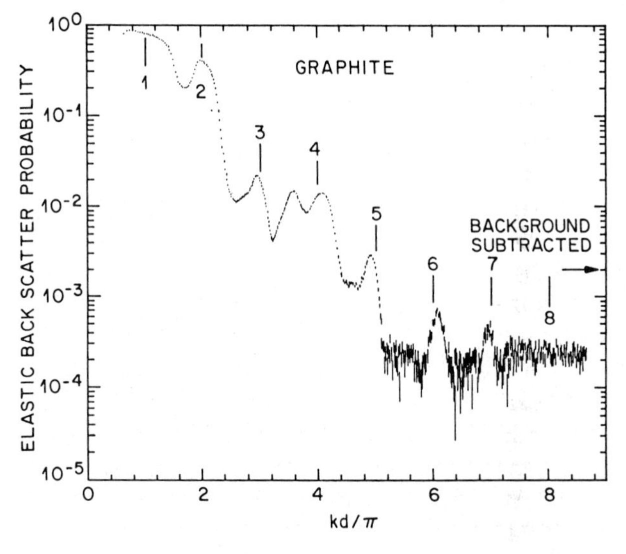

Fig. 9.22. Positron elastic backscattered intensity plotted vs positron momentum k in the crystal for a graphite $\{0001\}$ surface. A positron inner potential of $-1.4\,\text{eV}$ was used to compute k and $\text{d} = 3.29\,\text{Å}$. The integer markers denote the location of the calculated Bragg peaks, corrected for $V_0 = -1.4\,\text{eV}$ [9.78]

tened top hat shape of the lowest two peaks. The higher order peak positions are not in exact agreement with the simplistic interpretation, and there seems to be an half-order peak between $n = 3$ and $n = 4$. The exact shape of the diffraction curve could presumably be understood by a multiple scattering calculation. Nevertheless, the inner potential value was useful in deducing that positronium emission from graphite seems to occur by a phonon-assisted mechanism [9.79].

9.6 Conclusion

In the brief ten years of LEPD development, numerous differences between LEPD and LEED have been observed. These differences not only raise questions of fundamental interest regarding the physics of low energy diffraction but could also be viewed as raising the question: Does LEPD offer a more reliable structural determination than LEED? At present, a definitive answer would be premature. The early speculations pertaining to metals have yet to be subjected to experimental scrutiny with modern LEPD studies. The CdSe LEPD/LEED results suggest that LEPD provides a more reliable structure determination for the CdSe cleavage faces based on the improved R_x values, but the question of the reduced sensitivity of R_x to variations in ω has to also be addressed before this matter is fully resolved. Future LEPD studies carried out on other systems, i.e., metals, semiconductors and overlayers are now well warranted in view of the points discussed in this review. With the 1–2 orders of magnitude higher positron fluxes that can be anticipated in the near future, LEPD will take its place as a valuable complement to LEED for quantitative surface structure determination.

Acknowledgements. We gratefully acknowledge fellow researchers who collaborated in our individual and joint programs and who are indicated in the references pertaining to these programs. We particularly wish to acknowledge the collaboration of T. N. Horsky, A. Kahn, and D. L. Lessor in the CdSe studies. We would also like to thank F. Jona for providing the computer generated curves in Fig. 9.4. This work is supported in part by the National Science Foundation Grant #DMR-8820345. Batelle Memorial Institute is operated for the Department of Energy under Contract DE-AC06-76RLO 1830.

References

9.1 C. Davisson, L. H. Germer: Phys. Rev. **30**, 705 (1927)
9.2 C. D. Anderson: Science **76**, 238 (1932)
9.3 I. J. Rosenberg, A. H. Weiss, K. F. Canter: Phys. Rev. Lett. **44**, 1139 (1980)
9.4 J. M. Jauch, F. Rohrlich: *The Theory of Photons and Electrons* (Addison-Wesley, Reading, MA 1959)
9.5 For a review see: K. G. Lynn, P. J. Schultz: Rev. Mod. Phys. **60**, 701 (1988)
9.6 T. N. Rhodin, S. Y. Tong: Phys. Today **28 (10)**, 23 (1975)
9.7 C. B. Duke: Crit. Rev. Solid State Sci. **4**, 371 (1974)
9.8 J. B. Pendry: *Low Energy Electron Diffraction* (Academic, New York 1974)
9.9 C. B. Duke: Adv. Chem. Phys. **27**, 1 (1974)
9.10 M. B. Webb, M. G. Lagally: Solid State Phys. **28**, 301 (1973)
9.11 G. E. Laramore, C. B. Duke: Phys. Rev. B **2**, 3358 (1972)
9.12 R. J. Meyer, C. B. Duke, A. Paton, J. L. Yeh, J. C. Tsang, A. Kahn, P. Mark: Phys. Rev. B**21**, 4740 (1980)
9.13 D. W. Jepsen, P. M. Marcus, F. Jona: Phys. Rev. B**5**, 3933 (1972)
9.14 P. M. Marcus, F. Jona: Appl. Surf. Sci. **11/12**, 20 (1982)
9.15 M. A. Van Hove, S. Y. Tong: *Surface Crystallography by LEED* Springer Ser. Chem. Phys., Vol. 2 (Springer, Berlin, Heidelberg 1979)
9.16 A. H. Weiss, I. J. Rosenberg, K. F. Canter, C. B. Duke, A. Paton: Phys. Rev. B**27**, 867 (1983)

9.17 C. B. Duke, D. L. Lessor, T. N. Horsky, G. Brandes, K. F. Canter, P. H. Lippel, A. P. Mills, Jr., A. Paton, Y. R. Wang: J. Vac. Sci. Technol. A**7**, 2030 (1989)
9.18 T. N. Horsky, G. R. Brandes, K. F. Canter, C. B. Duke, S. F. Horng, A. Kahn, D. L. Lessor, A. P. Mills, Jr., A. Paton, K. Stevens, K. Stiles: Phys. Rev. Lett. **62**, 1876 (1989)
9.19 F. Jona, D. W. Jepsen, P. M. Marcus, I. J. Rosenberg, A. H. Weiss, K. F. Canter: Solid State Commun. **36**, 957 (1980)
9.20 R. Mayer, C-S. Zhang, K. G. Lynn, W. E. Frieze, F. Jona, P. M. Marcus: Phys. Rev. B**35**, 3102 (1987)
9.21 J. J. Lander: Prog. Solid State Chem **2**, 26 (1965)
9.22 P. Mark, S. C. Chang, W. F. Creighton, B. W. Lee: Crit Rev. Solid State Sci. **5**, 189 (1975)
9.23 A. R. Lubinsky, C. B. Duke, B. W. Lee, P. Mark: Phys Rev. Lett. **36**, 1058 (1976)
9.24 C. B. Duke: Crit Rev. Solid State. Mater. Sci. **8**, 69 (1978)
9.25 F. Jona: J. Phys. C: Solid State **11**, 4271 (1978)
9.26 P. Mark, A. Kahn, G. Cisneros, M. Bonn: Crit. Rev. Solid State Mater Sci. **8**, 317 (1979)
9.27 A. Kahn: Surf. Sci. Repts. **3**, 195 (1983)
9.28 C. B. Duke: In *Surface Properties of Electronic Materials*, ed. by D. A. King, D. P. Woodruff (Elsevier, Amsterdam 1988) pp. 69-118
9.29 C. B. Duke, A. Paton: Surf. Sci. **164**, L797 (1985)
9.30 C. B. Duke: Appl. Surf. Sci. **11/12**, 1 (1982)
9.31 J. M. Dale, L. D. Hullet, S. Pendgala: Surf. Interface Anal. **2**, 199 (1980)
9.32 A. P. Mills, Jr.: Appl. Phys. **23**, 189 (1980)
9.33 Alex H. Weiss: Ph.D. Dissertation, Brandeis University, 1982
9.34 R. H. Ritchie: Phys. Rev. **114**, 644 (1959)
9.35 J. Oliva: Phys. Rev. B **21**, 4909 (1980)
9.36 G. E. Laramore, C. B. Duke, N. O. Lipari: Phys. Rev. B**10**, 2246 (1974)
9.37 M. N. Read, D. N. Lowy: Surf. Sci. **107**, L313 (1981)
9.38 R. Feder: Sol. St. Commun. **34**, 541 (1980)
9.39 P. W. Zitzewitz, J. Van House, A. Rich, D. W. Gidley: Phys. Rev. Lett. **43**, 1281 (1979)
9.40 K. G. Lynn, J. R. MacDonald, R. A. Boie, L. C. Feldman, J. D. Gabbe, M. F. Robbins, E. Bonderup, J. Golovchenko: Phys. Rev. Lett. **38** 241 (1977)
9.41 E. Bonderup, J. U. Anderson, D. N. Lowy: Phys. Rev. B**20**, 883 (1979)
9.42 M. N. Read, G. J. Russel: Surf. Sci. **88**, 95 (1979)
9.43 P. A. Sturrock: *Static and Dynamic Electron Optics* (Cambridge University Press, London 1955)
9.44 G. R. Brandes, K. F. Canter, T. N. Horsky, P. H. Lippel, A. P. Mills, Jr.: Rev. Sci. Instrum. **59**, 228 (1988)
9.45 G. R. Brandes, K. F. Canter, A. P. Mills, Jr.: Phys. Rev. Lett. **61**, 492 (1988)
9.46 D. M. Chen, K. G. Lynn, R. Pareja, Bent Nielsen: Phys. Rev. **31**, 4123 (1985)
9.47 P. J. Schultz, E. M. Gullikson, A. P. Mills, Jr.: Phys. Rev. B**34**, 442 (1986)
9.48 K. G. Lynn, Alan Wachs: Appl. Phys. A**29**, 93 (1982)
9.49 K. F. Canter: *Positron Scattering in Gases*, ed. by J. W. Humberston, M.R. C. McDowell (Plenum, London 1984) pp. 219-225
9.50 W. E. Frieze, D. W. Gidley, K. G. Lynn: Phys. Rev. B **31**, 5628 (1985)
9.51 K. F. Canter, G. R. Brandes, T. N. Horsky, P. H. Lippel, A. P. Mills, Jr.: in *Atomic Physics with Positrons*, ed. by J. W. Humberston, E. A. G. Armour (Plenum, New York 1987) pp. 153-160
9.52 R. Mayer, C. S. Zhang, K. G. Lynn, J. Throwe, P. M. Marcus, D. W. Gidley, F. Jono: Phys. Rev. B **36**, 5659 (1987)
9.53 C. B. Duke, A. R. Lubinsky, S. C. Chang, B. W. Lee, P. Mark: Phys. Rev. B**15**, 4865 (1977)
9.54 C. B. Duke, R. J. Meyer, A. Paton, P. Mark: Phys. Rev. B**18**, 4255 (1978)
9.55 C. B. Duke, A. Paton, Y. R. Wang, K. Stiles, A. Kahn: Surf. Sci. **197**, 11 (1988)
9.56 C. B. Duke, Y. R. Wang: J. Vac. Sci. Technol. B**6**, 1440 (1988)

9.57 P. C. Stair: Rev. Sci. Instrum. **51**, 132 (1980)
9.58 E. Zannazi, F. Jona: Surf. Sci. **62**, 61 (1977)
9.59 C. B. Duke, S. L. Richardson, A. Paton: Surf. Sci. **127**, L135 (1983)
9.60 C. B. Duke, A. Paton, W. K. Ford, A. Kahn, J. Carelli: Phys. Rev. B**24**, 562 (1981)
9.61 Y. R. Wang, C. B. Duke: Phys. Rev. B**37**, 6417 (1988)
9.62 T. N. Horsky: PhD Thesis, Brandeis University (1988)
9.63 C. B. Duke, D. L. Lessor: submitted to Surf. Sci.
9.64 R. J. Meyer, C. B. Duke, A. Paton: Surf. Sci. **97**, 512 (1980)
9.65 S. Hara: J. Phys. Soc. (Japan) **22**, 710 (1967)
9.66 W. Kohn, L. J. Sham: Phys. Rev. **140**, A1133 (1965)
9.67 W. K. Ford, C. B. Duke, A. Paton: Surf. Sci. **112**, 195 (1982)
9.68 C. B. Duke, N. O. Lipari, U. Landman: Phys. Rev. B**8**, 2454 (1973)
9.69 E. G. McRae: Rev. Mod. Phys. **51**, 541 (1979)
9.70 C. Gaubert, R. Baudoing, Y. Gauthier: Surf. Sci. **147**, 162 (1984)
9.71 A. P. Mills, Jr.: In *Positron Solid State Physics*, ed. by W. Brandt, A. Dupasquier (North-Holland, Amsterdam 1983) p. 432
9.72 T. N. Horsky, G. R. Brandes, K. F. Canter, P. H. Lippel, A. P. Mills, Jr.: Phys. Rev. B**40**, 7898 (1989)
9.73 A. P. Mills, Jr., P. M. Platzman: Solid State Commun. **35**, 321 (1980)
9.74 D. R. Hamann: Phys. Rev. **146**, 277 (1966)
9.75 C. G. Darwin: Philos. Mag. **27**, 675 (1914)
9.76 E. M. Gullikson, A. P. Mills, Jr., E. G. McRae: Phys. Rev. B**37**, 588 (1988)
9.77 A. P. Mills, Jr., W. S. Crane: Phys. Rev. B**31**, 3988 (1985)
9.78 E. M. Gullikson, A. P. Mills, Jr.: Phys. Rev. B**36**, 8777 (1987)
9.79 P. Sferlazzo, S. Berko, K. G. Lynn, A. P. Mills, Jr., L. O. Roellig, A. J. Viescas, R. N. West: Phys. Rev. Lett. **60**, 538 (1988)

10. Time-of-Flight Scattering and Recoiling Spectrometry (TOF-SARS) for Surface Analysis

O. Grizzi, M. Shi, H. Bu, and J.W. Rabalais

Department of Chemistry, University of Houston,
Houston, TX 77204–5641, USA

Low energy (< 10 keV) ion scattering spectrometry [10.1] is becoming increasingly important as a surface analysis technique in three specific areas, i.e., surface elemental analysis [10.2–4], probing surface structure [10.5–16], and studying electronic transition probabilities [10.7, 7–19] between ions or atoms and surfaces. This is largely due to the following recent advances: (i) impact collision ion scattering spectrometry [10.6] (ICISS) in which the scattering angle is close to 180°, thus simplifying the scattering geometry and allowing experimental determination of the shadow cone radii, (ii) the use of alkali primary ions [10.9, 10] which have low neutralization probabilities, leading to higher scattered ion fluxes, (iii) time-of-flight (TOF) techniques [10.20–23] with detection of both neutrals and ions in a multichannel mode in order to enhance sensitivity, (iv) scattered ion fractions [10.7, 17] to probe the spatial distributions of electrons, and (v) the use of recoiling [10.24, 25] to determine the structure of light adsorbates on surfaces.

This chapter describes the technique of *time-of-flight ion scattering and recoiling spectrometry* (TOF-SARS). The technique is a nondestructive method of analysis due to collection of both neutrals plus ions in a multichannel TOF mode. It is sensitive to all elements (including hydrogen) due to the combination of scattering and recoiling with continuous angular variation. It can determine surface and adsorbate structures to an accuracy of < 0.1 Å, and it can delineate between structural effects and electronic exchange (neutralization) effects by collection of neutrals plus ions and neutrals only.

TOF-SARS has the following characteristics for surface analysis:

(i) *Elemental Analysis*. All elements can be analyzed by either scattering, recoiling, or both techniques, TOF peak identification is straightforward using classical scattering expressions, and collection of neutrals plus ions results in scattering and recoiling intensities that are determined by elemental concentrations, shadowing and blocking effects, and classical cross sections.

(ii) *Structural Analysis*. It provides "real space" information on the relative positions of all atoms in the surface region, including hydrogen, based on simple classical concepts, i.e., shadowing and blocking cones. The cone dimensions can be calculated and calibrated from known interatomic spacings and the analyses are not complicated by ion neutralization effects.

(iii) *Ion-Surface Electronic Transition Probabilities*. Since the scattered and recoiled ion fractions can be directly measured, electron exchange probabilities as a function of ion energy and type, surface type, crystallographic orientation and chemical state, and angle of incidence, ejection, scattering, and recoiling can be determined.

The chapter is organized as follows. Section 10.1 presents a brief historical review of the use of low energy ion scattering in surface structure determinations, for this is one of the most important applications of the technique. The experimental procedure is described in Sect. 10.2. Section 10.3 presents some of the capabilities and applications of TOF-SARS as follows: TOF spectra as a function of scattering angles, θ, and recoiling angles, ϕ; Comparison to LEED and AES; Surface structure determinations; Analysis of surface roughness; Surface semichanneling effects; Scattered ion fraction measurements; and Ion induced Auger electron measurements.

10.1 Historical Review

The technique of low energy scattering and recoiling is an outgrowth of conventional ion scattering spectrometry (ISS). In 1967, *Smith* [10.26] recognized that the scattering of low energy ions from surfaces was well-described by the binary collision approximation and that analysis of such spectra provided a sensitive surface elemental analysis. The technique did not reach the popularity level of the more conventional surface analysis methods, e.g., AES and XPS, because of its destructive nature stemming from the analysis of only scattered ions. In 1982, *Aono* et al. [10.5] proposed a technique called impact collision ion scattering spectroscopy (ICISS) in which θ is close to 180°, ca. 165°. This greatly simplified the scattering geometry and allowed experimental determination of shadow cone radii. These known shadow cones can be applied to surfaces with unknown or reconstructed structures in order to determine surface geometry. There are two difficulties with this technique: (i) It analyzes only the scattered ions; these are typically only a very small fraction ($< 5\%$) of the total scattered flux. Thus, high primary ion doses are required for spectral acquisition which are potentially damaging to the surface and adsorbate structures. (ii) Neutralization probabilities tend to increase as the ion-beam incidence angle with respect to the surface decreases. This is not a simple behavior since the probabilities depend on the distances of the ion to specific atoms. As a result, this can attenuate the features that arise from specific geometrical arrangements. In 1984 *Niehus* [10.9, 10] proposed the use of alkali primary ions which have low neutralization probabilities, leading to higher intensities and pronounced focusing effects. The contamination of the sample surface by the reactive alkali ions is a potential problem with this method. *Buck* et al. [10.8] used TOF methods for surface structure analysis in 1984 and demonstrated the capabilities and high sensitivity of the technique when both neutrals and ions are detected. In 1987, *van Zoest* et al. [10.11] demonstrated that TOF analysis of both the scattered and recoiled neutrals and ions provided much more information about surface structure. However, they did not have adequate resolution in the TOF mode to separate the scattered and recoiled particles. In 1989, *Rabalais* et al. [10.27,28] presented a TOF spectrometer system with sufficient resolution to separate scattered and recoiled particles and demonstrated its capabilities on clean [10.29] and adsorbate covered [10.30,31] surfaces.

Structure determinations for clean and adsorbate covered surfaces by low energy ion scattering are collected in Table 10.1. The table indicates whether scattering and/or recoiling was used and whether the particles detected were ions and/or neutrals. In these systems we note that there are only five cases where scattered plus recoiled particles and ions plus neutrals have been measured in the same experiment. Many of the structure determinations listed in Table 10.1

Table 10.1. Structural studies by low energy ion scattering

System	Adsorbates	Scattering and/or Recoiling	Particles detected: Ions (*I*) and/or Neutrals (*N*)	Ref.
Au{110}	–	S	I	[10.32]
Au{110}	–	S	I	[10.33]
Au{110}	–	S	I+N	[10.34]
Cu{001}	–	S	I	[10.35]
Cu{100} & {410}	–	S	I	[10.36]
Cu{100}	O_2	S+R	I	[10.37,38]
Cu{110}	–	S	I	[10.10]
Cu{110}	O_2	S+R	I	[10.39]
Cu{110}	O_2	S	I	[10.40,41]
Cu{410}	–	S	I	[10.42]
Cu{410}	O_2	S	I	[10.43]
Ir{110}	–	S	I	[10.44]
Ni{001}	–	S	I	[10.45]
Ni{001}	–	S	I	[10.46]
Ni{001}	S	S	I	[10.47]
Ni{100}	O_2,C	S	I	[10.48]
Ni{100}	O_2	S	I	[10.49,50]
Ni{110}	O_2	S	I	[10.51,52]
Ni{110}	S	S	I	[10.47]
Ni{110}	O_2	R	I	[10.53]
Ni{110}	O_2	S	I	[10.54]
Ni{111}	O_2,CO	S	I	[10.55]
Ni{111}	S	S	I	[10.47]
Fe{001}	–	S	I+N	[10.8]
Fe{100}	O_2	S+R	I+N	[10.11]
Pt{110}	–	S	I	[10.9,56]
Pt{111}	–	S	I	[10.57]
Pt{111}	–	S	I+N	[10.58]
Pt{111}	H_2	S+R	I	[10.25]

Table 10.1. (cont'd)

System	Adsorbates	Scattering and/or Recoiling	Particles detected: Ions (*I*) and/or Neutrals (*N*)	Ref.
Pd{110}	H_2	S	I+N	[10.59]
Si{001} & {111}	–	S	I	[10.60]
Si{001}	Ag	S	I+N	[10.61]
Si{100}	O_2,H_2O	S+R	I+N	[10.62]
Si{100}	–	S	I	[10.5]
Si{111}	Ag	S	I	[10.63]
Si{111}	O_2,CO	S	I	[10.64]
Si{111}	Ni	S	I	[10.65]
Ag{110}	O_2	S	I	[10.66]
W{100}	O_2	S	I	[10.12,67]
W{110}	O_2	S	I	[10.67]
W{110}	O_2	S	I	[10.68]
W{211}	–	S	I+N	[10.27,29]
W{211}	O_2	S+R	I+N	[10.27,30]
W{211}	H_2	S+R	I+N	[10.28,31]
W{211}	H_2	S	I	[10.69]
Cu_3Au	–	S	I+N	[10.70]
TiC{001}	O_2	S	I	[10.71,72]
UO_2{100}	O_2	S	I	[10.73]
ZnS{111}	–	S	I	[10.74]
CdS{001}	–	S	I	[10.75]
MgO{100}	–	S	I	[10.76]

used approximate model potentials to determine the shadow cone dimensions; this makes the structure determination model dependent and indirect. As a result, only some of these studies were capable of obtaining quantitative results. This problem can be avoided by calibrating the shadow cone calculations [10.29] against known interatomic distances.

10.2 Experimental Method

10.2.1 TOF-SARS Spectrometer

A schematic drawing of the TOF-SARS instrument is shown in Fig. 10.1 and the pulsed ion beam and timing electronics are shown in Fig. 10.2. It consists of the following components: (i) large flat vacuum chamber (radius of 1 m), (ii) pulsed primary ion beam, (iii) detector (channel electron multiplier) rotatable through

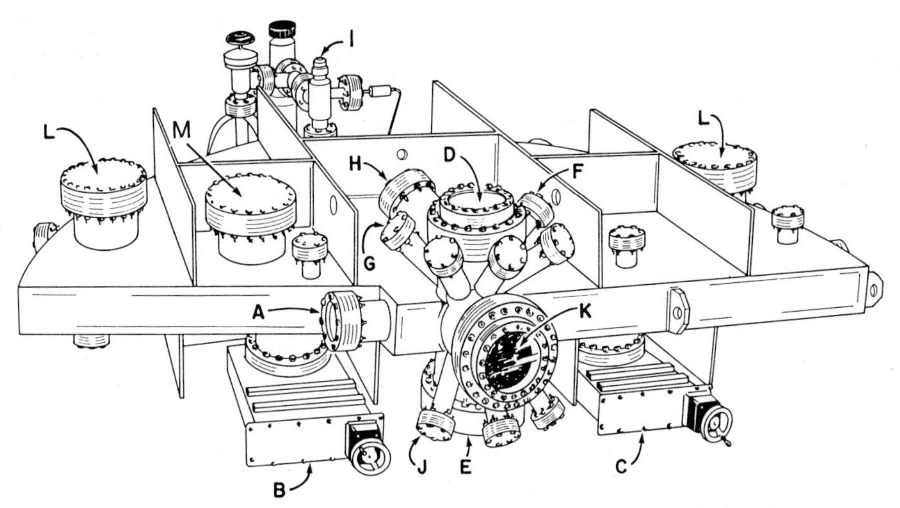

Fig. 10.1. Spectrometer system designed for TOF analysis of neutrals and ions and ESCA analysis of ions that are scattered and recoiled along with conventional AES, XPS, UPS, and LEED analysis. (*A*) pulsed ion beam; (*B*) turbomolecular pump; (*C*) ion pump; (*D*) sample manipulator; (*E*) detector precision rotary motion feedthru; (*F*) x-ray source; (*G*) electron gun; (*H*) 180° electrostatic hemispherical analyzer; (*I*) sorption pumps; (*J*) sputter ion gun; (*K*) viewport or reverse view LEED optics; (*L*) titanium sublimation pump; (*M*) cryopump

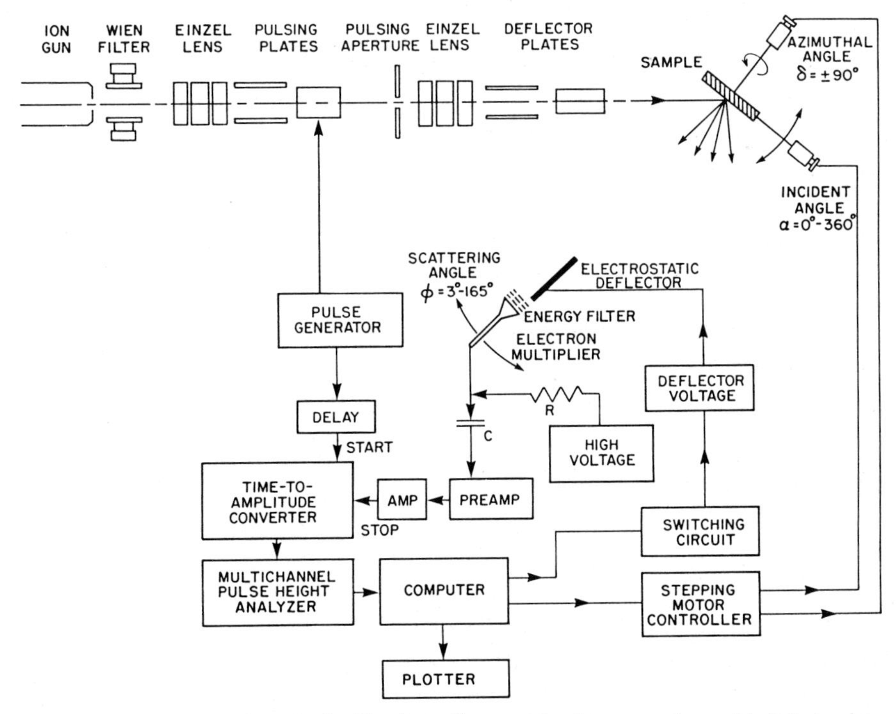

Fig. 10.2. Schematic drawing of pulsed ion beam line, sample, detector, and associated electronics

continuous scattering angles over the range $\theta = 0 - 165°$, (iv) $\sim 1\,\text{m}$ TOF drift region, (v) pulse generating, timing, detection, and control electronics, (vi) and conventional surface analysis techniques (AES, XPS, LEED). The instrument is described in detail elsewhere [10.77]. Typical experimental parameters are as follows: 2–5 keV He^+, Ne^+, or Ar^+ pulsed primary ion beam; pulse width 20–100 ns; pulse rate 10–50 kHz; average current density 0.05–0.1 nA/mm^2; signal detection rate up to $\sim 30,000$ c/s. A TOF spectrum can be acquired in ~ 20 s with a dose of $< 10^{-3}$ ions/target atoms.

10.2.2 Comparison to Rutherford Backscattering (RBS)

The primary difference between TOF-SARS and the independently developing technique of Rutherford backscattering spectrometry (RBS) is that for the former, the beam energy (E) is of the order of keV while in the latter E is of the order of MeV. This gives rise to two important differences. (i) In the low E range, ions are scattered by weak potentials and the radii of shadowing and blocking cones are comparable to interatomic spacings (≈ 1 Å). In the E range of RBS, ions are only scattered by strong potentials whose cone radii are very small (≈ 0.1 Å). (ii) the velocities of ions in the keV range are comparable to or smaller than those of valence electrons, while ions in the MeV range have velocities that are greater than those of valence electrons. As a result, low E ions with high ionization potentials pick up electrons near surfaces and are neutralized with high probability while neutralization of high E ions is negligible. Because of (i) and (ii), low E scattering is extremely sensitive to the first one or two atomic layers of a surface while the sampling depth of RBS is of the order of μm. By using shadow and blocking analysis, low E scattering and recoiling can be used for surface structure determinations whereas RBS is primarily a technique for bulk structural analysis.

10.3 Examples of Experimental Results

10.3.1 TOF Spectra

Example TOF spectra for 3 keV Ar^+ scattering from a Si sample which has some residual oxygen and hydrogen adsorbed on the surface are shown in Fig. 10.3 for several values of the polar beam incident angle α and scattering angle θ. These angles are defined in the inset of Fig. 10.3; for demonstrative purposes they have been adjusted to specular conditions in the range $\theta = 4.5°$-$85°$. The sharp, intense peak observed at low values of α and θ (near grazing incidence and exit) corresponds to Ar projectiles scattered with minor energy loss from combinations of single and multiple collisions with target atoms. In this angular region and for atomically flat surfaces, projectiles moving along azimuths of high

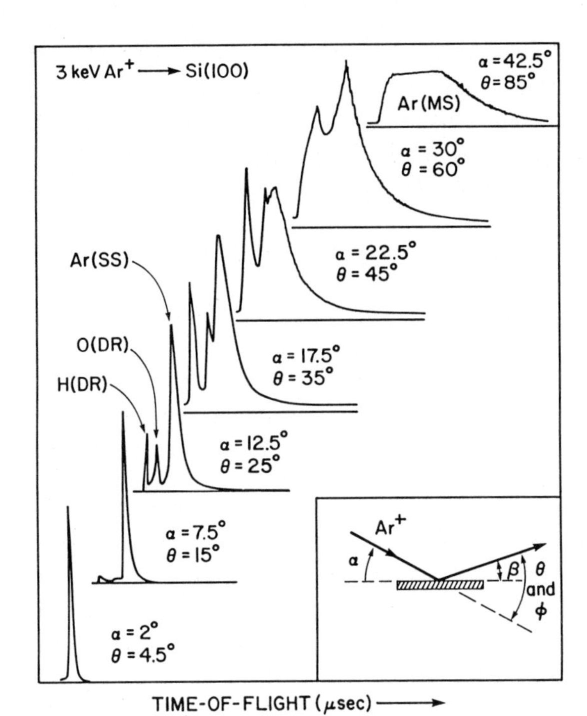

Fig. 10.3. TOF spectra for 3 keV Ar^+ scattering from a Si{100} surface with hydrogen and oxygen adsorbates at several α and θ values

symmetry collide with several target atoms before leaving the surface with exit angles similar to the incident angle. These surface channeling effects are useful for aligning the azimuthal angle of the sample, as will be shown in Sect. 10.3.6. The TOF for scattering of an incident ion of mass M_1 and energy E_0 from a target atom of mass M_2 into an angle θ can be calculated by the binary collision (BC) approximation [10.24, 78] as

$$t_{SS} = \frac{l(M_1 + M_2)}{(2M_1E_0)^{1/2}}\left[\cos\theta + \sqrt{(M_2/M_1)^2 - \sin^2\theta}\right] , \tag{10.1}$$

where l is the distance from the target to detector. For cases where $M_1 > M_2$, there is a critical angle $\theta_c = \sin^{-1}(M_2/M_1)$ above which only multiple scattering can occur. Since $\theta_c = 44.8°$ for Ar → Si collisions, a broad and featureless structure results at large θ as shown in Fig. 10.3.

Recoils ejected in single collisions with the projectile into an angle ϕ, i.e., direct recoils, have a TOF given by [10.24]

$$t_{DR} = \frac{l(M_1 + M_2)}{(8M_1E_0)^{1/2}} \cos\phi . \tag{10.2}$$

Both H and O(DR) peaks are present in the spectra of Fig. 10.3. For low α they appear at the short TOF side of the scattering peak, while for $\alpha > 35°$ the O(DR) and the scattering peak overlap. The angular range where this overlap occurs is determined by (10.1) and (10.2) and by the width of the peaks, which has a nonlinear relation with the recoiled energy E_r (it is proportional to $E_r^{-3/2}$). In general, the best resolution and sensitivity for different projectile energies and types of recoils are obtained experimentally by varying the observation angles, making continuous variation of the scattering angle, θ, a very useful feature in analysis of recoiling particles.

In order to illustrate the dependence of the scattering peak intensities I(S) on θ, 4 keV Ar^+ was scattered from a clean and well annealed (to 2200 K) W{211} single crystal. The incident angle was fixed at 20°, i.e., high enough to avoid shadowing effects [10.27-31], and θ was varied in the range 20°-150°, which corresponds to an exit angle range of $\beta = 0°$-130°. The experimental intensities I(S), obtained as the integrated area of the scattering peak, are shown in Fig. 10.4 together with the scattering cross sections calculated with the BC approximation and normalized to the experiment at θ = 125°. At high θ values both curves show the same angular dependence, indicating that the major contribution to the scattering peak originates from single collisions. In the range $\theta = 28°$-45°, I(S) is higher than the calculated BC cross sections because (i) multiple collisions, not taken into account in the BC approximation, have cross sections that approach those of single scattering collisions and (ii) focusing of the ion trajectories at the edge of the blocking cones cast by neighboring target atoms can considerably enhance the scattered intensities. For $\theta < 30°$ ($\beta < 10°$), there is a sharp decrease of I(S) because the scattered projectiles which have mass M_1 smaller than that of the target atom masses M_2 cannot penetrate the target atom blocking cones.

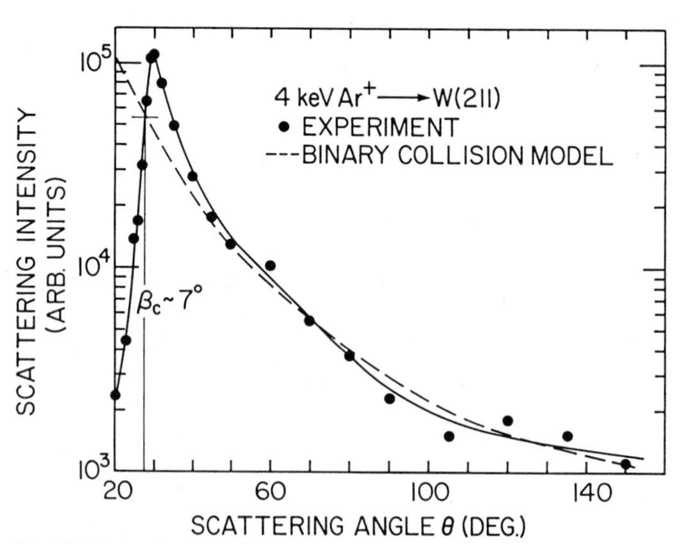

Fig. 10.4. Scattering intensities $I(S)$ as a function of Θ for 4 keV Ar^+ on a clean W{211} surface. The dashed line represents the scattering cross sections calculated from the binary collision approximation and normalized to the experimental values at θ = 125°. The sharp decrease in $I(S)$ for $\theta < 10°$ is due to blocking and β_c is the critical blocking angle

This blocks the exit trajectories of the projectile ions at such grazing angles. This sharp $I(\mathrm{S})$ decrease is a measure of the size of the blocking cones; it can be used to obtain the interatomic distance along the projectile direction.

10.3.2 Comparison of TOF-SARS to LEED and AES

It is highly desirable to have the complementary techniques LEED and AES available in the TOF-SARS instrument. Whereas LEED is capable of directly providing the symmetry of surface and adsorbate site structures (long-range structure), TOF-SARS is capable of directly measuring the interatomic spacings between substrate atoms and between adsorbate-adsorbate and adsorbate-substrate atoms, including hydrogen, (short-range structure) as will be shown in Sect. 10.3.4. Auger electron spectroscopy provides a fast, direct, quantitative analysis of all elements in the surface layers except hydrogen; TOF-SARS is sensitive to all elements, including hydrogen. However, it has difficulty in resolving spectral peaks of high mass elements, whose masses are similar, due to the low resolution of the TOF technique.

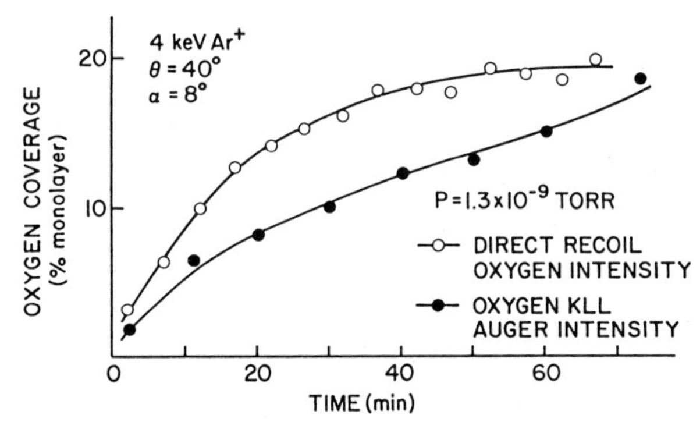

Fig. 10.5. Oxygen direct recoil intensity $I(R)$ and AES KLL intensity, calibrated in terms of oxygen coverage, as a function of exposure time to residual gases at 1.3×10^{-9} Torr

TOF-SARS has extremely high sensitivity to the outermost 1st- and 2nd-atomic layers. This can be observed from Fig. 10.5, in which the oxygen direct recoil intensity $I(\mathrm{R})$ and the AES oxygen KLL intensity on a W$\{211\}$ surface, calibrated in terms of oxygen coverage, are plotted as a function of exposure time to the residual gases of the vacuum chamber at a pressure of 1.3×10^{-9} Torr. $I(\mathrm{R})$ rises sharply and reaches a plateau at an oxygen coverage of $< 20\%$ of a monolayer, whereas the AES intensity has a more gradual slope and has not reached a plateau even after an exposure time of 75 min. Cleanliness of a surface with respect to TOF-SARS is more difficult to achieve than with respect to AES due to the high sensitivity of the former. The sensitivity of TOF-SARS is estimated to be < 0.01 monolayer of surface contamination; the "clean" surface condition was taken to be the case of absence of recoiled hydrogen, carbon, and oxygen signals in the TOF spectra.

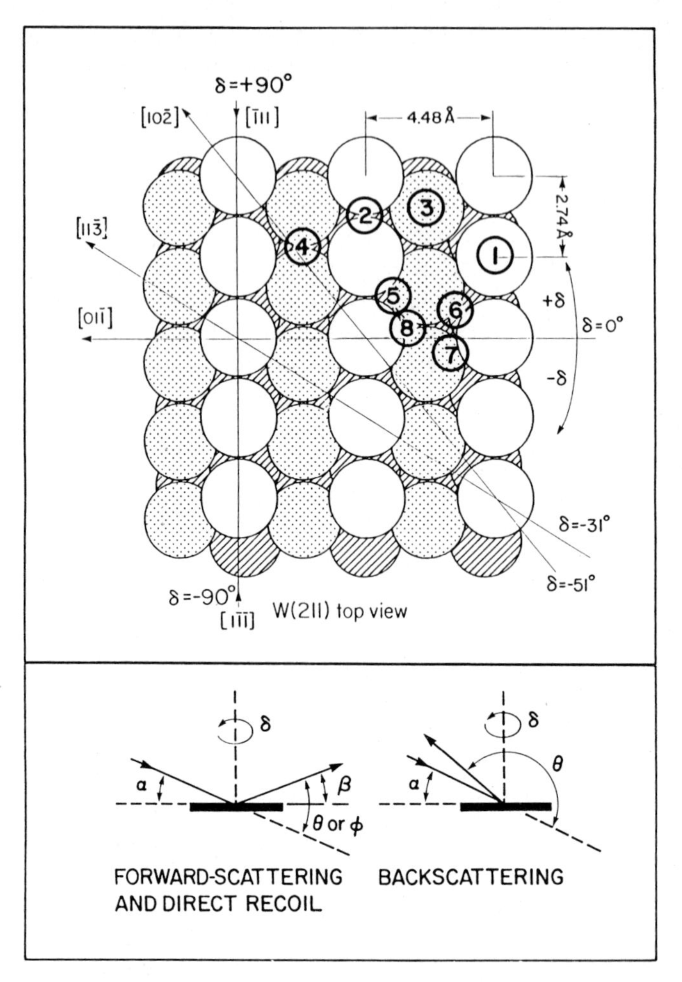

Fig. 10.6. Structure of the bulk truncated W{211} surface along with the definition of angles used herein. (*Open circles*) 1st-layer atoms; (*Dotted circles*) 2nd-layer atoms; (*Dashed circles*) 3rd- and 4th-layer atoms. Geometrically different adsorption sites are indicated

10.3.3 Structure Analysis

The ability to obtain direct information on surface structures from TOF-SARS is one of the most important applications of the instrument. In this section the technique [10.27, 28] is briefly described and examples of analysis of a clean W{211} surface [10.29] and oxygen [10.30] and hydrogen [10.31] chemisorbed on this surface are provided. A schematic drawing of the surface, possible adsorption sites, and the angle definition used herein are shown in Fig. 10.6. Analysis of both N + I scattered I(S) and recoiled I(R) intensities as a function of beam incident angle α and sample azimuthal angle δ provides direct information on surface atomic positions that is independent of the electron exchange (neutralization)

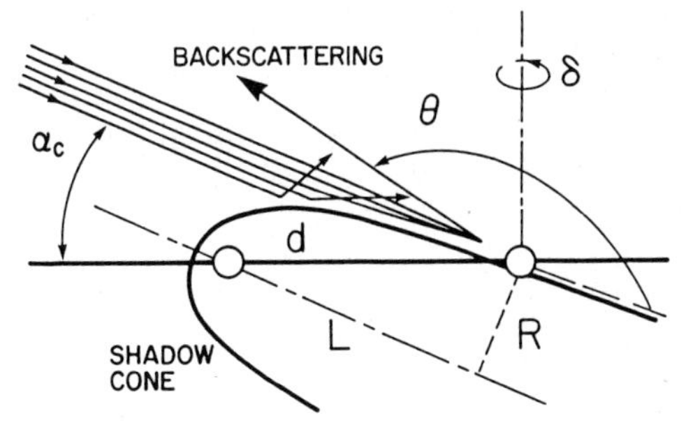

Fig. 10.7. Schematic illustration of backscattering events observed in TOF-SARS along with a representative plot of $I(S)$ vs α along the $[\bar{1}\,1\,1]$ azimuth of W(211) using 4 keV Ar^+

processes discussed in Sect. 10.3.6. Since a dose of $< 10^{-3}$ ions/surface atom is sufficient for spectral acquisition, structural analyses are relatively nondestructive and even analysis of adsorbates as a function of coverage is possible.

When an atomic projectile approaches a target atom, the trajectories are bent such that an excluded volume, i.e., a shadow cone, in the shape of a paraboloid is formed behind the target atom (Fig. 10.7). Since the radii of the cones are of the order of 1 Å, the technique is very sensitive to the outermost atomic layers. When an ion beam is incident on an atomically flat surface at grazing angles, each surface atom is shadowed by its neighboring atom such that only forward scattering (large impact parameter (p) collisions) is possible. As α increases, a critical value $\alpha^i_{c,sh}$ is reached each time the ith-layer of target atoms moves out of the shadow cone allowing for large angle backscattering (BS), i.e., small p collisions, from the specific ith-layer. If the BS intensity I(BS) is monitored as a function of α, steep rises with well defined maxima are observed when the focused trajectories at the edge of the shadow cone pass close to the center of neighboring atoms. These effects are shown in Fig. 10.7 for Ar^+ scattering at $\theta = 163°$ from a W$\{211\}$ surface along the $[1\,\bar{1}\,\bar{1}]$ azimuth. For this particular azimuth, the 1st- and 2nd-atomic layers are exposed to the beam at the same incident angle (see Fig. 10.6) giving rise to the peak observed at $\alpha = 29°$ with $\alpha^1_{c,sh} = 23°$); the underlying 3rd- and 4th-layers start to be accessible to the beam at $\alpha = 60°$, giving rise to the intense peak observed at $\alpha = 79°$. The experimen-

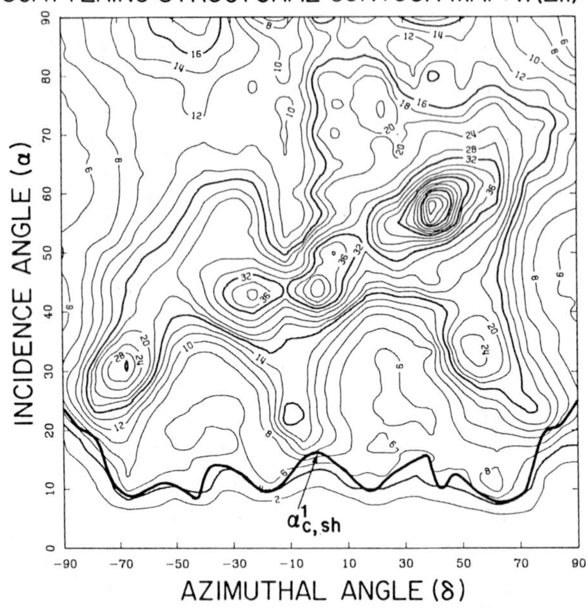

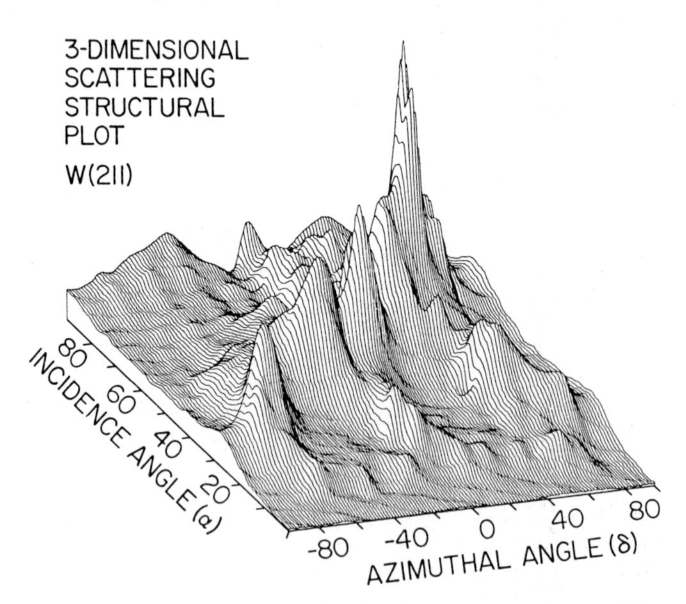

Fig. 10.8. (*Top*) Scattering structural contour map (SSCM) for the clean W(211) surface. Primary ion: 4 keV Ar^+; $\theta = 163°$; $\delta = 0°$ is the $[0\,1\,\bar{1}]$ azimuth; $\delta = -90°$ is the $[\,\bar{1}\,\bar{1}]$ azimuth; $\delta = +90°$ is the $[\bar{1}\,1\,1]$ azimuth. The critical value of α at low angles, $\alpha_{c,sh}$, is plotted as a heavy line. (*Bottom* Three-dimensional scattering structural plot (SSP) for the clean W(211) surface. Viewing directions are $\alpha = 35°$ and $\delta = -20°$

tal parameters used in this measurement were 4 keV Ar^+ primary beam; pulse width of ion beam 50 ns; pulse repetition rate 40 kHz; average current density 0.05 nA/mm^2; TOF path = 98.4 cm; and acquisition time for each spectrum 20 s.

If the shape of the cone [10.78–80], i.e., the radius (R) as a function of distance (L) behind the target atom is known, the interatomic spacing (d) can be directly determined from the I(BS) vs α plots. For example, by measuring $\alpha^1_{c,sh}$ along directions for which specific crystal azimuths are aligned with the projectile direction and using $d = R/\sin\alpha^1_{c,sh}$, interatomic spacings in the 1st-atomic layer can be determined. The 1st-2nd-layer spacing can be obtained in a similar manner from $\alpha^2_{c,sh}$ measured along directions for which the 1st- and 2nd-layer atoms are aligned.

I(BS) as a function of α along different crystal azimuths for $-90° < \delta < +90°$ can be presented in the form of a scattering structural contour map (SSCM) and a three-dimensional scattering structural plot (SSP) as shown in Fig. 10.8. Since the scans were taken at increments of $\alpha = 1°$ or $2°$ and $\delta = 6°$, an interpolation routine was used between the points from adjacent α scans along each δ. The I(BS) values are the experimental intensities multiplied by $\sin\alpha$ to correct for the fact that the area sampled varies as $I_0/\sin\alpha$ as α increases. The $\alpha^1_{c,sh}$ value is also plotted in Fig. 10.8.

The SSCM and SSP provide the following information. (i) They are a concise summary of all of the experimental BS data. (ii) They reveal the symmetry of the I(BS) data in α, δ space, thereby providing a fingerprint for a specific crystal face and type with minor perturbations due to relaxation and possible major perturbations due to reconstruction. The fingerprint is specific for the particular primary ion and energy. (iii) They show which general regions of α, δ-space contain interesting structures for more detailed investigation. (iv) Comparison of the clean surface SSCM and SSP to those of the adsorbate covered surface allows determination of adsorbate induced reconstruction or relaxation.

Consider the details of the SSCM and SSP of Fig. 10.8. The thick line at low α gives the value of $\alpha^1_{c,sh}$ vs δ corresponding to shadowing of 1st-layer atoms by their 1st-layer neighbors; this line is symmetrical about $\delta = 0°$, as is the 1st-atomic layer. Intense structures are observed as α increases above 20° due to subsurface-layer scattering; the high intensities are due to focusing and channeling of ion trajectories by 1st- and 2nd-layer W atoms onto 3rd- and 4th-layer W atoms and back out again. The asymmetry about $\delta = 0°$ is a result of the lack of symmetry between the 1st- and underlying layers, i.e., there is no mirror plane through the $\delta = 0°$ azimuth. The diagonal orientation of the line of intense peaks observed from $\alpha = 30°$, $\delta = -70°$ towards $\alpha = 75°$, $\delta = +80°$ results from the fact that focusing onto subsurface layers for $\delta < 0°$ occurs mainly at low α, while for $\delta > 0°$, this focusing occurs only at high α. The values of $\alpha^1_{c,sh}$ vs δ are consistent with 1st-layer interatomic spacings corresponding to the bulk truncated W{211} structure showing that the surface is not reconstructed. The 1st-layer is, however, relaxed vertically and laterally, as discussed elsewhere [10.27, 29] together with other details of the structures observed in the SSCM and SSP of Fig. 10.8.

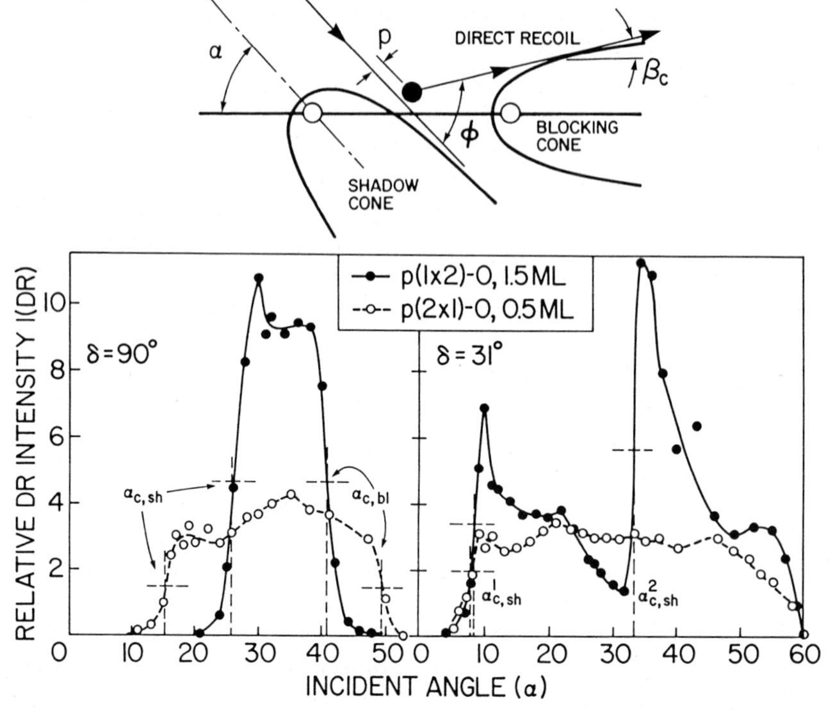

Fig. 10.9. Schematic illustration of direct recoiling events observed in TOF-SARS along with representative plots of $I(R)$ vs α for two different azimuths ($\delta = 31°$ and $90°$) and two different oxygen coverages, i.e., 0.5 and 1.5 ML, using 4 keV Ar^+

Light adsorbates can be efficiently detected [10.24] by recoiling them into forward scattering angles ϕ (Fig. 10.9). As α increases, the adsorbate atoms move out of their neighboring atom shadow cones so that direct collisions from incident ions are possible. When the p values necessary for recoiling of the adsorbate atom into a specific angle ϕ become possible in a single collision, adsorbate direct recoils are observed. Focusing at the edge of the shadow cone produces sharp rises in the recoiling intensity I (R) as a function of α. By measuring $\alpha_{c,sh}$ corresponding to the direct recoil event, the interatomic distance of the adsorbate atom relative to its nearest neighbors along the direction of the projectile can be directly determined from p and the shape of the shadow cone.

Example plots of oxygen recoiling intensities as a function of α obtained by bombardment of an oxygen covered W{211} surface with 4 keV Ar^+ ions are shown in Fig. 10.9 for two different azimuths and two different oxygen coverages; $p(1 \times 2)$-O (high dose) and $p(2 \times 1)$-O (low dose). All of the other experimental parameters were similar to those used above. Besides the sharp rises of I (R) determined by shadowing effects, sharp decreases in I (R) are observed at high α values. These result from the deflection of the recoiling trajectories by the blocking cones cast by neighboring atoms. The $\alpha_{c,bl}$, defined in Fig. 10.9, provide a measure of the size of the blocking cones and can also be used to determine interatomic distances. Along both of the $\delta = 90°$ azimuths, $\alpha_{c,sh} = 24°$ and $\alpha_{c,bl} = 42°$ at high coverage, which is considerably higher and lower, respectively, than the values $\alpha_{c,sh} = 16°$ and $\alpha_{c,bl} = 48°$ obtained at low coverage. Along other azimuths, such as $\delta = 31°$, the $\alpha^1_{c,sh}$ and $\alpha_{c,bl}$ values are nearly identical for different coverages. This data indicates that as coverage increases, both the shadowing and blocking effects become more severe along the $\delta = 90°$ azimuths than along other azimuths; this results from shadowing and blocking of O atoms by their neighboring O atoms. It is also noted that along the $\delta = 31°$ azimuth, a second peak and corresponding sharp $\alpha^2_{c,sh}$ value are observed only at high coverage. This second peak indicates that an additional adsorption site is being occupied which was previously unoccupied.

The features of I(R) in α, δ space can also be presented in the form of a recoiling structural contour map (RSCM) as shown in Figs. 10.10 and 10.11 for oxygen and hydrogen saturated W{211} surfaces, respectively. In both cases a 4 keV Ar^+ primary beam was used and the recoiling intensities were collected at $\phi = 65°$ for O and $\phi = 45°$ for H. These maps are a concise summary of the experimental data and reveal the symmetry of the recoil data in α, δ space, providing a fingerprint for O and H on the W{211} surface. For example, the symmetry of the O (RSCM) about $\delta = 0°$ indicates that the adsorption sites are symmetrical about this azimuth, thus eliminating the asymmetrical sites 3, 4, 6, 7, and 8 shown in Fig. 10.6. Comparison of the BS intensities for the clean and the

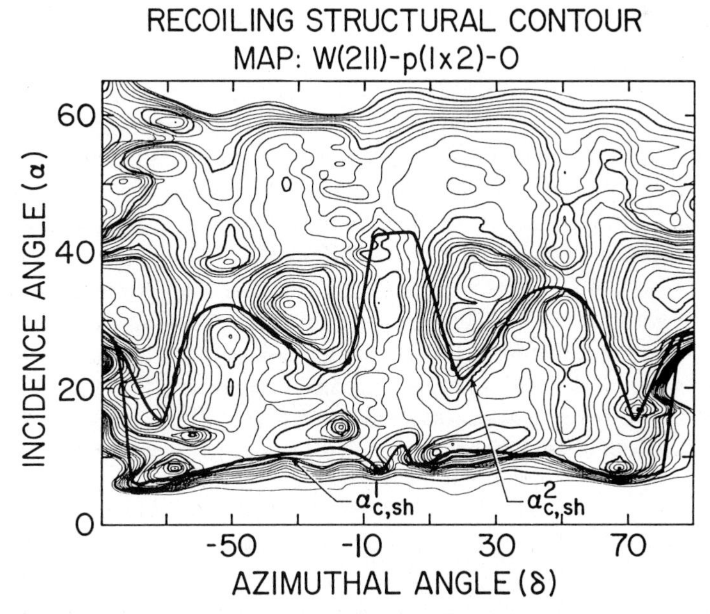

Fig. 10.10. Recoiling structural contour map (RSCM) for the W{211}-$p(1 \times 2)$-O surface. Azimuths are defined in Fig. 10.8. Contours of equal O recoil intensity $I(R)$ are plotted as a function of α and δ. The two critical shadowing angles, $\alpha_{c,sh}$ and $\alpha^2_{c,sh}$, and the critical blocking angle, β_c, are plotted on the map

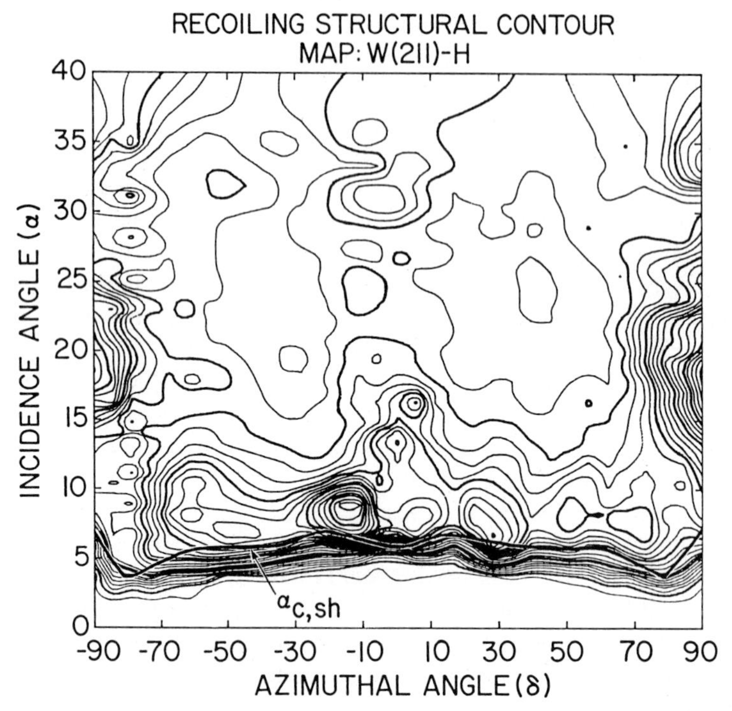

Fig. 10.11. Recoiling structural contour map (RSCM) for the W{211}-H surface. Contours of equal H recoil intensity $I(R)$ are plotted as a function of α and δ. The critical value of α, $\alpha_{c,sh}$, is plotted on the map

O covered surface indicates that only site 5, i.e., the three-fold site in which the O atom is bound to two 1st- and one 2nd-layer W atoms, is consistent with all of the experimental data. The O–W bond lengths are determined from detailed analysis [10.30] as 1.83 Å and 2.17 Å to the 1st- and 2nd-layer W atoms, respectively. In contradiction with the O case, the H (RSCM) shows only one $\alpha_{c,sh}$ at low α, which is approximately constant at 4° for $-85° < \delta < +85°$ and increases to $\alpha_{c,sh}$ = 10° for δ = 90°; a relatively flat region is observed in the center and background of the H (RSCM). This indicates that close packing along the $\delta = 90°$ direction occurs and that there is no H buried in subsurface layers that is accessible for recoiling. A detailed analysis presented elsewhere [10.31] allows determination of the hydrogen position as 0.58 ± 0.2 Å above the 1st-layer W plane and confined within a band that is centered above the $\langle 111 \rangle$ troughs.

10.3.4 Monitoring Sputtering Induced Damage

TOF-SARS is capable of monitoring the damge in the outermost atomic layers caused by ion collisions during sputtering. Such an experiment was carried out by bombarding a clean W{211} surface with 3 keV Ar^+ ions for a specific dose

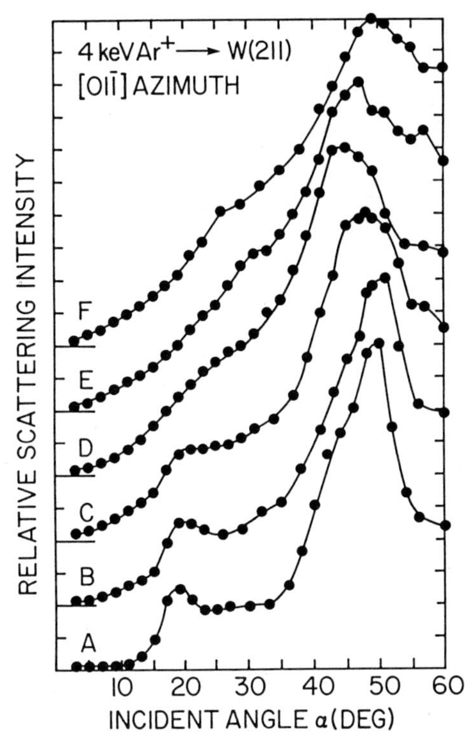

Fig. 10.12. Relative scattering intensity $I(S)$ vs α after bombardment with different 4 keV Ar^+ ion doses. (*A*) clean W{211}, (*B*) 9.4×10^{14}, (*C*) 2.8×10^{15}, (*D*) 7.5×10^{15}, (*E*) 3.8×10^{16}, and (*F*) 7.9×10^{16} ions/cm^2

and then collecting backscattering ($\theta = 163°$) spectral intensities as a function of incident angle α. After every such incident angle scan, the crystal was annealed to 2000°C for 10–15 min. in order to eliminate damage caused by the previous sputtering dose and to clean off any impurities that may have accumulated on the surface. The primary pulsed ion beam used for scattering was 4 keV Ar^+ at a flux density of 0.1 nA/mm^2; a dose of $\approx 10^{13}$ ions/cm^2 was used for each incident angle scan. This analysis dose is negligible compared to the doses used to create damage, i.e., $> 9 \times 10^{14}$ ions/cm^2.

The results of this experiment are shown in Fig. 10.12 as incident angle scans along the [0 1 $\bar{1}$] azimuth (perpendicular to the close-packed rows) for the clean surface and the clean surface after bombardment with specific Ar^+ ion doses. Due to the difficulty in calibrating the I (S) scales for the different scans, the individual scans were normalized by fixing the highest intensity point to a standard value. The clean surface scan exhibits two peaks due to shadowing and focusing effects of the 1st-layer atoms with their 1st-layer neighbors (low α) and 1st-layer atoms with their 2nd-layer neighbors (high α). The atomic density of the W{211} surface is 8×10^{14} atoms/cm^2. At the lowest Ar^+ dose of 9×10^{14} ions/cm^2, i.e., ≈ 1 ion/surface atom, the initial slope of the low α peak begins to decrease and the intensity at $\alpha < 15°$ begins to increase. The 1st-layer scattering peak is reduced to a minor inflection as the dose increases. The intensity of the valley between the 1st- and 2nd-peaks increases, the height of the plateau following the 2nd-peak increases, and the 2nd-peak broadens as the dose increases.

The disappearance of the low α peak and enhanced intensity at $\alpha < 15°$ as dose increases signifies the destruction of the 1st-atomic layer, i.e., the fixed interatomic spacings between 1st-layer atoms are replaced by a 1st-layer structure with random interatomic distances. The effects observed at higher α values associated with the 2nd-peak signify randomization of the 1st–2nd-layer interatomic spacings. Although quantification of these observations would require dynamic three-dimensional trajectory simulations, the results do illustrate the collision induced destruction of the 1st-atomic layer.

10.3.5 Surface Semichanneling

Sharp spatial anisotropy of scattered ion intensity has been observed [10.78] when an ion beam is incident at a glancing angle on a surface which has so-called surface semichannels. These semichannels are formed by close-packed rows in the 1st-atomic layer which serve as the "walls" of the channel and similar rows in the 2nd-atomic layer which serve as the "base" of the channel. Such surface semichannels are able to focus scattered ions effectively under certain conditions when the incident plane of the projectile is parallel to the channels. The W(211) surface has semichannels along the $[1\,\bar{1}\,\bar{1}]$ azimuthal direction of width 4.48 Å and depth 1.17 Å, considering the clean relaxed [10.29] (211) structure. TOF-SARS has been used to study the surface semichanneling behavior of ions scattering along the $\langle 111 \rangle$ channels.

The measurements were performed by directing a pulsed 5 keV Ne^+ or Ar^+ ion beam along the $[1\,\bar{1}\,\bar{1}]$ direction of the W(211) surface and collecting a series of TOF spectra as a function of surface azimuthal angle δ in a region $\approx \pm 25°$ about the $[1\,\bar{1}\,\bar{1}]$ axis (labeled $\delta = 90°$). Specific beam incident α and exit β angles were chosen and $I\,(\mathrm{S})$ was monitored as a function of δ. As an example of the results, Figs. 10.13 and 10.14 show plots of $I\,(\mathrm{S})$ vs δ at $\alpha = 10°$ and different β values. The scans are symmetric about the $\langle 111 \rangle$ troughs and their structure is very sensitive to β. A sharp peak (FWHM 4°–5°) is observed at low β due to surface semichanneling. As β increases above the specular condition, i.e., $\alpha = \beta = 10°$, the intensity of the focusing peak decreases until a minimum is finally observed. Similar semichanneling effects have been observed [10.78] on other surfaces.

The above measurements were made in order to determine the maximum incident angle α_{max} for which focusing along the semichannels still occurs. For 5 keV Ne^+, $\alpha_{\mathrm{max}} = 15°$, i.e., there is no focusing peak at $\delta = 90°$, but instead, a minimum is observed regardless of the value of β. Measurements with 4 keV He^+ and 5 keV Ar^+ exhibited structures similar to those of Ne^+, although with the values $\alpha_{\mathrm{max}}(He^+) = 12°$ and $\alpha_{\mathrm{max}}(Ar^+) = 18°$, respectively.

If the angles between the direction of motion of the scattering particles and the atomic rows are small enough, the scattering potential can be approximated as a multi-atom potential due to the chains of atoms along the close-packed rows. For such a case, focusing is enhanced at the specular condition, i.e., $\alpha = \beta$.

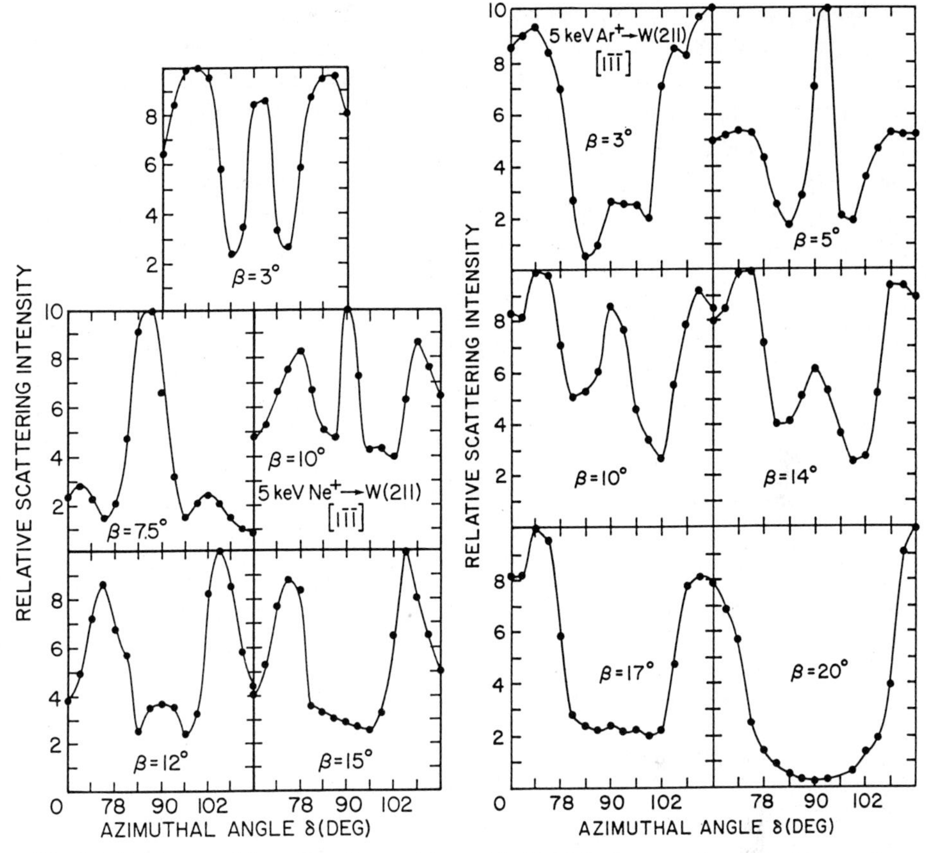

Fig. 10.13. Surface semichanneling illustrated by plots of $I(S)$ vs δ in the region near the W(211) $\delta = 90°$ $[1\,\bar{1}\,\bar{1}]$ azimuth with $\alpha = 10°$ and different β values using 5 keV Ne^+

Fig. 10.14. Surface semichanneling illustrated by plots of $I(S)$ vs δ in the region near the W(211) $\delta = 90°$ $[1\,\bar{1}\,\bar{1}]$ azimuth with $\alpha = 10°$ and different β values using 5 keV Ar^+

Linhard has defined a critical angle α_L above which the concept of continuous chains of atoms is no longer valid; for $\alpha > \alpha_L$, focusing is lost due to scattering from individual atomic centers rather than continuous multi-atom potentials. In order to estimate α_L, the expressions [10.78] for the Linhard inverse-square approximation to the Thomas-Fermi continuum potential, $(V(r))$, and the critical angle, given by

$$V(r) = \pi A/rd \tag{10.3}$$

and

$$\alpha_L = \left(\pi A/E_0 d^2\right)^{1/3} \tag{10.4}$$

were evaluated. In these expressions, r is the distance of closest approach of the projectiles to the rows, d is the lattice distance along the rows, $A = \varepsilon_v Z_1 Z_2 e^2 a/2$ (the potential constant), $\varepsilon_v = 2/(2.7183 \times 0.8853)$, Z_1 and Z_2 are the atomic numbers of the projectile and target atoms, respectively, and the screening length $a = 0.8853 a_0 (Z_1^{2/3} + Z_2^{2/3})^{-1/2}$ where a_0 is the Bohr radius. The resulting calculated α_L's are 12.3°, 19.0°, and 23.3° for 4 keV He^+, 5 keV Ne^+, and 5 keV Ar^+, respectively. These α_L's exhibit the same qualitative ordering as observed for the experimental α_{max} values, however, quantitatively they are all larger than the α_{max} values; this is because α_L provides the upper limit for the breakdown of the continuous chain approximation.

These results demonstrate the strong focusing effects occurring along the $[1\bar{1}\bar{1}]$ azimuth at glancing incident angles. The focusing peak is so sharp that it can be used to accurately align the surface crystal direction. Since TOF-SARS detects both neutrals and ions, it is asserted that the observed structures are due to geometric effects rather than ion neutralization effects.

10.3.6 Ion Fractions

TOF spectra of neutrals plus ions, neutrals only (obtained by deflection of ions), and ions only (obtained by subtraction of neutral from total spectra) are shown in Fig. 10.15 for 2 keV Ne^+ ions scattering from a clean magnesium sample. The

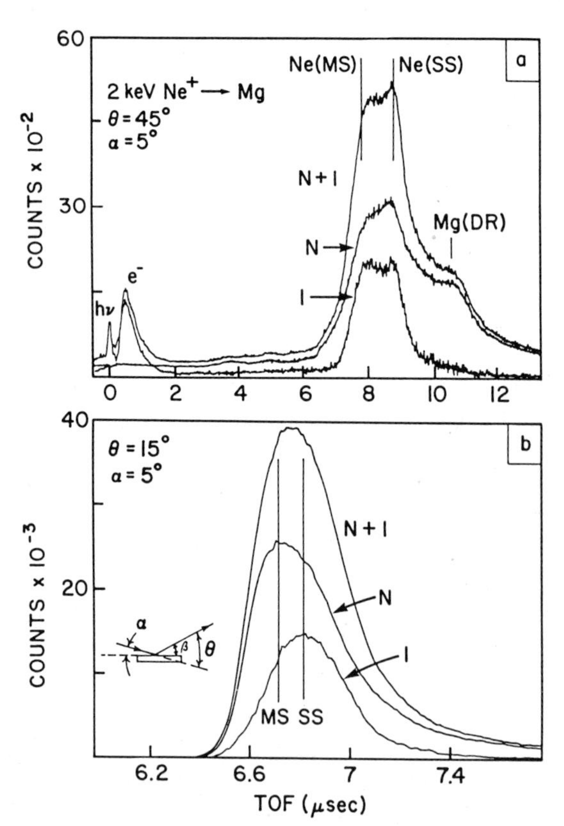

Fig. 10.15. TOF spectra of neutrals + ions (N + I), neutrals (N), and ions (I) for 2 keV Ne^+ scattering from a clean magnesium surface for $\alpha = 5°$ and (**a**) $\theta = 45°$ and (**b**) 15°. Ion induced photon ($h\nu$) and electron (e^-) peaks are indicated

spectra were acquired at an incident angle $\alpha = 5°$ to the surface and scattering angle $\theta = 45°$ (Fig. 10.15a) and 15° (Fig. 10.15b). The Mg recoiling peak can be clearly seen on the high TOF side of the scattering peak for $\theta = 45°$, but it is beneath the scattering peak for $\theta = 15°$. At the relatively large scattering angle of $\theta = 45°$, the multiple scattering (MS) sequences have cross sections of the same order as the single scattering (SS) events and their scattering energies differ by a few hundreds of eV (1165 eV for SS and 1540 eV for a double collision of $\theta = 22.5°$ each). This allows the observation of both the MS and SS contributions at $\theta = 45°$, while the $\theta = 15°$ spectrum is dominated by MS. The TOF scale in Fig. 10.15a has been extended in order to include the ion/surface collision induced photon peak (which defines the origin of the scale) and secondary electron peak (which has a maximum at approximately 10 eV).

Now consider the scattered and recoiled ion fractions (F) observed in the spectra of Fig. 10.15. First, note that Mg is recoiled predominately as a neutral species, i.e., there are no observable Mg^+ ions at the recoil position in Fig. 10.15a such that $F \approx 0$. According to the correlation diagram for the Ne–Mg collision, [10.81–83] electron excitations in the core of Mg atoms are improbable at 2 keV. Even though such excitations do occur, they are easily neutralized by resonant capture from the valence band along the outgoing trajectory.

The scattered Ne^+ ion fraction at $\theta = 45°$ is large, i.e., $F = 42\%$. This is due [10.83–85] to the strong electron promotion of Ne $2p$ electrons (by means of the $4f\sigma$ molecular orbital) taking place in the quasimolecule formed during the Ne–Mg collision. The minimum interatomic distance necessary for such a promotion is 0.6 Å. At 2 keV this interatomic distance is attained at a scattering angle of only $\approx 15°$ for SS sequences. At $\theta = 45°$ the distance of closest approach is < 0.6 Å for both MS and SS sequences, resulting in large ion fractions at both positions (Fig. 10.15a). Since electron promotion occurs only in the SS sequences for $\theta = 15°$, a decrease of the ion fractions is observed and a shift of the ion spectrum towards higher TOF (approximately to the position of the SS event) is observed in Fig. 10.15b). Other effects, such as different neutralization probabilities during the reflected part of the projectile trajectory for different exit angles and scattering energies, complicate the analysis of scattered ion fractions. Delineation of these different processes can be obtained by combining measurements of ion fractions and electron (both secondary and Auger) energy distributions as a function of projectile energy and scattering angle.

10.3.7 Ion Induced Auger Electron Spectra

The electron energy distributions from a clean magnesium surface stimulated by 2 keV Ne^+ bombardment are shown in Fig. 10.16 for two different sample orientations, one with a beam incident angle $\alpha = 15°$ and electron ejection angle $\beta = 90°$ and the other with $\alpha = 90°$ and $\beta = 15°$. The angle between the electron analyzer and the ion beam direction is fixed at 75°. Both spectra were acquired with a continuous ion current of 50 nA and an acquisition time of 5 min for

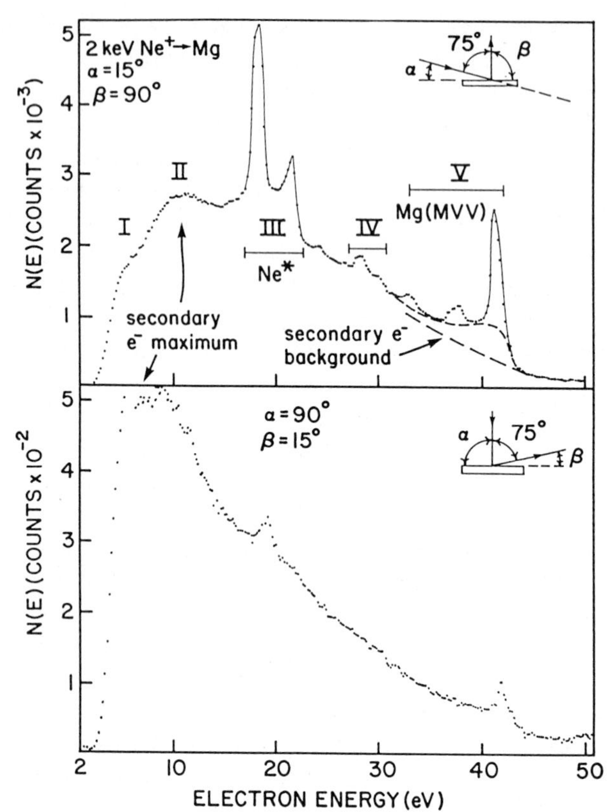

Fig. 10.16. Electron energy distributions from a clean magnesium surface stimulated by 2 keV Ne^+ ions under the conditions $\alpha = 15°$, $\beta = 90°$ and $\alpha = 90°$, $\beta = 15°$

$\alpha = 15°$ and 12 min for $\alpha = 90°$. The spectra were collected in the fixed retard-ratio mode and were not corrected for the transmission function of the analyzer, which in this mode is proportional to the electron kinetic energy.

Both electron distributions show a background of secondary electrons with a maximum at low energies (10–12 eV) and decaying towards higher energies. The intensity of the secondary electrons is ≈ 5 times higher for the $\alpha = 15°$, $\beta = 90°$ spectrum due to the short attenuation lengths of electrons in solids. At low α the collision cascade develops closer to the surface generating electrons with high escape probabilities. Also, with the geometry used, low α corresponds to detection close to the surface normal where electrons have the highest escape probability. Electrons ejected at grazing angles suffer, on the average, more inelastic collisions than those ejected near the surface normal and, additionally, they can be reflected back to the bulk when their velocity in the direction of the surface normal is not sufficient to traverse the surface barrier.

Several structures appear superimposed on the background of secondary electrons; they correspond to Auger electron emission from Ne and Mg atoms. The peaks labeled III have been previously assigned [10.86, 87] as electron emission from deexcitation of the autoionizing states $Ne^{**}\ 2p^4(^3P, {}^1D)3s^2$ formed in Ne–Mg collisions. The two Ne $2p$ electrons are promoted via the $4f\sigma$ molecular orbital, following Auger and resonant neutralization by electrons from the valence band of the solid. These excited electronic states decay in vacuum by electron emission along the outgoing part of the trajectory. Since the energies of the autoionizing states $2p^4(^3P, {}^1D)3s^2$ are higher than the work function of the solid, the reverse transition, resonant ionization of the projectile, is highly improbable [10.88]. At 2 keV, most of the collisions (> 90%) resulting in $4f\sigma$ promotion occur for scattering angles between 15° and 30°. In the case of α = 15°, excited Ne projectiles can be reflected from the solid in a single collision, decaying subsequently in vacuum. At α = 90°, most of the excited Ne atoms are inside the solid at the time of deexcitation. A small impact parameter collision or a MS sequence is required to eject the excited atom, resulting in a smaller cross section and, consequently, the lower Ne autoionizing peak observed at α = 90°. Neon Auger decay inside the solid also accounts for the broad structure observed beneath the sharp Ne autoionizing peak [10.86] at α = 90°. Structures I and IV have been tentatively ascribed [10.83] to Auger neutralization of Ne^{2+} to Ne^{*+} (excited ion) and Ne^+ (ground state ion), respectively.

Structure V appearing in the energy range 32–45 eV has been previously assigned [10.88–91] to Auger transitions from excited Mg atoms. Two different contributions are present in the Auger electron structure, a broad one, approximately twice the width of the valence band, arising from Mg(MVV) transitions inside the solid, and several sharp quasi-atomic structures, resulting from Auger transitions outside the solid in excited sputtered Mg* atoms. In this projectile energy range, i.e., below a few keV, most of the Mg excitations occur via $4f\sigma$ promotion in symmetric Mg–Mg collisions. For low incident angle bombardment, the collision cascade develops closer to the surface, enhancing the probability for ejection of both the excited Mg* atoms and Mg Auger electrons produced inside the solid.

The ability to rotate the sample position continuously for grazing or perpendicular incidence and electron detection angles, together with the possibility of changing the work function of the sample by deposition of different adsorbates, is useful in identification of the electronic transitions taking place in and near the solid surface. Coupling this with measurements of ion fractions of scattered and recoiled particles performed under the same experimental conditions can provide a better understanding of the electron exchange processes associated with ion-solid collisions.

10.4 Summary

The technique of time-of-flight ion scattering and recoiling (TOF-SARS) has been described and examples of its capabilities have been provided. A new spectrometer system has been designed which incorporates time-of-flight techniques and continuous angular movements with standard surface science techniques. TOF-SARS uses a pulsed low keV ion beam and TOF methods for velocity analysis of both scattered and recoiled neutrals and ions simultaneously with a flight path of ~ 1 m. The system contains a hemispherical electrostatic analyzer for kinetic energy measurements of ions and electrons. This provides the ability to perform nonstandard measurements (such as TOF scattering and recoiling, adsorbate and substrate structure determinations, scattered and recoiled ion fractions, and ion induced Auger electron emission) along with standard surface science measurements (such as LEED, XPS, UPS, and AES) in the same spectrometer system.

Acknowledgment This material is based on work supported by the National Science Foundation under Grant No. CHE-8814337 and the U.S. Air Force Research Office under Contract No. F19628-86-K-0013.

References

10.1 For a recent review, see Th. Fauster: Vacuum **38**, 129 (1988)
10.2 Y.S. Jo, J.A. Schultz, S. Tachi, J.W. Rabalais: J. Appl. Phys. **60**, 2564 (1986)
10.3 J.W. Rabalais, J.A. Schultz, R. Kumar: Nucl. Instrum. Meth. **218**, 719 (1983)
10.4 M.H. Mintz, U. Atzmony, N. Shamir: Phys. Rev. Lett. **59**, 90 (1987); M.H. Mintz, U. Atzmony, N. Shamir: Surface Sci. **185**, 413 (1987)
10.5 M. Aono, Y. Hou, C. Oshima, Y. Ishizawa: Phys. Rev. Lett. **49**, 567 (1982)
10.6 M. Aono, R. Souda: Jap. J. Appl. Phys. **24**, 1249 (1985)
10.7 R. Souda, M. Aono, C. Oshima, S. Otani, Y. Ishizawa: Nucl. Instrum. Meth. **B15**, 138 (1986)
10.8 L. Marchut, T.M. Buck, G.H. Wheatley, C.J. McMahon, Jr.: Surface Sci. **141**, 549 (1984)
10.9 H. Niehus: Surface Sci. **145**, 407 (1984)
10.10 H. Niehus, G. Comsa: Surface Sci. **140**, 18 (1984)
10.11 J.M. van Zoest, J.M. Fluit, T.J. Vink, B.A. van Hassel: Surface Sci. **182**, 179 (1987)
10.12 D.R. Mullins, S.H. Overbury: Surface Sci. **210**, 481 (1989; **210**, 501 (1989)
10.13 J.H. Huang, R.S. Williams: Surface Sci. **204**, 445 (1988); Phys. Rev. **B38**, 4022 (1988)
10.14 T.L. Porter, C.S. Chang, I.S.T. Tsong: Phys. Rev. Lett. **60**, 1739 (1988); T.L. Porter, C.S. Chang, U. Knipping, I.S.T. Tsong: Phys. Rev. **B36**, 9150 (1987)
10.15 J. Möller, K.J. Snowdon, W. Heiland, H. Niehus: Surface Sci. **178**, 475 (1986)
10.16 W. Hetterich, W. Heiland: Surface Sci. **210**, 129 (1989)
10.17 J.W. Rabalais, J.A. Schultz, R. Kumar, P.T. Murray: J. Chem. Phys. **78**, 5250–5259 (1983); J.N. Chen, M. Shi, J.W. Rabalais: J. Chem. Phys. **86**, 2403–2410 (1987); J.W. Rabalais, J.N. Chen, R. Kumar, M. Narayana: J. Chem. Phys. **83**, 6489–6500 (1985)
10.18 F.W. Meyer, C.C. Havener, S.H. Overbury, K.J. Snowdon, D.M. Zehner, W. Heiland, H. Hemme: Nucl. Instrum. Meth. **B23**, 234 (1987)
10.19 U. Umke, K.J. Snowdon, W. Heiland: Phys. Rev. B **34**, 41 (1986)
10.20 J.W. Rabalais, J.A. Schultz, R. Kumar: Nucl. Instrum. Meth. **218**, 719 (1983)
10.21 M.H. Mintz, J.A. Schultz, J.W. Rabalais: Phys. Rev. Lett. **51**, 1676–1679 (1983)
10.22 J.A. Schultz, J.W. Rabalais: Chem. Phys. Lett. **108**, 328–332 (1984)
10.23 M.H. Mintz, J.A. Schultz, J.W. Rabalais: Surface Sci. **146**, 457–466 (1984)
10.24 J.W. Rabalais: CRC Critical Rev. Sol. St. and Mat. Sci. **14**, 319 (1988)
10.25 B.J.J. Koeleman, S.T. deZwart, A.L. Boers, B. Polsema, L.K. Vereij: Phys. Rev. Lett. **56**, 1152 (1986)
10.26 D.P. Smith: J. Appl. Phys. **38**, 340 (1967)
10.27 J.W. Rabalais, O. Grizzi, M. Shi, H. Bu: Phys. Rev. Lett. **63**, 51 (1989)
10.28 O. Grizzi, M. Shi, H. Bu, J.W. Rabalais, R.R. Rye, P. Nordlander: Phys. Rev. Lett., submitted
10.29 O. Grizzi, M. Shi, H. Bu, J.W. Rabalais, P. Hochmann: Phys. Rev. B, submitted
10.30 H. Bu, O. Grizzi, M. Shi, J.W. Rabalais: Phys. Rev. B, submitted
10.31 M. Shi, O. Grizzi, H. Bu, J.W. Rabalais, R.R. Rye, P. Nordlander: Phys. Rev. B, submitted
10.32 S.H. Overbury, W. Heiland, D.M. Zehner, S. Datz, R.S. Thoe: Surface Sci. **109**, 239 (1981)
10.33 H. Derko, H. Hemme, W. Heiland, S.H. Overbury: Nucl. Instrum. Meth. **B23**, 374 (1987)

10.34 J. Möller, H. Niehus, W. Heiland: Surface Sci. **166**, L111 (1986)
10.35 R.S. Williams, J.A. Yarmoff: Nucl. Instrum. Methods **218**, 235 (1983)
10.36 A.J. Algra, S.B. Luitjens, E.P.Th.M. Suurmeijer, A.L. Boers: Nucl. Instrum. Methods **203**, 515 (1982)
10.37 Th.M. Hupkens: Nucl. Instrum. Methods **B9**, 277 (1985)
10.38 Th.M. Hupkens: Nucl. Instrum. Methods **B9**, 285 (1985)
10.39 R.P.N. Bronkers, A.G.J. DeWit: Surface Sci. **112**, 133 (1981)
10.40 A.G.J. DeWit, R.P.N. Bronckers, Th.M. Hupkens, J.M. Fluit: Surface Sci. **82**, 177 (1979)
10.41 J.A. Yarmoff, D.M. Cyr, J.H. Huang, S. Kim, R.S. Williams: Phys. Rev. B **33**, 3856 (1986)
10.42 A.J. Algra, S.B. Luitjens, E.P.Th.M. Suurmeijer, A.L. Boers: Surface Sci. **100**, 329 (1980)
10.43 A.J. Algra, E.P.Th.M. Suurmeijer, A.L. Boers: Surface Sci. **128**, 207 (1983)
10.44 W. Hetterich, W. Heiland: Surface Sci. **210**, 129 (1989)
10.45 Th. Fauster, M.H. Metzner: Surface Sci. **166**, 29 (1986)
10.46 J.A. Yarmoff, R.S. Williams: Surface Sci. **127**, 461 (1983)
10.47 Th. Fauster, H. Durr, D. Hartwig: Surface Sci. **178**, 657 (1986)
10.48 D.J. Godfrey, D.P. Woodruff: Surface Sci. **105**, 438 (1981)
10.49 D.J. Godfrey, D.P. Woodruff: Surface Sci. **95**, 76 (1979)
10.50 H.H. Brongersma, J.B. Theeten: Surface Sci. **54**, 519 (1976)
10.51 L.K. Verheij, J.A. Van Den Berg, D.G. Armour: Surface Sci. **84**, 408 (1979)
10.52 J.A. Van Den Berg, L.K. Verhey, D.G. Armour: Surface Sci. **91**, 218 (1980)
10.53 D.J. O'Connor, R.J. MacDonald, W. Eckstein, P.R. Higginbottom: Nucl. Instrum. Methods **B13**, 235 (1986)
10.54 H. Niehus, G. Comsa: Surface Sci. **151**, L171 (1985)
10.55 W. Englert, W. Heiland, E. Taglauer, D. Menzel: Surface Sci. **83**, 243 (1979)
10.56 H. Niehus, G. Comsa: Surface Sci. **152/153**, 93 (1985)
10.57 B. Poelsema, R.L. Palmer, G. Mechtersheimer, G. Comsa: Surface Sci. **117**, 60 (1982)
10.58 H. Niehus, G. Comsa: Nucl. Inst. Meth. **B15**, 122 (1986)
10.59 H. Niehus, C. Hiller, G. Comsa: Surface Sci. **173**, L599 (1986)
10.60 D.E. Eastman: J. Vac. Sci. Technol. **17**, 492 (1980)
10.61 H. Niehus, K. Mann, B.N. Eldridge, M.L. Yu: J. Vac. Sci. Technol. **A6**, 625 (1988)
10.62 J.A. Schultz, Y.S. Jo, S. Tachi, J.W. Rabalais: Nucl. Instrum. Method. *B*15, 134 (1986)
10.63 M. Saitoh, F. Shoji, K. Oura, T. Hanawa: Jpn. J. Appl. Phys. **19**, L421 (1980)
10.64 J. Onsgaard, W. Heiland, E. Taglauer: Surface Sci. **99**, 112 (1980)
10.65 T.L. Porter, C.S. Chang, U. Knipping, I.S.T. Tsong: Phys. Rev. B **36**, 9150 (1987)
10.66 W. Heiland, F. Iberl, E. Taglauer, D. Menzel: Surface Sci. **53**, 383 (1975)
10.67 H. Niehus, E. Bauer: Surface Sci. **47**, 222 (1975)
10.68 S. Prigge, H. Niehus, E. Bauer: Surface Sci. **65**, 141 (1977)
10.69 W.P. Ellis, R.R. Rye: Surface Sci. **161**, 278 (1985)
10.70 T.M. Buck, G.H. Wheatley, L. Marchut: Phys. Rev. Lett. **51**, 43 (1983)
10.71 M. Aono, Y. Hou, R. Souda, C. Oshima, S. Otani, Y. Ishizawa: Phys. Rev. Lett. **50**, 1293 (1983)
10.72 C. Oshima, M. Aono, T. Tanaka, S. Kawaii, S. Zaima, Y. Shibata: Surface Sci. **102**, 312 (1981)
10.73 T.N. Taylor, W.P. Ellis: Surface Sci. **107**, 249 (1981)
10.74 H.H. Brongersma, P.M. Mul: Chem. Phys. Lett. **10**, 217 (1973)
10.75 K. Matsuda, R. Shimizer: Jpn. J. Appl. Phys. **21**, L670 (1982)
10.76 H. Nakamatsu, A. Sudo, S. Kawai: Surface Sci. **194**, 265 (1988)
10.77 O. Grizzi, M. Shi, H. Bu, J.W. Rabalais: Rev. Sci. Instrum., submitted
10.78 E.S. Mashkova, V.A. Molchanov: *Medium Energy Ion Reflection From Solids* (North-Holland, Amsterdam 1985)
10.79 J.F. Ziegler, J.P. Biersack, U. Littmark: *The Stopping and Range of Ions in Solids* (Pergamon, New York 1985)
10.80 S.R. Kasi, M.A. Kilburn, H. Kang, J.W. Rabalais, L. Tavernini, P. Hochmann: J. Chem. Phys. **88**, 5902 (1988)
10.81 M. Barat, W. Lichten: Phys. Rev. **A6**, 211 (1972)
10.82 N. Stolterfoht: In *Structure and Collisions of Ions and Atoms*, ed. by I.A. Sellin (Springer Berlin, Heidelberg 1978)
10.83 G. Zampieri, F. Meier, R. Baragiola: Phys. Rev. **B11**, 3951 (1984)
10.84 J.W. Rabalais, Jie-Nan Chen, R. Kumar, M. Narayana: Chem. Phys. Lett. **120**, 406 (1985)

10.85 J.A. Schultz, C.R. Blakley, M.H. Mintz, J.W. Rabalais: Nucl. Instr. Meth. **B14**, 500 (1986)
10.86 S.V. Pepper: Surface Sci. **169**, 39 (1986)
10.87 G. Zampieri, R.A. Baragiola: Surface Sci. **114**, L15 (1982)
10.88 R. Whaley, E.W. Thomas: J. Appl. Phys. **56**, 1505 (1984)
10.89 J.A. Matthew: Phys. Scripta **T6**, 79 (1983)
10.90 J. Mischler, N. Benazeth: Scanning Electron Microsc. II, 351 (1986)
10.91 R. Baragiola: In *Inelastic Particle-Surface Collisions*, Springer Ser. in Chem. Phys. Vol. 17, ed. by E. Taglauer, W. Heiland (Springer, Berlin, Heidelberg 1981), p. 38

11. Scanning Electron Microscopy with Polarization Analysis: Studies of Magnetic Microstructure

J. Unguris, M.R. Scheinfein, R.J. Celotta and D.T. Pierce

National Institute of Standards and Technology, Gaithersburg, MD 20899, USA

When a beam of electrons with energies greater than several hundred eV is incident upon a ferromagnetic metal, spin polarized secondary electrons are emitted. The polarization of these secondary electrons is related to the polarization of the electrons in the ferromagnet. In the case of transition metal ferromagnets, the polarization of the secondary electrons is directly proportional to the magnetization. Spin polarization analysis of the secondary electrons, therefore, provides a direct measurement of the magnetization in the region probed by the incident electron beam. Scanning electron microscopy with polarization analysis (SEMPA), illustrated schematically in Fig. 11.1, combines the finely focused beam of the scanning electron microscope with secondary electron spin polarization analysis to obtain a technique that provides high resolution images of the surface magnetic microstructure of ferromagnetic materials. The purpose of this chapter is to review the SEMPA technique and to present several examples of magnetic microstructures that were studied using SEMPA.

SEMPA is the product of extensive research, which began in the late 1960s, studying the emission and scattering of spin polarized electrons from solid sur-

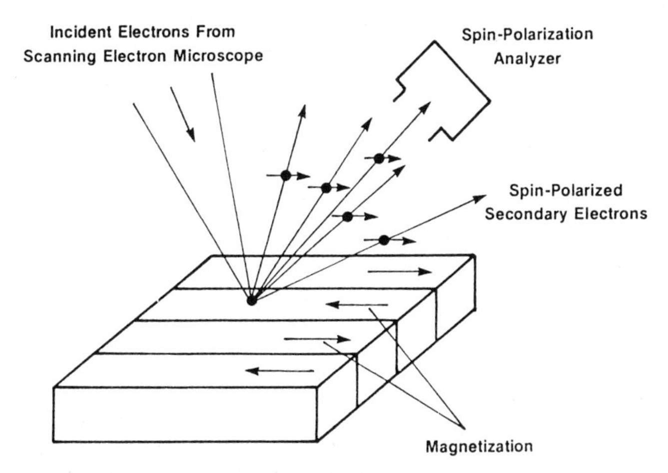

Fig. 11.1. Schematic diagram illustrating the SEMPA technique. A ferromagnetic specimen is scanned by the focused electron beam of a scanning electron microscope. A spin polarimeter is used to measure the polarization of the emitted secondary electrons. The spin polarization of the secondary electrons is directly proportional to the magnetization, so that a rastered image of the magnetic microstructure is produced

faces. Several reviews of this work are available [11.1–4]. The possibility of using the spin polarization of secondary electrons to image magnetic microstructure was initially discussed in the early 1980s [11.5–7]. By the mid 1980s the first SEMPA measurements had been made [11.8, 9]. Most of this early SEMPA work has already been reviewed [11.10–12].

Motivation for the development of SEMPA has come primarily from the need to look at very small, sub-micron, magnetic structures, such as domain configurations and domain walls. Interest in these magnetic microstructures ranges from fundamental research studying the physics of low dimensional magnetic systems to application oriented problems in magnetic recording and fine particle permanent magnets. Conventional magnetic imaging techniques such as domain wall decoration using the Bitter method [11.13] or polarized light magneto-optic Kerr microscopy [11.14, 15] are optical techniques with a resolution limit of about 0.5 micron. Higher resolution is achieved with electron microscope based techniques such as Lorentz microscopy in reflection [11.16] or transmission [11.17, 18], and electron holography [11.19]. Transmission Lorentz microscopy and electron holography offer the highest resolution, on the order of 10 nm, but both techniques require thin (less than 100 nm thick) unsupported specimens. The most recently developed technique, based on a scanned tip geometry, is magnetic force microscopy [11.20]. The ultimate resolution of the magnetic force microscope is uncertain because of interactions between the tip and sample. All of the magnetic observation techniques, except for magneto-optic Kerr microscopy, derive magnetic contrast from their sensitivity to magnetic fields, either inside or outside of the sample, rather than magnetization. In contrast, SEMPA is a high resolution, domain imaging technique in which the signal contrast is directly related to the direction and magnitude of the magnetization. In addition, SEMPA can be applied to both thick specimens or thin films grown on thick substrates.

11.1 Spin Polarization of Secondary Electrons

The secondary electron spin polarization from ferromagnetic materials has been extensively studied for the past ten years. A recent review of much of this work is available [11.21]. Measurements of the intensity, $N(E)$, and polarization, $P(E)$, energy distributions of secondary electrons generated from transition metal ferromagnets have general features that are illustrated schematically in Fig. 11.2. The polarization component along some direction is defined as

$$P = \frac{N_\uparrow - N_\downarrow}{N_\uparrow + N_\downarrow} , \qquad (11.1)$$

where $N_\uparrow(N_\downarrow)$ are the number of electrons with spins parallel (antiparallel) to the specified direction. The polarization distributions have the following common features: First, the direction of the polarization is exactly opposite to that of the magnetization. The polarization direction of the electrons with their negative magnetic moments is not changed by the emission process. Second, the

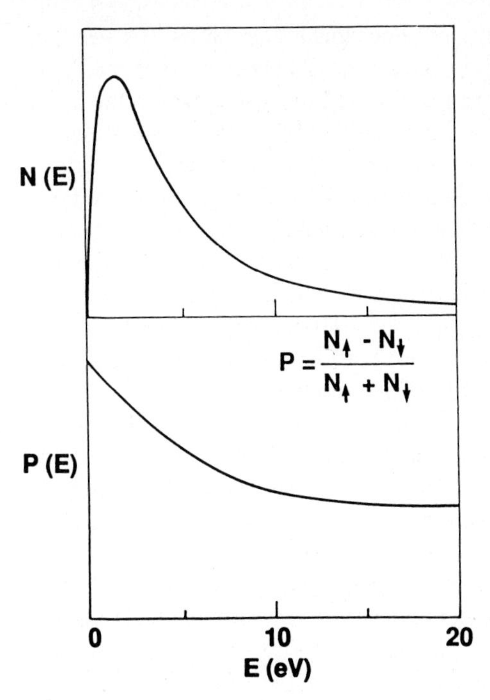

Fig. 11.2. Schematic drawing showing the energy dependence of the intensity, $N(E)$, and polarization, $P(E)$, of secondary electrons emitted from a simple transition metal ferromagnet

polarization is relatively constant for energies greater than 10 eV except for occasional characteristic loss features. And finally, the polarization increases as the secondary electron energy approaches zero.

Some quantitative understanding of the polarization distribution can be gained by assuming that only the valence electrons contribute to the secondary electron polarization. This would apply, for example, to the transition metals, Fe, Co, or Ni, in which the orbital magnetic moment is quenched. The magnetization, M, is then related to the next spin density, $n_\uparrow - n_\downarrow$, by

$$M = -\mu_{\mathrm{B}}(n_\uparrow - n_\downarrow) , \tag{11.2}$$

where μ_{B} is the electron magnetic moment or Bohr magneton. If the secondary electrons were simply valence electrons that were emitted without changing their polarization the secondary polarization should be equal to

$$P_{\mathrm{v}} = n_{\mathrm{b}}/n_{\mathrm{v}} , \tag{11.3}$$

where n_{b} is the magnetic moment per atom, and n_{v} the number of valence electrons per atom. This simple model predicts polarizations of 28%, 19%, and 5% for Fe, Co, and Ni, respectively. These values agree well with polarizations measured for energies above 10 eV for Fe [11.22], Co [11.22], and Ni [11.23].

The enhancement of the secondary polarization at low secondary electron energies is due to a spin dependent filtering of the slow secondary electrons. As the kinetic energy of the secondary electron becomes smaller, the probability that it will lose energy and drop down into an unoccupied state below the vacuum level increases. The spin polarization enhancement is the result of there being more empty minority states than majority states available to de-excite into. This spin filter model gives good quantitative agreement with the observed low energy polarization enhancement [11.24, 25]. A slightly different approach to describe the enhancement of secondary electron polarization at low energies has also been proposed which emphasizes the role of Stoner excitations in the spin filter mechanism [11.26–28]. In general, though, the relationship between the magnetization of a transition metal and the polarization of the secondary electrons emitted from it is relatively well understood. Note, however, that for ferromagnets with different electronic structures such as the rare earths, the relationship between polarization and magnetization is probably more complicated.

Another important feature of the secondary electrons is their surface sensitivity. The incident electron beam deposits its energy and, therefore, creates secondary electrons deep within the material, but only those that are close enough to the surface can escape before losing their energy [11.29]. The exact escape depth of the secondary electrons is still unresolved. If the inelastic mean free path [11.30] is used to determine the escape depth, then the secondaries are emitted from the top 2 to 3 nm of the sample. If the transport decay length [11.21] associated with a continuous slowing down of the electrons is used then escape depths of about 1 nm are predicted. In addition, there may be the further complication that the magnetic probing depth [11.31] may be different from the escape depth. More experimental work measuring the probing or escape depth of secondaries from ferromagnetic materials would be extremely valuable.

In summary, the features of polarized secondary electron emission that make SEMPA a useful tool are: First, the polarization is directly proportional to the magnetization of transition metal ferromagnets. Second, there are a lot of secondary electrons and they have relatively large polarizations so that there is a large signal to measure. And finally, the secondaries come from the outermost few atomic layers of the material.

11.2 Experimental

A schematic drawing of the SEMPA apparatus at the National Institute of Standards and Technology (NIST) is shown in Fig. 11.3. Versions of SEMPA instruments developed in different laboratories may vary in detail, but they all consist of the following basic components: 1) A scanning electron microscope column which produces the focused incident electron beam; 2) An ultra-high vacuum chamber with tools for surface preparation and analysis; 3) One or more spin polarization analyzers; 4) Electron optics for collecting and transporting the emitted secondary electrons to the detectors; And, 5) some form of image processing to transform the polarization measurements into magnetization images. All of these components are described in detail elsewhere [11.10, 32, 33] so that the instrumentation will only be briefly reviewed here.

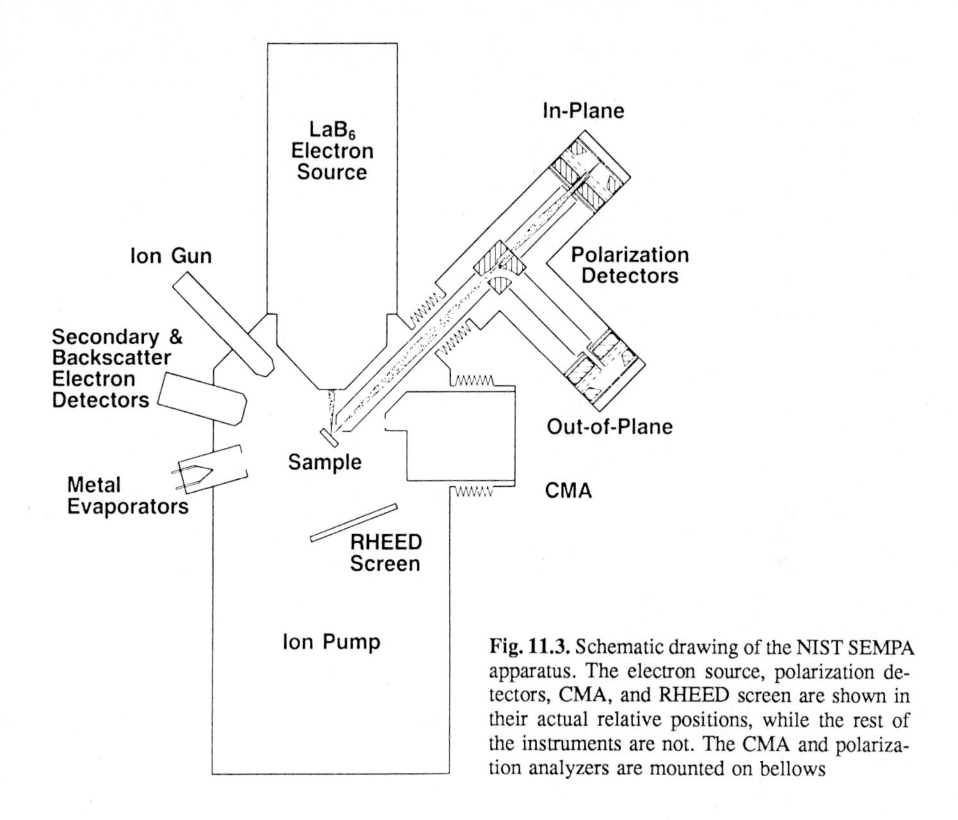

Fig. 11.3. Schematic drawing of the NIST SEMPA apparatus. The electron source, polarization detectors, CMA, and RHEED screen are shown in their actual relative positions, while the rest of the instruments are not. The CMA and polarization analyzers are mounted on bellows

11.2.1 Electron Microscope and Specimen Chamber

An electron microscope with a high brightness electron source is desirable, because of the inefficiency of existing electron spin polarization analyzers. The minimum beam current required to obtain a SEMPA image in a reasonable amount of time is about 1 nA. For a LaB_6 cathode, such as the one used on the NIST SEMPA apparatus, 1 nA corresponds to a nominal beam diameter of about 40 nm. Smaller probes can be formed using either cold or thermally assisted field emission sources [11.34]. SEMPA instruments using field emission sources should have usable probes that are 10 nm or less in diameter.

The incident beam energy used is a compromise between resolution and signal intensity. The secondary electron yield increases with decreasing beam energy [11.29], but the beam diameter also increases. In addition, a lower energy primary electron beam is more susceptible to deflections and distortions due to any electrostatic extraction fields or stray magnetic fields that may be present. Under typical operating conditions, a 10 keV incident electron beam is used which produces a secondary yield that is about 20% of the incident current.

Two final considerations concerning the microscope column are the working distance between the objective lens and the sample and the stray magnetic field coming from the objective lens. A large working distance is desirable in order to provide access to the sample by the polarization analyzers as well as other surface analysis and preparation instruments. Unfortunately, the incident beam diameter increases with increasing working distance. A long working distance is also desirable in order to remove the sample from the stray magnetic field of the objective lens. This stray magnetic field can affect the secondary electrons by deflecting their trajectories and rotating their spins. In addition, a stray magnetic field can change the domain structure of the magnetic specimen. A working distance of about 10 mm and a stray field at the sample of one gauss or less are typical.

As a result of its surface sensitivity, the SEMPA technique requires that the microscope have a bakeable, ultra-high vacuum specimen chamber that is equipped with various devices for the in situ preparation and characterization of samples. The base pressure of the NIST SEMPA apparatus shown in Fig. 11.3 is 6×10^{-8} Pa. The chamber has an ion gun for cleaning samples, a heated stage for annealing the samples after ion bombardment, and metal evaporators for depositing magnetic or nonmagnetic thin films. The chemical composition of the surface is monitored by Auger electron spectroscopy using the cylindrical mirror energy analyzer. This apparatus can, therefore, generate compositional maps from the same areas that are imaged using SEMPA. The surface order is measured by tilting the sample and analyzing the reflection high energy electron diffraction (RHEED) patterns from the surface. In addition, the microscope can generate standard SEM images using either secondary electrons, backscattered electrons, or absorbed sample current. One feature that has not been implemented, but is highly desirable for certain magnetic studies, is some method for applying a magnetic field to the sample such that the secondary electron trajectories are not disturbed.

11.2.2 Transport Optics and Polarization Analyzers

The function of the spin polarization analyzer input optics is to collect as many of the emitted secondary electrons as possible, transport them to the analyzer with minimum loss of intensity or change in polarization, and deliver the electrons with the correct energy and momentum for analysis. The design of the input electron optics depends critically upon the type of polarization analyzer used and the specimen chamber geometry. In the NIST SEMPA apparatus [11.32] the front end of the input optics is biased at a positive 1500 volts in order to collect most of the low energy secondary electrons emitted from the sample. One problem with this biasing arrangement is that the sample and sample holder become part of the extraction electron optics, so that tilting the sample or changing the sample geometry can change the secondary electron trajectories. On the positive side, rapid acceleration of the secondaries minimizes any adverse effects from magnetic stray fields from the lens or sample. The transport optics is designed to transmit all of the accelerated secondary electrons within a 8 eV wide energy window to the polarimeter. In practice, about 90% of these electrons are actually transmitted.

The input optics also contain deflection plates to descan the secondary electrons and a 90 degree electrostatic deflector to switch the electrons from one polarimeter to another. The purpose of the descan deflectors is to keep the motion of the incident beam at the sample from being transmitted to the polarimeter where the beam motion due to the scan can introduce false polarization signals especially at low magnifications (large scans). Descanning is accomplished by sensing the scan voltage of the SEM and driving the electrostatic deflection plates of the input optics with a voltage that is 180 degrees out of phase. A 90 degree deflector and two polarimeters are used so that all components of an arbitrary magnetization vector can be measured. Each polarimeter measures the two transverse polarization components of the electron beam. For the specimen-polarimeter geometry shown in Fig. 11.3, the undeflected, straight through analyzer measures the two in-plane components of the sample magnetization, while an orthogonal detector, accessed by activating the 90 degree deflector, measures the out-of-plane mgnetization and a redundant in-plane component. The redundant component is used to ensure that both detectors have identical polarization sensitivities. Alternate methods for measuring all of the magnetization components using just one analyzer involve rotating the specimen [11.33, 35] or rotating the polarization with a Wien filter [11.36].

Various electron spin polarization analyzers are currently available. The choices include traditional 100 keV Mott detectors [11.37], 30 keV retarding Mott analyzers [11.38], low energy electron diffraction (LEED) analyzers [11.4], low energy diffuse scattering (LEDS) analyzers [11.39, 40], and low energy absorbed current detectors [11.41, 42]. The features and relative merits of the various analyzers have been discussed in detail elsewhere [11.32, 43]. Fully operational SEMPA instruments have been constructed using a Mott analyzer [11.44], a LEED analyzer [11.33], and a LEDS analyzer [11.11].

The basis for the spin sensitivity of most of the polarization analyzers is the spin-orbit interaction [11.1]. When an electron scatters from the central potential of some high atomic number atom, there is an additional interaction between the electron's spin and its orbital angular momentum about the central potential. The spin-orbit interaction has the effect of making the scattering cross sections different for electrons with spins parallel or anti-parallel to the scattering plane normal. The scattering plane is the plane in which both the incident and scattered electron trajectories lie. The simplest polarization analyzer, therefore, consists of a high-Z target and two electron detectors, one measuring the number of electrons scattered to the right, the other to the left. An incident beam of electrons with a polarization component P perpendicular to the scattering plane would result in a scattering asymmetry A

$$A = \frac{N_{\mathrm{L}} - N_{\mathrm{R}}}{N_{\mathrm{L}} + N_{\mathrm{R}}} = PS\,, \tag{11.4}$$

where $N_{\mathrm{L}}(N_{\mathrm{R}})$ are the number of electrons scattered to the left (right) detector and S, the Sherman function, is a parameter which describes the analyzer's sensitivity to polarization.

The performance of a polarization analyzer is characterized by the Sherman function and the fraction of incident electrons scattered into the detectors, I/I_0. The figure of merit F for a spin polarization analyzer in a measurement limited by counting statistics is given by

$$F = (I/I_0)S^2 . \tag{11.5}$$

For optimized analyzers, F is approximate 1×10^{-4}. In practice, this means that it will take 10^4 times as long to acquire a polarization measurement as an intensity measurement with the same statistics. For comparison, a typical Auger measurement takes about 10^6 times as long as an intensity measurement [11.45].

In selecting a particular spin polarization analyzer, features other than the figure of merit must also be considered. Among these are proper electron optical coupling with the phase space of the electrons to be analyzed, immunity from false apparatus asymmetries, minimum disturbance to the functioning of the SEM, and ease of use. No single detector fully satisfies all of these requirements, so that some compromises have to be made. For example, high energy Mott detectors have the largest electron acceptance phase space, but the low energy polarimeters are much more compact and, therefore, easier to attach to an SEM. Further polarimeter comparisons can be found in the review papers [11.32, 43].

A schematic drawing of the low energy diffuse scattering (LEDS) analyzer used on the NIST SEMPA apparatus is shown in Fig. 11.4. It is based on the scattering of 150 eV electrons from an evaporated polycrystalline Au target. The Au films are evaporated in situ and are stable for at least a week in ultrahigh vacuum. The incident electrons are scattered diffusely from the poycrystalline surface. A negatively biased electrode E_1 focuses the scattered electrons into a pair of retarding grids, G_1 and G_2, which filter out the low energy secondary electrons generated at the Au target. After passing the grids, the electrons proceed to a microchannel plate electron multiplier. The amplified signal is then detected by an anode which is divided into four equal quadrants as shown in the inset of Fig. 11.4. The two orthogonal transverse polarization components of the incident beam are measured simultaneously and are given by

$$P_x = \frac{1}{S} \frac{N_A - N_C}{N_A + N_C} , \tag{11.6}$$

$$P_y = \frac{1}{S} \frac{N_B - N_D}{N_B + N_D} , \tag{11.7}$$

where N_i is the number of electrons counted by quandrant "i".

A common problem of electron spin polarimeters is the elimination of any false polarization signals due to instrumental asymmetries. Instrumental asymmetries can be the result of intrinsic nonuniformities in the detector such as differences in gain and zero signal levels between different detector channels and mechanical deviations from a symmetric scattering geometry. These instru-

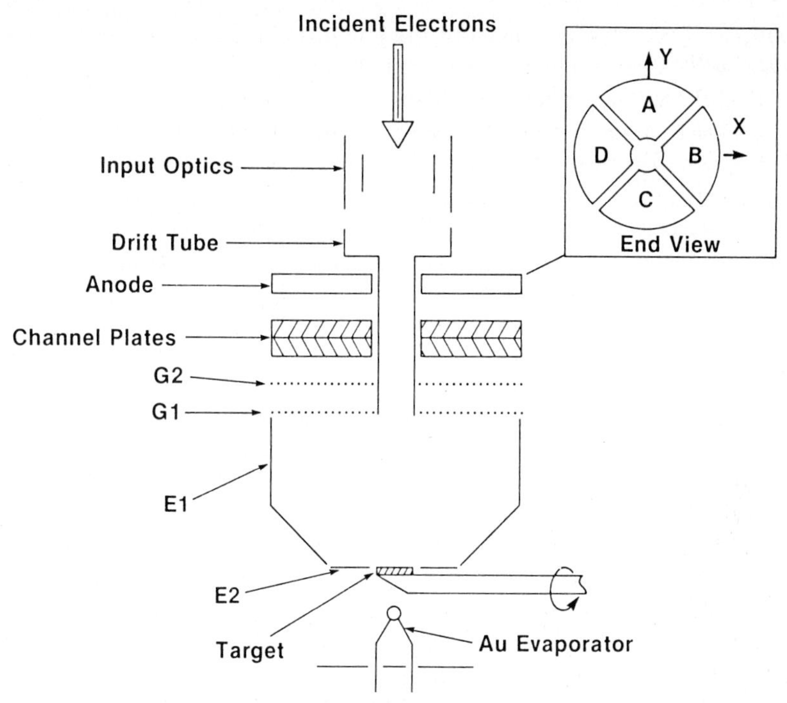

Fig. 11.4. Schematic drawing of a low energy diffuse scattering spin polarization analyzer. The inset shows how the anode is divided into quadrants so that polarization components along both the x and y directions may be measured simultaneously

mental asymmetries are relatively easy to eliminate by using the unpolarized electrons from a non-magnetic sample for calibration and standard electronic and mechanical design practices. Another source of false asymmetry is the sensitivity of the polarimeter to incident electron trajectories. If the position or angle of the incident electron beam at the high-Z scattering target changes, then opposing electron counters will record different relative intensities and, therefore, measure a false polarization. Deviations of the electron beam trajectories at the detector can be caused by scanning of the incident electron beam at the sample, variations in the electrostatic extraction field due to sample geometry, topography, or work function, and stray magnetic fields from the objective lens and the specimen. These instrumental asymmetries are much more difficult to eliminate because they can be different for each polarization measurement.

In the NIST SEMPA apparatus, trajectory related instrumental asymmetries are reduced using several techniques. First, the scan of the primary electron beam is removed by descanning the secondary electron beam in the analyzer input optics. Second, the detector uses an electron optical compensation scheme which balances spatial and angular components of the instrumental asymmetry [11.40]. In the compensation scheme, electrostatic lenses are set such that a change in position of the beam at the target is balanced by a change in the incident angle and the angular change in the asymmetry is equal and opposite to the spatial one. These first two methods can eliminate instrumental asymmetries from most SEMPA measurements. Any remaining false asymmetries, such as those associated with very rough specimens, can be removed by using a low-Z graphite reference target in the polarimeter so that the spin dependence of the polarimeter can be turned off without changing the instrumental asymmetries. In these cases, a reference measurement is made using the graphite target for every polarization measurement using the Au target, and the difference between them is the true polarization. An example of this process will be described in a later section.

11.2.3 Image Processing

The final product of SEMPA is a picture of the direction and magnitude of the magnetization in the area scanned by the SEM. Unlike conventional microscopies, which display a scalar quantity such as the secondary electron intensity in the standard SEM image, SEMPA produces an image of a vector quantity, the magnetization. Therefore, the SEMPA image processing system must not only be able to perform standard image processing tasks such as data storage, display, filtering, and background subtraction, but, in addition, the system must be able to combine the individual polarization measurements into maps of the magnetization vector field [11.46]. For example, the polarimeters can measure two in-plane magnetization components, M_x and M_y, and an out-of-plane component, M_z. In order to determine the direction and magnitude of the magnetization, the image processing system must be able to remove any residual instrumental asymmetry offsets, check for registration of the images by cross correlation techniques, and produce images of the magnitude

$$|M| = \sqrt{M_x^2 + M_y^2 + M_z^2} \tag{11.8}$$

and the direction of magnetization within the surface plane

$$\Theta_{\text{ip}} = \tan^{-1} \frac{M_y}{M_x} \tag{11.9}$$

or out of the surface

$$\Theta_{\text{op}} = \tan^{-1} \frac{M_z}{\sqrt{M_x^2 + M_y^2}} \,. \tag{11.10}$$

In addition, the image processing system should be able to generate quantitative information such as, line scans, histograms, and scatter plots [11.47] from the SEMPA data.

11.3 SEMPA Measurement Examples

11.3.1 Iron Crystals

Because of their large intrinsic secondary polarization and well known bulk magnetic properties, Fe crystals are good specimens for demonstrating various SEMPA features [11.3, 8, 9, 10, 11, 33, 46, 48–53]. Figure 11.5 shows SEMPA measurements of the domain structure of the {100} surface of an Fe-3%Si crystal. The surface was prepared by ion sputtering with 1 keV Ar ions followed by annealing to 700°C. The small amount of Si present in the sample makes cleaning easier by suppressing the bcc to fcc phase transition of pure Fe. Figure 11.5a shows the horizontal in-plane magnetization component, M_x. The brightness of the image is linearly proportional to the magnetization component. In the grey scale of the M_x image, positive magnetization points to the right and is mapped to white while negative magnetization points to the left and is mapped to black. Figure 11.5b shows the corresponding vertical magnetization component, M_y, with white corresponding to magnetization pointing up and black pointing down. The secondary intensity image, which is simply equal to the denominator,

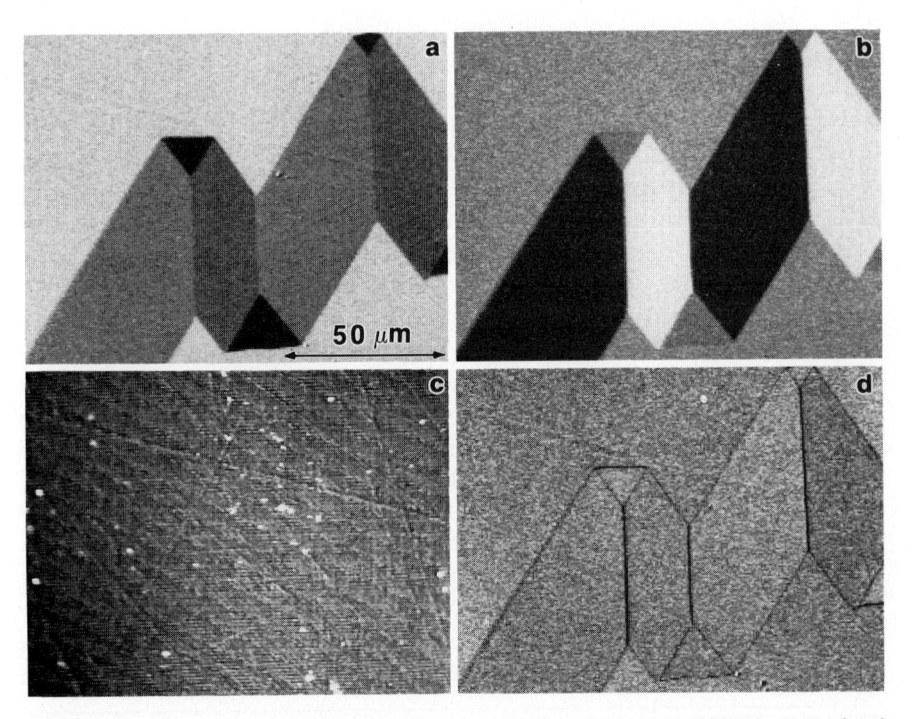

Fig. 11.5a–d. An example of a SEMPA measurement showing the domain structure of the {100} surface of an Fe-3%Si crystal. The (**a**) horizontal, M_x, and (**b**) vertical, M_y, in-plane magnetization components are shown along with (**c**) the simultaneously measured intensity topograph. The magnitude of the magnetization computed from the components is shown in (**d**)

$N_A + N_C$, of the polarization in (6), is shown in Fig. 11.5c. M_x, M_y and the intensity are all measured simultaneously by one detector.

The 256×192 pixel images in Fig. 11.5 are 140 μm across. The measurements were made using an incident beam current of 15.5 nA and a dwell time of 4.0 ms/pixel, so that the images took about four minutes to acquire. No out-of-plane, M_z, component was observed. The domain structure is primarily determined by the fact that Fe, a material with cubic anisotropy, has two orthogonal easy magnetization axes that lie in the $\{100\}$ surface and one which is perpendicular to the surface. Any magnetization component pointing out of the plane, however, is associated with a large magnetostatic energy. The magnetization therefore remains in-plane and lies along one of the two orthogonal easy axis.

One feature of the Fe SEMPA data that is common to all SEMPA measurements is that the magnetic and topographic images can be separated. The secondary electron intensity and polarization are measured simultaneously, but they are completely independent measurements. SEMPA is therefore a useful technique for studying the relationships between topographic and magnetic structure. For example, SEMPA has been used to observe the pinning of domain walls by point defects within a ferromagnet [11.12].

The individual magnetization components can be used to calculate the magnitude and direction of the in-plane magnetization. First, the instrumental asymmetry offset is removed by subtracting a plane from the image so that domains of opposite magnetization corresponded to equal but opposite values of polarization. This is based on assuming a constant magnetization magnitude for the entire image. The magnitude of the magnetization is then calculated, as shown in Fig. 11.5d. It is essentially constant which verifies that the instrumental asymmetry was correctly removed. There appears to be some magnetization missing at the domain walls, but this is purely an instrumental artifact due to the finite size of the beam. When the incident electron beam diameter, in this case about 200 nm, is larger than the domain wall, the polarization measurement averages over domains with magnetization components of opposite direction, leading to a reduced value for the magnitude of the magnetization [11.32]. The direction of the magnetization is shown by the in-plane angle image in Fig. 11.9a. The direction of the magnetization is displayed by calculating the angles from the value of the magnetization components and mapping these angles into color using the color wheel shown in the inset of Fig. 11.9. In this representation, the two easy magnetization axes are obvious.

In some cases, specifically, when large topographic features are present, spurious instrumental asymmetries can only be removed by using magnetization independent reference images obtained with the graphite target in the analyzer. This procedure is illustrated by SEMPA images from single crystal Fe whiskers shown in Fig. 11.6. The whiskers were cleaned by Ar ion bombardment and

Fig. 11.6a–h. SEMPA measurements of magnetic domains in an Fe whisker which demonstrate how spurious instrumental asymmetries can be removed using a graphite reference target in the detector. (**a**) and (**b**) SEMPA measurements of M_x and M_y are made using a Au target; (**c**) and (**d**) using a graphite target; (**e**) and (**f**) after subtracting the graphite data from the Au. (**g**) The magnetization magnitude and (**h**) intensity are also shown

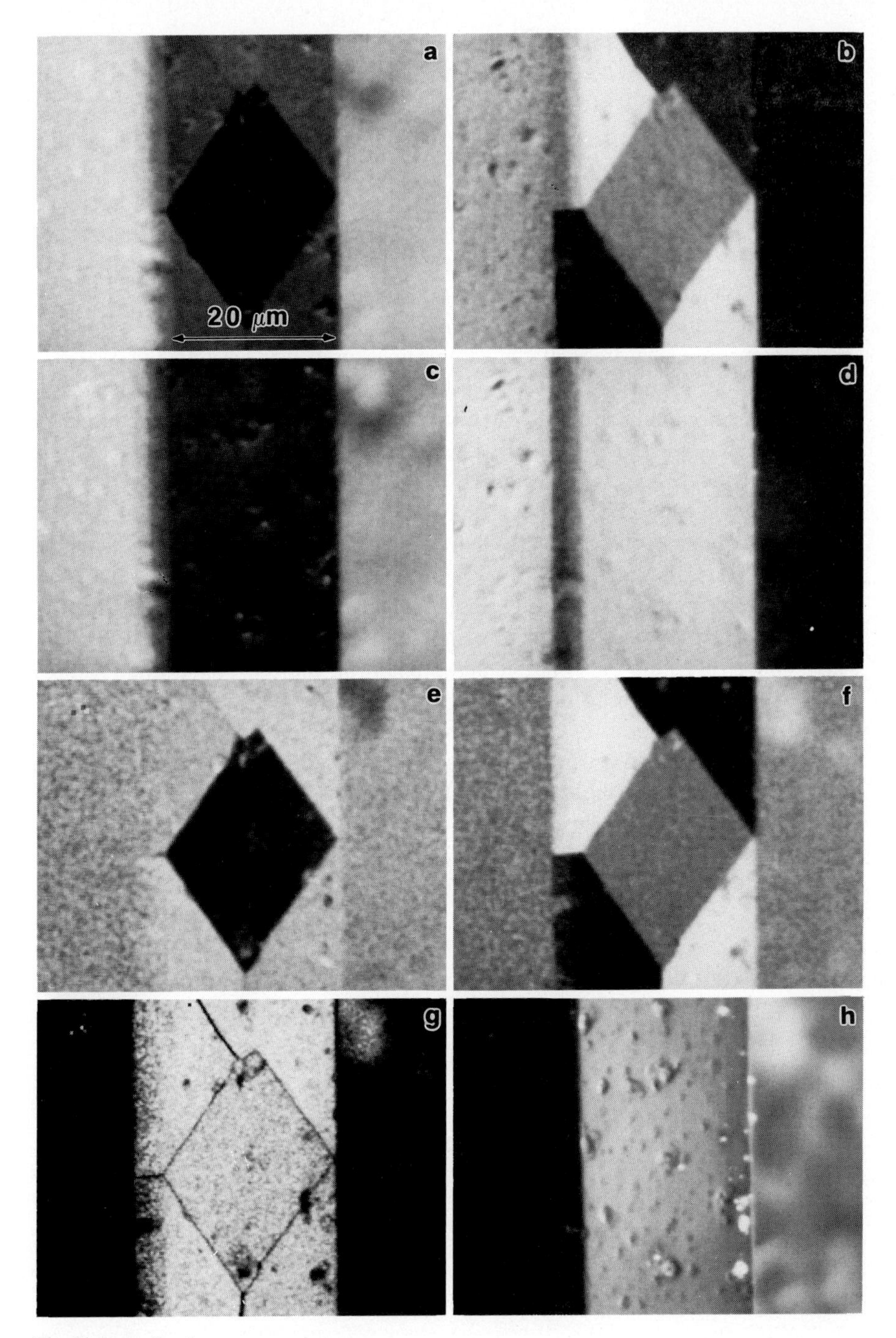

Fig. 11.6a–h. Caption see opposite page

700°C annealing. The whiskers are grown along the ⟨100⟩ direction with sides that are {100} surfaces. The whiskers are about 20 μm wide by several mm long. They are mounted on a nonmagnetic sample holder which is visible to the right and left of the whisker. Unprocessed SEMPA images of the in-plane magnetization components are shown in Figs. 6a and b. The intensity is shown in Fig. 6h. Although the domain structure of the whisker is clearly visible, the unprocessed magnetization images contain several artifacts due to sample topography. Examples of these artifacts include the non-zero polarization signals from the sample holder, the polarization level from the top of the sample differs from the side, and surface roughness "feed through" into the magnetization signal. Figures 6c and d show the same areas of the sample measured using a graphite target in the polarimeter. Figures 6e and f show the magnetization components after subtracting the graphite data. Inspection of these images shows that the instrumental artifacts have been greatly reduced. This can also be observed in the magnitude of the magnetization shown in Fig. 6g; the magnetization is essentially constant over the top and side of the whisker and equal to zero on the sample holder. The remaining regions of low magnetization on the sample are due to nonmagnetic contamination.

A final example of a SEMPA measurement of an Fe whisker is shown in Fig. 11.7. Two important features of SEMPA are illustrated by this measurement, which shows both the diamond shaped domain on the top of the tilted sample and the associated zig-zag domain wall running down the side. First, because SEMPA uses an electron microscope to form the electron probe, the technique has excellent depth of focus making it useful in non-planar geometries. Secondly, this measurement emphasizes that SEMPA, like all of the other domain imaging techniques, only looks at the surface closure domain structure. The bulk domain structure of this whisker is the diamond domain observed from the top. This bulk structure is not at all obvious from the zig-zag closure domain structure observed along the side.

Fig. 11.7a,b. A SEMPA measurement showing (**a**) M_y and (**b**) the intensity from an Fe single crystal whisker demonstrating the technique's depth of focus. Domains are clearly visible on the top and side of this tilted sample which has a rectangular cross section

11.3.2 Cobalt Crystals

The domain structure of cobalt is primarily determined by its uniaxial crystalline anisotropy. The easy magnetization axis of a *hcp* cobalt crystal is along the c axis. SEMPA has been used to investigate domains on Co surfaces with the c axis lying in plane [11.10, 44, 48, 51] and normal to the surface plane [11.46, 54]. SEMPA measurements of the Co{0001} surface are especially interesting because the strong c axis crystalline anisotropy results in a magnetization component that is perpendicular to the surface. Figure 11.8 shows all three magnetization components and the intensity for the Co{0001} surface. A positive perpendicular magnetization component, M_z, corresponding to magnetization pointing out of the surface, is colored white in Fig. 11.8. Black corresponds to magnetization pointing into the surface. The images are 10 μm across and took about 25 minutes to acquire. The Co surface was cleaned by Ar ion bombardment followed by annealing to 400°C. As can be seen from the SEMPA images the domain microstructure is rather complex [11.54]. Even with an easy magnetization axis perpendicular to the surface, the large amount of magnetostatic energy associated with a perpendicular magnetization forces the magnetization to primarily lie in plane. The result is a surface magnetic microstructure, which consists of

Fig. 11.8a–d. Domain structure of the {0001} surface of Co. The magnetization has in-plane components, (**a**) M_x and (**b**) M_y, and (**c**) an out-of-plane component, M_z. (**d**) Intensity topograph

Fig. 11.9a–d. SEMPA images showing the direction of the in-plane magnetization in (**a**) Fe-3%Si {100}, (**b**) Co {0001}, (**c**) a stressed ferromagnetic glass, and (**d**) a NiFe thin film with a cross tie domain wall. Magnetization directions are represented by colors as shown in the color wheel in the inset

narrow, branched regions of perpendicular magnetization separated by regions of in-plane magnetization. The magnetization direction varies continuously as the magnetization comes to the surface, flows along the surface, and returns into the bulk. The relationship between the in-plane and out-of-plane domain structures is easier to visualize by comparing the M_z image in Fig. 11.8c with the map of the in-plane magnetization direction which is shown in Fig. 11.9b. The map of the in-plane magnetization angle also shows that the in-plane magnetization has a domain substructure that appears to reflect the sixfold symmetry of the Co{0001} surface.

11.3.3 Ferromagnetic Metallic Glasses

Because of the long range nature of magnetic interactions, measurements of the surface domain microstructure of a magnetic sample can, in some cases, yield useful information about the internal bulk properties of the material. One example of this use of domain imaging is in the determination of internal stresses in ferromagnetic metallic glasses [11.55, 56]. Because crystalline order is missing in the ferromagnetic glasses, they do not have an easy magnetization axis associated with magnetocrystalline anisotropy. This lack of strong anisotropy results in their useful magnetic properties such as, low coercivity, low losses, and high permeability. Magnetic domains in these materials are generally large with weak shape anisotropies determining the orientation of the domains. When the glasses are strained, however, magnetostrictive interactions cause anisotropies which lead to the development of a fine domain pattern. Investigations of the surface domain structure, therefore, yield information about the strains within the ferromagnetic glass and about how different kinds of stresses affect their magnetic properties.

SEMPA measurements of the domain structure in a stressed $Fe_{81}B_{13.5}Si_{3.5}C_2$ metallic glass (Allied 2605SC) are shown in Figs. 11.9 and 11.10. The glass was plastically deformed by bending which produced the black and white shear bands

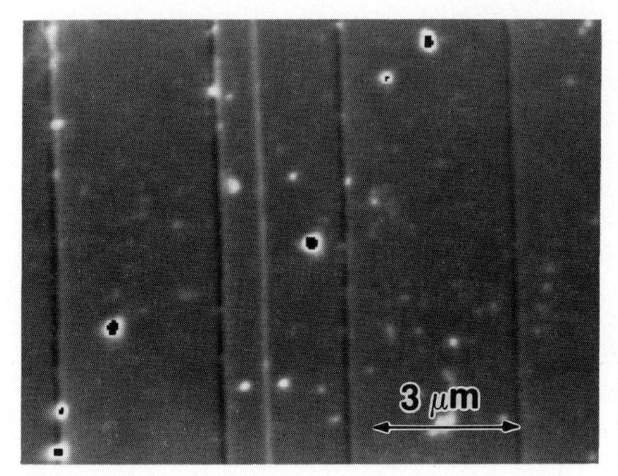

Fig. 11.10. Intensity topograph of a stressed ferromagnetic glass. The vertical lines are shear bands. The corresponding domain structure is shown in Fig. 11.9c

that are visible in the topograph in Fig. 11.10. The domain structure from the same region is shown in Fig. 11.9c. Detailed analysis of the domain structure reveals information about the microstructural properties of the shear bands [11.57].

11.3.4 Domain Walls

A particularly useful feature of the SEMPA technique, when compared with other methods of imaging domains, is that SEMPA can not only produce pictures of the domain structure but also provide quantitative information about the magnetization. SEMPA is, therefore, an appropriate measurement technique for testing various theoretical predictions of magnetic microstructure. One example of this work is the study of the width and the internal structure of domain walls at surfaces.

Within the bulk of a thick ferromagnet the boundary between antiparallel domains is a 180° Bloch wall, in which the magnetization rotates in the plane of the wall. If the Bloch wall were terminated by a surface, the magnetization would point out of the surface causing a large, energetically unfavourable stray field. In a thin film, this surface magnetostatic energy is sufficient to force the magnetization to rotate totally within the plane of the film, forming a Néel wall [11.58]. Because of the magnetostatic energy contribution associated with the surface, it has been predicted [11.59, 60] that, even in a thick ferromagnet, Bloch-like domain walls in the bulk would terminate as Néel walls at the surface. These Néel wall "caps" have been observed using magneto-optic Kerr microscopy [11.61] and SEMPA [11.10, 11, 33, 62]. SEMPA images of domain walls in a Co-based ferromagnetic glass are shown in Fig. 11.11 [11.62]. Only in-plane magnetization components are observed, clearly showing Néel wall behavior at the surface. A Bloch wall would show contrast in the out-of-plane component, which is not observed. Instead, contrast in the horizontal in-plane component is observed. Note that there can be two opposite senses to the rotation of the magnetization within the walls, which results in either black or white contrast in the horizontal component image, Fig. 11.11d.

Because of its spatial resolution, SEMPA measurements provide a stringent test of model calculations of surface domain wall configurations. The calculations are described in detail elsewhere [11.62, 63]. Basically, the calculations involve minimizing the total magnetic energy, which includes contributions from exchange, anisotropy and magnetostatic energy. Only well-known, bulk magnetic properties were used as input parameters for these calculations. The results for the calculation of the magnetization distribution in a cross section of an Fe crystal are shown schematically in Fig. 11.12. The cross section shows a bulk Bloch wall which turns over into a Néel wall at the surface. Domain wall profiles were measured for a 20 μm thick single crystal Fe whisker, a 1200 Å thick $Ni_{80}Fe_{20}$ Permalloy film, and a 2400 Å thick $Ni_{80}Fe_{20}$ Permalloy film. Calculated and measured wall profiles of the in-plane component of the magnetization perpendicular to the wall are shown in Fig. 11.13. The agreement between the measured and calculated wall profiles is impressive, especially, considering the large variation

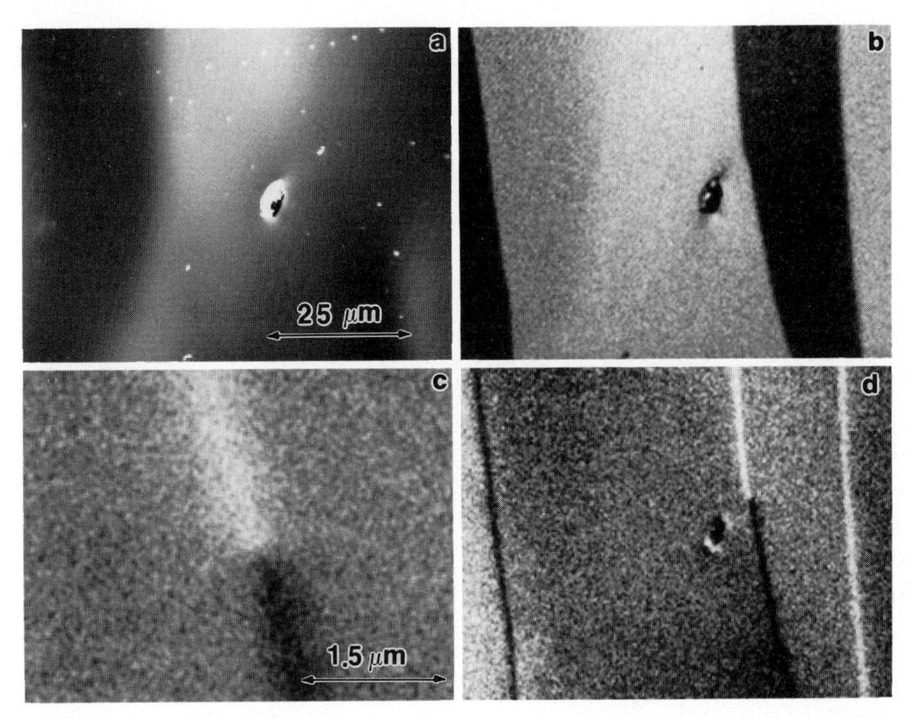

Fig. 11.11a–c. SEMPA images of domain walls in a Co based ferromagnetic glass. Shown are (**a**) the intensity, (**b**) M_y, (**d**) M_x, and (**c**) a magnified image from the region where the wall changes chirality

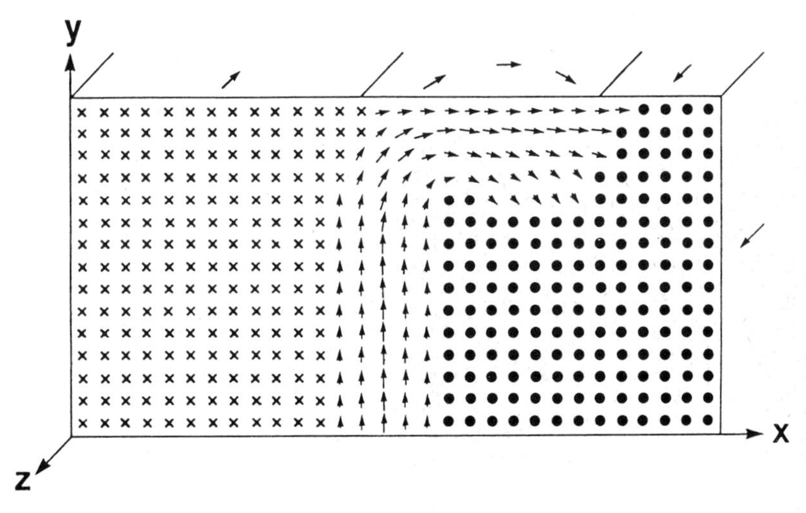

Fig. 11.12. Schematic representation of a calculated magnetization distribution in the upper 0.2 μm of the cross section through an Fe sample. The domain wall is a Bloch wall in the solid, but rotates into a Néel wall at the surface

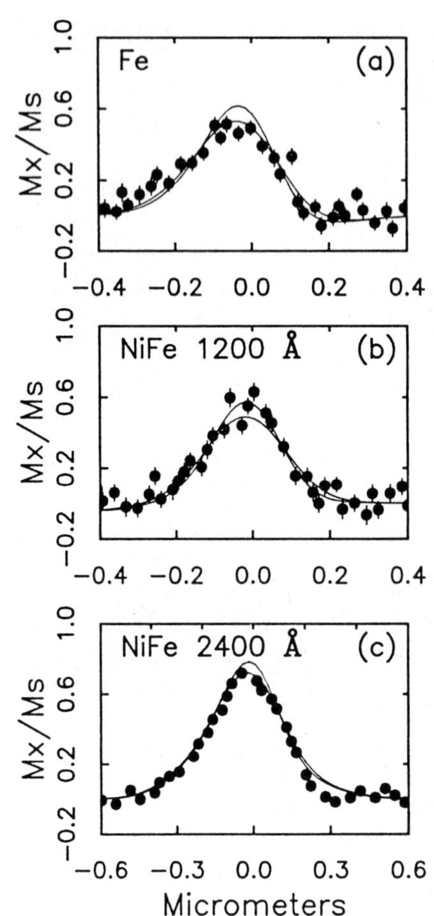

Fig. 11.13a–c. Comparison of calculated and measured profiles of the magnetization component perpendicular to the wall for (**a**) an Fe {100} single crystal, (**b**) a 1200 Å thick $Ni_{80}Fe_{20}$ Permalloy film, and (**c**) a 2400 Å thick Permalloy film

in magnetic properties between these various materials. Similar calculations and measurements were also made for other $Ni_{80}Fe_{20}$ films as a function of film thickness, and again there was good agreement between calculated and measured wall widths [11.64]. Because of this excellent agreement, the magnetic models can be used with confidence to predict magnetization distributions in other systems. For example, the model can be used to interpret the results of TEM Lorentz measurements which average over the thickness of a sample, or understand the interactions between the tip and sample in magnetic force microscopy measurements [11.63].

In addition to measurements of domain wall widths, SEMPA can be used to study other magnetic microstructural phenomena associated with domain walls. An example of such a structure is the magnetic singularity shown in Fig. 11.11c. The singularity is associated with a change in the chirality of the domain wall. The in-plane magnetization swirls about this singularity in a clockwise direction. At the center of the singularity the topology does not permit a non-zero in-plane magnetization component, so that the magnetization must either go to zero or be forced into or out of the plane. Either case would be extremely interesting to observe, but the current spatial resolution of SEMPA only permits placing an upper limit of 65 nm on the radius of the singularity on a Co based ferromagnetic glass [11.65]. This is still about three times as large as predicted by model calculations [11.66]. Another question is whether the singularity occurs only at the surface. Then the surface Néel wall changes chirality but the underlying Bloch wall does not. If it penetrates into the bulk, then both the surface Néel and the bulk Bloch wall switch, forming a Bloch line [11.67]. Both cases can occur and SEMPA can be used to differentiate between them by measuring the lateral displacement of the surface Néel wall at the singularity. When only the surface wall changes chirality, there is a shift in the position of the domain wall at the singularity. This is the case in Fig. 11.11c. The shift is simply a result of the surface Néel wall being offset slightly from the bulk Bloch wall as shown in Fig. 11.12. When both the surface and bulk domain walls change chirality, no lateral shift in the surface domain wall position is observed.

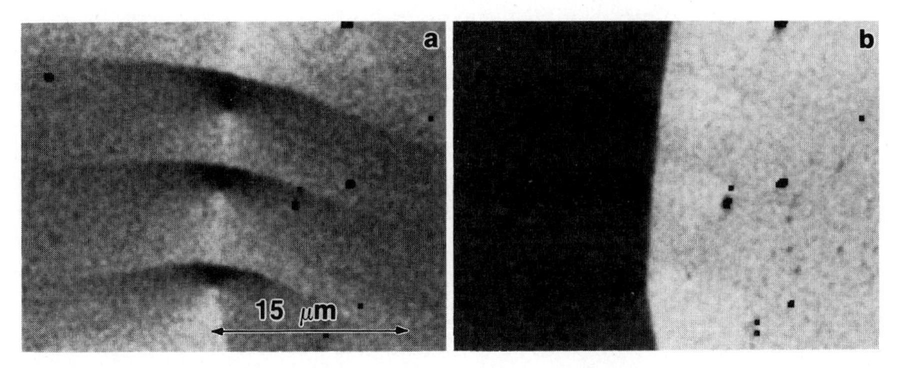

Fig. 11.14. SEMPA images of "Cross Tie" domain walls in a NiFe thin film showing (**a**) M_x and (**b**) M_y magnetization components

As a final example of how complicated domain wall structures can become, Figs. 11.14 and 11.9d show SEMPA measurements of "cross-tie" domain walls in a 40 nm thick $Ni_{80}Fe_{20}$ film. A single cross-tie, of which there are three in the figure, consists of two changes in domain wall chirality and hence two singular points [11.58]. The pair of singularities differ in that the magnetization forms a vortex about one but diverges from the other. The existence of cross-ties is a function of the film thickness and the applied magnetic field. Cross-ties occur in Permalloy films that are between 10 nm and 100 nm thick. Besides their scientific interest, most of the domain wall structures, such as cross-ties and Bloch lines, have potential uses in various magnetic memory devices where their small size makes them ideal for high density storage.

11.3.5 Magnetic Storage Media

One of the major driving forces behind magnetics technology today is the search for ways to store the maximum amount of information in the smallest space. One result of this push toward high density magnetic storage is that the physical size of the basic unit of information, a logical bit, has become smaller than can be imaged by conventional domain observation techniques. Bit dimensions are currently on the order of a μm, an will tend toward a tenth of a μm in the near future. In order to be a useful analytical tool for examining magnetic recording media, a domain observation technique must be able to image the recorded bits with sufficient resolution to not only see the bit but also resolve the transition region between bits. The quality of the recording media, in terms of the signal-to-noise during reading, is a sensitive function of the sharpness of the transition between bits. For example, in a conventional hard disc memory unit, bits written with poorly resolved transitions will yield lower signals when read by an inductive or magneto-resistive read/write head. Because of the high spatial resolution and the ability to look at as-deposited opaque samples, SEMPA is a useful technique for examining the magnetic microstructure of written information in conductive recording media. This information can be correlated with measurements of macroscopic magnetic properties in order to understand the recording characteristics of the media on a microscopic level.

SEMPA measurements of bits written on commercial memories have so far been limited to hard disc recording media composed primarily of Co-Ni alloy thin films with in-plane magnetization [11.10, 11.32, 11.68]. An example of how SEMPA can be used to understand the difference in recording characteristics between various media is illustrated in Fig. 15. The intensity topograph and the component of the magnetization approximately along the recorded track are shown for two different media compositions. Figure 11.15a, b correspond to a $Co_{26}Cr_{12}Ta_2$ film in which the bits were read with good signal-to-noise ratios. Figure 11.16c, d correspond to $Co_{75}Ni_{25}$ films that were noisier to read. The SEMPA images show that the reason for the poorer performance of the $Co_{75}Ni_{25}$ film is that the bits are not separated into distinct domains, but instead are randomly bridged together. In the good media the bits are more clearly separated.

Information can also be stored in magnetic media by using a focused laser beam to write and read the bits [11.69]. Briefly, writing a bit involves heating the media in an applied field. Bits are read by sensing the rotation of polarized light reflected from the surface. The materials involved are not the simple transition metal ferromagnets as discussed so far, but are transition metal-rare earth ferrimagnets in which the transition metal and the rare earth are magnetized in opposite directions. At a particular temperature, referred to as the compensation point, the two magnetic subsystems have equal magnetizations and, therefore, the net magnetization of the material is zero. Since compensation points are usually near room temperature, imaging domains with observation techniques that are sensitive to the net magnetization or field can be difficult. SEMPA has the useful feature that it is primarily sensitve to the magnetization of the transition metal. A

Fig. 11.15a–d. SEMPA images of bits written on thin film hard disc media of two different componsitions. The topography and M_y are shown for a sample with good recording properties in (**a**) and (**b**), and for a sample with goord recording properties in (**a**) and (**b**), and for a sample with noisier, less well resolved bits in (**c**) and (**d**)

possible explanation is that only the weakly bound polarized valence electrons of the transition metal contribute to the secondary electron cascade while the more localized magnetic $4f$ electrons of the rare earth do not. This feature allows SEMPA to be used for imaging magnetic structures in rare earth–transition metal ferrimagnets even at the compensation point.

Figure 11.16 shows SEMPA measurements of magnetic domains written using a focused laser beam in a magneto-optic storage media, $Tb_{23.6}Fe_{67.6}Co_{8.8}$ (214 K compensation temperature) [11.70]. In this case, as with most magneto-optic media, the magnetization of the bulk material is perpendicular to the surface. The SEMPA measurements, however, show that there is also a sizable component of the magnetization lying in the plane of the surface. In fact, the magnetization makes an angle of about 45° with respect to the surface. The angle of the magnetization at the surface was found to depend upon the amount of Fe in the surface. As the amount of Fe is increased, the magnetization becomes more in plane.

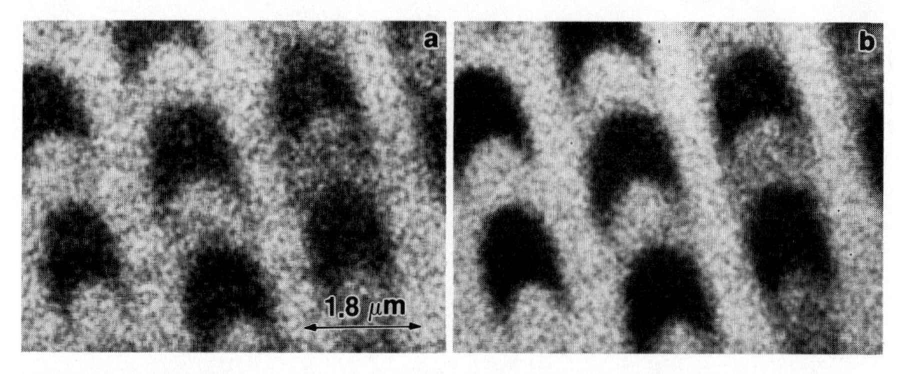

Fig. 11.16a, c. SEMPA images of bits written by a laser in a $Tb_{23.6}Fe_{67.6}Co_{8.8}$ magneto-optic recording media. The (**a**) M_x and (**b**) M_z magnetization components are shown

11.4 Summary and Future Directions

There are several features of SEMPA that make it an extremely powerful technique for the investigation of surface magnetic microstructure. First, the most important feature is that, because the polarization of secondary electrons emitted from a transition metal ferromagnet is directly proportional to the magnetization, SEMPA can directly measure the direction of the magnetization in a ferromagnet. SEMPA can, therefore, not only provide images showing the domain structure, but also quantitative information about the magnetization. Second, because SEMPA is based on a measurement of the electron spin polarization, a quantity that is independent of intensity, the magnetic structure can be separated from the topography. This separation permits investigations of the relationships between magnetic, topographic and, in some cases, chemical structures. Third, the technique is surface sensitive with a probing depth on the order of a nanometer. Surface sensitivity is an asset for studying thin films and surfaces, but the ultra-high vacuum requirements and extensive surface preparation are a drawback when only bulk magnetic structures are of interest. Finally, because SEMPA uses a scanning electron microscope as the incident probe, magnetic structures in opaque ferromagnetic samples can be imaged over a long depth of field with high spatial resolution (about 40 nm at present). The use of an electron probe does require that samples be electrically conducting in order to reduce charging and that stray magnetic fields near the sample must be held to a minimum (less than a few gauss) so that the secondary electron trajectories and polarizations are not disturbed.

Future improvements in SEMPA instrumentation will primarily be directed towards improving the spatial resolution and developing more efficient electron spin polarization analyzers. In the near future, the use of field emission electron sources along with optimized electron optics should push the spatial resolution of the technique to about 10 nm. Further major improvements in resolution may be difficult because of multiple scattering within the specimen which increase the sampling volume. On the other hand, there is room for four orders of magnitude of improvement in the efficiency of the electron spin polarimeters. More efficient polarimeters would effectively improve the spatial resolution of SEMPA since less current would be needed and image acquisition times could be significantly shortened. There is sufficient contrast and intensity in the secondary electron signal so that television rate domain imaging would be possible if polarimeter efficiencies could be significantly improved. Unfortunately, only minor improvements in efficiency appear to be possible with the analyzers that are currently available.

Another potential area for improvement of the SEMPA technique involves the absolute quantification of the measurement; in effect, turning SEMPA into a surface magnetometer. Currently SEMPA measurements are proportional to the total magnetization. Absolute magnetization measurements will require some improvements in detector calibration, but primarily quantification will require more information about the secondary electron emission features such as sampling depth, effects of magnetic and non-magnetic adsorbates, and the energy and angle dependence of secondary electron polarization for various materials. Perhaps a short term solution will involve the use of calibration standards, such as those used in x-ray and Auger compositional analysis, in order to make more meaningful relative magnetization measurements.

Future work using SEMPA will be influenced by the fact that SEMPA has become a tested, routine method for the imaging of magnetic microstructures. One can, therefore, expect to see less work involving the development of the SEMPA technique and more emphasis on applying SEMPA to the myriad of technological and fundamental magnetics problems. One potentially fertile area of research that SEMPA is especially well suited for and that has barely been touched upon is the study of magnetic structures in films that are only a few monolayers thick. Because the films must be grown under ultra high vacuum conditions, SEMPA can be used to study the layer-by-layer development of the magnetic microstructure during film growth. Preliminary work in which domains in three monolayer thick Fe films were easily observed using SEMPA clearly show that SEMPA has the necessary surface sensitivity to observe magnetic structures in thin films [11.71]. SEMPA could be used to answer questions about whether domains exist in monolayer films, how layered thin film structures are magnetically coupled, and how the interlayer coupling is influenced by roughness and defects.

Acknowledgements. We gratefully acknowledge the assistance of many collaborators, especially M.H. Kelley, M. Aeschliman, M. Hart and A. Hubert. Samples were supplied by A. Arrott (Fe whiskers), J.C. Li (Co metglasses), P. Ryan (NiFe films), M. Khan (CoNi hard disc media), and S. Klahn (TbFe magneto-optic media). This work was supported in part by the Office of Naval Research.

References

11.1 J. Kessler: *Polarized Electrons*, 2nd Ed. (Springer, Berlin, Heidelberg 1985)
11.2 R. Feder: *Polarized Electrons in Surface Physics* (World Scientific, Singapore 1985)
11.3 R.J. Celotta, D.T. Pierce: Science **234**, 333 (1986)

11.4 J. Kirschner: *Polarized Electrons at Surfaces* (Springer, Berlin, Heidelberg 1985)
11.5 T.H. DiStefano: IBM Tech. Discl. Bull. **20**, 4212 (1978)
11.6 J. Unguris, D.T. Pierce, A. Galejs, R.J. Celotta: Phys. Rev. Lett. **49**, 72 (1982)
11.7 J. Kirschner: Scanning Electron Microsc. 1984; III: 1179 (1984)
11.8 K. Koike, K. Hayakawa: Jpn. J. Appl. Phys. **23**, L187 (1984)
11.9 J. Unguris, G.G. Hembree, R.J. Celotta, D.T. Pierce: J. Mircroscopy **139**, RP1 (1985)
11.10 K. Koike, H. Matsuyama, K. Hayakawa: Scanning Micro. Intern., Supp. **1** 241 (1987)
11.11 G.G. Hembree, J. Unguris, R.J. Celotta, D.T. Pierce: Scanning Micro. Intern., Supp. **1**, 229 (1987)
11.12 D.T. Pierce, J. Unguris, R.J. Celotta: MRS Bulletin **13**(6), p. 19 (1988)
11.13 H.S. Williams, R.M. Bozorth, W. Shockley: Phys. Rev. **75**, p. 155 (1949)
11.14 W. Rave, R. Schafer, A. Hubert: J. Magn. Magnet. Mat. **65**(7) (1987)
11.15 B.E. Argyle, B. Petek, D.A. Herman, Jr.: J. Appl. Phys. **61**, 4303 (1987)
11.16 D.E. Newbury, D.C. Joy, P. Echlin, C.E. Fiori, J.I. Goldstein: (Plenum New York, 1986) p. 147
11.17 J.P. Jakubovics: *Electron Microscopy in Materials Science* Part IV, ed. by E. Ruedl, U. Valdre (Commission of the European Communities, Brussels, 1973) p. 1305
11.18 J.N. Chapman, S. McVitie, J.R. McFadyen: Scanning Electron Microscopy Suppl. **1**, 221 (1987)
11.19 A. Tonomura: J. Appl. Phys. **61**, 4297 (1987)
11.20 Y. Martin, D. Rugar, H.K. Wickramasinghe: Appl. Phys. Lett. **52**, 244 (1988)
11.21 J. Kirschner: In *Surface and Interface Characterization by Electron Optical Methods*, ed. by A. Howie, U. Valdre, (Plenum, New York, 1987) p. 267
11.22 E. Kisker, W. Gudat, K. Schröder: Solid State Commun. **44**, 591 (1982)
11.23 H. Hopster, R. Raue, E. Kisker, G. Guntherodt, M. Campagna: Phys. Rev. Lett. **50**, 70 (1983)
11.24 D. Penn, S.P. Apell, S.M. Girvin: Phys. Rev. Lett. **55**(5), 518 (1985)
11.25 D.R. Penn, S.P. Apell, S.M. Girvin: Phys. Rev. B **32**(12), 7753 (1985)
11.26 J. Kirschner, D. Rebenstorff, H. Ibach: Phys. Rev. Lett. **53**, 694 (1984)
11.27 J. Glazer, E. Tosatti: Solid State Commun. **52**, 905 (1984)
11.28 H. Hopster, R. Raue, R. Clauberg: Phys. Rev. Lett. **53**, 695 (1984)
11.29 H. Seiler: J. Appl. Phys. **54**, R1 (1983)
11.30 M.P. Seah, W.A. Dench: Surf. Interface Anal. **1**, 2 (1979)
11.31 D.L. Abraham, H. Hopster: Phys. Rev. Lett. **58**, 1352 (1987)
11.32 M.R. Scheinfein, J. Unguris, M.H. Kelley, D.T. Pierce, R.J. Celotta: Rev. Sci. Inst. (to be published)
11.33 H.P. Oepen, J. Kirschner: Phys. Rev. Lett. **62**(7), 819 (1989)
11.34 J. Orloff: Ultramicr. **28**, 88 (1989)
11.35 K. Koike, H. Matsuyama, H. Todokoro, K. Hayakawa: Scanning Micr. **1**(1), 31 (1987)
11.36 M.R. Scheinfein: Optik **82**(3), 99 (1989)
11.37 E. Kisker, R. Clauberg, W. Gudat: Rev. Sci. Instr. **53**, 1137 (1982)
11.38 L.G. Gray, M.W. Hart, F.B. Dunning, G.K. Walters: Rev. Sci. Instr. **55**(1), 88 (1984)
11.39 J. Unguris, D.T. Pierce, R.J. Celotta: Rev. Sci. Instr. **57**(7), 1314 (1986)
11.40 M.R. Scheinfein, D.T. Pierce, J. Unguris, J.J. McClelland, R.J. Celotta: Rev. Sci. Instr. **60**(1), 1 (1989)
11.41 D.T. Pierce, S.M. Girvin, J. Unguris, R.J. Celotta: Rev. Sci. Instr. **52**(10), 1437 (1981)
11.42 K. Koike, H. Matsuyama, K. Hayakawa: Jap. J. Appl. Phys. **27**(7), L1352 (1988)
11.43 D.T. Pierce, R.J. Celotta, M.H. Kelley, J. Unguris: Nuc. Inst. Meth. **A266**, 550 (1988)
11.44 K. Koike, K. Hayakawa: J. Appl. Phys. **57**(1), 4244 (1985)
11.45 J.A. Venables, A.P. Janssen: Proc. of the 9th Int. Conf. on Electron Microscopy, Toronto (1978)
11.46 M.H. Kelley, J. Unguris, M.R. Scheinfein, D.T. Pierce, R.J. Celotta: Proc. of the Microbeam Analysis Society – 1989, ed. by P.E. Russell, (San Francisco Press, San Francisco 1989) p. 391
11.47 T. VanZandt, R. Browning, C.R. Helms, H. Poppa, M. Landolt: to be published in Rev. Sci. Inst.
11.48 K. Koike, H. Matsuyama, H. Todokoro, K. Hayakawa: Jap. J. Appl. Phys. **24**(8), 1978 (1985)
11.49 K. Koike, H. Matsuyama, H. Todokoro, K. Hayakawa: Jap. J. Appl. Phys. **24**(10), L833 (1985)
11.50 H. Matsuyama, K. Koike, H. Todokoro, K. Hayakawa, IEEE Translation Journal on Magnetics in Japan, **TJMJ-1**(9), 1071 (1985)

11.51 K. Hayakawa, K. Koike, H. Matsuyama: Hitachi Inst. News **14**, 11 (1988)
11.52 G.G. Hembree, J. Unguris, R.J. Celotta, D.T. Pierce: Proc. of the 44th EMSA Meeting, ed. by G.W. Bailey, (San Francisco Press, San Francisco, 1986) p. 634
11.53 J. Unguris, G.G. Hembree, R.J. Celotta, D.T. Pierce: J. Vac. Sci. Technol. **A5**(4), 1976 (1987)
11.54 J. Unguris, M.R. Scheinfein, D.T. Pierce, R.J. Celotta: Appl. Phys. Lett. **55**(24), 2553 (1989)
11.55 H. Kronmuller, W. Ferengel: Phys. Stat. Sol. (a) **64**, 593 (1981)
11.56 J.D. Livingston: J. Appl. Phys. **57**, 3555 (1985)
11.57 V. Lakshmanan, J.C.M. Li: Proc. of the Sixth Int. Conf. on Rapidly Quenched Metals, Mat. Sci. Eng. **98**, 483 (1988)
11.58 B.D. Cullity: *Introduction to Magnetic Materials*, (Addison-Wesley, Reading, MA 1972) p. 429
11.59 A. Hubert: Phys. Status Solidi B **32**, 519 (1969)
11.60 A.E. LaBonte: J. Appl. Phys. **40**, 2450 (1969)
11.61 F. Schmidt, W. Rave, A. Hubert: IEEE Trans. Mag. **21**, 1596 (1985)
11.62 M.R. Scheinfein, J. Unguris, R.J. Celotta, D.T. Pierce: Phys. Rev. Lett. **63**, 668 (1989)
11.63 M.R. Scheinfein, J. Unguris, D.T. Pierce, R.J. Celotta: Proc. of 34th Conf. on Magnetism and Magnetic Materials, Boston, MA.
11.64 M.R. Scheinfein, J. Unguris, P. Ryan, R.J. Celotta, D.T. Pierce: unpublished
11.65 M.R. Scheinfein, J. Unguris, R.J. Celotta, D.T. Pierce: International Workshop on the Magnetic Properties of Low Dimensional Systems, San Luis Potosi, Mexico, ed. by L.M. Falicov, J.L. Moran-Lopez, (Springer, Berlin, Heidelberg, November 1989)
11.66 A. Hubert: J. de Phys. Coll. C8, **49**, 1865 (1988)
11.67 R. Schafer, W.K. Ho, J. Yamasaki, A. Hubert, F.B. Humphrey: Proc. of Intermag 1989, Washington, DC, to be published in IEEE Trans. Magn.
11.68 D.T. Pierce, M.R. Scheinfein, J. Unguris, R.J. Celotta: Materials Research Society Symposium Proceedings **151**, 49 (1989)
11.69 M. Hartmann, B.A.J. Jacobs, J.J.M. Braat: Phillips Techn. Review **42**, 37 (1985)
11.70 M. Aeschlimann, M.R. Scheinfein: private communication
11.71 J.L. Robins, R.J. Celotta, J. Unguris, D.T. Pierce, B.T. Jonker, G.A. Prinz: Appl. Phys. Lett. **52** (22), (1988)

12. Low Energy Electron Microscopy

E. Bauer

Physikalisches Institut, Technische Universität Clausthal,
D-3392 Clausthal-Zellerfeld
und Sonderforschungsbereich 126, Göttingen-Clausthal, Fed. Rep. of Germany

Low energy electron microscopy (LEEM) is a surface imaging technique in which the surface is illuminated by an approximately parallel electron beam at near normal incidence. The image is formed with those electrons which are elastically backscattered into a small angular region around the surface normal. The limitation to a small angular region is necessary because of the large aberrations of the objective lens which produces the primary image. This lens is a so-called cathode lens which not only has imaging properties but at the same time decelerates the fast electrons of the illuminating beam to the desired low energy at the specimen and re-accelerates the backscattered electrons to high energies again. In order to achieve this, the specimen is at a high negative potential which differs from the potential of the emitter of the electron gun of the illumination system by $V_0 = E_0/e$, where E_0 is the energy of the electrons at the specimen. Typical energy values are $E = eV = 15\text{–}20\,\text{keV}$ for the fast electrons and $0 < E_0 < 50\,\text{eV}$ at the specimen. There are three fundamental quantities which are important in LEEM: resolution, intensity and contrast. These will be discussed in Sect. 12.1. Section 12.1 also describes how LEEM can be combined with other surface characterization techniques, such as low energy electron diffraction (LEED), photoemission electron microscopy (PEEM) and other emission microscopies. Section 12.2 illustrates the applications of LEEM and of the associated techniques to the study of clean surfaces, while Sect. 12.3 presents examples of the power of LEEM in the study of surface layers. Section 12.4 gives an outlook for possible future developments. A brief summary (Sect. 12.5) concludes this chapter.

12.1 Fundamentals of LEEM

12.1.1 Resolution and Intensity Transmission

The theoretical resolution limit of LEEM is determined by the cathode lens. This lens may be divided, in principle, into an acceleration region and into a lens which produces a real image. Reasonable imaging lenses have aberrations smaller than those of the acceleration region so that the latter limits the resolution. To a first approximation, the accelerating field may be considered to be homogeneous and the radius d of the disk of confusion referred to the object plane which is caused by diffraction at the angle limiting aperture, by chromatic and spherical aberration may be approximated by the geometric mean of these quantities.

$$d^2 = d_C^2 + d_S^2 + d_D^2 \,. \tag{12.1}$$

With the approximation $r_A/L = R_A = 2\sqrt{\varrho}\sin\alpha$ (r_A α-limiting aperture, L field length, $\varrho = V_0/V$) and neglecting terms with $\varrho^{3/2}$, $\varepsilon^{3/2}$ and higher order terms in ϱ and ε ($\varepsilon = \Delta V_0/V$), one obtains for the relative radii $\delta = d/L$:

$$\delta_C = -\varepsilon\sin\alpha \,, \quad \delta_S = \varrho\sin^3\alpha \,, \quad \delta_D = 1.2\lambda/R_A = 0.6\lambda/(\sqrt{\varrho}\sin\alpha) \,, \tag{12.2}$$

with $\lambda = \Delta/L$ and $\Delta = \sqrt{1.5/\mathrm{V}}\,\mathrm{nm}$ [12.1]. Thus,

$$\delta^2 = \varepsilon^2\sin^2\alpha + \varrho^2\sin^6\alpha + \frac{0.36\lambda^2}{\varrho\sin^2\alpha} \,. \tag{12.3}$$

If intensity is of no concern but only resolution is to be optimized, we obtain for the angular aperture which gives minimim δ^2

$$\sin^4\alpha_{\mathrm{opt}} = \frac{\varepsilon^2}{6\varrho^2}\left[\sqrt{1 + \frac{4.32\lambda^2\varrho}{\varepsilon^4}} - 1\right] \approx 0.36\frac{\lambda^2}{\varrho\varepsilon^2} \,. \tag{12.4}$$

The approximation is valid when $\lambda^2\varrho/\varepsilon^4 \ll 1$, a condition which is usually fulfilled in LEEM. With this approximation

$$\delta^2 = 1.2\lambda\varepsilon/\sqrt{\varrho} \,. \tag{12.5}$$

Expressed in ΔV_0, V_0 and field strength $F = V/L$, this gives for the absolute resolution

$$d = 3.83 \times 10^{-5}(\Delta V_0)^{1/2} V_0^{-1/4} F^{-1/2}\mathrm{m} \,. \tag{12.5a}$$

For $V = 15\,\mathrm{kV}$, $V_0 = 1.5\,\mathrm{V}$, $\Delta V_0 = 0.3\,\mathrm{V}$, $L = 3\,\mathrm{mm}$, equations (12.5, 5a) yield $d \approx 8.5\,\mathrm{nm}$. An increase of V_0 by a factor of n decreases d by a factor of $n^{-1/4}$. For example, for $V_0 = 150\,\mathrm{V}$, $d \approx 2.7\,\mathrm{nm}$. Thus, resolutions between about 3 and 10 nm are achievable at start energies between 150 eV and 1.5 eV for small energy widths. Other resolution considerations, which also take the angular distribution of the electrons into account, can be found in Refs. [12.2, 3].

When intensity becomes a critical quantity then the quality of the microscope is not only determined by the resolution but also by the transmitted intensity I. We may then define a quality-factor $Q(\alpha) = I(\alpha)/\delta^2(\alpha)$ which has to be optimized. For the particularly simple case of a cosine emitter, $I(\alpha) = I_0\cos\alpha$, independent of energy within the energy window $\Delta E = e\Delta V_0$ transmitted, one obtains with a limited angular aperture α_0

$$I(\alpha_0) = I_0 2\pi \int_0^{\alpha_0} \cos\alpha\,\sin\alpha\,d\alpha\,\Delta E = I_0 V\varepsilon\pi\sin^2\alpha_0 \,, \tag{6}$$

and for the quality factor $Q(\alpha_0)$, using (12.3)

$$Q(\alpha_0) = I_0 V \pi \frac{\varepsilon}{\varepsilon^2 + \varrho^2 \sin^4 \alpha_0 + 0.36\lambda^2/\varrho \sin^4 \alpha_0} . \tag{12.7}$$

Optimization with respect the α_0 leads to

$$\sin^4 \alpha_{0,\text{opt}} \equiv \sin^4 \alpha_1 = 0.6\lambda/\varrho^{3/2} , \tag{12.8}$$

which yields

$$Q(\alpha_1) = I_0 V \pi \frac{\varepsilon}{\varepsilon^2 + 1.2\lambda\varrho^{1/2}} = Q(\alpha_1, \varepsilon) . \tag{12.9}$$

It is obvious that $Q(\alpha_1, \varepsilon)$ can still be optimized with respect to the energy window ε, which yields

$$\varepsilon_{\text{opt}}^2 \equiv \varepsilon_1^2 = 1.2\lambda\varrho^{1/2} . \tag{12.10}$$

Thus

$$Q_{\max} = Q(\alpha_1, \varepsilon_1) = \frac{I_0 V \pi}{2\sqrt{1.2}} \frac{1}{\lambda^{1/2}\varrho^{1/4}} = \frac{I_0 V \pi}{2\varepsilon_1} . \tag{12.11}$$

For the radius δ_1 of the disc of confusion which is connected with ε_1 and α_1, one obtains

$$\delta_1^2 = 4 \times (0.6)^{3/2} \lambda^{3/2} / \varrho^{1/4} . \tag{12.12}$$

Similar results are obtained from signal to noise considerations [12.4]. If the reduced quantities $\delta = d/L$, $\lambda = \Delta/L$ and $\varrho = V_0/V$ are expressed by V_0 and the field strength $F = V/L$, then the actual radius d_1 of the minimum disc of confusion is given by

$$d_1 \approx 1.6 F^{-1/4} V_0^{-1/8} \text{m} . \tag{12.12a}$$

Thus, there is only a weak dependence on field strength and an even weaker one on starting energy. For $E = 15000\,\text{eV}$ and $L = 3\,\text{mm}$, the values of α_1, ΔE_1 and d_1 are 12.2°, 0.10 eV and 5.7 nm for $E_0 = 1.5\,\text{eV}$ and 2.1°, 0.29 eV and 3.2 nm at $E_0 = 150\,\text{eV}$. The better resolution at 1.5 eV compared to that obtained from δ^2-optimization is due to the lower ΔE value used here. If the optimum ε value in (12.10) which maximizes Q is inserted into the expressions (12.3, 4) connected with the optimization of δ^2, then

$$\sin^2 \alpha_{\text{opt}} = \tfrac{1}{3} \sin^4 \alpha_1 \tag{12.13}$$

and

$$\delta_{\text{opt}}^2 = \frac{4}{3\sqrt{3}} \delta_1^2 \quad \text{or} \quad \delta_{\text{opt}} = 0.877\,\delta_1 . \tag{12.14}$$

Thus, not much is gained in resolution by choosing the optimum aperture obtained by minimizing δ^2. In LEEM, the angular distribution is not cosine-like but peaked in the diffraction beams of which usually only one, generally the (00) beam, passes the aperture, so that Q-optimization is not very relevant. In emission microscopy, in particular with Auger electrons or characteristic photoelectrons, however, it is of fundamental importance because of the low intensity available from the specimen.

Before concluding the section on resolution, the influence on the resolution of the imaging lens beyond the accelerating field has to be discussed briefly. Several lens configurations have been used or considered: i) electrostatic triodes, ii) electrostatic tetrodes, iii) magnetic lenses and iv) magnetic "triodes" in which the magnetic pole pieces have different potentials. An electrostatic triode is used in the present LEEM instruments. Its resolution is worse by about a factor of two than that of the homogeneous field [12.5, 6]. Magnetic lenses have been used very successfully with resolutions as low as 12 nm in UV photoemission [12.7]. A comparison of various lens types has been made recently [12.8] with the result that at low energies, the magnetic triode is slightly better than the electrostatic tetrode which in turn is significantly better than the electrostatic triode. At high energies, the magnetic triode is significantly better than the electrostatic lenses, which is not surprising because the imaging lens increasingly determines the resolution with increasing energy. At low energy, the resolution of the magnetic triode approaches the value of the homogeneous field. A resolution ranging from 7.5 nm at 2 eV to 3.5 mn at 200 eV for $e\Delta V_0 = 0.5$ eV and $V = 20$ kV is predicted for such a lens and should become available in the near future.

12.1.2 Intensity

The intensity I_0 discussed above is determined by the intensity of the illumination of the specimen and by the response of the specimen to the illumination by photons, electrons or other particles. Here we are only interested in the elastic backscattering of slow electrons. Other processes such as Auger electron or photoelectron emission are discussed elsewhere [12.9]. The backscattering of slow electrons differs vastly from the forward scattering of fast electrons, which is used in conventional transmission and reflection microscopy and which is determined mainly by the nuclear charge Z. The backscattering of slow electrons is nonmonotonic in Z and in energy E [12.10]. In crystalline samples the E-dependence is, of course, dominated by diffraction effects, but the intensity of the diffracted beams is subjected to the general E-dependence of the backscattering cross-sections. Of fundamental importance for LEEM is the fact that the backscattering cross-section can be very large at low energies so that a large fraction of the incident intensity is available for imaging.

A good example is shown in Fig. 12.1, which illustrates not only the high specular reflectivity $R(E) = I_{\text{reflected}}/I_{\text{incident}}$ at normal incidence at low energies but also the strong surface orientation dependence caused mainly by diffraction effects [12.11, 12]. In any case, reflectivities of more than 50% are obtainable at low energies. This is not due to the large Z of the material but follows from backscattering calculations for many other materials and has also been demonstrated experimentally for lower Z materials [12.13]. A particularly impressive case is H. Figure 12.2 shows the change $\Delta R(E)$ of the reflectivity of a

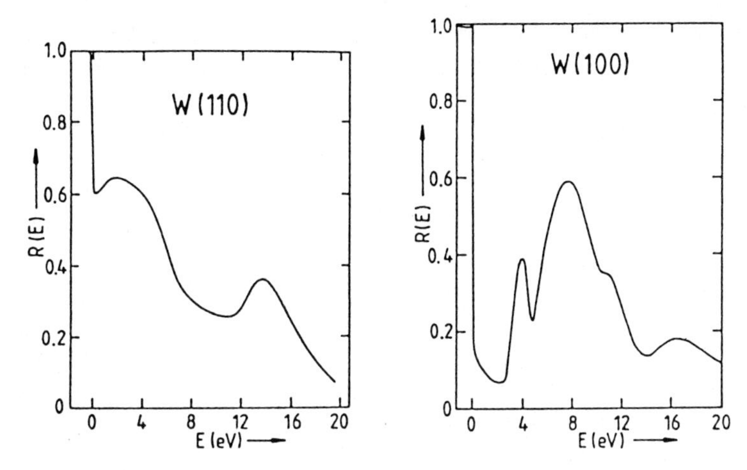

Fig. 12.1. Specular reflectivity $R(E)$ of W$\{110\}$ and W$\{100\}$ for slow electrons at normal incidence [12.12]

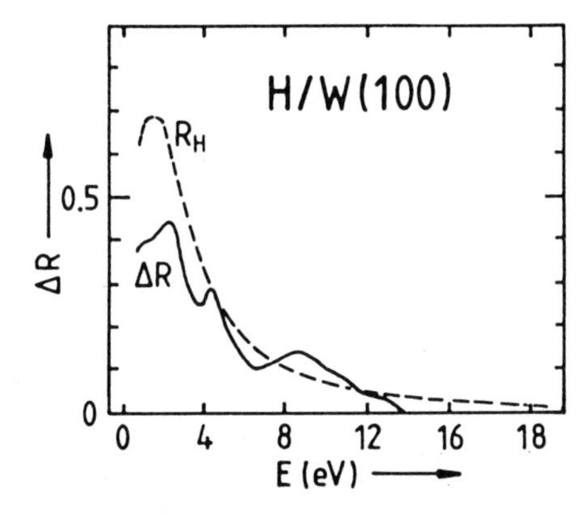

Fig. 12.2. Change $\Delta R(E)$ in specular reflectivity of W$\{100\}$ due to adsorption of $\frac{3}{4}$ monolayers of hydrogen (*solid line*). The dashed line shows the calculated $\Delta R(E)$ [12.12, 14]

W$\{100\}$ surface caused by adsorption of $\frac{3}{4}$ monolayers of hydrogen [12.12, 14]. For comparison, the "reflectivity" of a free H monolayer is shown as obtained by neglecting diffraction effects and simply adding up the backscattering cross-sections of the individual H atoms. The measured reflectivity change increases with decreasing E qualitatively in the same manner as the H monolayer reflectivity, the deviations being caused by diffraction and surface effects [12.12, 14]. At the usual LEED energies ($E > 20\,\mathrm{eV}$), H is below the detection limits. This example clearly demonstrates the usefulness of very low energies in LEEM.

It should be noted, however, that reflectivity changes, ΔR, due to surface layers are not necessarily caused by the R contribution of the layer but frequently are produced by adsorbate-induced relaxation, reconstruction or deconstruction of the substrate or by constructive or destructive interference between layer and substrate. A good example is oxygen on W$\{100\}$ which causes a very strong R decrease, in particular at the R maxima seen in Fig. 12.1 [12.12]. Whatever the mechanism may be, adsorbates can change the specular reflectivity of the surface significantly so that adsorbate islands with dimensions above the resolution limit can be imaged clearly.

12.1.3 Contrast

The preceding intensity considerations lead immediately to the all-important question of contrast. It is obvious from Fig. 12.1 that surfaces with different orientations, e.g., in a polycrystalline foil, differ strongly in reflectivity at certain energies, which produces strong contrast. Strong contrast is also available between clean and adsorbate-covered regions of a single crystal surface, or more generally between surface regions with different structures, e.g., reconstructed and unreconstructed regions. This type of contrast may be called diffraction contrast, which dominates whenever it can occur.

When no structurally different regions with dimensions above the resolution limit are present on the surface, there may still be contrast due to interference effects which we call interference contrast. Two causes of this type of contrast are shown in Fig. 12.3, a) geometric phase contrast [12.15] and b) quantum size

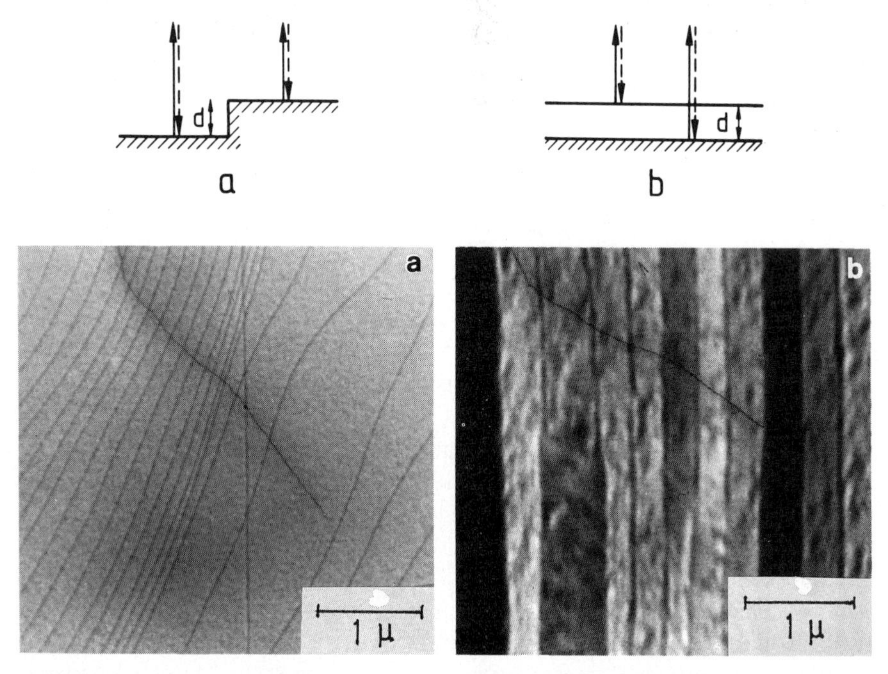

Fig. 12.3a, b. Interference contrast mechanisms. **(a)** Geometric phase contrast. **(b)** Quantum size contrast. The micrograph (a) is from a Mo{110} surface and shows monoatomic steps (E_0 = 4 eV) [12.15], (b) is from an epitaxial Cu layer on Mo{110} with an average thickness of about 10 monolayers and shows thickness fluctuations of a few monolayers from terrace to terrace (E_0 = 4 eV) [12.16]. The fine dark line which starts on the left side of the upper figure edge and runs about diagonally towards the right of the micrograph is a crack in the channel plate

contrast [12.16]. Optical path differences between waves reflected from terraces bordering a step (a) or from the front and back face of a thin layer (b) can lead to constructive or destructive interference between the two waves depending upon the ratio d/λ_0. Varying the wavelength $\lambda_0 = \sqrt{1.5/V_0}$ nm at the specimen by changing V_0 causes maxima and minima of the local $R(E_0)$ from which d [and in case (b) also the mean inner potential in the layer] may be determined with high accuracy, providing atomic depth resolution.

In addition to diffraction and interference contrast there is also topography contrast due to the field distortion by surface roughness. The presently available experience with single crystal and polycrystalline surfaces shows, however, that diffraction contrast is dominating in general.

12.1.4 Instrumental Aspects

Cathode lenses, as used in LEEM, were originally developed for emission microscopy [12.9, 17]. Therefore, a LEEM instrument is simultaneously an emission electron microscope, provided the specimen can be heated for thermionic emission or irradiated with photons for photoemission, or with ions, fast neutrals or electrons for secondary electron emission. Side entry ports pointing at the specimen allow these various emission modes. In addition, mirror electron microscopy (for reviews see [12.18, 19]) is possible by making the specimen more negative than the electron gun.

An important aspect of a LEEM instrument is the fact that structural analysis arises as a by-product: in the back focal plane of the objective lens, all electrons which are reflected at the specimen in the same direction are focussed in a point, so that the LEED pattern can be observed by imaging the back focal plane onto the final screen. The LEED pattern is in many cases sufficient to identify the chemical composition of the imaged surface region and, therefore, complements the LEEM image. A variable aperture in one of the intermediate image planes allows, in addition, selection of small regions on the surface for LEED analysis ("selected area diffraction").

The first LEEM instrument built and presently still the only fully operational one is shown schematically in Fig. 12.4 [12.15]. The specimen (1) which is at a high negative potential $-V + V_0$ of about 20 kV is illuminated through the electrostatic triode consisting of the electrodes (1–3) by a parallel beam of electrons with energy $E_0 = eV_0$. The parallel beam is obtained by focussing the high energy (E = eV) beam from the electron gun (4) with a condensor (5) into the back focal plane (6) of the objective lens. The incident electron beam is separated from the imaging beam by a magnetic prism (7). The intermediate lens (8) and the projective (9) provide the desired magnification of the first intermediate image (in the center of the prism) on the microchannel plate image intensifier (10). The final image on the fluorescent screen (11) is recorded by a still or video camera (12). The LEED pattern can be imaged by changing the focal length of the intermediate lens and removing the objective aperture

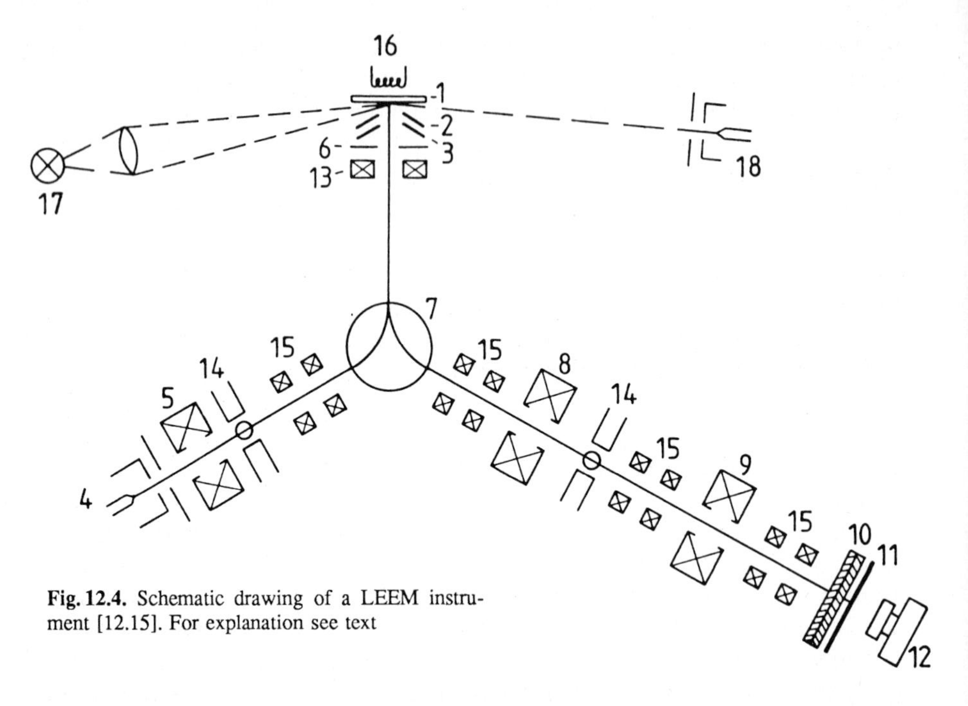

Fig. 12.4. Schematic drawing of a LEEM instrument [12.15]. For explanation see text

(which limits the angle α_0 of the imaging beam) from the back focal plane of the objective (4). The quadrupole (13) corrects the astigmatism of the objective lens, the quadrupoles (14) the astigmatism of the prism (7). Double deflection systems (15) are used for beam alignment.

A specimen heater (16) allows observations over a wide temperature range and also thermionic emission microscopy. On the side entry ports, a UV lamp (17) and an auxiliary electron gun (18) are mounted for PEEM and secondary electron emission microscopy, respectively. Other ports are equipped with an ion gun, mainly for in situ sputter cleaning, and evaporators for thin film growth studies. Directed gas supply via a side entry port is possible, too. The complete system is built of stainless steel, is bakeable up to 200°C and can reach a base pressure in the high 10^{-11}, low 10^{-10} Torr range after bakeout. Large Helmholtz coils surrounding the complete microscope compensate DC and AC magnetic fields. Air springs reduce vibrations to an acceptable level. The results below were all obtained with this instrument.

12.2 LEEM Studies of Clean Surfaces

Because of the good vacuum, studies of clean surfaces are quite feasible with LEEM. While many surfaces such as Au{100} or Pb{110} require ion bombardment cleaning, some surfaces such as Si{111} or Si{100} can be cleaned simply by heating in UHV, or – e.g., in the case of Mo{110} or W{100} –

by heating in O_2 followed by flash desorption of the adsorbed oxygen. Most of the work done up to now concentrated on surfaces which can be cleaned by heating. One of the fundamental questions in surface science is that of the influence of surface imperfections on a wide range of surface phenomena. LEEM is very well suited to the study of surface imperfections because of its high depth resolution, provided that their distances are above the lateral resolution. A variety of imperfections have been studied such as monatomic steps formed during sublimation, glide lines formed during thermal cycling of the crystal, screw dislocations, mosaic grain boundaries, hillocks and etch pits [12.20, 21]. On many surfaces, such as Mo{110}, W{110}, Pb{110} or Au{100}, steps can be imaged by geometric phase contrast as illustrated in Fig. 12.3a. On other surfaces such as Si{111} and Si{100} this contrast is too weak to be useful so that other contrast mechanisms have to be used, the most generally applicable one being decoration contrast. Steps are preferred nucleation sites for phase transformations and film growth. If these processes are stopped at an early stage, only a narrow band is formed along the steps. Figure 12.5 shows two examples. In (a), the monoatomic steps on Si{111} are decorated by small (7×7) structure nuclei formed in the early stages of the $(7 \times 7) \leftrightarrow (1 \times 1)$ phase transition, in (b), the monoatomic steps on Mo{110} are decorated by narrow bands of a Cu monolayer, which formed during Cu deposition at elevated temperature. The same decoration also allows step imaging by PEEM (c) because a monolayer of Cu reduces the work function of Mo{110} sufficiently [12.22] to cause a strong local increase of the photoemission yield, which gives excellent decoration contrast [12.23].

Another indirect step-imaging method is available on surfaces with superstructures whose azimuthal orientation changes from one monoatomic terrace to the next one. An example is the Si{100} surface which has a (2×1) superstructure caused by dimerization, as indicated in Fig. 12.6a by the dotted lines. At normal incidence of the electron beam, the two domain orientations are equivalent but a beam tilt towards one of the superperiodicity directions destroys this equivalence resulting in contrast between the two domains (Fig. 12.6b) [12.24]. The domain boundaries delineate the step position. It is obvious from Fig. 12.6b, which was

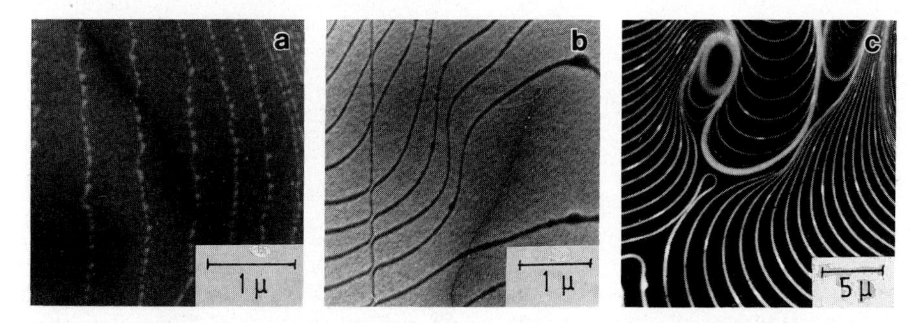

Fig. 12.5a-c. Step decoration in LEEM, (**a**) by (7×7) nuclei on a Si{111} surface at a few degrees below the transition temperature (E_0 = 10.5 eV) [12.37], (**b**) by Cu deposition on a Mo{110} surface at about 700 K, (E_0 = 4 eV) [12.20]; (**c**) is a PEEM image of Cu decorated steps [12.23]

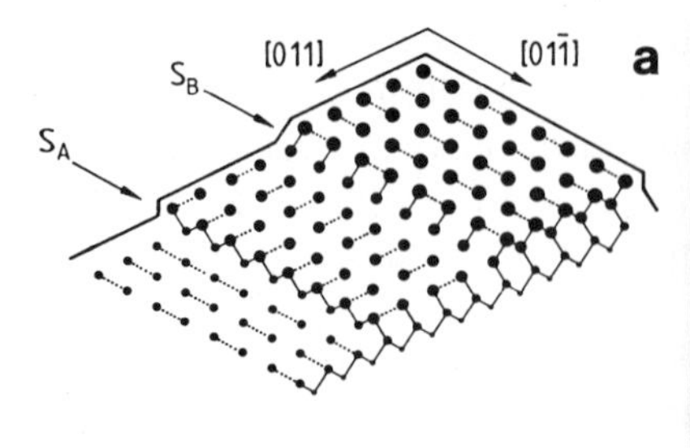

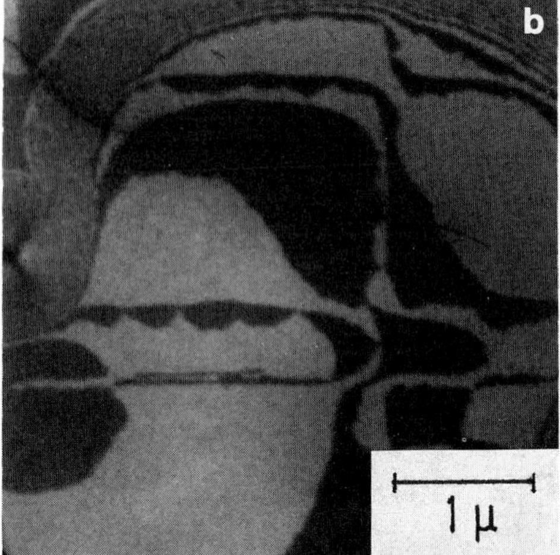

Fig. 12.6a, b. Si{100} surface. (a) Surface model. The S_A steps are parallel, the S_B steps normal to the dimer chains. (b) Tilted bright field image of a well-oriented surface (E_0 = 6 eV) [12.24]

obtained from a surface heated to about 1300 K and cooled slowly, that one step is smooth, the other rough. Closer inspection shows that the smooth steps are S_A steps, the rough ones S_B steps in good agreement with theory, which predicts that the formation energy for S_B steps is by a factor of 15 larger than that of S_A steps [12.25], a fact which is also reflected in the equilibrium shape of two-dimensional (12.2d) islands.

Such islands can be produced by deposition at temperatures at which the crystal does not grow by step flow growth but rather by $2d$ nucleation and at high temperatures by sublimation. At high temperatures, entropy makes such a large contribution to the specific edge free energy that the anisotropy of the specific edge energy is lost and the islands are round. At low temperatures,

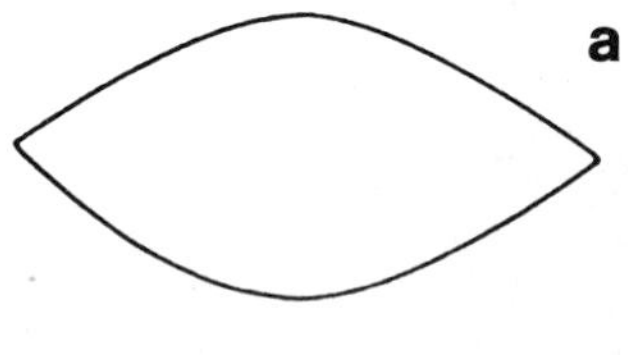

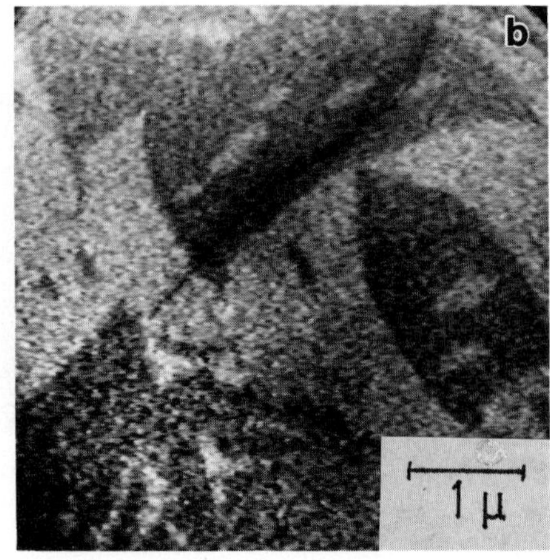

Fig. 12.7a, b. Two-dimensional Si islands on Si{100}. (a) Model of the equilibrium shape [12.26]. (b) Tilted bright field image of islands grown at very low supersaturation (E_0 = 5 eV) [12.27]

the island shape is determined by the strong anisotropy of the specific edge energy. A 2d Wulff construction gives the cusped shape schematically shown in Fig. 12.7a, in which no step orientations appear beyond a critical angle with the S_A step orientation [12.26]. The shape observed during Si growth on Si{100} at very low supersaturation at low temperatures (Fig. 12.7b) [12.27] is in good agreement with this theoretical prediction.

LEEM is particularly well suited to the study of many step-related processes on clean surfaces. Examples are: the interaction of sublimation steps and slip steps at elevated temperatures [12.28], the step generation during sublimation and growth, or the formation of sublimation hillocks and etch pits.

It should be kept in mind that even a well-oriented low-index single crystal surface has a high step density. For a monolayer thickness of 2 Å and a misorientation of 0.5° as used in many experiments, the mean step distance is 230 Å. There are always small fluctuations in the local surface orientation, which produce corresponding fluctuations in the step spacings. Very large step spacings can be produced by localized step pile-up which leaves large atomically smooth areas up to several μm in diameter in between. Step pile-up occurs, for example, during sublimation by localized sublimation-depressing impurities, which produce sublimation hillocks and large flat areas behind them (Fig. 12.8). This process is described in detail in Ref. [12.21]. On these large terraces, the elementary process of sublimation, the "Lochkeim" formation, has been observed via the monolayer contrast in tilted illumination. The growth of the Lochkeime, as well as the monolayer-by-monolayer sublimation, has been studied in real time [12.27].

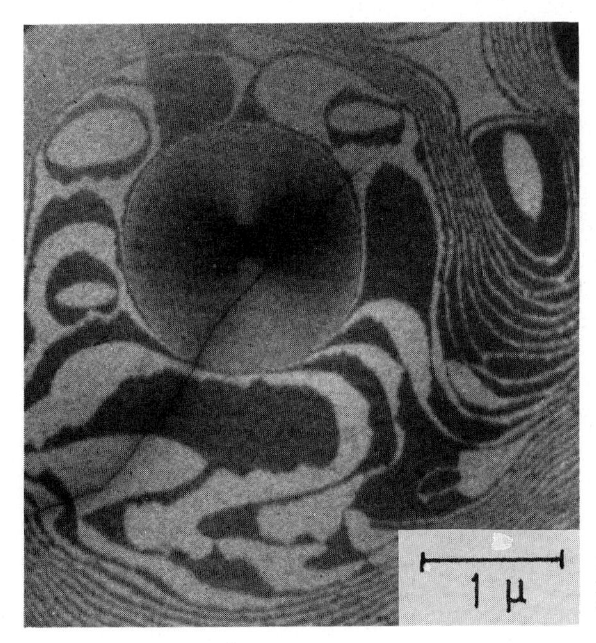

Fig. 12.8. Sublimation hillock and large atomically flat areas left behind during sublimation (E_0 = 7 eV) [12.21]

The homoepitaxy of Si on Si{100} is of considerable technological importance and has, therefore, been studied in recent years extensively by RHEED, LEED, STM and other techniques. LEEM has the advantage of good lateral resolution when compared with RHEED and LEED; when compared with STM, it has the advantage of real time observation of the growth process. In situ studies of the growth at a constant low deposition rate as a function of temperature clearly show the various growth modes: step flow growth, occasional $2d$ nucleation on terraces followed by step flow growth, and frequent $2d$ nucleation causing a rapid roughness increase at high, intermediate and low temperatures, respectively. In the intermediate case, the faster growth of the $2d$ islands in the cusp direction (Fig. 12.7) compared to the smooth direction causes double step formation because cusped and smooth direction alter from terrace to terrace [12.27]. These steps are D_B steps which have a much lower formation energy than D_A steps [12.25].

These examples illustrate the information on step kinetics, which can be obtained with LEEM. Improvements in specimen stability, temperature and deposition rate control will allow the direct extraction of atomistic parameters from video recordings. Another important application of LEEM is the study of surface phase transitions. The Si{111} $(7 \times 7) \leftrightarrow (1 \times 1)$ phase transition at 1110 K [12.29], the Au{100} "(5×1)"$\leftrightarrow (1 \times 1)$ phase transition at 1050 K [12.30, 31] and the Pb{110} $c(2 \times 4) \leftrightarrow (1 \times 1)$ phase transition between 390 K and 440 K [12.32] have been studied up to now. These transitions appear in LEED as continuous transitions. The continuous nature of the transitions has been questioned later [12.33] for Si{111} on the basis of a more precise LEED study, which did not show the critical scattering typical for continuous transitions. Reflection electron microscopy (REM) studies [12.34, 35] were inconclusive because of the difficulties in image interpretation caused by the strong foreshortening due to the grazing incidence of the electron beam. For Pb{110} the continuous nature appeared questionable at the very beginning because of the unphysically large critical exponent of the order parameter [12.32].

Even the early LEEM studies [12.36, 37] of the Si{111} $(7 \times 7) \leftrightarrow (1 \times 1)$ phase transition revealed unambiguously that the transition was first order and showed that the microstructure of the surface and the details of the transition were strongly influenced by the thermal history of the crystal. Later, extensive studies for a wide range of surface treatments confirmed the earlier studies and extended the variety of phenomena considerably. Examples have been published in some reviews [12.38, 39]. Here, only two extreme cases are reproduced in Fig. 12.9: (a) is from a surface which had been heated briefly to 1450 K, (b) from a surface which was annealed for several hours slightly above the transition temperature. The bright regions have the (7×7) structure, the dark regions the (1×1) structure. (a) converts completely from the (1×1) to the (7×7) structure upon slight undercooling, (b) converts only in the bright regions into the (7×7) structure and requires undercooling by about 100 K for complete conversion, in agreement with the LEED work. (a) can be converted into (b) and vica versa by the appropriate heat treatment. The apparent continuous transition observed in

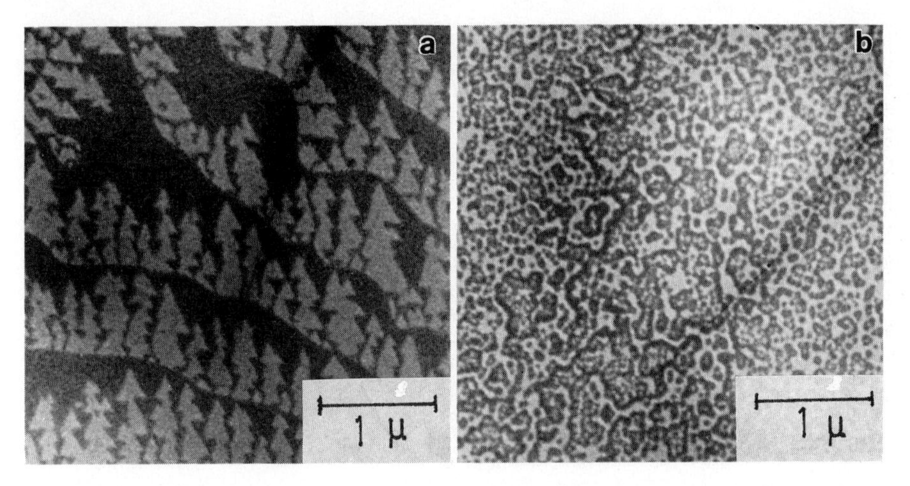

Fig. 12.9a, b. Si{111} $(7\times7) \leftrightarrow (1\times1)$ phase transition. (**a**) After heating to 1450 K (E_0 = 10.5 eV), (**b**) after long annealing at about 1150 K (E_0 = 11 eV) [12.24]

LEED is thus a first-order transition with locally varying transition temperatures, T_t. The T_t range depends strongly upon pretreatment and may approach 0 K as in case (a). Similarly, the wide transition range on Pb{110} is caused by local variations of the transition temperature of a first order transition [12.40].

12.3 LEEM Studies of Surface Layers

Surface layers are ideal subjects for LEEM studies because their formation, transformation or disappearance may be studied in real time over a wide temperature range. The first step in any surface science experiment is the cleaning process, e.g., the removal of the oxide from Si, of the carbonaceous layer from Au, of the carbide from Mo or the complex surface compounds on Pb by various techniques. Examples of the cleaning of Au are shown in [12.31]. Figure 12.10 gives another example for Mo{110}. The LEED pattern (a) identifies the carbide but says nothing about its lateral distribution while the LEEM image (b) shows that the carbide is distributed as $2d$ islands whose size is limited by steps. Following the cleaning process in O_2 by LEEM is much more sensitive than observing the LEED pattern because even a few small islands can be detected. Carbide may form on the surface not only by segregation from the bulk but also by dissociation of CO from the residual gas. The latter process produces the same LEED pattern but quite different islands [12.61, 41] so that LEEM can distinguish between segregation and dissociation as the cause of the surface contamination.

Although every adsorbate changes the reflectivity of the surface, no local contrast differences can be expected unless the adsorbate forms islands. Therefore, most of the work done up to now concentrated on systems which are known to have attractive lateral interactions in the layer. These layers may either be

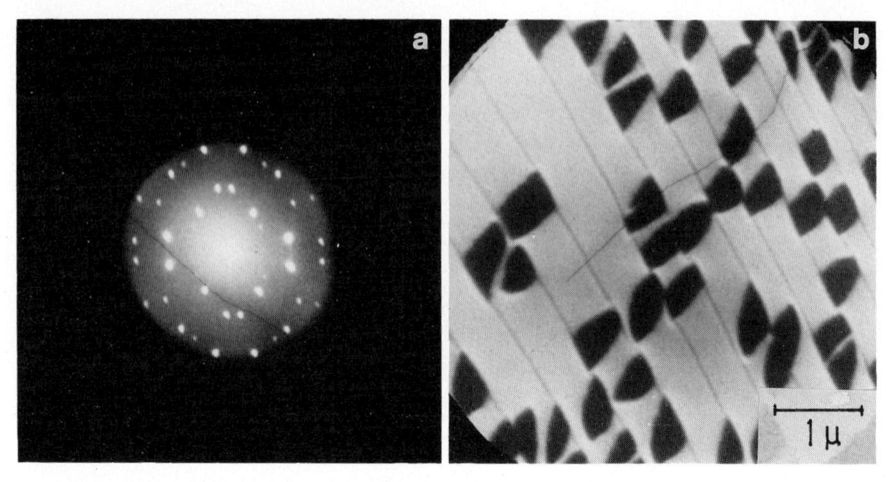

Fig. 12.10a, b. Carbon segregation into Mo carbide on the Mo{110} surface. (**a**) LEED pattern (E_0 = 5 eV), (**b**) LEEM image (E_0 = 4 eV)

pure or reaction layers. Examples of pure layers are Cu and Au on Mo{110}, of reaction layers Cu and Co on Si{111} while Au on Si{111} is a border case.

Cu and Au on Mo{110} are two typical non-alloying metal-metal systems. Up to 1 monolayer (ML) Cu grows pseudomorphically while Au forms a misfitting layer [12.42]. This has a profound influence on the stability of the layers: Except at low coverages, Cu remains in the $2d$ crystalline state up to the desorption temperature [12.43] while Au transforms into a $2d$ gas below this temperature [12.44]. LEEM shows additional differences during growth at elevated temperatures at which the layer can approach the equilibrium configuration [12.16]: Cu decorates the substrate steps ("step flow growth"), i.e., it wets the steps (Fig. 5b, c) [12.20, 23, 45, 46] while Au forms islands on the terraces which are elongated in the direction of the densely packed rows in layer and substrate (Fig. 12.11a), i.e., it does not wet the steps [12.46].

With increasing thickness, the layers rearrange structurally into the structure of the bulk materials, with small distortions to accommodate the misfit at the interface. Cu is an interesting case because the double layer has two structures, a high and a low temperature structure, which convert reversibly and with large hysteresis into each other [12.47]. In this transition, the direction of the misfit dislocation changes from ⟨110⟩ at low to ⟨100⟩ at high temperatures ("misfit flip" transition). The laterally averaging studies [12.47] had suggested vacancies in the dislocation cores of the low temperature structure in order to accommodate the density difference between the two structures. LEEM [12.48], however, indicates that the density difference is accommodated by crack formation during cooling through the transition. LEEM also shows the kinetics of the transition: during heating, the high temperature structure nucleates in the low temperature layer in long narrow ribbons along the zero misfit direction of the high temperature structure. In the reverse transition the nuclei are more isometric.

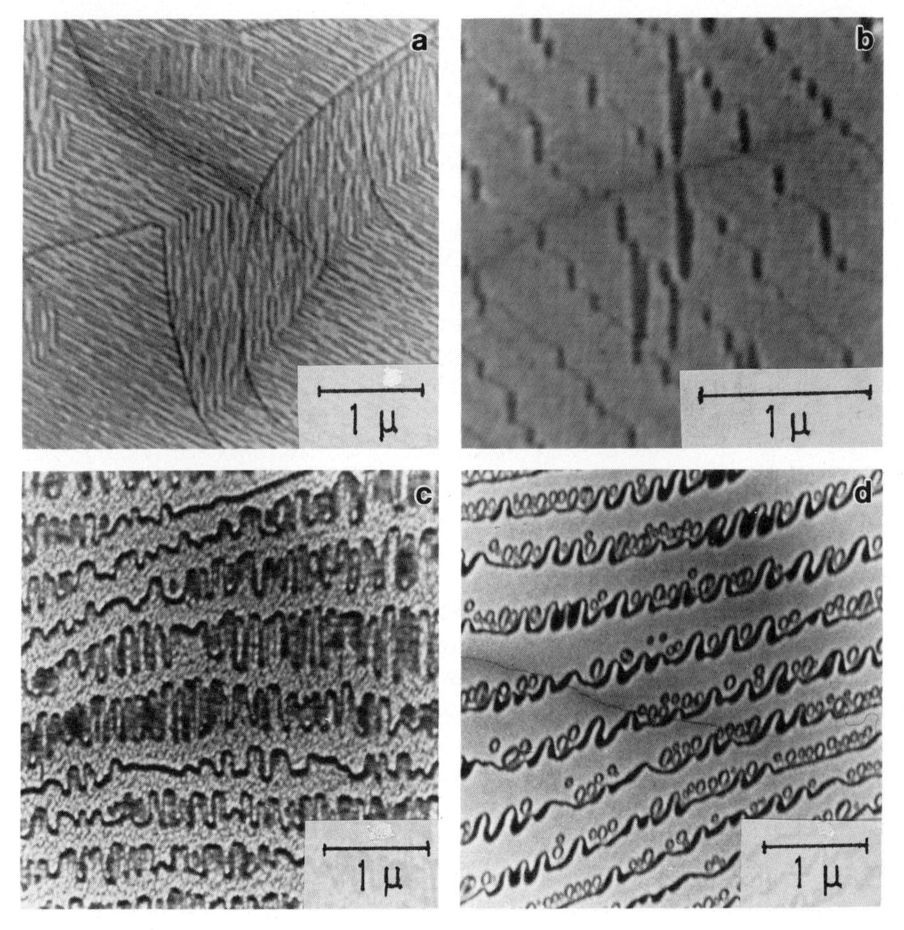

Fig. 12.11a-d. Metal layers on Mo{110} at elevated temperatures. (**a**) Au at sub-monolayer coverage (E_0 = 14 eV) [12.46]. (b-d) Cu double layers; (**b**) step facetting above 700 K (E_0 = 3 eV), (**c**) step break-up above 900 K (E_0 = 3 eV), (**d**) step rounding during desorption (E_0 = 3 eV) [12.16, 41, 48]

When a Mo{110} surface covered with a Cu double layer is annealed at temperatures above about 700 K, the steps initially facet (Fig. 12.11b) and finally develop deep crevices with walls along the zero misfit direction ⟨100⟩. With increasing temperature, this process leads to a nearly complete break-up of the terraces (Fig. 12.11c). During desorption, the rectangular step structure becomes increasingly rounder and smoothes out (Fig. 12.11d) so that finally the wavy steps seen in Fig. 12.5b, c appear again. During desorption, the steps fluctuate considerably [12.41, 48]. Desorption occurs from the two-phase region, both in the second and in the first ML. In Au on Mo{110}, this is true only for the second ML. The first ML desorbs from the single-phase region, in agreement with the results from laterally averaging studies [12.43, 44].

At low temperatures (below about 400–500 K) when the layer formation process is diffusion-limited, both Cu and Au grow via $2d$ nucleation and growth on the terraces. At the beginning of the second ML, spontaneous recrystallization into the double layer structure occurs locally. The ensuing growth is monolayer-by-monolayer ("Quasi-Frank-van der Merwe growth mode") up to many monolayers with steps acting as growth barriers as seen in Fig. 12.3b [12.16, 45, 46]. At higher temperatures, only 2 MLs grow layer-by-layer while material in excess of this coverage forms flat $3d$ crystals ("Stranski-Krastanov growth mode").

The growth of metals on Si$\{111\}$ is quite different from that on Mo$\{110\}$. Although quasi-Frank-van der Merwe growth can be obtained by deposition at low temperatures [12.49, 50], the quasi-equilibrium growth at elevated temperatures is more complex due to the possibility of silicide formation. Cu is known to form a $2d$ silicide with an approximate "(5×5)" superstructure, which is followed by the growth of large $3d$ crystals. (For references see [12.51]). Au forms two superstructures, (5×1) and $(\sqrt{3} \times \sqrt{3})R30°$, initially. With further increasing coverage, the $2d$ layer saturates below about 650 K with a (6×6) structure. Excess Au forms $3d$ Au crystals or a $3d$ Au–Si eutectic depending upon temperature [12.52]. Cobalt finally forms various silicides which tend to grow three-dimensionally. (For references see [12.53]). Although much is known about these systems from laterally averaging studies, little is known about their microstructure. LEEM provides this information.

The $2d$ Cu silicide nucleates with a temperature-dependent incubation period, apparently because of partial diffusion of the condensing Cu into the bulk. Once nucleated, the "(5×5)" structure spreads rapidly along the terraces until the surface is completely covered with the $2d$ silicide. The $3d$ Cu silicide which forms subsequently grows in various crystal habits: in long roofs, triangular or hexagonal platelets or small polyhedra with hexagonal bases. During the growth of these particles, a considerable restructuring of the $2d$ silicide-covered surface takes place, resulting in facetted steps with step heights up to several MLs. An example is shown in Fig. 12.12a. While the particle has a structureless flat surface in the PEEM image, LEEM clearly shows a stepped surface. The LEED patterns of the larger particles allow their identification as η"-Cu_3Si. The smaller particles migrate over long distances during their reaction with the substrate, leaving reaction trails behind them (Fig. 12.12b).

Above 850-900 K, the $3d$ and $2d$ silicide dissolves in the bulk of the Si crystal. Upon cooling below 850 K, the $2d$ silicide nucleates and grows again, frequently to both sides of the steps. The "(5×5)" islands in the (7×7) "sea" have preferentially boundaries along the Si $\langle 1\bar{1}0 \rangle$ direction. Apparently, the strain field of the step aids the nucleation process in segregation. Pure Cu layers can be grown only at low temperatures, e.g., at 300 K after formation of the $2d$ silicide layer [12.46, 51].

The quasi-equilibrium growth of Au on Si$\{111\}$ [12.27] starts with the formation of islands with a (5×1) structure with three equivalent orientations preferentially at steps and (7×7) domain wall boundaries. The islands are narrow in the direction of the fivefold periodicity and grow until the surface is com-

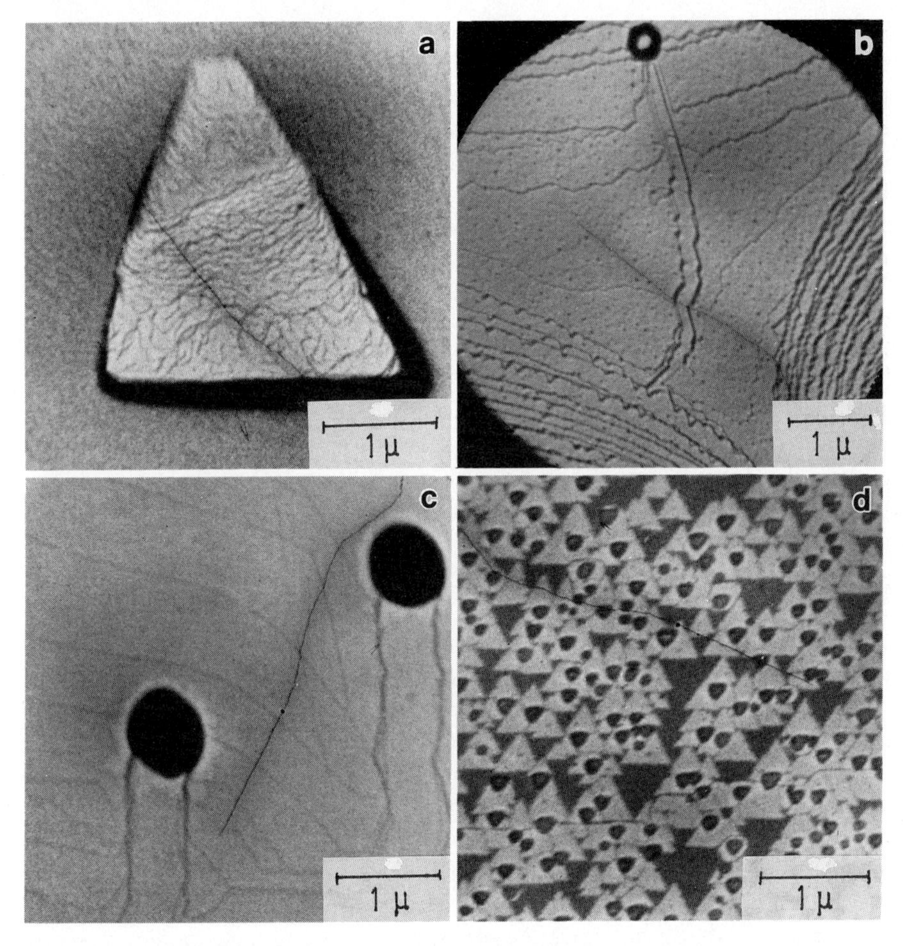

Fig. 12.12a-d. Metal layers on Si{111}. (**a, b**) A large and a small three-dimensional Cu silicide crystal on the two-dimensional Cu silicide layer on Si{111} (E_0 = 4 eV). Note the migration trail in (b) [12.46, 51]. (**c**) Au particles and migration trails on a Si{111} $(\sqrt{3} \times \sqrt{3})R30°$ surface (E_0 = 7 eV) [12.27] (**d**) $CoSi_2$ crystals (*small dark triangles*) on Si{111} surface with (7 × 7) (*bright triangles*) and (1 × 1) (*dark areas in between*) structure (E_0 = 10 eV) [12.48]

pletely covered by the (5 × 1) structure (see for example Fig. 12.12 in [12.46]). The $(\sqrt{3} \times \sqrt{3})R30°$ islands nucleate then with little delay and grow without preferred growth direction apparently unimpeded over very large areas. After the completion of this structure, $3d$ Au–Si droplets form above about 650 K, while at lower temperatures, Au particles and (6 × 6) structure appear simultaneously. The Au particles are quite mobile already at 500 K and undergo interesting reversible volume and shape changes during heating via the eutectic temperature to high temperatures. The (5 × 1) ↔ (1 × 1) phase change, which appears continuous in LEED, is shown by LEEM to be a first-order transition with locally varying transition temperature. The (6 × 6) ↔ $(\sqrt{3} \times \sqrt{3})R30°$ transition, on the other hand, occurs over a very narrow temperature range. Similarly to Cu, at high temperatures Au causes considerable surface rearrangement and the Au–Si particles leave pronounced migration trails behind them (Fig. 12c). The details of the Au/Si{111} system will be reported elsewhere [12.27].

The system Co on Si{111} was studied with the specific goal of understanding why it is necessary to first grow a very thin continuous "template" $CoSi_2$ layer at low temperature before a pinhole-free thick film can be grown at high temperatures. In other words: does $CoSi_2$, which has a very low lattice mismatch with Si, really not wet Si, so that one has to resort to kinetic limitations in order to grow a quasi-$2d$ layer? LEEM gives a clear answer as seen in Fig. 12d. When Co is deposited at high temperatures, (dark) triangular and hexagonal $CoSi_2$ islands grow on the otherwise clean Si{111} surface, which shows large (bright) triangular (7×7) islands intermixed with (dark) untransformed (1×1) regions. This shows that $CoSi_2$ does not wet Si, i.e., $CoSi_2$ has a significantly larger specific surface free energy than Si.

12.4 Outlook

The results reported in Sects. 12.2 and 12.3 illustrate the present capabilities and limitations of LEEM. The best resolution obtained to date (15 nm at a few eV) is approximately what is expected for the electrostatic triode objective lens [12.5, 9]. A magnetic lens should allow resolutions close to the resolution limit of the homogeneous field, that is by about a factor of 3 better [12.7, 9]. Improvements in the illumination system which have already been made [12.4] increase the intensity by orders of magnitude – at the expense of coherence – so that images can be taken at higher energies which increases the resolution even more. Nevertheless, LEEM will not reach the resolution of reflection electron microscopy (REM) normal to the electron beam direction, which is at least good enough to resolve the 5-fold superperiodicity of the Si{111} (5×1) Au structure, that is 1.7 nm [12.54]. The lateral atomic resolution of STM is completely beyond the range of LEEM.

What is then the outlook for LEEM? The strong foreshortening in REM has already led to false conclusions and makes image interpretation much more difficult than in LEEM. The limited field of view and the sequential image acquisition in STM makes LEEM superior to STM when a large field of view or rapid image acquisition is necessary; for example, in the study of surface processes or in studies at elevated temperatures. Thus, each of the methods has its pitfalls and virtues so that – inspite of partial competition – they complement each other.

Where do we go from LEEM, once this technique has been perfected? There is an immediate extension whose experimental feasibility still has to be tested: spin-polarized LEEM (SPLEEM). The scattering of electrons, also of slow electrons, is spin-sensitive [12.55, 56] so that the different reflectivities of surface regions differing in spin orientation produces spin contrast. Thus SPLEEM should allow the study of the magnetic structure of clean surfaces and surface layers. It remains to be seen whether or not the available spin-polarized electron sources have enough luminosity simultaneously with high spin polarization and small energy width so that SPLEEM images with resolutions comparable to ordinary LEEM images can be obtained.

One of the main disadvantages of present day LEEM compared to scanning electron microscopy (SEM) is the lack of chemical characterization. In SEM this is a byproduct if an energy analyzer is added to detect the Auger electrons, thus creating a scanning Auger microprobe (SAM). A similar capability can be incorporated into a LEEM instrument by adding an imaging energy analyzer. The Auger electrons can be excited either by a side entry electron gun or [12.4] by using higher energies in the incident beam which requires a more complex magnetic prism (12.7) in Fig. 12.4. There is an important difference between such an Auger electron emission microscope (AEEM) and a SAM: in the former image, detection is parallel as in LEEM, in the latter, sequential as in SEM. The image acquisition time is, therefore, much shorter in AEEM than in SAM but, if a complete spectrum of a small area is to be obtained, SAM is superior to AEEM.

The same instrument which is capable of AEEM can, of course, also be used for imaging with characteristic photoelectrons excited by X-rays or monochromatic synchrotron radiation (XPEEM). With sufficient photon flux from wigglers or undulators a much better resolution than that of present fine spot ESCA instruments should be possible. For a more detailed discussion of the possibilities and limitations of AEEM and XPEEM, see [12.8].

12.5 Summary

Although LEEM was conceived more than 25 years ago [12.57] and much of the first instrument was already completed in the 1960s, the technique is still far less developed than REM, SEM and the much younger STM. This is reflected in the mainly qualitative character of the data presented. However, even these results demonstrate the power of LEEM in surface studies for which a lateral resolution of 15-20 nm is sufficient. The main virtues of LEEM are the parallel image acquisition with a large field of view, without foreshortening, at high rate (video rate) and the good access to the specimen which allows a large variety of in situ studies. The LEEM instrumentation can still be significantly improved in resolution, intensity, flexibility, specimen temperature control and other experimental aspects. Once these improvements are made, quantitative surface studies with high lateral resolution should open a new era of surface science.

Acknowledgements. This chapter is dedicated to the late Wolfgang Telieps who completed the first LEEM instrument and made it operational. He was able to build on the pioneering work of George Turner who developed most of the instrument in the 1960s when ultra high vacuum technology was still young. The experimental data reported here were obtained by Wolfgang Telieps, Michael Mundschau, Waclaw Swiech and Michael Altman. Support by the Deutsche Forschungsgemeinschaft and the Volkswagenstiftung is gratefully acknowledged.

References

12.1. E. Bauer: J. Appl. Phys. **35**, 3079 (1964); Ultramicroscopy **17**, 51 (1985)
12.2. H. Liebl: Optik **80**, 4 (1988) and references therein
12.3. G.F. Rempfer, O.H. Griffith: Ultramicroscopy **27**, 273 (1989)
12.4. L. Veneklasen: private communication
12.5. D.R. Cruise, E. Bauer: J. Appl. Phys. **35**, 3080 (1964) and unpublished work. The results of these calculations are compiled in Ref. [12.6]
12.6. W. Telieps: Ph.D. thesis, TU Clausthal (1983)
12.7. W. Engel: Ph.D. thesis, Berlin (1968); reproductions of emission images can be found in Ref. [12.9]
12.8. J. Chmeli'k, L. Veneklasen, G. Marx: Optik **83**, 155 (1989)
12.9. E. Bauer, W. Telieps: In *Surface and Interface Characterization by Electron Optical Methods*, ed. by A. Howie, U. Valdre (Plenum Press, New York 1988) p. 195
12.10. E. Bauer: J. Vacuum Sci. Technol. **7**, 3 (1970)
12.11. H.-J. Herlt, R. Feder, G. Meister, E. Bauer: Solid State Commun. **38**, 873 (1981)
12.12. H.-J. Herlt: Ph.D. thesis, TU Clausthal (1982)
12.13. S. Andersson: Surface Sci. **18**, 325 (1969); **25**, 273 (1971)
12.14. H.-J. Herlt, E. Bauer: Surface Sci. **175**, 336 (1986)
12.15. W. Telieps, E. Bauer: Ultramicroscopy **17**, 57 (1985)
12.16. E. Bauer, M. Mundschau, W. Swiech, W. Telieps: Ultramicroscopy **31**, 49 (1989)
12.17. R.A. Schwarzer: Microscopica Acta **84**, 51 (1981)
12.18. A.B. Bok: Ph.D. thesis, Delft 1968
12.19. A.B. Bok, J.B. LePoole, J. Roos, H. de Lang: Adv. Opt. Electron Microscopy **4**, 161 (1971)
12.20. M. Mundschau, E. Bauer, W. Swiech: Phil. Mag. **A59**, 217 (1989)
12.21. M. Mundschau, E. Bauer, W. Telieps, W. Swiech: Surface Sci. **223**, 413 (1989)
12.22. E. Bauer, H. Poppa, G. Todd, F. Bonczek: J. Appl. Phys. **45**, 5164 (1974)
12.23. M. Mundschau, E. Bauer, W. Swiech: Surface Sci. **203**, 412 (1988)
12.24. W. Telieps: Appl. Phys. **A44**, 55 (1987)
12.25. D.J. Chadi: Phys. Rev. Lett. **59**, 1691 (1987)
12.26. C. Herring: In *Structure and Properties of Solid Surfaces*, ed. by R. Gomer, C.S. Smith (University of Chicago Press, Chicago 1953), p. 5;
M. Wortis: In *Chemistry and Physics of Solid Surfaces VII*, ed. by R. Vanselow, R. Howe (Springer, Berlin, Heidelberg 1988), p. 367
12.27. W. Swiech, M. Mundschau, E. Bauer: to be published
12.28. M. Mundschau, E. Bauer, W. Telieps, W. Swiech: Phil. Mag. **A61**, 257 (1989)
12.29. P.A. Bennett, M.W. Webb: Surface Sci. **104**, 74 (1981)
12.30. W. Telieps, M. Mundschau, E. Bauer: Optik **77**, 93 (1987)
12.31. W. Telieps, M. Mundschau, E. Bauer: Surface Sci. **225**, 87 (1990) and references therein
12.32. A. Pavlovska, E. Bauer: Europhys. Lett. **9**, 797 (1989)
12.33. W. Witt: Ph.D. thesis, TU Clausthal (1984)
12.34. N. Osakabe, Y. Tanishiro, K. Yagi, G. Honjo: Surface Sci. **109**, 353 (1981)
12.35. Y. Tanishiro, K. Takayanagi, K. Yagi: Ultramicroscopy **11**, 95 (1983)
12.36. W. Telieps, E. Bauer: Surface Sci. **162**, 163 (1985)
12.37. W. Telieps, E. Bauer: Ber. Bunsenges. Phys. Chem. **90**, 197 (1986)
12.38. E. Bauer, W. Telieps: Scanning Microscopy Suppl. **1**, 99 (1987)
12.39. E. Bauer, M. Mundschau, W. Swiech, W. Telieps: In *Evaluation of Advanced Semiconductor Materials by Electron Microscopy*, ed. by D. Cherns (Plenum, New York 1989), p. 283
12.40. M. Altman, E. Bauer: to be published

12.41. M. Mundschau, E. Bauer, W. Swiech: Catalysis Lett. **1**, 405 (1988)
12.42. E. Bauer, H. Poppa: Thin Solid Films **121**, 159 (1984)
12.43. M. Paunov, E. Bauer: Appl. Phys. **A44**, 201 (1987)
12.44. A. Pavlovska, H. Steffen, E. Bauer: Surface Sci. **195**, 207 (1988)
12.45. M. Mundschau, E. Bauer, W. Swiech: J. Appl. Phys. **65**, 581 (1988)
12.46. M. Mundschau, E. Bauer, W. Telieps, W. Swiech: Surface Sci. **213**, 381 (1989)
12.47. M. Tikhov, M. Stolzenberg, E. Bauer: Phys. Rev. **B36**, 8719 (1987)
12.48. M. Mundschau, W. Swiech, E. Bauer: to be published
12.49. M. Jalochowski, E. Bauer: J. Appl. Phys. **63**, 4501 (1988)
12.50. M. Jalochowski, E. Bauer: Phys. Rev. **B37**, 8622 (1988)
12.51. M. Mundschau, E. Bauer, W. Telieps, W. Swiech: J. Appl. Phys. **65**, 4747 (1989)
12.52. S. Ino: Japan. J. Appl. Phys. **16**, 891 (1977)
12.53. J.M. Phillips, J.L. Batstone, J.C. Hensel, M. Cerullo, F.C. Unterwald: J. Mater. Res. **4**, 144 (1989) and references therein
12.54. Y. Tanishiro, K. Takayanagi: Ultramicroscopy **31**, 20 (1989)
12.55. E. Bauer: In *Techniques of Metals Research*, ed. by R.F. Bunshah (Interscience, New York 1969) Vol. II part 2, p. 559
12.56. J. Kessler: *Polarized Electrons* (Springer, Berlin 1976)
12.57. E. Bauer: In *Electron Microscopy*, ed. by S.S. Breese, Jr. (Academic, New York 1962) Vol. 1, D-11

13. Atomic Scale Surface Characterization with Photoemission of Adsorbed Xenon (PAX)

K. Wandelt

Institut für Physikalische und Theoretische Chemie,
Universität Bonn, Wegelerstr. 12, D-5300 Bonn 1, Fed. Rep. of Germany

13.1 Introduction

The physical and chemical properties of solid surfaces are strongly influenced – if not dominated – by structural and chemical heterogeneities. For example, structural surface defects like steps, kinks and vacancies act as heterogeneous nucleation centers and thereby affect the growth mode and the final structure of epitaxial films and adsorbed gas layers. Chemical defects like heteroatoms are the origin of local surface structure modifications or may even trigger structural phase transitions and surface reconstruction. Both surface topography and composition are decisive parameters for the chemical reactivity and catalytic activity of solid surfaces. Surface defects and dopants (heteroatoms) create surface states in the band gap of semiconductors, which in turn determine the electronic properties of semiconductor devices. A full understanding of all kinds of surface properties and surface processes ultimately requires a characterization of real surfaces on an atomic scale. A Kossel model of a real surface is depicted in Fig. 13.1, showing schematically a binary surface with several kinds of structural defects.

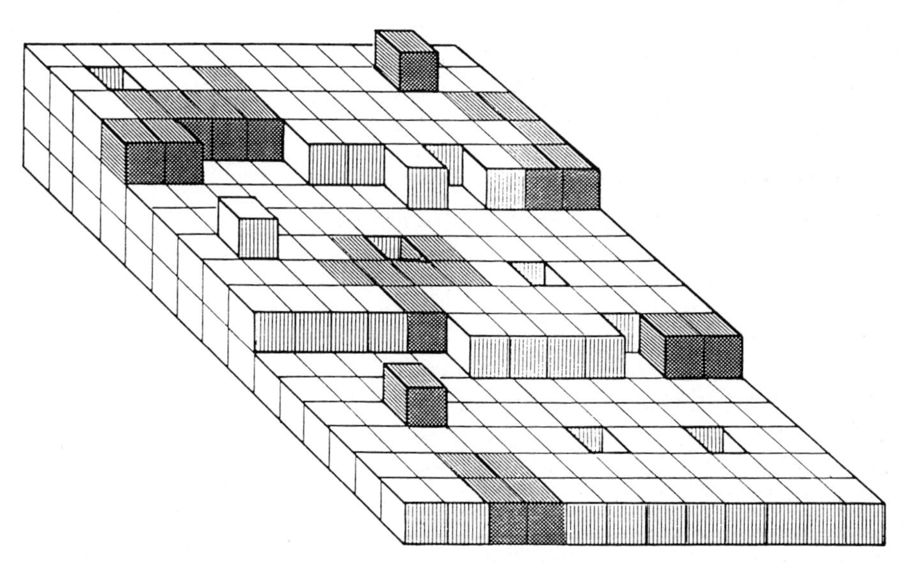

Fig. 13.1. Schematic Kossel model of a binary surface with structural defects like steps, kinks, adatoms, vacancies, etc

After two decades of detailed model studies with well prepared single crystal samples, surface heterogeneities begin to dictate a new trend in surface science. Several new analytical techniques have been developed recently, which can provide information about structural and chemical surface disorder on different scales of resolution, like low energy electron microscopy (LEEM) [13.1], high-resolution electron microscopy [13.2], photoelectron microscopy [13.3] and scanning tunneling microscopy (STM) [13.4–6]. All these techniques provide real space images of surface disorder. They can be applied to extended heterogeneous surfaces and do not require fine tips as the field ion microscope (FIM) does [13.7].

Photoelectronspectroscopy of Adsorbed Xenon (PAX) [13.8, 9] is different from the aforementioned techniques in that this method does not provide real space images of surface heterogeneities. PAX, in turn, may be considered as a reversible, non-destructive decoration and titration technique, which provides information about the number density of specific kinds of surfaces and, in particular, about the local surface potential at these sites. For many purposes it does not seem as important to know where the defects are as to know how many of them there are. The lateral "resolution" of the PAX-technique is also of the order of a few Ångstroms, which means that sites of different surface potential which are only ~ 5 Å apart may be distinguished by PAX. In particular with respect to the capability of providing surface potential differences on an atomic scale, PAX seems to complement the STM in a unique way.

This chapter presents a review of the present understanding and capability of the PAX-technique. Section 13.2 describes and discusses the physical principles of the technique and addresses some methodical aspects. Section 13.3 presents some case studies which illustrate the applicability of the method. This work concentrates on metallic surfaces; a forthcoming review will also include studies with semiconductor surfaces [13.10]. Section 13.4 gives a short summary and addresses some implications of the results presented.

13.2 Principles of the PAX-Technique

At temperatures sufficiently below 80 K (that is below liquid nitrogen temperature), xenon can be adsorbed dosewise on any solid surface. Both the photoemission spectrum of the adsorbed Xe and the subsequent thermal desorption spectrum provide information about important properties of the underlying surface, such as the number density and the distribution of structural and chemical defects as well as the local surface potential at specific surface sites. Section 13.2.1 first describes the relevant features of the Xe(5p) valence and Xe(4d) core level photoemission spectra of Xe adsorbed on different homogeneous as well as heterogeneous surfaces. Based on these features, Section 13.2.2 describes the theoretical model underlying the PAX-technique, which then is supported by further experimental evidence in Sect. 13.2.3. Section 13.2.4 extends the PAX-model from homogeneous surfaces to heterogeneous surfaces and introduces the con-

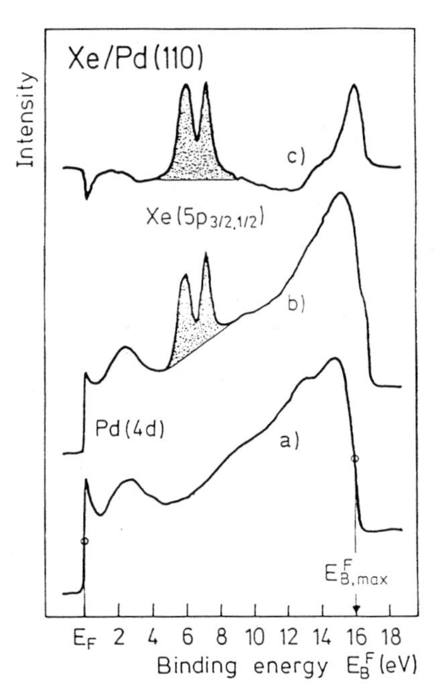

Fig. 13.2. UV photoemission spectra excited with He I ($h\nu = 21.22$ eV) radiation of (*a*) a clean Pd{110} surface and (**b**) a Pd{110} surface covered with one complete monolayer of xenon. The Xe induced $5p_{3/2,1/2}$ extra emission (*shaded*) is accentuated in the difference spectrum (c) = (b) − (a). The total width of the spectra $E^{\mathrm{F}}_{\mathrm{B,max}} - E_{\mathrm{F}}$ serves to determine the average work function of the surface using (13.1)

cept of the "local work function". Section 13.2.5 addresses some methodological aspects.

13.2.1 Photoemission Spectra of Adsorbed Xe

The 5p-valence band photoemission spectrum, excited with UV-radiation from a He-discharge lamp (He I = 21.22 eV) and detected with an angle-integrating hemispherical electron analyzer (angle of acceptance $\sim 45°$ around the direction $\sim 45°$ off surface normal), of a complete monolayer of Xe on a well defined Pd{110} surface is shown in Fig. 13.2. Spectrum a) first shows the integral UPS spectrum of the valence band region of the bare palladium substrate with dominant emission from the $4d$ band close to the Fermi level E_{F}. The rapidly increasing background of secondary electrons at higher binding (low kinetic) energies falls abruptly to zero at $E_{\mathrm{kin}} = h\nu - E^{\mathrm{F}}_{\mathrm{B,max}} - \varphi = 0$, with $h\nu$ the primary energy of the photons and φ the work function. Rearranged, this equation

$$\varphi = h\nu - E^{\mathrm{F}}_{\mathrm{B,max}} \tag{13.1}$$

provides the basis for the photoelectric determination of the average work function of the substrate surface. Spectrum b) in Fig. 13.2 shows the UPS spectrum of the same Pd{110} surface covered with one complete monolayer of xenon. The two sharp and intense extra peaks (shaded) between 5 eV and 8 eV below E_{F} arise from the $5p_{3/2}$ and $5p_{1/2}$ final states of the photoionized adsorbed xenon atoms. In order to accentuate this extra emission features, curve c) in Fig. 13.2

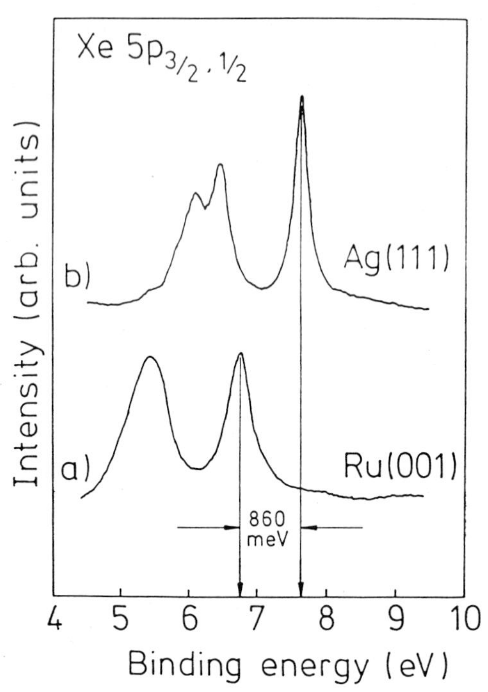

Fig. 13.3. Expanded Xe($5p_{3/2,1/2}$) photoemission spectra from complete monolayers of Xe on a Ag{111} and a Ru{001} surface, respectively. The different shape of the $5p_{3/2}$ signal (at lower binding energy) compared to that of the sharp $5p_{1/2}$ peak and, in particular, the peak shift of ΔE_B^F = 860 meV between the two surfaces is discussed in the text

shows the difference between spectrum b) and spectrum a). Figure 13.3 shows only the spectral range of the $5p_{3/2,1/2}$ emission of a monolayer of Xe adsorbed on an Ag{111} and a Ru{001} surface, respectively. Two important differences become immediately apparent. Firstly, the $5p_{3/2}$ signal is not only broader than the $5p_{1/2}$ signal on both surfaces but it is even clearly split on the Ag{111}. Secondly, both spectra are shifted with respect to each other by 860 meV. This shift is clearly seen between the $5p_{1/2}$ signals from both surfaces, it occurs also in the case of the $5p_{3/2}$ signals, however, it is less easy to recognize, because of the more complicated structure of the $5p_{3/2}$ signal. While the overall line width of all 5p-peaks of adsorbed Xe is determined by a combination of inhomogeneous broadening [13.11–13], phonon-broadening [13.14] and dispersion effects [13.15–18], the selective, further increase of the $5p_{3/2}$ line width results from a lifting of the energetic degeneracy of the $5p_{3/2}$, $m_j = \pm 3/2$ and $5p_{3/2}$, $m_j = \pm 1/2$ subpeaks into two separate lines due to interaction with the substrate [13.19–21] and, in particular, lateral Xe–Xe interaction accompanied by two-dimensional band structure formation within the adsorbed xenon-layer [13.15–18]. The 5p spectrum of adsorbed Xe is, therefore, a composite of three (Lorentzian) lines as shown in Fig. 13.4 for Xe on Ag{111}. The actual separation of the two $5p_{3/2}$ states depends on the nature of the substrate as well as on the Xe coverage [13.22, 23]. For more details concerning these two spectral features, the interested reader is referred to the original literature. In the context of the present work, we will concentrate mainly on the energy shift between the 5p spectra as a whole between the two metal surfaces. Henceforth, this shift

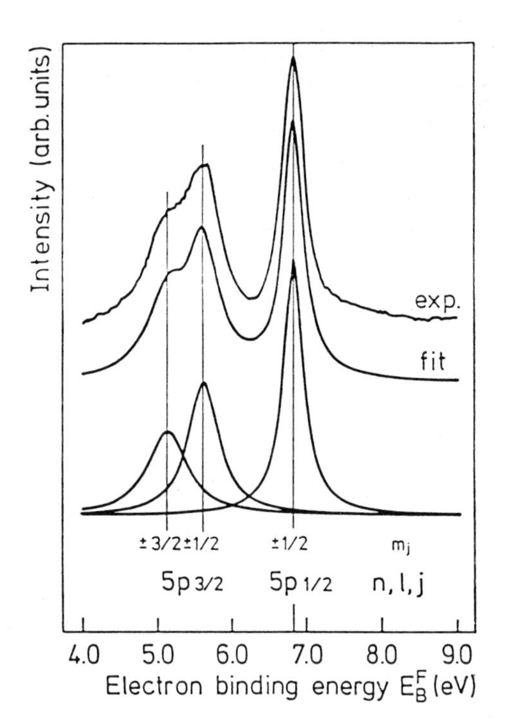

Fig. 13.4. Illustration of the decomposition of a typical Xe($5p_{3/2,1/2}$) spectrum of adsorbed Xe into the three component lines $5p_{3/2}(m_j = \pm 3/2)$, $5p_{3/2}(m_j = \pm 1/2)$ and $5p_{1/2}(m_j = \pm 1/2)$ which can be represented by Lorentz lines

will only be discussed with the $5p_{1/2}$ signal, because it is easier to locate on the energy axis, as mentioned previously. As shown in earlier publications, this shift corresponds to the difference in work function between the two substrates:

$$\Delta E_B^F(5p_{1/2})_{ads} \cong -\Delta\varphi \,. \tag{13.2}$$

The larger $E_B^F(5p_{1/2})$ value with respect to the Fermi level (superscript F) belongs to the surface with the lower work function. This empirical relationship has now been verified with some twenty different metal and semiconductor surfaces. An updated compilation of in situ measured work function and E_B^F values is given in Tables 13.1 and 13.2.

A consequence of (13.2) is that the ionization potential $E_B^F(5p_{1/2})_{ads}$ with respect to the vacuum level (superscript V) is nearly independent of the substrate:

$$E_B^V(5p_{1/2})_{ads} = E_B^F(5p_{1/2})_{ads} + \varphi \cong \text{const} \,. \tag{13.3}$$

On the basis of the data given in Tables 13.1 and 13.2, $E_B^V(5p_{1/2})_{ads}$ acquires a value of

$$E_B^V(5p_{1/2})_{ads} = 12.3 \pm 0.1\,\text{eV} \,. \tag{13.4}$$

This value, however, is still ~ 1.1 eV lower than the corresponding gas phase ionization potential $(E_B^V 5p_{1/2})_{gas} = 13.41$ eV [13.24] due to extra-atomic relaxation effects [13.25]. The 5p-hole left behind on an *adsorbed* photoionized Xe

Table 13.1. Compilation of macroscopic work functions, φ, and electron binding energies, E_B^F (with respect to the Fermi level) of the $5p_{1/2}$ level of Xe adsorbed on the various homogeneous metal single crystal surfaces. The φ-values are determined according to (13.1). Note the approximate invariance of the ionization potential, $E_B^V = E_B^F + \varphi$, compared to φ and E_B^F. The significance of the value $E_B^F \approx 12.3$ eV is discussed in the text, ΔE_R denotes the extra atomic relaxation at the surface. On Cu{111} and Ag{111}, adsorbed Xe was found to undergo a $2d$ gas $\leftrightarrow$ $2d$ solid phase transition [13.22, 23] which affects E_B^F. The lower E_B^V values correspond to the $2d$ gas phase; the higher E_B^V values to the $2d$ solid phase on these two surfaces

Sample	φ [eV]	E_B^F [eV]	$E_B^V + E_B^F + \varphi$ [eV]
Pd {110}	5.20	7.03	12.23
Pd {100}	5.65	6.75	12.40
Pd {111}	5.95	6.47	12.42
Pt {111}	6.40	5.90	12.30
Ir {100}(1 × 1)	6.15	6.24	12.39
Ir {100}(5 × 1)	6.00	6.38	12.38
Re {0001}	6.65	5.4	12.15
Ru {0001}	5.52	6.76	12.28
W {001}	4.50	7.90	12.40
W {110}	5.10	7.15	12.25
Ni {110}	4.65	7.75	12.40
Ni {100}	5.30	6.90	12.20
Ni {111}	5.40	6.80	12.20
Cu {110}	4.48	7.80	12.28
Cu {111}	4.90	7.10	12.0 *)
		7.30	12.2
Ag {111}	4.72	7.27	11.99 *)
		7.62	12.34
Ag {100}	4.65	7.68	12.33
Au {111}	5.68	6.72	12.40
Al {111}	4.48	7.74	12.22
Al {110}	4.45	7.8	12.25
K (poly)	2.20	10.21	12.41
Cs (poly)	1.80	10.50	12.30
Xe gas			13.40
ΔE_R	= const. =		1.1 eV

atom is screened more efficiently due to the vicinity of the highly polarizable charge in the metal substrate, than a hole on a free Xe atom which undergoes only intra-atomic relaxation. A 5p photoelectron from an *adsorbed* Xe atom benefits from this extra-atomic relaxation energy, ΔE_R, and leaves the Xe atom with $\sim$ 1.1 eV higher kinetic energy, which converts into an apparently $\sim$ 1.1 eV lower binding energy.

Both values, i.e., $E_B^V(5p_{1/2}) = 12.3$ eV and $\Delta E_R = 1.1$ eV, have been debated in the literature and shall, therefore, be discussed here in more detail. The exact energy position E_B^F of the $5p_{1/2}$-signal is not only determined by the nature of the substrate but to some extent also by the structure of the adsorbed Xe layer itself. In particular, it has been shown that the lateral interactions between adsorbed Xe atoms not only cause the splitting of the $5p_{3/2}$-signal (see Fig. 13.4) but may also

Table 13.2. Compilation of macroscopic work functions, φ, and electron binding energies, E_B^F (with respect to the Fermi level) of the $5p_{1/2}$ level of Xe adsorbed on the various semiconductor surfaces. E_B^V and ΔE_R have the same meaning as in Table 13.1. Note that all φ values are determined photoelectrically at low temperatures, that is, under flat band conditions (Sect. 13.2.3). E_B^V has the same rather constant value as in Table 13.1 for metallic substrates

Surface	φ [eV]	E_B^F [eV]	$E_B^V = E_B^F + \varphi$ [eV]
ZnO (000$\bar{1}$) O	4.35	8.00	12.35
ZnO (0001) Zn	3.75	8.70	12.45
ZnO (10$\bar{1}$0)	4.05	8.25	12.30
n-Si {111}	4.65	7.70	12.35
p-Si {111}	5.55	6.80	12.35
n-Si {100}	4.60	7.70	12.30
Xe gas			13.40
ΔE_R	= const. =		1.1 eV

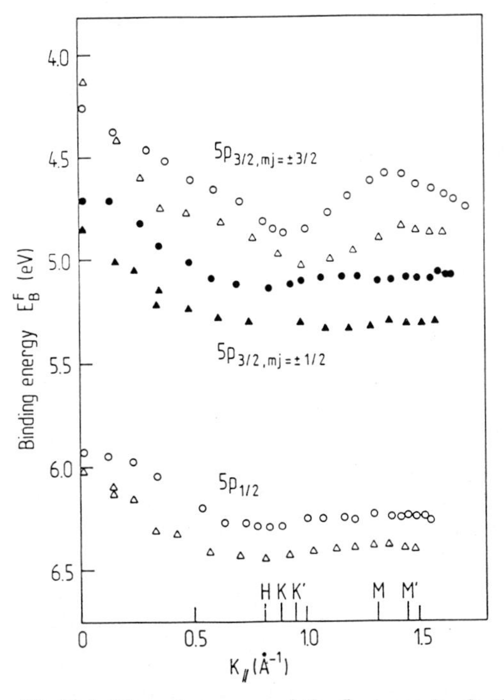

Fig. 13.5. Dispersion curves of the 5p xenon levels along the ΓKMK direction of the surface Brillouin zone of a Xe layer adsorbed on Pt{111}. Angle of incidence = 45°. (•, ○) refer to a $\sqrt{3} \times \sqrt{3}$R30° commensurate layer; with a Xe–Xe distance of 4.8 Å; (▲, △) refer to an incommensurate rotated phase with a Xe–Xe distance of 4.38 Å. After [13.18]

lead to a two-dimensional band structure formation with substantial dispersion effects. Within a close-packed, {111}-like Xe monolayer, the dispersion of the $Xe(5p_{1/2})$ level may be as large as ~ 0.5 eV. As an example, Fig. 13.5 reproduces the dispersion curves for a $(\sqrt{3} \times \sqrt{3})$ R 30° structure as well as for a hexagonally close-packed incommensurate structure of Xe on Pt{111} from *Cassuto* et al. [13.26]. In the $\sqrt{3}$-structure, the Xe–Xe separation is 4.8 Å and the $5p_{1/2}$ dispersion is 0.32 eV in the $\Gamma M-$ and 0.35 eV in the ΓK-direction of the surface Brillouin zone. These values increase to ~ 0.5 eV within the close-packed layer with a Xe–Xe interatomic distance of 4.38 Å, in agreement with theoretical predictions [13.27]. These dispersion effects, in fact, may already occur at rather low coverages if the Xe layer undergoes a $2d$ gas $\leftrightarrow$ $2d$ solid phase transition with the formation of $2d$ close-packed Xe islands. At the commonly used adsorption temperatures, this is, in fact, true for many substrates, see e.g. [13.22, 23, 28–31]. As a consequence, the experimentally determined binding energy $E_{\mathrm{B}}^{\mathrm{F}}$ of the $5p_{1/2}$ signal depends on the Xe-coverage and the detection geometry. Those measurements which led to the value of $E_{\mathrm{B}}^{\mathrm{V}}(5p_{1/2}) = 12.3$ eV given above, integrate in a somewhat undefined way over a broad region of the first Brillouin zone of a Xe layer. Obviously, more care has to be exercised, and the spectra need to be taken at a well defined point of the Brillouin zone using an angle-resolving analyzer. Since this detection mode also affects the resolution and the shape of the zero kinetic energy cut-off at $E_{\mathrm{B,max}}^{\mathrm{F}}$ (Fig. 13.2) of a spectrum, it is no wonder that work function values calculated from angle resolved spectra using (13.1) are also slighly different from those obtained from angular integrated spectra [and used in (13.4)]. Table 13.3 summarizes some $E_{\mathrm{B}}^{\mathrm{F}}(5p_{1/2})$ values and work function data from angle resolved spectra taken at the Γ point of the surface Brillouin zone, i.e., in normal emission. The sum

$$E_{\mathrm{B}}^{\mathrm{F}}(5p_{1/2}, \Gamma) + \varphi = E_{\mathrm{B}}^{\mathrm{V}}(5p_{1/2}, \Gamma) = 11.77 \pm 0.12\,\mathrm{eV} \tag{13.5}$$

is again constant proving that the basic conclusion from Tables 13.1 and 13.2 and of (13.3) is correct. Obviously, the final state relaxation then becomes $\Delta E_{\mathrm{R}} = 1.6$ eV. The difference in the absolute values, i.e., $E_{\mathrm{B}}^{\mathrm{V}} = 11.8$ eV versus 12.3 eV, or $\Delta E_{\mathrm{R}} = 1.6$ eV versus 1.1 eV, is purely a consequence of the dispersion effects (and the angle-resolved detection mode), which had not been taken into account in the earlier angle-integrated measurements. The difference certainly does not invalidate the basic concept of the PAX-technique as a local surface potential probe (Sect. 13.2.4), i.e. (13.3). As pointed out correctly by *Cassuto* et al. [13.18], the dispute about the absolute value of $E_{\mathrm{B}}^{\mathrm{V}}(5p_{1/2})$ brought into the literature by several authors was based on an improper comparison of data. The small scatter of the $E_{\mathrm{B}}^{\mathrm{V}}$ values in Table 13.3 (as well as Tables 13.1 and 13.2) may still be the consequence of structural differences of the Xe overlayer on different substrates, namely, in the Xe–Xe interatomic distance, and, hence, in the dispersion amplitude.

Another point of controvery concerning (13.4) and (13.5) lies in the choice of the proper work function ϕ. In both cases, the work function of the bare surface

Table 13.3. Compilation of macroscopic work functions, φ, and $E_B^F(5p_{1/2})$ of adsorbed Xe at the Γ-point of the surface Brillouin zone. Note the rather constant value of $E_B^V(5p_{1/2}) = E_B^F(5p_{1/2}) + \varphi$. The Pt{100}(1 × 1) surface was prepared by hydrogen adsorption at low temperatures

Surface	φ [eV]	$E_B^F(5p_{1/2})$at Γ [eV]	$E_B^V(5p_{1/2})$ at Γ [eV]
Pd {100}	5.65	6.2	11.85
Al {111}	4.30	7.55	11.85
Cu {110}	4.5	7.3	11.8
Pt {111}	5.85	5.95	11.8
Pt {100}(1 × 1)*)	5.77	6.08	11.85
Pt {100}(hex)	5.80	6.10	11.90
Pb {111}	4.05	7.60	11.65

(without Xe adsorbed) was used. Other authors, e.g. [13.32] tend to use the work function of the surface covered by one complete monolayer of Xe. This so-called reference-problem, which is as old as photoelectron spectroscopy of adsorbates itself, still appears not to be solved even for the simplest case, namely rare gas adsorption. Probably the "truth" lies between the two extremes mentioned above [13.33], but the proper fraction of φ to be used is unknown. This uncertainty may also contribute to the scatter of the E_B^V-values in Table 13.1–4, but the basic "reference-problem" is, in principle, an open one and calls for further theoretical work.

Table 13.4. Compilation of macroscopic work functions, φ, and values of $4d^F$ and $4d^V = 4d^F + \varphi$ for Xe adsorbed on the different single crystalline metal surfaces. Note the almost constant value of the binding energy of $4d^V$ compared to φ and E_B^F

Surface	φ [eV]	$4d^F_{5/2}$ [eV]	$4d^F_{3/2}$ [eV]	$4d^V_{5/2}$ [eV]	$4d^V_{3/2}$ [eV]
Pd {111}	5.95	60.22		66.17	
Pd {100}	5.65	60.5		66.15	
Gd {0001}	3.3	62.8		60.10	
Ru {001}	5.44	60.74		66.18	
1 ML Cu/Ru {001}	4.86	61.33		66.19	
Al {111}	4.48			66.00	68.00

13.2.2 The $\Delta\varphi$-Model

Despite the complications addressed in the previous section, we would like to stress, however, that (13.3) expresses an empirical result, which is based on a large number of Xe(5p) measurements. This result may be summarized by the model illustrated in Fig. 13.6. The left half of Fig. 13.6 represents a potential

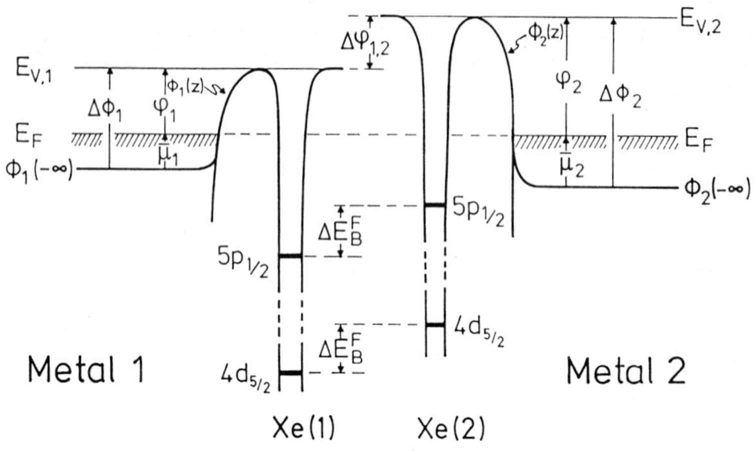

Fig. 13.6. Potential energy diagram for Xe atoms adsorbed on two homogeneous and infinitely extended metal surfaces with different work functions φ_1 and φ_2. $\bar{\mu}$ is the chemical potential and $\Delta\phi$ the surface dipole barrier. All other terms are explained in the text. Note that the Xe potential well floats with the respective surface potential

energy diagram for a Xe atom being adsorbed on the (large) homogeneous surface of metal 1, with $E_{V,1}$ the vacuum potential near this surface, E_F the Fermi level and φ_1 the work function of the (bare) metal surface. The curve $\phi_1(z)$, which connects $E_{V,1} = \phi(+\infty)$ outside the surface with the electrostatic potential, $\phi(-\infty)$, inside the metal corresponds to the variation of the electrostatic potential (*without* image effects) due to the surface dipole layer. $E_{V,1} - \phi_1(-\infty) = \Delta\phi_1$ depends on the charge density at the surface and is the surface contribution to the total work function

$$\varphi_1 = E_{V,1} - E_F = \Delta\phi_1 - \bar{\mu}_1 \tag{13.6}$$

where $\bar{\mu}_1$ is the bulk chemical potential [with respect to $\phi_1(-\infty)$] of the metal [13.34]. $\phi_1(z)$ varies rather steeply and reaches its saturation value $E_{V,1}$ at only ~ 2 Å outside the surface [13.35]. Typical Xe-metal interatomic distances as derived from LEED investigations, however, are larger than ~ 3 Å. Consequently, (at least the center of) an adsorbed xenon atom sits outside of the surface potential barrier, as depicted in Fig. 13.6 for Xe(1). This picture has received strong theoretical support by self-consistent charge density calculations for a single Xe atom on a jellium surface and by potential curves derived herefrom by *Lang* and *Williams* [13.35]. Therefore, the potential well of the *adsorbed* Xe atom on metal 1 is "pinned" to that part of $\phi_1(z)$ which levels off to form $E_{V,1}$, and which is the characteristic vacuum level of the surface of metal 1. This is taken to be the first important reason for the validity of (13.3).

Another important point is the weak bonding between Xe and any substrate, which manifests itself in the requirement of the low adsorption temperatures. Typical adsorption energies range between 5 and 8 kcal/mol on different substrates. Again, self-consistent charge-density calculations show [13.36] that this interaction, which includes bonding and polarization of the Xe atom, leads to absolute initial state shifts of the Xe(5p) valence levels of less than 0.3 eV. Consequently, *changes* of this values when going from one metal substrate to another are much smaller, of the order of ~ 0.1 eV. This is the second reason, why $E_B^V(5p_{1/2})_{ads}$ is practically substrate independent.

All the above arguments hold likewise for the Xe(2) atom which is adsorbed on the metal 2 on the right half of Fig. 13.6. This metal has the work function φ_2. Since the Fermi levels E_F of both metals are aligned through their contact to the spectrometer, the potential well of the Xe(2) atom is shifted up, together with $E_{V,2}$ by the amount $\Delta\varphi_{1,2}$; and because of the constancy of $E_B^V(5p_{1/2})_{ads}$ the $5p_{1/2}$ level shifts by precisely the same amount. This model provides a straightforward explanation for the correlation $\Delta E_B^F(5p_{1/2}) = -\Delta\varphi$, as obtained experimentally (13.2). One may visualize this correlation as if the potential well of an adsorbed Xe atom as a whole "floats" with the vacuum level $E_{V,i}$ of the respective surface, i, ($\Delta\varphi$-model).

13.2.3 Further Experimental Evidence for the $\Delta\varphi$-Model

Obviously, the floating of the potential well of an adsorbed Xe with the vacuum level in front of the respective substrate surface should not only manifest itself with corresponding $\Delta\varphi$-shifts of the 5p *valence* levels, but should hold for all *core* levels as well. In fact, the core levels should comply even better with (13.3) because they are more contracted than the 5p valence orbitals [13.35] and, hence, are located even further outside the surface dipole potential $\Delta\phi$ (Fig. 13.6). High resolution Xe(4d) core level measurements from different metal surfaces are scarce because they require the use of synchrotron radiation. But the few data given in Table 13.4, convincingly show that the $E_B^V(4d_{5/2})$ values are very close for Xe adsorbed on all surfaces although the respective $E_B^F(4d_{5/2})$ and φ values deviate by as much as 2.6 eV. As an example, Fig. 13.7 shows the $4d_{5/2,3/2}$

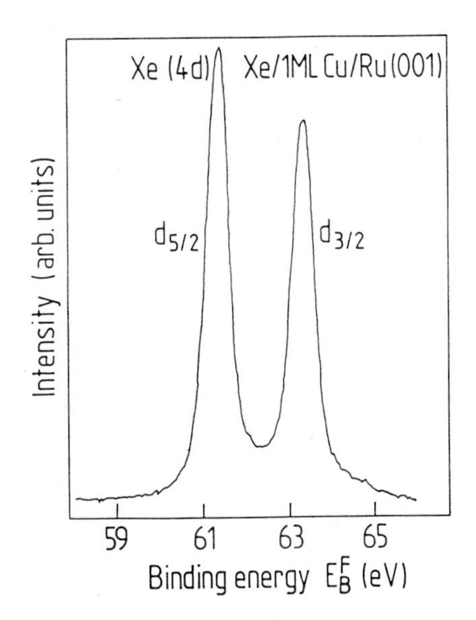

Fig. 13.7. Xe(4d) core level spectra of one monolayer of Xe adsorbed on a Ru{001} surface which, prior to Xe adsorption, was covered with precisely one complete monolayer of copper

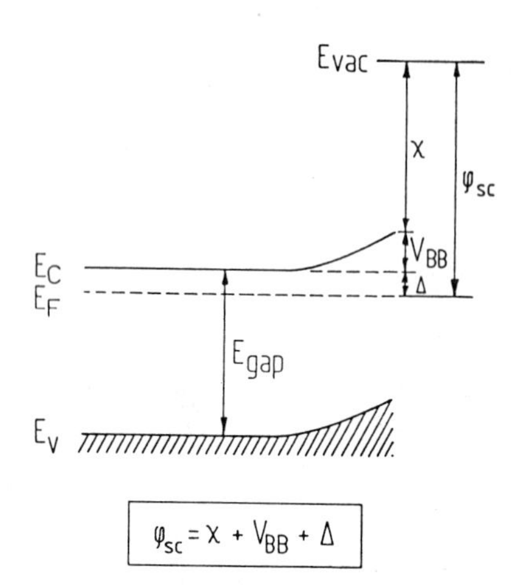

Fig. 13.8. Potential energy diagram near the surface of an n-doped semiconductor to illustrate the various contributions to the work function φ_{sc} of a semiconductor surface: χ is the electron affinity, V_{BB} the band bending, Δ the position of the Fermi level E_F in the band gap with respect to the bottom of the conduction band E_C. E_V is the top of the valence band and E_{vac} the vacuum level

spectrum of one monolayer of xenon on a complete monolayer of Cu on Ru{001} [13.37]. Also Xe(3d) data, though obtained with simple, non-monochromatized XPS, reflect reasonable constancy of $E_B^V(3d)$ over several different metal surfaces [13.12].

Another very elegant way of testing the $\Delta\varphi$-model is made possible through the use of a semiconductor substrate, e.g. Si{111}7 × 7. Due to a high density of surface states and the extra surface charge associated with these states, all the electronic bands exhibit the well-known band bending near the surface of the semiconductor. Therefore, the work function of a semiconductor, which, of course, is still the total energy difference between the Fermi level inside the sample and the vacuum level just outside the surface, is somewhat more complex than that of a metal. According to Fig. 13.8, the work function, φ_{sc}, of a semiconductor is given by

$$\varphi_{sc} = \chi + V_{BB} + \Delta \ , \tag{13.7}$$

where χ is the electron affinity, V_{BB} the band bending and Δ the energy separation between the Fermi level, E_F, within the band gap (E_{gap}) and the bottom of the conduction band, E_C. χ is an intrinsic property of the semiconductor, e.g., silicon and Δ is a bulk property but depends on the doping of the sample. The term V_{BB} is a surface property and depends on the amount of extra surface charge accommodated in the intrinsic, defect induced or adsorbate induced surface states. In principle there are two independent ways to change the work function of one and the same semiconductor surface. The first way is to change the bulk doping of the sample and, hence, the quantity Δ; this leaves V_{BB} unaffected. The second way is to change V_{BB} for a given Δ. Both approaches are easily practicable if the experiments are carried out at the low temperatures needed for xenon adsorption as described in the following.

Under the conditions of a PAX-experiment, that is under UV irradiation and at low sample temperatures ($T \lesssim 50$ K), the band bending at a semiconductor surface can be lifted. The creation of electron-hole pairs across the band gap leads

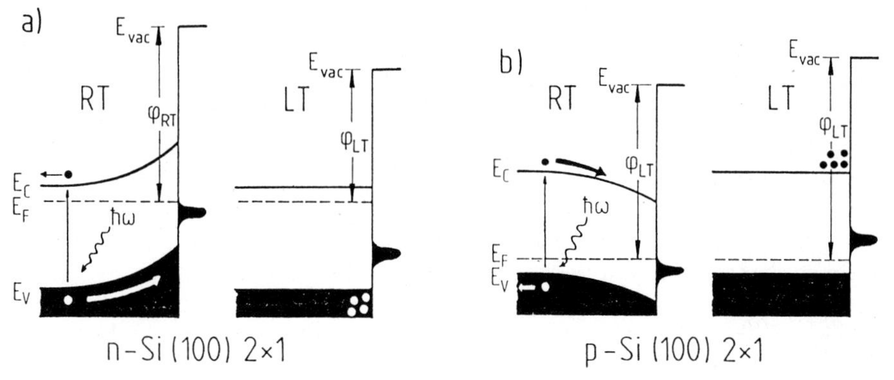

Fig. 13.9. Schematic representation of the surface voltaic effect at an n-doped and a p-doped Si{100}(2 × 1) surface. At low temperature ($\lesssim$ 60 K) the photon induced production of electron (•) hole (○) pairs under He I irradiation from a He-resonance lamp can lead to accumulation of holes (n-doped) or electrons (p-doped) at the surface driven by the respective bending and, hence, in the steady state to a band flattening (see text)

to a compensation of the surface charge in the surface states and a corresponding flattening of the bands (surface photovoltaic (SPV) effect) [13.38]. The low temperatures are required to reduce the recombination rate of the electron-hole pairs and to build up the compensation charge. This is illustrated schematically in Fig. 13.9 (upper half) for an n-doped Si surface. The photo-generated *holes in the valence band*, which under the influence of the upward band bending drift to the surface, eventually (under steady state conditions) lead to a compensation of the excess charge in the surface states and to a corresponding shift of the valence band states (Fig. 13.9 right hand side). Figure 13.10 shows the entire He I-excited valence band spectrum of an n-doped Si{111}7 × 7 surface measured at room temperature (RT) and at 60 K (LT). All spectral features, including the valence band onset (enlarged in the inset), the Si valence band emission and the secondary electron cut-off, shift by the same amount V_D to higher binding energies. V_D is given by:

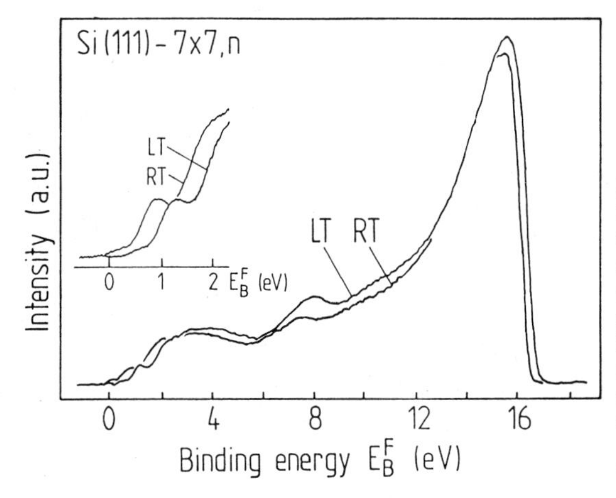

Fig. 13.10. Total He I excited valence band spectra of an n-doped Si{111}(7×7) surface measured at room temperature (RT) and at 60 K (LT), respectively. Note that at LT *all* spectral features are shifted to higher binding energy due to the operation of the surface photovoltaic effect (Fig. 13.8). The inset enhances the spectral region near $E_F = 0$

$$V_D = V_{BB} + \Delta_{RT} - \Delta_{LT} \approx V_{BB} \ , \tag{13.8}$$

where V_{BB} is the band bending at room temperature (Fig. 13.9), and $\Delta_{RT} - \Delta_{LT}$ is a very small temperature induced shift of the bulk Fermi level within the band gap (which converts into a corresponding shift of the gap edges since E_F is fixed with respect to the Fermi level of the spectrometer). Since the secondary electron cut-off edge of the spectrum also shifts by V_D, (13.1) suggests a corresponding change of the work function. In fact, for an n-doped Si{111}7 × 7 surface (1 Ω cm) we have measured work function values of $\varphi_{RT}(n) = 5.05$ eV at room temperature and $\varphi_{LT}(n) = 4.65$ eV at 60 K [13.39,40]. The difference indicates a corresponding shift of the vacuum level outside this surface (with respect to the Fermi level of the spectrometer).

Of course, the same arguments also hold for a p-doped Si{111}7 × 7 surface (1 Ω cm). Since here, at room temperature, the bands are bent downwards, at low temperature, the photo-excited *electrons in the conduction band* drift to the surface and lead to an upwards shift of all levels, including the vacuum level (Fig. 13.9, lower half). This is, indeed, again reflected in the work function data, which yield $\varphi_{RT}(p) = 5.05$ eV and $\varphi_{LT}(p) = 5.55$ eV, respectively. Thus, under the conditions of a PAX-experiment, that is under He I-irradiation at $\lesssim 60$ K, the work function difference,

$$\Delta\varphi_{SC}(p,n) = \varphi_{LT}(p) - \varphi_{LT}(n) = 0.9\,\text{eV} \ , \tag{13.9}$$

is equivalent to a corresponding shift of the Fermi level within the band gap or, vice versa, to a corresponding shift of the vacuum level with respect to the (fixed) Fermi level of the spectrometer.

Figure 13.11 now shows the He I excited valence band spectra of the same p- and n-doped Si{111}7 × 7 surfaces covered with a monolayer of adsorbed xenon

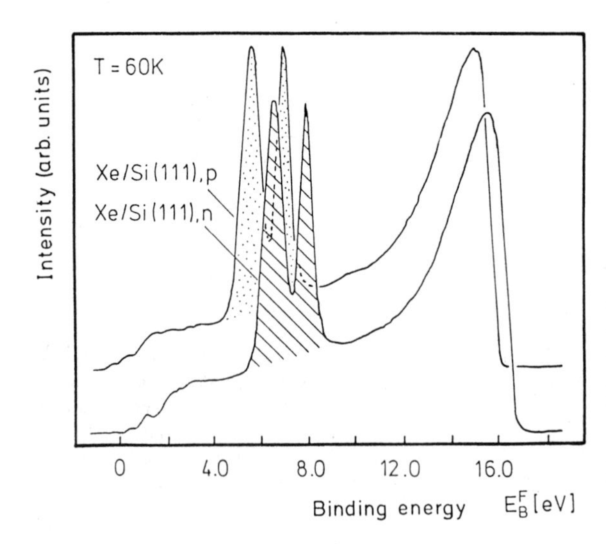

Fig. 13.11. He I excited valence band spectra of a p-doped Si{111} (7 × 7) and an n-doped Si{111} (7 × 7) surface covered with a complete monolayer of Xe. The Xe induced $5p_{3/2,1/2}$ extra emission (*shaded*) is shifted by the difference in the position of the Fermi level in the two samples (see text)

at 60 K. The shaded double-peak structure in both spectra corresponds to the Xe(5p) emission (non-angular-resolved). It is shifted by $\Delta E_B^F = 0.9$ eV between the two surfaces in full agreement with (13.2) and (13.9). The origin of this shift and of the same work function difference reported above is the different energetic position Δ of the Fermi level within the band gap of the Si{111} sample due to the different doping. V_{BB} is zero under these flat band conditions.

The second way of changing the work function of the silicon surface is to change V_{BB} for a given Δ. Starting from the flat band conditions ($V_{BB} = 0$) during a PAX-experiment at 60 K described above, V_{BB} can be changed by increasing the temperature of the sample. This is expected to increase the recombination rate of the electron-hole pairs and, hence, to reduce the available compensation charge at the surface.

Figure 13.12a shows the valence band edge of the p-doped Si{111}7 × 7 surface at room temperature, at 60 K and at 100 K [13.39, 40]. Indeed, there is a successive shift to higher binding energies with rising temperature due to an increasing downwards bending of the bands. The shift between 60 K and 100 K amounts to $\Delta V_{BB} = 230$ meV. The same change is observed in the work function as well as in the energetic position of the Xe(5p) emission. Figure 13.12b displays Xe($5p_{3/2,1/2}$) spectra from the p-Si{111}7 × 7 surface at 60 K and 100 K

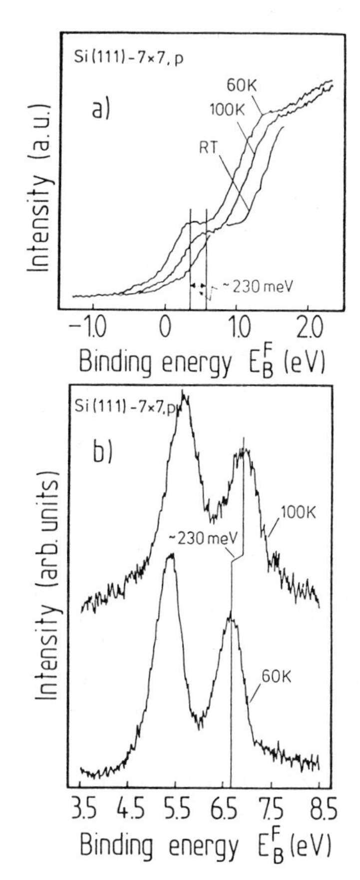

Fig. 13.12a,b. Illustration of the influence of different temperatures on the spectral valence band features of a Xe covered, p-doped Si{111}(7 × 7) surface. Increasing temperature shifts all features, e.g., the valence band edge (**a**) and the Xe(5p) peaks (**b**) to higher binding energies compared to the flat band situation at low temperatures (see Fig. 13.9 and text)

[13.39,40]. At 100 K, a steady state pressure of xenon gas is needed to establish a detectable coverage on the surface. Figure 13.12b compares the corresponding spectrum with that of the *same* coverage measured at 60 K. The spectra are, in fact, shifted by $\Delta E_B^F(5p_{1/2}) = \Delta V_{BB} = \Delta\varphi\,(60\,K{-}100\,K) = 230\,meV$, again in agreement with (13.2) and (13.3) and the $\Delta\varphi$-model displayed in Fig. 13.6.

In summary, this substrate independence of the $5p_{1/2}$ ionization potential of an adsorbed xenon atom (13.3) is easily understandable on the basis of the weak Xe/metal interaction and the large Xe diameter, and holds likewise for all other electron levels on the atom. The potential well of an adsorbed xenon atom as a whole floats with the vacuum level outside different homogeneous surfaces, and, as a consequence, the corresponding E_B^F values (with respect to E_F) reflect the surface potential (contact potential) differences between different surfaces. This is the basis for PAX being a sensitive "local work function" probe as discussed in the next section.

13.2.4 The "Local Work Function" Concept

As pointed out in the previous section, the work function of the surface of a homogeneous, one-component single crystal is determined by the sum of the bulk chemical potential, $\bar{\mu}$, and the surface barrier, $\Delta\phi$ (Fig. 13.6). Both depend on the charge density, ϱ, of the particular material [13.34], but $\bar{\mu}$ is constant throughout an equilibrated solid, while $\Delta\phi$ depends on the charge density of the specific crystal face of the material, that is, on the atomic packing density of the surface. Consequently, different crystallographic faces of one and the same single crystal exhibit different work functions (Table 13.1). More important, however, is the fact that the *local* charge density varies across any real atomistic (non-jellium) surface and, of course, in particular, across surfaces which are chemically and topographically inhomogeneous. Therefore, we are led to conclude that also $\Delta\phi$ and, hence, φ varies across heterogeneous surfaces. This is schematically shown in Fig. 13.13 by a three-dimensional potential energy diagram for a "patchy" surface, e.g., a bimetallic surface. Depending on the nature of the surface site (patch) i, the local surface potential $E_{V,i}$ near the surface is energetically different "high above" the common Fermi level, E_F, throughout the (equilibrated) solid. The respective energy difference

$$\varphi_i = E_{V,i} - E_F = \Delta\phi_i - \bar{\mu} \tag{13.10}$$

has the same physical meaning, as has $\varphi = \Delta\phi - \bar{\mu}$ for a perfect single crystal surface [see (13.6)]. The term φ_i is the *local* equivalent of φ and may therefore be called the "local work function". This term, in fact, appears particularly justified because it will be shown in the following sections that φ_i of surface patches as small as a few nanometers in diameter may have the same values as φ of an extended single crystal surface of the same material and structure.

The important feature of the photoemission of adsorbed Xe atoms is that it provides an experimental method for determining "local work functions", φ_i,

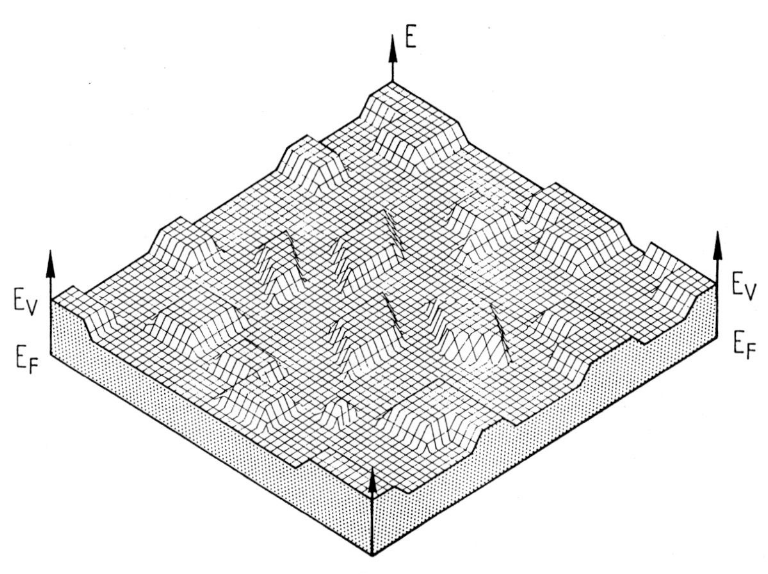

Fig. 13.13. Schematic representation of the modulated surface potential ("local vacuum potential" E_V) near a heterogeneous surface. This modulation flattens out with increasing distance from the surface. The Fermi level E_F is constant throughout the sample

because, as has been demonstrated in a large number of publications [e.g. 13.41–52], (13.2) is also applicable to Xe atoms adsorbing onto different sites (patches) on one and the same but *heterogeneous* surface. In this case, the potential well of the adsorbed Xe atoms is pinned to the local surface potential $E_{V,i}$ of the respective surface site. As a consequence, a superposition of different Xe(5p) as well as Xe(4d) states is observed in the photoemission spectrum, which are shifted by

$$\Delta E_B^F(i,j)_{ads} = -\Delta\varphi_{i,j} \,, \tag{13.11}$$

with respect to each other, $\Delta\varphi_{i,j}$ being the *local* work function difference between the sites (patches) i and j. This is sketched in Fig. 13.14. Considering further, that the adsorption energy E_{ad} of a Xe atom is also site specific, then depending on the chemical nature and the geometry (coordination) of the respective adsorption site, different kinds of surface sites can be populated selectively so that their local surface potential can be determined in isolation before other, weaker-bonding surface sites become populated with Xe leading to new emission peaks, etc. This, for instance, becomes immediately evident from Fig. 13.15, which shows a set of Xe(5p) spectra of xenon adsorbed on a bimetallic Cu/Ru{001} surface. The Xe(Cu) state is populated after the Xe(Ru) state. A detailed discussion of these spectra is given in Sect. 13.3.1.3. Likewise, Fig. 13.16 shows the Xe(4d) spectrum from a Ru{001} surface which was covered with 0.8 monolayers of Cu. Again the two sets of Xe($4d_{5/2,3/2}$) peaks are indicative of Xe atoms on free Ru surface sites [Xe(Ru)] and on Cu-islands [Xe(Cu)], respectively

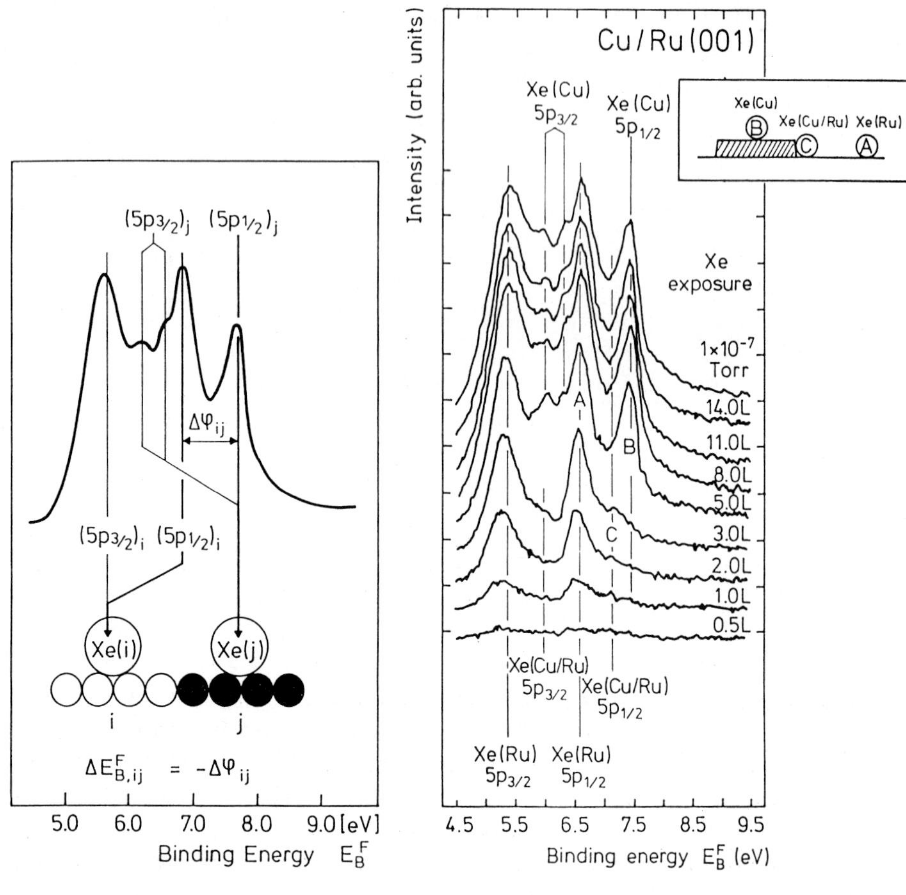

Fig. 13.14. Superposition of $5p_{3/2,1/2}$ spectra from two coexisting Xe adsorption states on two kinds of surface patches i and j on a heterogeneous surface (see Fig. 13.13). The $(5p_{3/2})_i$ signal is only broadened, while the $(5p_{3/2})_j$ peak is split (see, for example, Fig. 13.3). The shift $\Delta E^F_B(5p_{1/2})_{i,j}$ is a measure of the "local work function" difference $\Delta\varphi_{i,j}$ between both patches

Fig. 13.15. Xe($5p_{3/2,1/2}$) spectra from a Ru{001} surface covered with ~ 0.4 monolayers of Cu in the form of well annealed 2*d* Cu islands. Three different Xe(5p) spectra (*A*, *B*, *C*) develop with increasing Xe coverage, which can be associated with the three different kinds of adsorption sites on this surface, namely *A* on free Ru sites, *B* on top of the Cu islands and *C* at the Cu island boundaries, as illustrated in the inset. The highest spectrum corresponds to a complete Xe monolayer on this bimetallic surface

(Sect. 13.3.3). The successive population of the two kinds of states is the same as with the Xe(5p) spectra in Fig. 13.15.

13.2.5 Methodological Aspects

The two "spectroscopic observables" detected with the PAX technique are thus: a) the Xe photoelectron spectrum, namely the binding energies, shapes and intensities of the Xe(5p) or Xe(4d) peaks and b) the relative adsorption energies

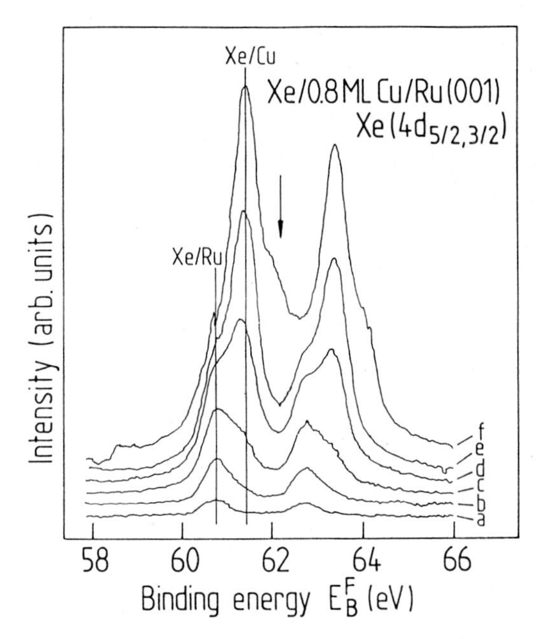

Fig. 13.16. Xe(4d) photoemission spectra for different coverages of Xe on a Ru{001} surface which, prior to Xe adsorption, was covered with ~ 0.8 monolayers of Cu in the form of well annealed 2*d* Cu islands. Spectra a–f correspond to Xe exposures of 0.2, 0.5, 1.0, 2.0, 3.0 and 4.0 L (L ≡ Langmuir ≡ 1 Torr s), respectively. Note the sequential population of the Ru-sites and Cu-sites on this surface

E_{ad} of different site-specific Xe adsorption states, which manifest themselves in the sequence of appearance of the corresponding 5p or 4d peaks (see Figs. 13.15 and 13.16). Due to the weak bonding in the initial state and the nearly substrate independent relaxation in the photoionized final state, the role of an adsorbed Xe atom can be compared to the deposition of a test-electron very near to the surface (~ 2.5 Å) at a particular surface site. Via photoemission, this (5p or 4d) test electron provides the information about the local surface potential at this site. The energetic difference between this local surface potential and E_F may be regarded as the "local work function". This terminology receives its justification from the observation that the "local work function" of for example, a small surface patch, often has the same value as the work function of a macroscopic surface of the same nature as the patch. Experimental results, in particular with vicinal surfaces (Sect. 13.3.1), strongly suggest that the lateral "resolution" of the PAX technique is of the order of the xenon diameter, namely ~ 5 Å. The site specificity of E_{ad} governs the location of the test-electron on the surface, in that different sites on a heterogeneous surface are populated sequentially in the order of decreasing adsorption energy. At completion of an adsorbed Xe monolayer on a heterogeneous surface, the relative intensities of the corresponding Xe($5p_{1/2}$) or Xe(4d) states are a measure of the relative abundances of the different kinds of surface sites. These relative abundances also provide some insight into the distribution of the different surface sites (see for example sect. 13.3.5 and [13.51]). Most often, however, the adsorption energies E_{ad} at different surface sites are not sufficiently different in order to ensure strictly sequential population. In this case, due to an equilibrium distribution of the Xe atoms on the surface, the more or less simultaneous population of different surface sites and, hence, the simultaneous growth of the corresponding (mostly overlapping) photoemission peaks makes a decompositon of the total photoemission spectrum into the partial site specific contributions necessary. For instance, each site specific Xe(5p) spectrum is then represented by the set of three Lorentzian lines as shown in Fig. 13.4. The energetic position and the intensity (area) of the Xe($5p_{1/2}$) Lorentzian line are the desired quantities which determine the local surface potential and the surface concentration of the particular type of surface site. The same applies to the analysis of the Xe(4d) spectra.

Although, in this way, the Xe(5p) PAX spectrum of a complete Xe monolayer on a surface with three distinguishable sites needs to be fitted with the large number of nine Lorentzians, it has to be pointed out that the position, width and intensity of these nine lines are *not* free variables, but that they are subject to physical constraints. Firstly, the energetic separations and the relative intensities within one and the same set of three Lorentzians per Xe state vary only within narrow limits and may be further confined by measurements on suitable reference surface. Secondly, difference curves between spectra from successive Xe coverages (so-called incremental spectra) accentuate the sequential population of different surface sites and consist of one or two Xe adsorption states only. Incremental spectra are displayed, for instance, in Figs. 13.23 and 13.29. After all, only the Xe(5p) signal of each Xe state is considered for the surface characterization. Under these conditions, the fits become rather reliable. The situation is somewhat simpler with the Xe(4d) spectra because each Xe state contributes only with two peaks of given shape and separation. For more details on the fitting procedure, see [13.9]. A PAX fitting program based on physically realistic parameters is described in [13.53].

Summarizing this section, PAX may be considered as a non-destructive, reversible "decoration" and "titration" technique, which not only provides information about the relative abundance of specific surface sites, but also about their "local work function". In this latter respect, the PAX technique appears to provide complementary information to scanning tunneling microscopy (STM). In particular, the adsorbed Xe probe atoms "sense" the local properties of the respective adsorption site in a way more like other adsorbing molecules than does the STM tip. Furthermore, because of the weak interaction between a Xe atom and any substrate and the low temperatures involved, no adsorption induced modifications of the surface (such as displacement reactions, surface reconstruction or segregation effects), influence the measurements as do strong adsorbates like O_2, H_2, CO, etc. in "chemisorptive titration" [13.54]. The Xe probe atoms may be desorbed at temperatures below $\sim$ 120 K where still no thermally activated alteration of the surface is expected.

In the following sections, selected case studies will show best what kind of information can be obtained from PAX studies from heterogeneous metallic surfaces. The examples are presented in order of increasing complexity, that is, of increasing number of distinguishable kinds of surface sites.

13.3 Selected Case Studies of Metallic Surfaces

13.3.1 Surface Steps: Vicinal Surfaces

Steps and kinks are ubiquitous surface structure defects as has become clear from scanning tunneling microscopy. The influence and the properties of these step defects at surfaces are studied best with well prepared vicinal surfaces as model systems [13.55] because they enable a control of the average step density and the crystallographic step direction. Vicinal surfaces are prepared by cutting a single crystal under a small angle ($\alpha \lesssim 10°$) off a low index crystal plane. After polishing and cleaning in ultrahigh vacuum, such an equilibrated vicinal surface assumes a rather regular step structure, which can be characterized very accurately by STM [13.56] and LEED [13.55] with respect to the step height (which is mostly monoatomic), the crystallographic direction, γ, of the step ledges and the mean width, W, of the terraces between the steps. The terraces between two adjacent steps have the same structure as the parent low-index plane ($\alpha = 0°$), even if this surface tends to reconstruct [13.56]. Figure 13.17 shows schematically an fcc $\{100\}$ surface with three monoatomic steps of different orientations $\langle uvw \rangle$. Steps with $\langle 110 \rangle$ direction are formed from close-packed rows of atoms resulting in smooth ledges. Steps with directions γ off the $\langle 110 \rangle$ direction include kinks. In the $\langle 100 \rangle$ direction ($\gamma = 45°$) the step should be fully kinked.

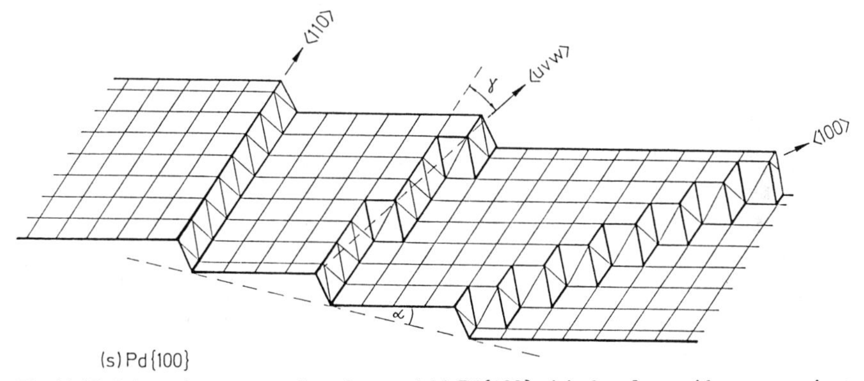

Fig. 13.17. Schematic representation of stepped (s) Pd$\{100\}$ vicinal surfaces with monoatomic steps of different orientation $\langle uvw \rangle$. $\alpha \equiv$ vicinal angle, $\gamma \equiv$ step direction off the $\langle 110 \rangle$ direction

Step wedges (including kinks) provide surface sites with high coordination to the substrate, and it is, therefore, safe to assume that physisorbed xenon atoms will preferably occupy (down-) step and kink sites before the terrace sites are populated. This assumption is particularly justified in the case of physisorbed gases which are held at the surface by van der Waals forces. Figure 13.18 shows adsorption isobars of xenon on a stepped Pd(s) [8(100) × (110)] surface [13.57]. On this surface, the steps run along the $\langle 100 \rangle$-directions as sketched in Fig. 13.17 and are fully kinked. At very low equilibrium pressures only the step (= kink) sites are populated (curves a–c in the inset). The saturation coverage of these

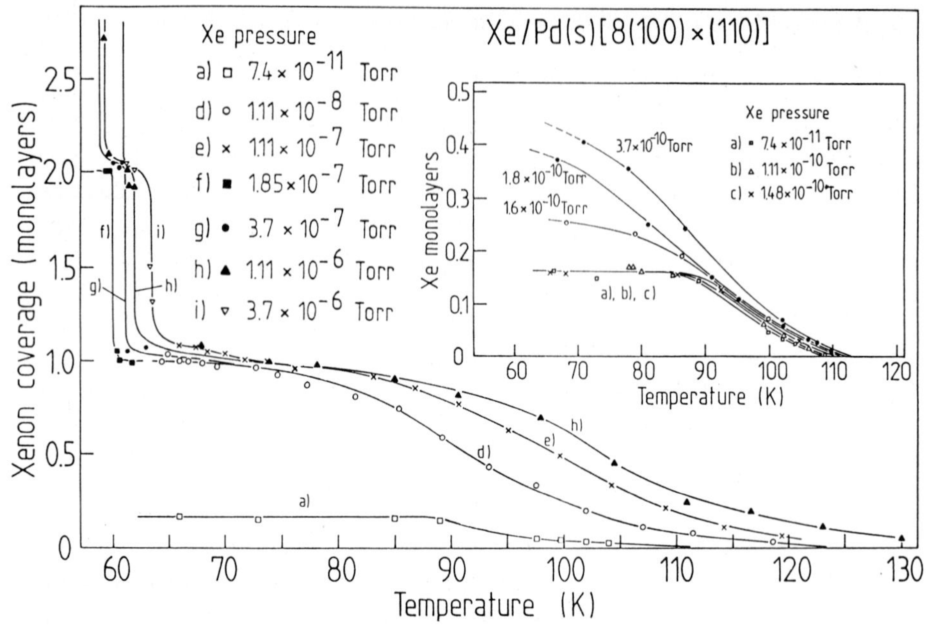

Fig. 13.18. Adsorption isobars of Xe on a Pd(s)[8(100) × (110)] vicinal surface. The Xe coverage was determined with AES. The inset shows the very low pressure curves which lead to preferential population of the step sites. From [13.57]

isobars (curve a in Fig. 13.18) agrees with the relative abundance of step sites on this surface, as determined from the LEED-derived ideal terrace-ledge-kink model of this surface [13.57, 58]. At higher Xe pressures, the terraces of this surface (curves d and e in Fig. 13.18) and even higher Xe layers (curves f–i) are also populated with Xe. A Clausius-Clapeyron analysis of these isobars yields extrapolated initial heats of adsorption of 42.6 kJ/mol at step sites and 34.3 kJ/mol at terrace sites [13.57]. These results are in close agreement with corresponding thermal desorption experiments for Xe on this and other stepped Pd{100} surfaces (Table 13.5) and support the preferred initial occupation of high-coordination (step) sites [13.58].

Xe($5p_{3/2,1/2}$) UPS spectra from a stepped Pd{100} surface are displayed in Fig. 13.19. One can see clearly that at low Xe coverages obtained after Xe exposures $< 4L$ Xe (spectra a–c), the $5p_{1/2}$ peak is located at somewhat higher electron binding energy than a second $5p_{1/2}$ peak which first becomes visible in spectrum d ($4L$) and which finally overgrows the low coverage signal. Spectrum g corresponds to a complete monolayer of Xe on this surface (corresponding to the monolayer plateau in Fig. 13.18); its intensity serves as a reference for the assignment of the lower coverages. A careful analysis of the $5p_{1/2}$ intensity of the spectra from Fig. 13.19 using the procedure described in Sect. 13.2.5 and [13.9], yields two $5p_{1/2}$ lines which are separated by ~ 350 meV and the intensities of which reflect the relative abundances of step and terrace sites on this

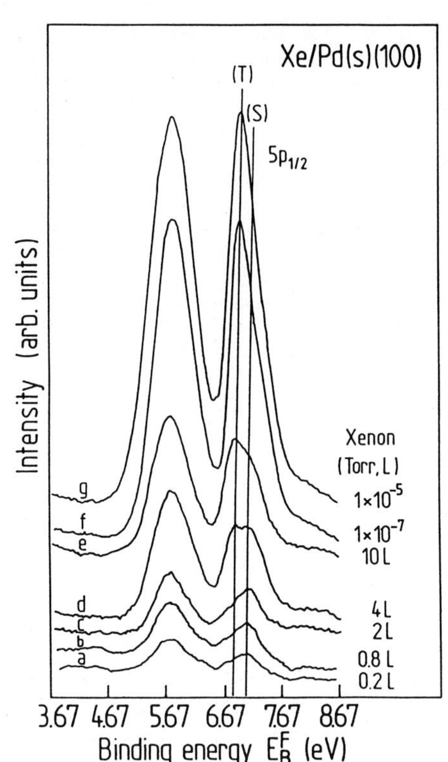

Fig. 13.19. Xe($5p_{3/2,1/2}$) spectra for Xe adsorption on a stepped Pd{100} vicinal surface. Note the sequential population of the step (*S*) and terrace sites (*T*) with xenon. The highest spectrum corresponds to a complete Xe monolayer on this surface

surface which are known from an LEED analysis. Very similar results have been obtained on other Pd{100} vicinal surfaces with different step density and step orientation. Table 13.5 summarizes these results as follows: The step direction ⟨uvw⟩ and the vicinal angle α are defined by Fig. 13.17 and are determined from LEED images of these surfaces. The average step density and the average terrace width (perpendicular to the respective step ledge) are a simple geometrical consequence of the vicinal angle α if only steps of monoatomic height are considered. The numbers of Xe atom rows which can be accommodated on the terraces parallel to the steps are calculated assuming a hexagonal structure for the Xe layer [13.57–59]with a Xe–Xe interatomic distance of 4.48 Å[13.59]. E^0_{ad} (step) and E^0_{ad} (terrace) are the initial heats of adsorption of Xe at step and terrace sites, respectively, as determined from thermal desorption spectra. In particular, the E^0_{ad} (step) values are in close agreement with those derived from the isobars in Fig. 13.18. The terms $E^F_{B,0}(5p_{1/2}$, step) and $E^F_{B,0}(5p_{1/2}$, terrace) are the Xe($5p_{1/2}$) electron binding energies as obtained by careful fits with Lorentzian curves and extrapolated to zero coverage *of the respective Xe adsorption state*. Finally, the last two lines in Table 13.5 give the intensity ratios $(\mathrm{Xe(T)} : \mathrm{Xe(S)})_{\mathrm{model}}$ and $(\mathrm{Xe(T)} : \mathrm{Xe(S)})_{\mathrm{exp}}$, which are obtained theoretically from a ball model (model) and experimentally from the partial Xe($5p_{1/2}$) UPS intensities (exp) at Xe monolayer saturation of the respective surface. In the case of the surfaces with ⟨100⟩

Fig. 13.20. Structure model of Xe adsorption on a Pd(*s*)[8(100) × (110)] vicinal surface as derived from LEED observations [18.57]. The steps in ⟨100⟩ direction on this surface are "fully kinked". At low Xe coverage only every second kink site is populated. At higher Xe coverages a hexagonally close-packed Xe layer is formed. Both orientations I and II with respect to the step edges are observed in the LEED picture. The uptake data put emphasize on structure II, which grows exclusively beyond Xe monolayer saturation. From [18.57]

step direction, it is considered that with exclusive population of the steps with Xe atoms, only every other kink site is occupied, as illustrated in Fig. 13.20.

This adsorption structure is confirmed by direct LEED evidence, namely, by a Xe induced diffraction pattern which exhibits double periodicity in the ⟨100⟩ direction [13.57]. Only when all steps are saturated in this way do higher Xe coverages lead to the formation of a hexagonal Xe layer on the terraces [13.57]. Along smooth step ledges of [110] direction (see Fig. 13.17) the Xe atoms form close-packed rows.

Several observations become clear from the results in Table 13.5. The adsorption energy at the step sites is approximately 10% higher than at the terrace sites, in agreement with theoretical calculations based on the summation over Lennard-Jones pair-potentials between an adsorbed Xe atom and substrate atoms [13.60]. The adsorption energy at the fully kinked steps with ⟨100⟩ direction is reproducibly higher than at a smooth step of ⟨110⟩ direction (see Fig. 13.17). The (average) adsorption energy at a step with the intermediate ⟨430⟩ direction falls between the other two limiting cases. All E^0_{ad} (step) values are very similar to the value found for Xe adsorption on a perfect Pd{110} surface, namely 41.0 kJ/mol. Indeed, a step wedge resembles very much the trough structure of the Pd{110} surface. Similarly, all E^0_{ad} (terrace) values in Table 13.5 are very close to the corresponding value found on the perfect Pd{100} surface ($\alpha = 0°$ in Table 13.5).

The initial selective population of the step sites clearly determines the assignment of the two Xe(5p) photoemission states seen in Fig. 13.19, in that the higher binding energy peak (at low coverage) corresponds to Xe atoms at step sites [Xe(S)]. On average $E^{\mathrm{F}}_{\mathrm{B},0}(5\mathrm{p}_{1/2}$, step) is higher by 320 meV than $E^{\mathrm{F}}_{\mathrm{B},0}(5\mathrm{p}_{1/2}$,

Table 13.5. Summary of experimental results for Xe adsorption on Pd{100} vicinal surfaces. The various quantities are explained in the text. E^0_{ad} and $E^F_{B,0}$ denote the initial adsorption energy (at the limit of zero coverage of the respective Xe adsorption state) and the electron binding energy of the $5p_{1/2}$ level of Xe adsorbed at step and terrace sites of the vicinal surfaces, respectively. The last two lines correspond to intensity ratios for Xe adsorbed at terrace and step sites of the vicinal surfaces expected on the basis of ideal terrace-ledge-kink models (model) and determined experimentally (exp) from PAX spectra

Step direction ⟨uvw⟩	–	⟨100⟩			⟨110⟩	⟨430⟩
vicinal angle α	0°	2.5°	7°	8°	5.6°	3.7°
step density [10^6/cm]	–	2.25	6.37	7.25	5.05	3.25
terrace width [Å]	–	44.5	15.7	13.8	19.8	30.4
Xe rows per terrace	–	11	4	3	5	7–8
E^0_{ad} (step) [kJ/mol]	–	41.8	41.0	41.4	39.8	40.6
E^0_{ad} (terrace) [kJ/mol]	36.8	37.2	36.4	37.6	36.4	36.4
$E^F_{B,0}(5p_{1/2}$/step) [eV]	–	6.98	6.98	6.94	6.97	–
$E^F_{B,0}(5p_{1/2}$/terrace) [eV]	6.75	6.70	6.64	6.61	6.63	–
(Xe(T): Xe(s)) model	–	20	6	4	4	6–7
(Xe(T): Xe(S)) exp.	–	24	5.4	4	3.8	5.1

terrace). All $E^F_{B,0}(5p_{1/2}$, step) values are again close to the corresponding value found on a perfect Pd{110}, namely 7.03 eV (Table 13.1). In turn, all $E^F_{B,0}(5p_{1/2}$, terrace) values are near the value found on a perfect Pd{100} surface ($\alpha = 0°$ in Table 13.5).

Finally, the experimentally determined intensity ratios $(Xe(T) : Xe(S))_{exp}$ agree within ~ 15% with the $(Xe(T) : Xe(S))_{model}$ value estimated on the basis of an ideal terrace-ledge-kink model of the respective vicinal surface populated with hard-sphere Xe atoms as described above and shown in Fig. 13.20

Several important conclusions can be drawn from the above observations.

1. As regards the adsorption energy E^0_{ad} and the electron binding energy $E^F_{B,0}(5p_{1/2})$ of adsorbed Xe probe atoms, the step sites on the Pd(s){100} vicinal surfaces have similar properties to the adsorption sites in the troughs of a perfect Pd{100} surface [13.61]. Likewise, the properties of the {100} terraces are similar to those of a perfect Pd{100}, practically *independent of the terrace width.* This is in agreement with STM observations.

2. Using (13.11), the local surface potential at the step sites is $\Delta\varphi_{S,T} = \Delta E^F_{B,0}(S,T) = 320$ meV lower than at the terrace sites. This is a consequence of the so-called Smoluchowski smoothing effect [13.62]. The electronic charge near the surface does not follow the abrupt step geometry but flows from the upper step edge into the lower step wedge (Fig. 13.21). This creates a local extra dipole which is antiparallel to the normal surface dipole layer. Indeed, systematic studies with regularly stepped surfaces [13.58, 63, 64] revealed a linear decrease of the (macroscopic) work function of these surfaces as a function of step density. This

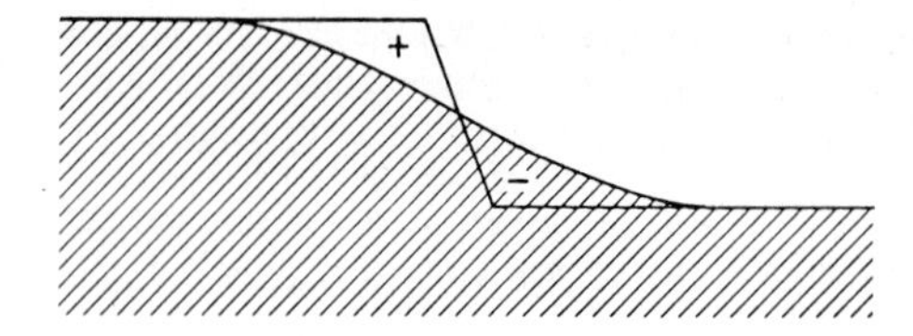

Fig. 13.21. Schematic representation of the Smoluchowski electron smoothing effect near a surface step. Charge flows from the upper step edge into the low step wedge, thereby creating a localized extra dipole which counteracts the normal surface dipole layer

linearity led to the *indirect* conclusion, that the step induced charge redistribution and, hence, the associated extra dipoles which counteract the normal surface dipole is really confined to the immediate vicinity of the steps. Each step adds an incremental dipole which leads to a linear decrease of the macroscopic work function with increasing step density. The present PAX results provide *direct* evidence that the surface potential is *locally* lowered by 320 meV at a step site compared to a terrace site.

3. Moreover, from the fact that the $(Xe(T) : Xe(S))_{exp}$ intensity ratios determined at Xe monolayer saturation always agree with the expectation based on the model derived numbers of available step and terrace sites, it follows conclusively that only those Xe atoms "feel" the lowered surface potential which are in immediate contact with the step; even the next-nearest row of Xe atoms away from the step does not contribute to the Xe(S) signal. Consequently, PAX proves directly that the work function decrease is strictly *localized* along the steps. This, in turn, provides the argument that the lateral "resolution" of the PAX technique is of the order of one Xe atom diameter, ~ 5 Å.

4. Obviously, the Xe($5p_{1/2}$, step) intensity varies linearly with the step density of the vicinal surfaces. Compared to the $5p_{1/2}$ intensity of a complete Xe monolayer, the Xe atom density of which is approximately known from LEED observations [13.57, 59, 61], this provides a *quantitative measure* of the absolute number of step sites on a surface. As yet, attempts to distinguish between step and kink sites by means of PAX have not been successful.

Taken together, the conclusions 1–4 make PAX an important tool for the characterization of surface steps. Of course, both the number of step sites and the local surface potential at a step site are basic ingredients for an understanding of the catalytic activity of structural surface defects. In fact, the local surface

Table 13.6. Local surface potential difference between step (s) and (sputter-induced) defect (d) sites and perfect sites on the different metal surfaces

Surface	$\Delta\varphi_{loc}$ [meV]
(s)Pt {111}	980
(s)Ru {001}	700
(d)Ru {001}	~ 500
(s)Pd {111}	500
(s)Pd {100}	300
(d)Pd {110}	~ 0
(d)Ag {111}	150
(d)Cu {111}	~ 150

potential decrease at step sites may be enormous as has been shown with PAX measurements on stepped surfaces of other metals. Table 13.6 summarizes PAX derived local surface potential decreases from stepped (s) and defected (d) metal surfaces; they reach the value of ~ 1 eV at steps on a Pt{111} surface. A consequence of this enormous surface potential difference which decays over a distance of a few Ångstroms only and of the associated local fields is addressed in Part 4.

13.3.2 Metallic Ruthenium Powder

Of course, single crystal vicinal surfaces are rather ideal model systems for defect surfaces. Yet, on the *atomic scale* a step defect on a realistic surface is not expected to be different from a step defect on a vicinal surface. This is already suggested by the *short-range* effect of a step defect which was the main conclusion from the experiments with the vicinal surfaces presented in the previous section. In this section we turn to the other extreme, namely, to Ru powder particles of ~ 500 Å diameter [13.46]. The PAX results from these highly defective nano-crystallites are, in fact, very similar to those obtained from a sputter-roughened Ru{001} surface, thereby supporting the expectation about the transferability of results from defective (stepped or sputtered) single crystal model surfaces to more realistic surfaces.

Figure 13.22a shows Xe(5p) PAX spectra from the cleaned [13.46] Ru powder which was pressed onto a Pt-foil by means of a sapphire pestle. For comparison,

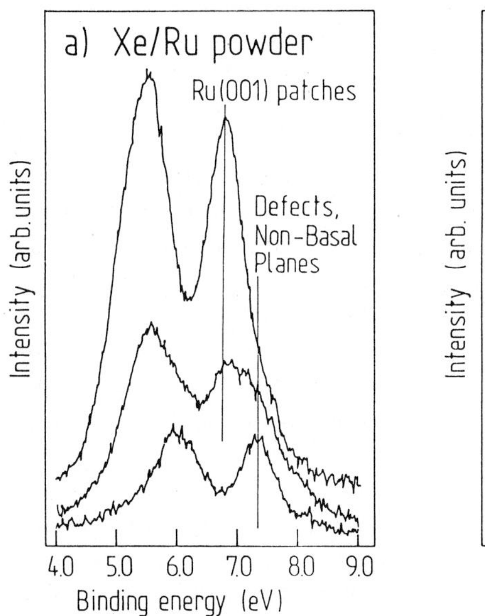

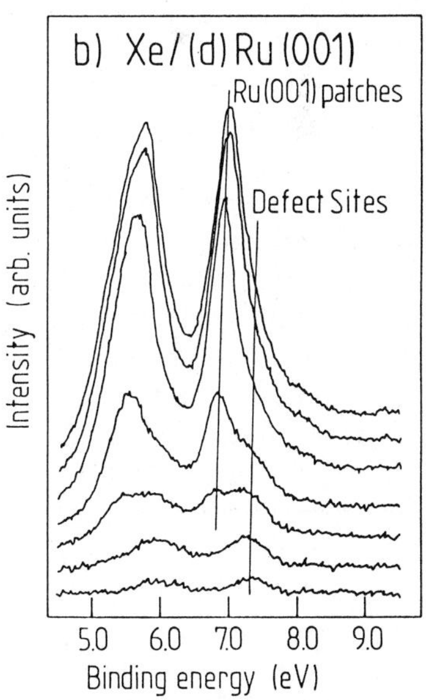

Fig. 13.22a,b. Xe($5p_{3/2,1/2}$) spectra from Xe adsorbed on metallic Ru powder particles (**a**) and a sputter-roughened Ru{001} surface (**b**). In both panels the highest spectrum corresponds to a complete Xe monolayer. Note the sequential population of step (defect) sites and intact Ru{001} patches

Figure 13.22b shows Xe(5p) PAX spectra from a defective Ru{001} surface which was prepared by bombarding the single crystal surface with 2 keV argon ions. The ion dose was equivalent to about one-half of a monolayer. Both sets of spectra are surprisingly similar, in that both show the successive population of surface defects at higher binding energy and of intact Ru{001} patches at lower binding energy. The value $E_B^F(5p_{1/2}) = 6.8\,eV$ from the Ru{001} patches in both cases is identical to the value from a perfect Ru{001} single crystal surface (Table 13.1 and [13.65]). The value of $E_B^F(5p_{1/2}) = 7.3\,eV$ for the defect sites is identical on both the Ru powder particles and the sputter-roughened Ru{001} surface (but slightly smaller than for a step site on a Ru{001} vicinal surface, see Table 13.6). It suggests a local surface potential decrease of $\sim 500\,meV$ near the defect sites.

The highest spectrum in both panels of Fig. 13.22 corresponds to completion of the first monolayer of Xe on the respective sample. The partial intensity arising from the Xe atoms at defects sites is larger on the powder sample than on the sputter-roughened Ru{001} surface. According to a quantitative analysis of the spectra in Fig. 13.22a (as described in Sect. 13.2.5 and [13.9]) about 70% of the total powder surface still consists of perfect Ru{001} sites.

The main conclusion which can be drawn from these experimental results is that PAX can be applied to realistic surfaces such as from metal powder catalysts. The information obtained from the PAX spectra about the surface defect concentration and the local surface potential at the defects can be correlated with adsorption and catalytic reaction data of reactive gases on the same surfaces, and, therefore, appears indispensable for an investigation of the fundamental parameters in heterogeneous catalysis (see also Sect. 13.4).

13.3.3 Initial Stages of Film Growth: Bimetallic Surfaces

Epitaxial metal monolayers have attracted increasing attention recently because of their potential technological importance. For instance, epitaxial monolayers of a magnetic material on a non-magnetic substrate like Fe/Cu{100} [13.66], Fe/Au{100} [13.67] may exhibit new magnetic behavior due to an altered geometrical structure and their electronic interaction with the substrate. Furthermore, the magnetic moment of thin Gd films deposited on Fe, for example, is found to couple antiparallel to that of the substrate [13.68]. These magnetic properties together with the advantages of thin films, like miniaturization and low material consumption, make these structures potentially useful for magnetic storage devices.

More generally, the occurrence or monolayer surface films of different composition than the bulk beneath are more or less a ubiquitous phenomenon as a consequence of segregation effects in multicomponent materials. These films most probably will also exhibit geometric and electronic properties different from both the substrate and from a lattice layer of the same material (as the film) in three-dimensional form. These modified films, in turn, dominate the surface chemistry and physics of the whole sample. Even when these films do not cover the whole substrate surface, the same arguments apply at least to corresponding surface patches. Vapor deposited monolayer metal and alloy films have, therefore, become a subject of intensive investigations.

The nucleation and growth (condensation) of thin metal films is governed by atomistic processes. Individual atoms are accommodated on the substrate. They may, thermally activated, diffuse across the surface, recombine with other atoms forming stable homogeneous nuclei of a few atoms, or may be incorporated into the atomic edges of already existing immobile islands of larger size. Eventually, the growing islands may coalesce giving rise to atomic mismatch and defect sites along the island (grain) boundaries. The case of heterogeneous nucleation *per se* is controlled by atomic scale defects like steps, kinks and point defects in the substrate surface. Any such structural (and chemical) surface disorder is accompanied by a corresponding variation of the surface electronic structure and energy, and results in a distribution of surface sites of different adsorption and reaction behavior. Hence, the ultimate goal of understanding the growth and the properties of thin films calls for techniques which allow the characterization of the initial stages of nucleation and of surface morphology on an atomic scale.

The PAX-technique can also provide quantitative information about the concentration and the lateral distribution of heteroatoms on a metal surface. In part, the PAX results complement STM reliefs in a unique way, in that (1) the PAX intensities are a quick and direct measure ("titration") of the density of specific surface sites, and (2) the $E_B^F(5p_{1/2})$ or $E_B^F(4d_{5/2})$ values of the adsorbed Xe probe atoms yield "local work functions" (as defined in Sect. 13.2.4), which do not seem to be obtainable by STM.

This section describes the evaluation of PAX spectra from Ru surfaces which were covered with submonolayer amounts of Cu. The Cu/Ru system is known as a *bimetallic* catalyst, in which the Cu is present only on the surface of the Ru because both metals are completely immiscible in the bulk. Earlier model studies with submonolayer deposits of Cu on a Ru{001} single crystal surface pointed to the formation of Cu islands of monoatomic thickness because Cu-TDS spectra from the Ru{001} substrate indicated a quasi-zeroth order desorption behavior as well as stronger Cu–Ru bonds than Cu–Cu bonds [13.69]. Here we discuss PAX spectra from submonolayer deposits of Cu on a perfectly flat Ru{001} single crystal surface and a Ru{001} surface which, prior to Cu deposition, was slightly sputter-roughened with Ar^+ ion bombardment. The results provide a rather direct insight into the completely different growth mode of the Cu layers on both substrates [13.46]. The results will serve as a reference for corresponding studies with bimetallic Cu/Ru powder particles presented in the following section.

Figure 13.15 shows a series of Xe($5p_{3/2,1/2}$) spectra from the perfect Ru{001}, which prior to Xe adsorption, was covered with ~ 0.4 monolayers of Cu by vapor deposition and was well annealed at 520 K. Up to an exposure of $3L$ Xe, the spectra exhibit two $5p_{1/2}$ states (A and C) at $E_B^F = 6.7$ eV and $E_B^F \approx 7.3$ eV, respectively. This becomes particularly evident from Fig. 13.23a which displays difference curves between consecutive spectra from Fig. 13.15. These difference

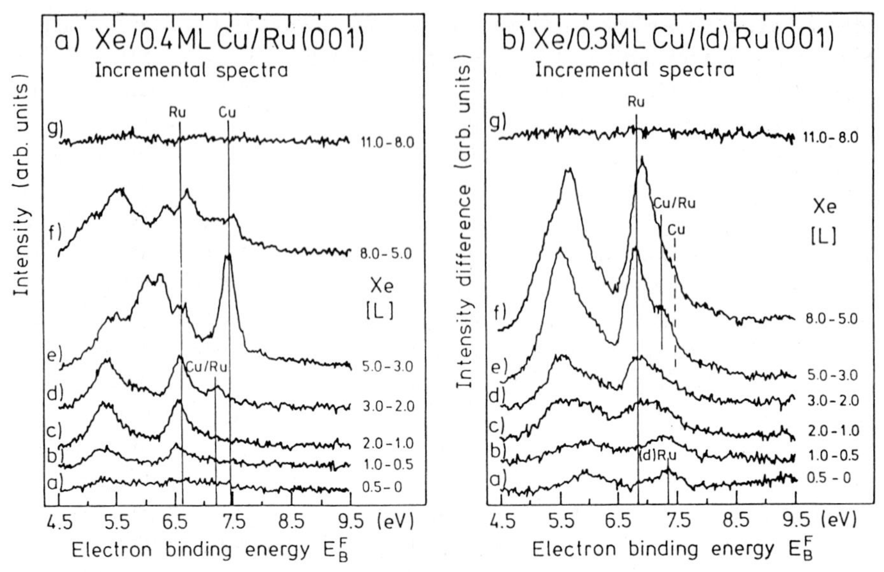

Fig. 13.23a,b. Difference curves between subsequent spectra from Fig. 13.15 (**a**) and from Fig. 13.22b (**b**). These so-called "incremental spectra" accentuate the sequential population of surface sites in the order of decreasing Xe adsorption energy E_{ad}

curves accentuate the intensity added by each new Xe dose and are, therefore, termed "incremental spectra". The dominant peak A at 6.7 eV, which appears first, is close to the position as on clean Ru, and is, therefore, assigned to Xe atoms on bare Ru patches [Xe(Ru)]. Above $3L$ Xe exposure, a new $5p_{1/2}$ peak B emerges at $E_B^F \approx 7.4$ eV (Fig. 13.15). Its $5p_{3/2}$ counterpart grows between the $5p_{3/2}$ and $5p_{1/2}$ signals of the Xe(Ru) state. This is again best seen in the incremental spectra of Fig. 13.23a. Signal B is very close to the (high-coverage) position on Cu{111} and is, therefore, assigned to Xe atoms *on* the deposited Cu [Xe(Cu)]. The fact that these Cu sites are populated with Xe after the Ru sites is in agreement with the lower Xe adsorption energy on Cu compared to Ru [13.23]. The Xe(Cu)$5p_{3/2}$ signal (between the Xe(Ru) peaks) is clearly split into two peaks (Figs. 13.15 and 13.23a). As stated in Sect. 13.2.1, it is generally accepted that this splitting is mainly a consequence of the formation of a two-dimensional (2d) electronic band structure within the adsorbed Xe layer [13.15–18], which in the present case is only conceivable if the deposited and well annealed Cu layer forms flat islands so that the Xe atoms on top can also form a densely packed overlayer of sufficient lateral extension. Hence, the structure of the Xe(Cu) spectrum itself, namely the $5p_{3/2}$ splitting, carries the unambiguous information that submonolayers of Cu on a flat Ru{001} surface form islands, which should be monoatomically thick because the Cu–Ru interaction is stronger than the Cu–Cu interaction, as mentioned above [13.69]. This, together with the fact that Cu and Ru are immiscible, and that, therefore, the Cu islands are *on* the Ru surface, leads to the assignment that the peak C in Figs. 13.15 and 13.23a corresponds to Xe atoms at the Cu/Ru step sites along the Cu island boundaries, as illustrated in the inset of Fig. 13.15. Since here the Xe atoms are in contact with both Ru and Cu atoms, these sites are henceforth called "mixed" sites [Xe(Cu/Ru)]. They are populated rather early because here they have the form of high-coordination step sites (see also Sec. 13.3.5). Similar results have also been obtained with Ag and Au submonolayers on Ru{001} [13.45, 50–52]. The partial intensities of all three Xe states; Xe(Ru), Xe(Cu) and Xe(Cu/Ru), are a quantitative measure of the relative abundance of these three kinds of surface sites [13.45], and the local surface potential at the mixed Cu/Ru boundary sites is intermediate between those *on* Ru and *on* Cu patches, respectively. The latter two are in agreement with the macroscopic work functions of a clean Ru{001} surface and a Ru{001} surface covered with one complete Cu monolayer. The above PAX results obtained from the Xe(5p) valence levels are supported by corresponding Xe(4d) core level measurements using synchrotron radiation. Xe(4d) spectra from a Ru{001} surface covered with 0.8 monolayers of Cu in the form of well annealed islands are shown in Fig. 13.16. The sequential population of the Xe(Ru) and the Xe(Cu) state is obvious. However, here the resolution is not good enough to resolve the Xe(Cu/Ru) state from the mixed Cu/Ru boundary sites.

The assignment of peak C in Figs. 13.15 and 13.23a, as being due to Xe at the mixed Cu/Ru boundary sites, is strongly supported by PAX spectra from

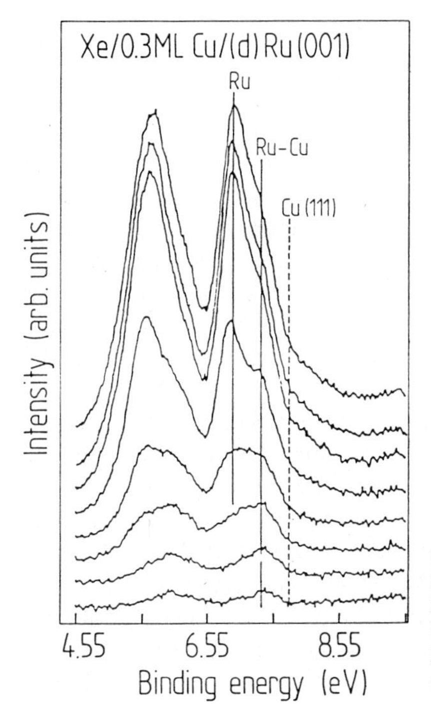

Fig. 13.24. Xe($5p_{3/2,1/2}$) spectra of Xe adsorbed in a bimetallic Cu/Ru surface. Before Cu deposition, the Ru surface was subjected to ion bombardment. Note the lack of any resolved Xe(Cu) peak compared to Fig. 13.15, suggesting a highly dispersed distribution of the Cu due to the Ru surface defects

a Ru{001} surface which prior to Cu deposition was slightly sputtered with Ar^+ ions of 2 keV (Sect. 13.3.2). The corresponding integral and incremental PAX spectra are shown in Figs. 13.24b and 13.23b. Although the overlayer of 0.3 ML Cu on the defected Ru{001} surface was again annealed for 10 min at 520 K (as on the flat Ru surface described above) prior to Xe adsorption, *no* Xe(Cu)($5p_{1/2}$) signal can be resolved on this surface. Instead, the Xe(Cu/Ru) intensity is much more pronounced, indicating a higher concentration of mixed Cu/Ru sites as a consequence of a more uniform dispersion of the Cu particles across the whole surface. Obviously, in this case, the formation of larger 2d Cu islands is prevented due to the high concentration of sputter-induced defect sites which act as heterogeneous nucleation centers. Note also that at no stage does the $5p_{3/2}$ signal from this surface show any obvious splitting, suggesting Cu clusters on the surface too small for a $2d$ electronic band structure to form within the Xe on top.

This Cu/Ru example shows that PAX cannot only distinguish qualitatively between the three possible surface sites for Xe adsorption, namely Ru, Cu and Cu/Ru sites, but also provides a quantitative measure of their relative surface concentration, of their local surface potential by virtue of their $5p_{1/2}$ or $4d_{5/2}$ electron binding energies using (13.11), as well as of the distribution of the Cu deposit, namely $2d$ island formation on the flat Ru{001} substrate versus nearly atomic dispersion on a sputtered Ru substrate [13.46].

In particular, these latter observations will serve as a reference for the following section, in which the corresponding surface properties of bimetallic Cu/Ru powder particles are examined by means of PAX.

13.3.4 Bimetallic Cu/Ru Powder Catalyst

The 5p spectra of Xe adsorbed to different coverages on bimetallic Cu/Ru powder particles are displayed in Fig. 13.25. This sample was prepared by deposition of approximately 0.5 ML of copper on the Ru powder characterized in Sect. 13.3.2. During copper deposition the powder substrate was kept at a temperature of 100 K and was subsequently annealed at 650 K prior to Xe adsorption. The highest spectrum in Fig. 13.25 is taken to represent saturation of the first Xe monolayer. These spectra bear a much closer resemblence to the spectra from Fig. 13.24 than to those shown in Fig. 13.15, in that they show no resolved Xe(Cu) state. The latter would indicate the existence of extended Cu islands on the Ru powder particles, which after all are $\sim$ 500 Å in diameter. Consequently, the Cu is highly dispersed over the surface of the Ru grains trapped by the many defects.

This information is again very important in order to explain the catalytic properties of bimetallic Cu/Ru powder catalysts. If a reaction required a site consisting of an extended array of bare Ru atoms, and the Ru catalyst was heavily loaded by copper (say 50% on the surface), the probability of finding such a copper free array is expected to be much lower if the copper is randomly distributed as atoms or small aggregates than if it is present as large islands on the

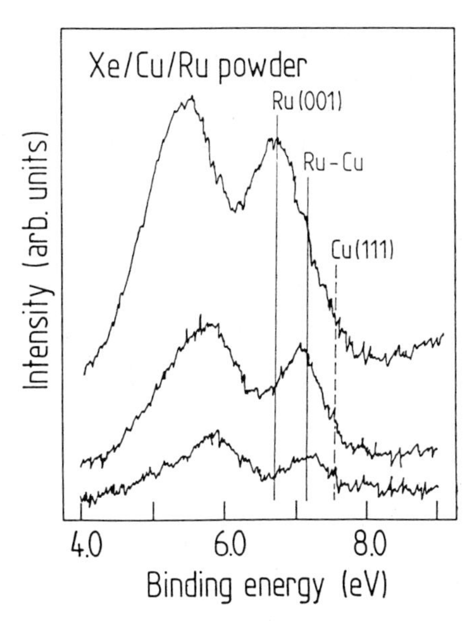

Fig. 13.25. Xe($5p_{3/2,1/2}$) spectra of Xe adsorbed on bimetallic Cu/Ru powder particles. As in Fig. 13.24, the lack of any resolved Xe(Cu) peak suggests a highly dispersed distribution of the Cu atoms on the surface

surface. The random distribution of the copper (if it were inactive for the reaction) would, therefore, be much more effective in inhibiting the supposed reaction. This has to be kept in mind when comparing results from model experiments with single crystal surfaces with realistic catalysts [13.70].

13.3.5 Thermal Stability of Metal/Metal Interfaces

Beyond the mere geometrical and electronical structure of epitaxial metal/metal interfaces, their stability with respect to temperature effects is also of great importance for the design and the technological application of layered structures. Analytically, it is not trivial to demonstrate (or exclude) the onset of site-exchange processes and intermixing across an atomically sharp interface between two metals. The high surface- and site-specificity of the PAX technique can again be taken advantage of to study this problem very successfully. This section describes model studies, in which the thermally activated site exchange between atoms across atomically sharp Ag/Au- and Au/Ag-interfaces supported on a Ru{001} substrate has been studies with PAX. Of course, these investigations with the silver/gold system must be regarded as a test study, which can be carried over to other metal/metal systems.

The atomically sharp interfaces between Ag on Au and Au on Ag, were formed by first preparing just one complete, well annealed monolayer of Au or Ag on the clean and perfect Ru{001} surface, and then depositing at 60 K a submonolayer amount of the other metal on top. At this low substrate temperature, no intermixing and significant atom mobility is expected. Both Ag and Au form well ordered epitaxial monolayers of uniaxially compressed {111} structure on

Ru{001} [13.71, 72]. The Ag and Au coverages, Θ, were best determined by the thermal desorption traces of both metals recorded after completion of each experiment [13.73]. Like copper, Au and Ag do not form alloys with Ru [13.45, 73]; the Ru surface acts as an infinite diffusion barrier. Hence, after desorption of the respective Ag + Au overlayer, the Ru surface is clean and ready for the next experiment.

First, Fig. 13.26 shows reference spectra for complete Xe monolayers on five different surfaces, namely, on bulk Ag{111} and bulk Au{111}, on one complete, well annealed monolayer of Ag or Au on Ru{001}, as well as on the {111} surface of a $Ag_{0.5}Au_{0.5}$ bulk alloy [13.52]. The sharpness of the $5p_{1/2}$ peak on all five samples is an indication for the perfectness of each surface (as will become clear from this section). The obvious splitting of the $5p_{3/2}$ signal on all five surfaces is again a fingerprint for lateral Xe–Xe interactions and the formation of a two dimensional electronic band structure in the Xe overlayer,

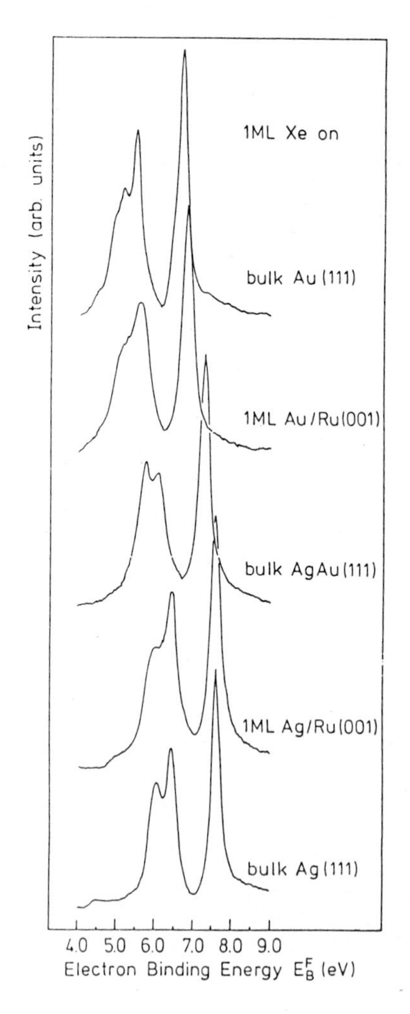

Fig. 13.26. Xe($5p_{3/2,1/2}$) reference spectra from Xe monolayers on the various indicated single crystal surfaces. Note the different energy positions

which occurs only on perfectly flat surfaces (Sect. 13.2.1). The electron binding energy $E_B^F(5p_{1/2})$ is 6.8 eV on the Au monolayer on Ru{001} and 7.55 eV on the Ag monolayer on Ru{001}. Both values are also very similar to those found on the bulk Au{111} and the bulk Ag{111} surface, as evidenced by Fig. 13.26. The difference $\Delta E_B^F(5p_{1/2}) = 7.55\,eV - 6.8\,eV = 0.75\,eV$, is due to the work function difference between these two surfaces: $\Delta\varphi = \varphi_{MLAu} - \varphi_{MLAg} = 5.6\,eV - 4.75\,eV = 0.85\,eV$ [13.52], in agreement with (13.2). On the bulk $Ag_{0.5}Au_{0.5}$ {111} surface with random distribution of the two components, the large Xe atoms are always bound to occupy "mixed" sites, being in contact with both Ag and Au atoms. Similarly, on ordered CuAu [13.74] and ordered NiAl [13.75] alloy surfaces, the large Xe atoms "integrate" over the surface constituents. In this case, the position of the $5p_{1/2}$ peak lies between the positions of $Xe(5p_{1/2})$ on pure Au and pure Ag (see Fig. 13.26). Of course, these "mixed" sites can be of variable composition (Ag : Au ratio), but also of different structure [Ag and Au forming a perfect alloy plane (see Fig. 13.31); Ag adsorbed *on* Au; Au adsorbed *on* Ag, etc.]. Careful reference studies with differently prepared binary Ag + Au surfaces have shown that $E_B^F(5p_{1/2})$ of Xe atoms adsorbed at all these various "mixed" sites ranges from 7.0 eV to 7.35 eV [13.51]. Consequently, from a heterogeneous surface which contains pure Ag-patches, pure Au-patches and "mixed" sites, and which is covered with a complete monolayer of Xe, the measured PAX-spectrum is a superposition of $5p_{3/2,1/2}$-spectra, each of which is characteristic for one kind of surface site. The partial spectra, as obtained by the decomposition procedure described in Sect. 13.2.5 and [13.9], are shifted with respect to each other by the difference in "local work function" between inequivalent sites, and the $5p_{1/2}$-intensities are a measure of the relative abundance of each kind of surface site.

On the basis of the above reference data, a rather detailed discussion of the PAX spectra from differently prepared and treated binary (Ag + Au) overlayers on Ru{001} is possible. The first series of experiments concerns a Ru{001} surface which was first covered with a complete, well annealed monolayer of Ag, which then was covered with 0.3 ML Au at 60 K and heated to successively increasing temperatures for 10 minutes each time. Figure 13.27 displays four sets of PAX-spectra. The spectra from Fig. 13.27a, were obtained after the 0.3 ML Au/1 ML Ag overlayer ("/" stands for "on") was warmed up to 275 K for 10 min and then recooled to 60 K. Note the lack of any $5p_{1/2}$-intensity at $E_B^F = 6.8\,eV$ (vertical line) which excludes unambiguously the existence of pure Au-sites (patches) on the surface. The single $5p_{1/2}$ peak is at most somewhat asymmetric and its position shifts from 7.1 eV at low Xe coverage to 7.45 eV at Xe monolayer saturation (highest spectrum). The evolution of this peak asymmetry (arrow) at intermediate Xe coverage suggests the growth of new Xe states at the higher binding energy side, which leads to the overall shift of the peak maximum. The $5p_{3/2}$ peak splits only visibly at higher Xe coverages. Since the total coverage of $\Theta_{Ag} + \Theta_{Au} = 1.3\,ML$ implies only a fractional second layer, the surface will comprise step sites, namely the boundary sites of islands or small 2*d* clusters. These step sites are "decorated" first by Xe atoms and prevent the formation of large 2*d* Xe islands, which explains the poor splitting of the

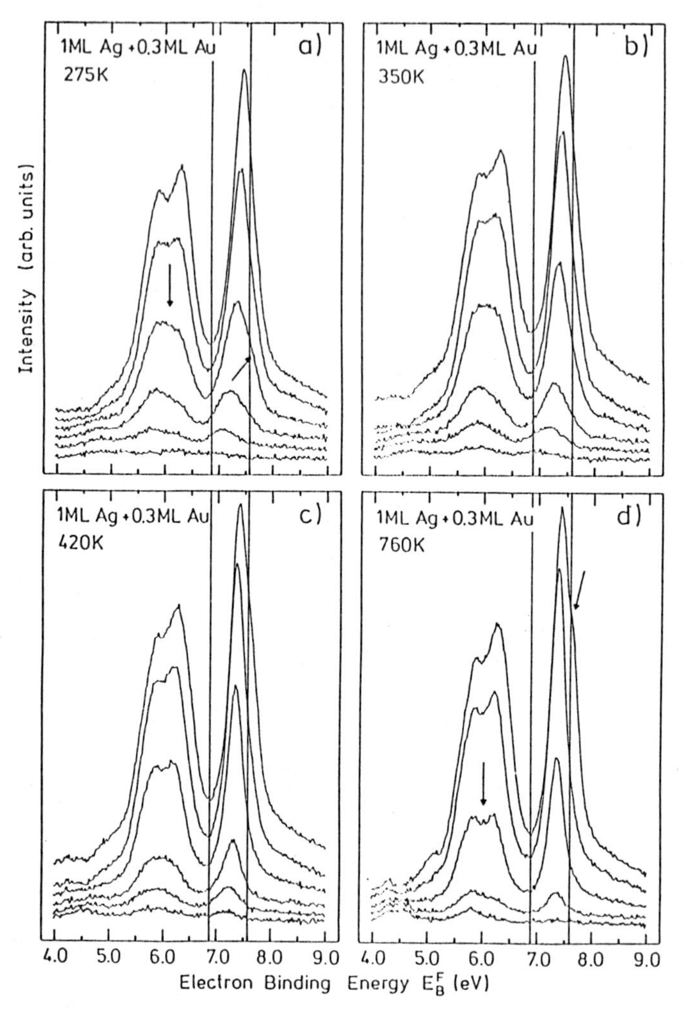

Fig. 13.27a–d. Xe($5p_{3/2,1/2}$) (= PAX-)spectra for different coverages of Xe adsorbed at 60 K on a complete monolayer of Ag (on a Ru{001} substrate) which was covered with 0.3 ML Au at 60 K followed by annealing for 10 minutes at (**a**) 275 K, (**b**) 350 K, (**c**) 420 K and (**d**) 760 K. Note the spectral changes as a function of annealing treatment in particular at low Xe coverages, which are related to structural transformations at the surface. The highest spectrum in each panel corresponds to a saturated Xe monolayer

$5p_{3/2}$-peak at lower Xe coverages. Furthermore, these step sites obviously involve Au-atoms as suggested by the low binding energy E_B^F of the $5p_{1/2}$-signal. Only with increasing Xe coverage are the remaining pure Ag patches from the original Ag monolayer populated, causing the $5p_{3/2}$-splitting and the $5p_{1/2}$ shift to higher binding energies. Further annealing of the 0.3 ML Au/1 ML Ag-film for 10 min at 350 K (Fig. 13.27b) and 420 K (Fig. 13.27c), respectively, causes only minor changes of the resultant PAX-spectra. The $5p_{1/2}$ peak sharpens up; at low Xe coverages, $E_B^F(5p_{1/2})$ is higher than in Fig. 13.27a, and the $5p_{3/2}$-signals begin to split at lower Xe coverages. All three trends support a progressive homogenization of the surface sites populated first by Xe (low coverages) as well as a growth of the metal islands in the second layer. Even annealing at 760 K for 10 min (Fig. 13.27d) causes hardly any further change, except one important detail: At Xe monolayer saturation (highest spectrum) the $5p_{1/2}$-peak exhibits a well resolved shoulder (arrow) near the binding energy characteristic of Xe on pure Ag. *Qualitatively*, the PAX-spectra from Figs. 13.27a–d support the notion that already at 275 K, the Au atoms (originally deposited on top of a complete Ag monolayer) have started to penetrate into the Ag underlayer because no Xe($5p_{1/2}$) peak from pure Au-sites is observed. An equivalent number of Ag atoms is displaced into the second layer, thereby producing a rough surface with high-coordination (step) sites. With an increase in annealing temperature, more Au atoms are dissolved homogeneously in the first layer (in contact with Ru) and more Ag atoms form increasingly growing Ag islands in the second layer.

Before drawing more detailed conclusions from a careful *quantitative* decomposition of the spectra shown in Figs. 13.27a–d, we shall first describe the PAX-spectra observed from the second series of experiments in which the order of Ag and Au deposition was reversed. Here, a complete and well annealed Au monolayer was prepared onto which 0.5 ML Ag were deposited at 60 K. At this low substrate temperature, the Ag atoms are basically immobile and remain statistically distributed across the Au underlayer, thereby producing a large number of high-coordination sites around them. The resultant PAX-spectra are displayed in Fig. 13.28a. The initial $5p_{1/2}$ peak is rather broad and grows *between* the positions for Xe on pure Au or pure Ag, respectively (vertical lines), which is characteristic of "mixed" sites. These sites correspond to the high-coordination, nearest neighbor sites of the Ag atoms because they are populated first (low Xe coverage). At intermediate Xe coverage, a $5p_{1/2}$ peak suddenly emerges at 6.8 eV characteristic of Xe on pure Au sites, while the original intensity from "mixed" sites remains visible as a shoulder. Further increase of the Xe coverage causes the appearance of a second $5p_{1/2}$ peak in the position for Xe on pure Ag with $E_B^F = 7.55$ eV. At no stage in Fig. 13.28a, does the $5p_{3/2}$ peak show any splitting. (The $5p_{3/2}$ double peak structure at high Xe coverages is not due to an m_j-splitting shown in Fig. 13.4 but arises from Xe on Au- and Ag-sites, respectively, as does the $5p_{1/2}$ double peak structure.) Warming this 0.5 ML Ag/1 ML Au film to 275 K for 10 min results in the PAX-spectra from Fig. 13.28b. There is only very little intensity from "mixed" sites left (arrow). Instead, a rather sharp $5p_{1/2}$ peak first grows at 6.8 eV (Xe on Au) followed by a second also rather sharp $5p_{1/2}$ peak at 7.5 eV (Xe on Ag). The $5p_{3/2}$ peak begins to show indications of a double peak structure (shoulder at low binding energies). These spectra from Fig. 13.28b are in marked contrast to those shown in Fig. 13.27a. Both figures relate to films which were annealed at 275 K. In the first case (0.3 ML Au/1 ML Ag), this temperature is enough to activate penetration of Au atoms into the Ag underlayer. In the second case (0.5 ML Ag/1 ML Au), however, the Ag atoms do not yet exchange site with Au atoms from the first layer, but rather coalesce and form extended $2d$ Ag islands *on* the Au underlayer as suggested by the negligible

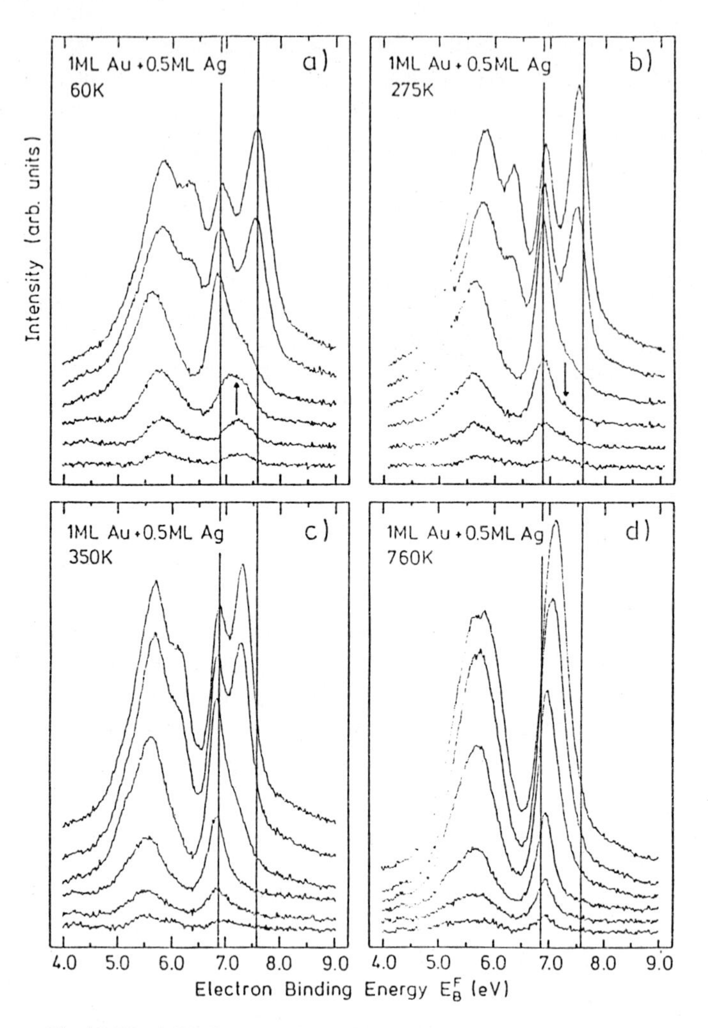

Fig. 13.28a–d. Xe($5p_{3/2,1/2}$) (= PAX_-)spectra for different coverages of Xe adsorbed at 60 K on a complete monolayer of Au (on a Ru{001} substrate) which was covered with 0.5 ML Ag at 60 K (**a**) followed by annealing for 10 mintes at (**b**) 275 K, (**c**) 350 K and (**d**) 760 K. Note the strong spectral changes of the $5p_{1/2}$ signal as a function of annealing temperature for both low and high Xe coverages. These changes are indicative of thermally activated structural and chemical redistributions at the surface. The highest spectrum in each panel corresponds to Xe monolayer saturation

number of ("mixed") boundary sites (arrow) and the improved order in the Xe overlayer ($5p_{3/2}$-splitting).

Further annealing of the 0.5 ML Ag/1 ML Au system at 350 K leaves the $5p_{1/2}$ peak at 6.8 eV (Xe on Au) unaffected (Fig. 13.28c). The second $5p_{1/2}$-signal now peaks at 7.3 eV, and can no longer be assigned to Xe on pure Ag islands but corresponds to a large number of "mixed" sites, which, however, are now populated at higher Xe coverages, in contrast to the "mixed" sites (6.8 eV $< E_B^F(5p_{1/2}) <$ 7.5 eV) in Fig. 13.28a. Therefore, these "mixed" sites from Fig. 13.28c must be assigned to *flat* alloy-sites created by intermixing between the Ag islands (concluded above from Fig. 13.28b) and the Au underlayer.

Finally, annealing at 760 K for 10 min causes the formation of a more or less homogeneous AgAu alloy throughout both layers as judged by the single $5p_{1/2}$ signal in Fig. 13.28d. The continuously increasing $5p_{1/2}$ peak width with increasing Xe coverage may be taken as an indication for the successive population of more and more Ag-rich sites (see below).

The above observations obtained from the (integral) PAX-spectra in Figs. 13.27 a–d and 13.28a–d can be refined somewhat by taking difference curves between consecutive measured spectra. Examples of such "incremental spectra" as obtained from Figs. 13.28a–d are shown in Figs. 13.29a–d. These incremental spec-

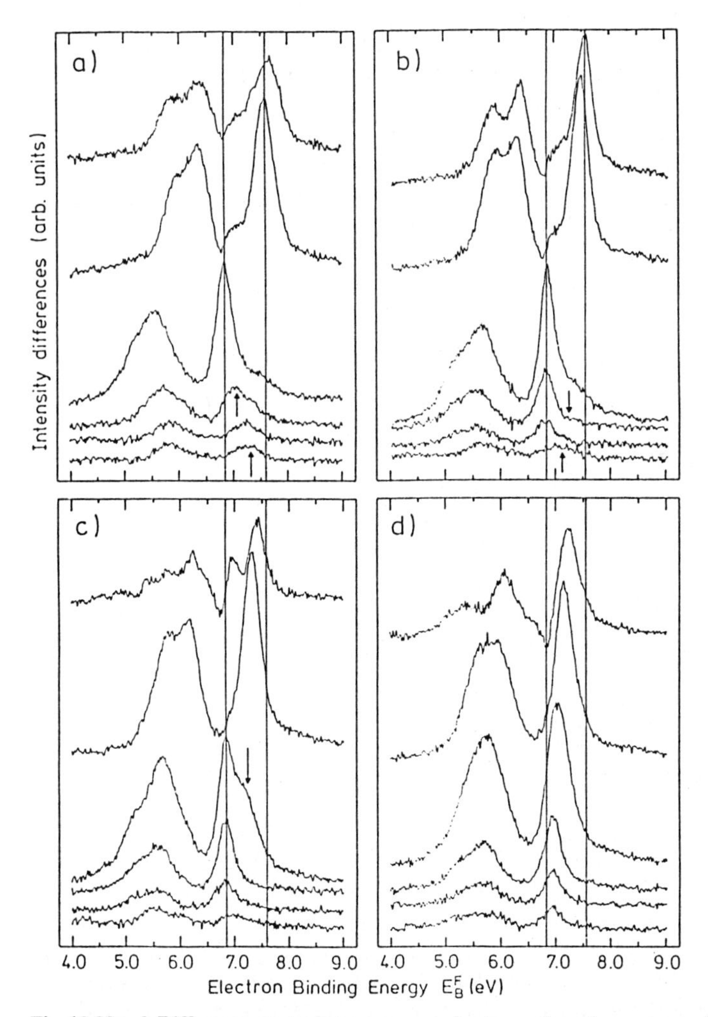

Fig. 13.29a–d. Difference spectra between successive traces from the corresponding panel in Fig. 13.28. Each difference spectrum corresponds to an addition to the Xe coverage and is, therefore, termed "incremental spectrum". These incremental spectra accentuate the successive population of inequivalent surface sites

tra accentuate the successive population of different surface sites in order of decreasing Xe adsorption energy. For instance, Fig. 13.29a shows more clearly the initial saturation of the "mixed" step-like sites on this surface before the Au-sites, then the Ag-sites are populated. Obviously, the adsorption energy E_{ad} of the Xe follows the sequence $E_{ad}(step) > E_{ad}(\mathrm{Au}) > E_{ad}(\mathrm{Ag})$. In Fig. 13.29b (obtained from 13.28b), the obvious splitting of the $5p_{3/2}$ signal of Xe on Ag (upper two increments) supports much more clearly the extended island structure of the Ag than Fig. 13.28b. Also Fig. 13.29c improves the splitting of the $5p_{3/2}$ signal, and, finally, Fig. 13.29d explains the continuous $5p_{1/2}$ peak *broadening* in Fig. 13.28d as a continuous *peak shift* to higher binding energy, probably due to a successive population of more Ag-rich sites, in agreement with the relative adsorption energies (see above).

Since the incremental spectra have the tendency to isolate the spectral properties of Xe atoms adsorbed at individual types of surface sites, they can help to establish the fit-parameters for the related partial $5p_{3/2,1/2}$ spectra by first fitting the incremental spectra in order of increasing Xe coverage with a set of three Lorentzian functions per Xe state, as depicted in Fig. 13.4 and discussed earlier. The thus determined *electron binding energies* per Xe state are then used as input to repeat the fit procedure with the *integral* spectra shown in Figs. 13.27a–d and

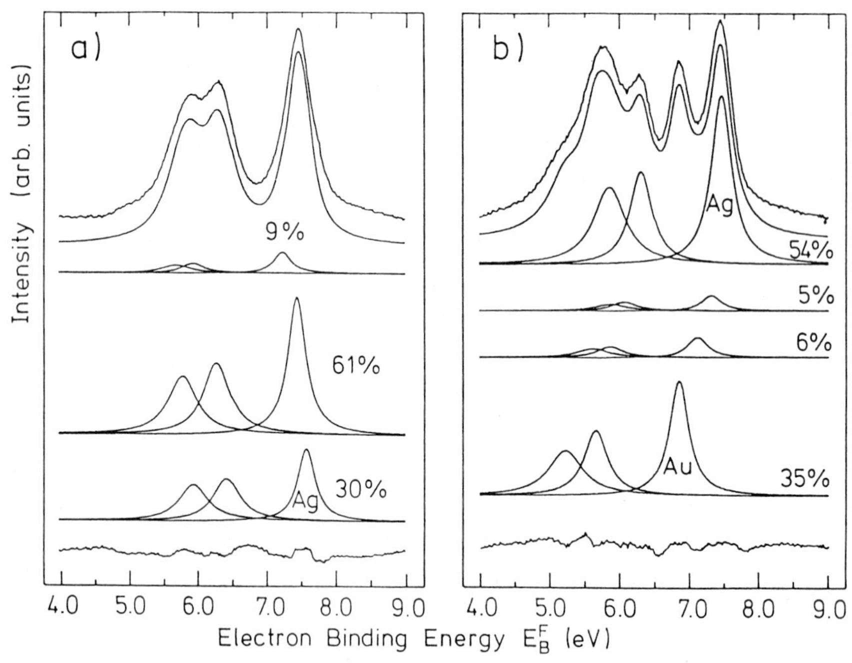

Fig. 13.30a,b. Examples for the best fits of Xe monolayer PAX spectra (using triplets of Lorentzian functions as defined in Fig. 13.4) for (**a**) a 0.3 ML Au/1 Ml Ag film on Ru{001} annealed for 10 minutes at 275 K, and (**b**) a 0.5 ML Ag/1ML Au film on Ru{001} annealed for 10 minutes at 275 K likewise. Note the clear difference in the $5p_{1/2}$ peak shape, namely, a single peak in panel (a) and a double peak in panel (b). The different sets of three Lorentzian lines correspond to Xe atoms adsorbed at different surface sites, and the percentages represent their relative surface concentration (see text)

Figs. 13.28a–d in order to find the *partial intensities* for all Xe states contributing to the respective Xe monolayer spectrum. As explained previously, the Xe($5p_{1/2}$) electron binding energy per Xe state is a measure of the local work functions at the corresponding kind of surface site while the partial intensity per Xe state provides the relative concentration of this particular type of surface site.

Examples of the finally obtained best fits of Xe monolayer PAX spectra are displayed in Fig. 13.30 for the two most important situations in the present study, namely the 0.3 ML Au/1 ML Ag system annealed at 275 K (Fig. 13.27a) and the 0.5 ML Ag/1 ML Au system annealed at 275 K (Fig. 13.28b). Qualitatively, the spectral distinction between the two systems could hardly be more obvious: In the first case, the single $5p_{1/2}$ peak suggests intermixing between Ag and Au, while in the second case the two $5p_{1/2}$ peaks indicate the persistence of pure Au and pure Ag. Quantitatively, the best fit to the spectrum in Fig. 13.30a results in three Xe states, which, based on their $5p_{1/2}$ electron binding energies and intensities, suggest that 30% of the Xe atoms (of a complete Xe monolayer) sit on pure Ag-sites, while (60% + 9%) sit on "mixed" sites influenced by Ag *and* Au. Certainly no Xe state characteristic of pure Au is found (according to our long experience, these percentages are accurate to within $\pm 5\%$). The best fit to the spectrum in Fig. 13.30b requires four Xe states, with 35% of the atoms emitting from pure Au sites, 54% of the Xe atoms sitting on pure Ag sites, and (5% + 6%) of the intensity coming from "mixed" sites. Although the exact assignment of the various "mixed" states is certainly not unique (two states were simply needed in order to account for the broad emission at these intermediate energies, reflecting a distribution in structure and composition of these sites), it is clearly possible to distinguish quantitatively between the three *classes* of sites on the respective surface, namely, pure Au-, pure Ag- and "mixed" sites.

A compilation of the fit results from the Xe monolayer PAX-spectra shown in Figs. 13.26–28a–d is given in Fig. 13.31. At the top the electron binding energy scale is divided into three regimes, namely two vertical lines at ~ 6.8 eV and ~ 7.6 eV indicating the positions of the Xe($5p_{1/2}$)-signal on a pure Au (Fig. 13.26) and a pure Ag-monolayer (Fig. 13.26) on Ru{001}, respectively, and a broad regime between 7.0 eV and 7.4 eV, which encompasses the Xe($5p_{1/2}$) binding energies from all kinds of "mixed" sites including the substitutional AgAu alloy surface from Fig. 13.26 (see AgAu line in the center of Fig. 13.31). The black bars mark the $5p_{1/2}$ positions as they came out from the decomposition of the experimental Xe monolayer spectra. As mentioned above, no further attempt is made to assign the various "mixed" states from one surface. The percentages in Fig. 13.31 denote the relative intensity of each decomposed state. The small atomistic models are to help visualize the conclusions on the chemical distribution and structure of the differently treated (Ag+Au)-films.

A careful inspection of the percentage numbers in Fig. 13.31 leads to several very interesting conclusions which go much beyond the qualitative observations described earlier. As pointed out previously, a temperature of 275 K suffices to activate intermixing between the 0.3 ML Au overlayer and the 1 ML Ag underlayer. The numbers in the upper half of Fig. 13.31, however, further suggest that

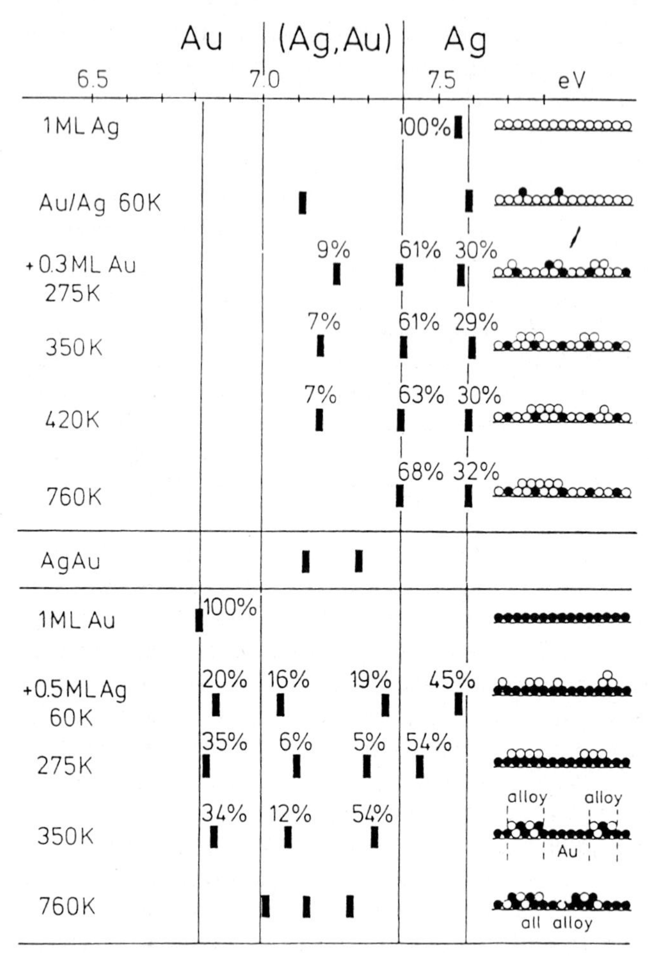

Fig. 13.31. Compilation of $5p_{1/2}$ electron binding energies (*black bars*) and relative intensities (percentages) of coexisting Xe adsorption states on differently annealed (Ag + Au) films on a Ru{001} substrate, as obtained from a decomposition of all the Xe monolayer PAX-spectra shown in Figs. 13.27a–d and 13.28a–d. These data support the small structure models shown on the right-hand side

always ~ 30% of the surface are Ag-like, even after annealing at the very high temperature of 760 K (see shoulder in spectrum 13.27d). This 30% agrees with the amount of 0.3 ML Au originally desposited on top of the Ag monolayer. This strongly suggests that the adsorbed Au atoms have *completely* penetrated into the Ag underlayer, thereby displacing an equivalent number of Ag atoms into the second layer where they then form *monatomically* thick islands of pure Ag. Furthermore, the corresponding relative intensities of the "mixed" states are compatible with a lateral growth of these pure Ag islands with an increase in annealing temperature (see the atomic models), because the number of island boundary sites decreases from 9% to ~ 0%. The high value of the $Xe(5p_{1/2})$ electron binding energy (7.4 eV) of the final "mixed" state (68% after 760 K) is suggestive of a Ag-rich alloy in the first layer. Also, in the second experiment with the Au underlayer two interesting details may be concluded from the percentage numbers compiled in the lower half of Fig. 13.31. Firstly, upon annealing from 60 K to 275 K, the relative intensity of the Xe state on pure Ag sites *increases* from 45% to 54%, the latter being in good agreement with the actual coverage, as determined by Ag-TDS after the experiment. This increase may be indicative of the existence of some small Ag clusters at 60 K which spread out upon warming to 275 K, as depicted in the small atomistic models. Secondly, annealing from 275 K to 350 K does *not* change the intensity of the Xe state on pure Au sites (34%), while on the other hand the Xe(Ag) signal vanishes. This is a strong indication that the *initial AgAu alloy formation* proceeds via site-exchange *only within the area of the original Ag islands* which were produced on top of the underlayer at 275 K. This suggests also that the activation energy for this site-exchange normal to the surface is smaller than the 2*d* heat of sublimation of the Ag islands on Au. Only very high temperature treatment at 760 K leads to a homogenization of the AgAu alloy parallel to the surface.

In summary, an Ag/Au interface (on Ru{001}) is thermally more stable than an Au/Ag interface. In the first case (initial Au-underlayer) Ag does not penetrate into the Au-layer below 275 K. Converely, in the second case (initial Ag-underlayer) the Au atoms are already exchanging sites with underlying Ag atoms by 275 K, leading to the formation of a monolayer of an $Ag_{0.7}Au_{0.3}$ alloy covered with pure, monoatomically thick Ag-islands. This remarkable difference is easily conceivable in terms of the energy terms involved: (1) The adsorption energy of Au on Ru{001} is significantly higher than that of Ag [13.73]. (2) The specific surface free energy of Ag is lower than that of Au [13.76]. (3) The lateral interaction energy between adsorbed Ag atoms (on Ru {001} is attractive [13.73]. (4) The heat of formation of AgAu alloys is slightly exothermic [13.77]. The first two terms favor Au atoms contacting the Ru substrate and Ag atoms being displaced in the second (outer) layer. This process, however, will compete with the exothermicity of the AgAu alloy formation which tends to maximize the number of Ag–Au bonds. Finally, the attractive Ag–Ag interaction explains the formation and the growth of the 2*d* Ag islands in the second layer of the first system (initial Ag-underlayer).

These results, once again, demonstrate the kind of detailed information that can be obtained by means of the PAX method for characterization of the structure and structural transformations at heterogeneous surfaces on an atomic scale. Of course, the (Ag + Au) system chosen is only a case study; such investigations can be carried over to other metal/metal combinations.

13.4 Summary and Implications

The intention of this work was to give an updated review of Photoelectron-spectroscopy of Adsorbed Xenon (PAX) as a method for local characterization of heterogeneous surfaces which comprise structural and chemical defects. The observables taken advantage of in PAX are: (1) the relative adsorption energy of Xe atoms at surface sites of different chemical nature and/or of different geometrical structure (coordination); (2) the electron binding energies of Xe(5p) or Xe(4d) electrons (with respect to the Fermi level of the spectrometer), and (3) the relative intensities of Xe adsorption states at different surface sites. The site specific adsorption energies E_{ad} govern the sequential population of surface sites in order of decreasing adsorption energy. The electron binding energies E_B^F carry information about the local surface potential at the respective adsorption site, and the relative intensities (determined at Xe monolayer saturation) provide information about the relative abundance of each kind of surface site. All in all, the PAX method as a local surface potential probe can be visualized as follows. Because of the weak bonding in the initial state and the nearly substrate independent extra atomic relaxation in the photoionized final state, the role of an adsorbed Xe atom is to deposit a test-electron (5p or 4d) at a particular site (controlled by E_{ad}) very near to the surface (~ 2.5 Å). Via photoemission, this 5p or 4d electron provides the information about the local surface potential at this site. Compared to an earlier review [13.9] this view is now also supported by elegant experiments with semiconductor surfaces. Experimental results from stepped metallic vicinal surfaces suggest that surface sites which are only ~ 5 Å apart can be distinguished with PAX by virtue of their difference in E_B^F, that is, their difference in local surface potential. With respect to the fact that PAX yields local surface potentials, this technique seems to complement scanning tunneling microscopy (STM) in a unique way.

Local surface potential differences of as much as 1 eV have been found between step and terrace sites on a stepped Pt$\{111\}$ surface. Since this potential difference decays over a distance of only a few Ångstroms, the strong localized fields associated with these surface irregularities are expected to exert a strong influence on the properties of adsorbed molecules at these sites. These strong *lateral* fields have not been taken into account so far in order to explain the different reactivity of adsorbed molecules on heterogeneous surfaces, e.g., at step sites or near heteroatoms. A recent theoretical investigation of the effect of these lateral electric fields on the degenerate $2\pi^*$ derived valence levels of chemisorbed CO molecules [13.78], for example, indicated, indeed, a strong (predominantly linear) dependence of level shifts and level broadening on the existing field. This feature, which can be considered as a local surface Stark effect, provides a mechanism which may be a clue to the understanding of some elementary processes in heterogeneous catalysis.

References

13.1 W. Telieps, E. Bauer: Surf. Sci. **162**, 163 (1985), see also E. Bauer in this volume
13.2 Consult the journal "Ultramicroscopy" for high-resolution electron microscopy work; see also D.J. Smith in Vol. VI of this series
13.3 H. Bethge, Th. Krajewski, O. Lichtenberg: Ultramicroscopy **17**, 21 (1985)
13.4 G. Binnig, H. Rohrer: Surf. Sci. **126**, 236 (1983): Surf. Sci. **152/153**, 17 (1985)
13.5 R.J. Behm, W. Hösler: in: Physics and Chemistry of Solid Surfaces VI, ed. by R. Vanselow and R. Howe (Springer, Berlin, Heidelberg, 1986)

13.6 IBM Journal of Research and Development, Vol. 30, No. 4+5 (1986)
13.7 E.W. Müller, T.T. Tsong: "Field Ion Microscopy" (Elsevier, New York 1969)
13.8 K. Wandelt: J. Vac. Sci. Technol. **A2**, 802 (1984)
13.9 K. Wandelt: in Thin Metal Films and Gas Chemisorption, ed. by P. Wissmann (Elsevier, Amsterdam 1987) p. 280–369
13.10 K. Wandelt: Progr. Surf. Sci., in preparation
13.11 K. Wandelt, J. Hulse: J. Chem. Phys. **80**, 1340 (1984)
13.12 R.J. Behm, C.R. Brundle, K. Wandelt: J. Chem. Phys. **85**, 1061 (1986)
13.13 K. Markert, P. Pervan, W. Heichler, K. Wandelt: Surf. Sci. **211/212**, 611 (1989)
13.14 J.W. Gadzuk, S. Holloway, C. Mariani, K. Horn: Phys. Rev. Lett. **48**, 1288 (1982)
13.15 K. Horn, M. Scheffler, A.M. Bradshaw: Phys. Rev. Lett. **44**, 822 (1978)
13.16 M. Scheffler, K. Horn, A.M. Bradshaw, K. Kambe: Surf. Sci. **80**, 69 (1979)
13.17 T. Mandel: PhD-Thesis, Free University Berlin, 1985
13.18 A. Cassuto, J.J. Ehrhardt, J. Cousty, R. Riwan: Surf. Sci. **194**, 579 (1988)
13.19 B.J. Waclawski, J.F. Herbst: Phys. Rev. Lett. **35**, 1594 (1975)
13.20 R.P. Antoniewicz: Phys. Rev. Lett. **38**, 374 (1977)
13.21 S.I. Ishi, Y. Ohno: J. Electron Spectr. **33**, 85 (1984)
13.22 A. Jablonski, S. Eder, K. Markert, K. Wandelt: J. Vac. Sci. Technol. **A4**, 1510 (1986)
13.23 S. Eder, K. Markert, A. Jablonski, K. Wandelt: Ber. Bunsenges. Phys. Chem. **90**, 825 (1986)
13.24 D.W. Turner, C. Baker, D.A. Baker, C.R. Brundle: "Molecular Photoelectron Spectroscopy" (Wiley-Interscience, New York 1970)
13.25 J.W. Gadzuk: J. Vac. Sci. Technol. **12**, 289 (1975)
13.26 A. Cassuto, J.J. Ehrhardt: J. Phys. France **49**, 1753 (1988)
13.27 K. Hermann, J. Noffke, K. Horn: Phys. Rev. **B22**, 1022 (1980)
13.28 K. Kern, R. David, R.L. Palmer, G. Comsa: Appl. Phys. **A41**, 91 (1986)
13.29 K. Kern: Phys. Rev. **B35**, 8265 (1987)
13.30 K. Kern, R. David, P. Zeppenfeld, R. Palmer, G. Comsa: Solid State Commun. **62**, 391 (1987)
13.31 K. Kern, R. David, P. Zeppenfeld, G. Comsa: Surf. Sci. **195**, 353 (1988)
13.32 G. Kaindl, T.-C. Chiang, D.E. Eastman, F.J. Himpsel: Phys. Rev. Lett. **45**, 1808 (1980) and in: "Ordering in Two Dimensions", ed. by Sinha (Elsevier, Amsterdam 1980) p. 99
13.33 N.D. Lang: private communication
13.34 N.D. Lang, W. Kohn: Phys. Rev. **B3**, 1215 (1971)
13.35 N.D. Lang, A.R. Williams: Phys. Rev. **B25**, 2940 (1982)
13.36 N.D. Lang: Phys. Rev. Lett. **46**, 842 (1981)
13.37 K. Kalki, B. Pennemann, K. Wandelt, B. Eisenhut, Ch. Schug, J. Stober, W. Steinmann, W. Heichler: BESSY-Report, 1989
13.38 J.E. Demuth, W.J. Thompson, H.J. DiNardo, R. Imbihl: Phys. Rev. Lett. **56**, 1408 (1986)
13.39 K. Markert: PhD-Thesis, Free University Berlin, 1988
13.40 K. Markert, P. Pervan, K. Wandelt: Phys. Rev. Lett., submitted
13.41 J. Küppers, K. Wandelt, G. Ertl: Phys. Rev. Lett. **43**, 928 (1979)
13.42 J. Küppers, H. Michel, F. Nitschke, K. Wandelt, G. Ertl: Surf. Sci. **89**, 361 (1979)
13.43 J. Hulse, J. Küppers, K. Wandelt, G. Ertl: Appl. Surf. Sci. **6**, 453 (1980)
13.44 K. Markert, K. Wandelt: Surf. Sci. **159**, 24 (1985)
13.45 A. Jablonski, S. Eder, K. Wandelt: Appl. Surf. Sci. **22/23**, 309 (1985)
13.46 K.S. Kim, J.H. Sinfelt, S. Eder, K. Markert, K. Wandelt: J. Chem. Phys. **91**, 2337 (1987)
13.47 Th. Berghaus, Ch. Lunau, H. Neddermeyer, V. Rogge: Surf. Sci. **182**, 13 (1987)
13.48 M. Alnot, V. Gorodetskii, A. Cassuto, J.J. Ehrhardt: Surf. Sci. **162**, 886 (1985); Thin Solid Films **151**, 251 (1987)
13.49 D. Fargues, J. J. Ehrhardt, M. Abon, J.C. Bertolini: Surf. Sci. **194**, 149 (1988)
13.50 K. Wandelt, K. Markert, P. Dolle, A. Jablonski, J.W. Niemantsverdriet: Surf. Sci. **189/190**, 114 (1987)
13.51 K. Markert, P. Dolle, J.W. Niemantsverdriet, K. Wandelt: J. Vac. Sci. Technol. **A5**, 2849 (1987)
13.52 K. Wandelt, J.W. Niemantsverdriet, P. Dolle, K. Markert: Surf. Sci. **213**, 612 (1989)
13.53 J.W. Niemantsverdriet, K. Wandelt: J. Vac. Sci. Technol. **A7**, 1742 (1989)
13.54 W.M.H. Sachtler: J. Vac. Sci. Technol. **9**, 828 (1971)
13.55 H. Wagner: in Springer Tracts in Modern Physics, Vol. 85 (Springer, Berlin, Heidelberg 1979)
13.56 R.J. Behm, W. Hösler, E. Ritter, G. Binnig: Phys. Rev. Lett. **56**, 228 (1986)
13.57 R. Miranda, S. Daiser, K. Wandelt, G. Ertl: Surf. Sci. **131**, 61 (1983)

13.58 S. Daiser: MS-Thesis, University of München, 1981
13.59 P.W. Palmberg: Surf. Sci. **25**, 598 (1971)
13.60 J. Küppers, U. Seip: Surf. Sci. **119**, 291 (1982)
13.61 J. Küppers, F. Nitschke, K. Wandelt, G. Ertl: Surf. Sci. **87**, 295 (1979)
13.62 R. Smoluchowski: Phys. Rev. **60**, 661 (1941)
13.63 K. Besocke, H. Wagner: Surf. Sci. **52**, 653 (1975)
13.64 K. Besocke, H. Wagner: Phys. Rev. **B8**, 4597 (1973)
13.65 K. Wandelt, J. Hulse, J. Küppers: Surf. Sci. **104**, 212 (1981)
13.66 A. Clarke, P.J. Rous, M. Arnott, G. Jennings, R.F. Willis: Surf. Sci. **192**, L843 (1987)
13.67 W. Dürr, R. Germar, D. Pescia, M. Taborelli, O. Paul, M. Landolt: to be published
13.68 M. Taborelli, R. Allenspach, G. Boffa, M. Landolt: Phys. Rev. Lett. **56**, 2869 (1986)
13.69 K. Christmann, G. Ertl, H. Shimizu: J. Catal. **61**, 397 (1980)
13.70 C.H.F. Peden, D.W. Goodman: in Catalyst Characterization Science: Surface and Solid State Chemistry, Amer. Chem. Soc., Washington, DC, 1985; ACS Symp. Ser. No. 288, p. 185
13.71 B. Konrad, MS-Thesis, University of München, 1984
13.72 B. Konrad, F.J. Himpsel, W. Steinmann, K. Wandelt: in: Proceedings of the International Seminar on Surface Structure Determination by LEED and Other Methods, University of Erlangen-Nürnberg, F.R.G., 1985, p. 109
13.73 J.W. Niemantsverdriet, P. Dolle, K. Markert, K. Wandelt: J. Vac. Sci. Technol. **A4**, 875 (1987)
13.74 P. Dolle, K. Markert, J.W. Niemantsverdriet, K. Wandelt: unpublished results
13.75 U. Schneider, H. Isern, M. Stöcker, G.R. Castro, K. Wandelt: in preparation
13.76 A.R. Miedema, J.W.F. Dorleijn, Surf. Sci. **95**, 447 (1980)
13.77 P. Hultgren, R.L. Orr, P.D. Anderson, K.K. Kelley: in "Selected Values of Thermodynamic Properties of Metalls and Alloys" (Wiley, New York 1963), p. 341
13.78 B. Gumhalter, K. Hermann, K. Wandelt: Vacuum, in press; Phys. Rev. Lett., submitted

14. Theoretical Aspects of Scanning Tunneling Microscopy

J. Tersoff

IBM Research Division, T.J. Watson Research Center,
Yorktown Heights, NY 10598, USA

Since its invention by *Binnig, Rohrer,* and coworkers [14.1], scanning tunneling microscopy (STM) has established itself as a remarkable tool for studying surfaces. This chapter reviews the present theoretical understanding of STM, with emphasis on the interpretation of atomic-resolution STM images. The basic ideas and instrumentation have already been described in detail elsewhere [14.1].

The primary task in any theoretical analysis of STM is to understand how the tunneling current J varies with voltage V and tip position, $\boldsymbol{r}_t$, i.e., to determine the function $J(\boldsymbol{r}_t, V)$. Most (though not all) issues in the theory of STM reduce to determining this function. Some general aspects of the tunneling problem, and of the determination of $J(\boldsymbol{r}_t, V)$, are discussed in Sect. 14.1.

Section 14.2 discusses STM imaging, i.e., operation at constant V. It includes sub-sections on 1. modes of imaging, 2. imaging of metals, 3. imaging of semiconductors, and 4. imaging band-edge states. Section 14.3 briefly discusses STM spectroscopy, i.e., operation at fixed position and variable voltage.

Finally, Sect. 14.4 introduces an issue which is crucial for STM, but which is outside the scope of the tunneling problem. This is the effect of mechanical interactions between tip and surface on the operation of STM. Some concluding remarks are offered in Sect. 14.5.

14.1 General Tunneling Theory

14.1.1 Non-perturbative Treatment

The tunneling problem may be viewed as a special case of the more general problem, of the current-voltage characteristic of an interface. Unfortunately, the latter problem is extremely difficult in general. Therefore, it is desirable to take advantage of the fact that, in tunneling, the coupling between surface and tip is weak. Such a perturbative approach is discussed in Sect. 14.1.2 below.

Nevertheless, it is sometimes necessary to treat the problem non-perturbatively, in particular, when the tip approaches very close to the surface. For example, *Gimzewski* and *Möller* [14.2] measured the resistance between tip and sample as the tip is brought down to the surface. They found that the resistance decreased exponentially with decreasing distance, as expected from simple arguments, until at very small distances the resistance reached a plateau, presumably where the tip and surface are touching. At this point, the interaction between surface and tip certainly cannot be assumed to be weak.

To understand these results, *Lang* [14.3] calculated the resistance between a model surface (jellium), and a model tip (a single atom adsorbed on jellium). (Jellium is simply a metal in which the positive ions are replaced by a uniform positive background.) The simplification of using jellium made it feasible to calculate the resistance "exactly", i.e., non-perturbatively, so the results are valid even when the surface and tip are in contact. The results describe the experiment very well, showing that the plateau corresponds to the point where the resistance is dominated by a one-atom contact, rather than a vacuum gap.

For real systems, however, the matching of wave functions across the interface becomes extremely difficult. Even for an atomically perfect planar interface, such a calculation represents a *tour de force* [14.4]. For the STM problem, non-perturbative calculations have not proven feasible to date except by using models which ignore the atomic character of the electrodes. Thus, while some early calculations [14.5] for STM were non-perturbative, this approach has now generally been abandoned except for cases where the tip and surface interact strongly.

14.1.2 Tunneling-Hamiltonian Treatment

Under typical tunneling conditions, the interaction between the two electrodes is sufficiently weak that it may be treated in first order perturbation theory, giving

$$J = \frac{2\pi e}{\hbar} \sum_{\mu,\nu} \{f(E_\mu)[1 - f(E_\nu)] - f(E_\nu)[1 - f(E_\mu)]\} \times |M_{\mu\nu}|^2 \delta(E_\nu + V - E_\mu) \, , \tag{14.1}$$

where $f(E)$ is the Fermi function, V is the applied voltage (in units of energy, i.e., eV), $M_{\mu\nu}$ is the tunneling matrix element between states ψ_μ and ψ_ν of the respective electrodes, and E_μ is the energy of ψ_μ. For most purposes, the Fermi functions can be replaced by their zero-temperature values, i.e., unit step functions, in which case one of the two terms in braces becomes zero. In the limit of small voltage, this expression further simplifies to

$$J = \frac{2\pi}{\hbar} e^2 V \sum_{\mu,\nu} |M_{\mu\nu}|^2 \delta(E_\mu - E_F) \delta(E_\nu - E_F) \, . \tag{14.2}$$

These results are quite simple. The real difficulty is in evaluating the tunneling matrix elements. *Bardeen* [14.6] showed that, under certain assumptions, the tunneling matrix element could be expressed as

$$M_{\mu\nu} = \frac{\hbar^2}{2m} \int d\mathbf{S} \cdot (\psi_\mu^* \nabla \psi_\nu - \psi_\nu \nabla \psi_\mu^*) \, , \tag{14.3}$$

where the integral is over any surface lying entirely within the barrier region.

If we choose a plane for the surface of integration, and neglect the variation of the potential in the region of integration, then the surface wave function at this plane can be conveniently expanded in the generalized planewave form

$$\psi = \int d\boldsymbol{q}\, a_q \exp\left[-\sqrt{\kappa^2 + |\boldsymbol{q}|^2} z\right] \exp(\mathrm{i}\boldsymbol{q} \cdot \boldsymbol{x}) \tag{14.4}$$

where z is measured from a convenient origin at the surface, $\kappa = \hbar^{-1}(2m\phi)^{1/2}$, and ϕ is the "local workfunction" (or, more precisely, the potential in the region of interest, relative to the Fermi level). A similar expansion applies for the other electrode, replacing a_q with b_q, z with $z_t - z$, and $\boldsymbol{x}$ with $\boldsymbol{x} - \boldsymbol{x}_t$. Here $\boldsymbol{x}_t$ and z_t are the lateral and vertical components of the position of the tip. Then, substituting these wavefunctions into (14.3), one obtains

$$M_{\mu\nu} = -\frac{4\pi^2\hbar^2}{m} \int d\boldsymbol{q}\, a_q b_q^* \sqrt{\kappa^2 + |\boldsymbol{q}|^2} \exp\left[-\sqrt{\kappa^2 + |\boldsymbol{q}|^2} z_t\right] \exp(\mathrm{i}\boldsymbol{q} \cdot \boldsymbol{x}_t) \,. \tag{14.5}$$

Thus, given the wave functions of the surface and tip separately, i.e., a_q and b_q, one has a reasonably simple expression for the matrix element and tunneling current. Note that $\phi \sim 4\,\mathrm{eV}$, so evaluating κ and its role in the matrix element (14.5), one finds that the current decreases by roughly an order of magnitude for each angstrom increase in the surface-tip separation.

14.1.3 Modeling the Tip

In order to calculate the tunneling current, and hence the STM image or spectrum, it is first necessary to have explicitly the wave functions of the surface and tip. Unfortunately, the actual atomic structure of the tip is generally not known [14.7]. Even if it were known, the very low symmetry would probably make accurate calculation of the tip wave functions infeasible.

One must, therefore, adopt a reasonable but somewhat arbitrary model for the tip. The most realistic model which has been used to date is that of *Lang* [14.8, 9] an atom adsorbed on jellium. However, to facilitate the treatment of real surfaces, an even more severe approximation for the tip is convenient. The simplest model which has been used for the tip is that of *Tersoff* and *Hamann* [14.10, 11]. Because this model leads to a particularly simple interpretation of the STM image, it is worth describing in a little detail.

To motivate the simplest possible model for the tip, [14.11] considered what would be the ideal STM. First, one wants the maximum possible resolution, and, therefore, the smallest possible tip. Second, one wants to measure the properties of the bare surface, not of the more complex interacting system of surface and tip. Therefore, the ideal STM tip would consist of a mathematical point source of current. In that case, Eq. (14.2) reduces to [14.11]

$$J \propto \sum_{\nu} |\psi_\nu(\boldsymbol{r}_t)|^2 \delta(E_\nu - E_\mathrm{F}) \equiv \varrho(\boldsymbol{r}_t, E_\mathrm{F}) \,. \tag{14.6}$$

Thus, the ideal STM would simply measure $\varrho(\boldsymbol{r}_t, E_F)$ of the bare surface. This is a familiar quantity, being simply the local density of states at E_F, i.e., the charge density from states at the Fermi level, at the position of the tip. Thus, within this model, STM has quite a simple interpretation.

It is important to see how far this interpretation can be applied for more realistic models of the tip. Reference [14.11] showed that (14.6) remains valid, regardless of tip size, so long as the tunneling matrix elements can be adequately approximated by those for an s-wave tip wave function. The tip position r_t must then be interpreted as the center of the tip, i.e., the origin of the s-wave which best approximates the tip wave function.

Even for a realistic model of a one-atom tip, *Lang* verified [14.8, 9] that the image corresponds quite closely to a contour of constant $\varrho(\boldsymbol{r}_t, E_F)$, confirming the applicability of (14.6). Of course, the s-wave model must break down when there is appreciable tunneling to several tip atoms at once. However, in view of the reduced resolution, and the lesser accuracy which is consequently acceptable for the theory in such a case, (14.6) is probably still an adequate approximation, if r_t is interpreted [14.11] as an effective center of curvature of the tip.

14.2 STM Images and their Interpretation

14.2.1 Modes of Imaging

In its most general form, STM measures the function $J(\boldsymbol{r}_t, V)$. While there is great interest in measuring the dependence on r and V simultaneously, [14.12, 13] it is usually more convenient to consider $J(\boldsymbol{r}_t)|_V$, i.e., the tunneling current as a function of tip position, at fixed voltage.

Even this simplification, however, is not enough, since it is not generally practical to map out the full three-dimensional function. The elegant approach of *Binnig* et al. [14.1] was to map out the two-dimensional surface $z(\boldsymbol{x})$ defined implicitly by the condition $J(\boldsymbol{r}_t) = J_f$, where z and $\boldsymbol{x}$ are the vertical and lateral components of the tip position $\boldsymbol{r}_t$. A feedback circuit is used to constantly adjust the tip height z, as x and y are varied, so as to maintain a constant current J_f.

This constant-current mode of imaging has two crucial advantages over any other approach. One is that this mode avoids the need to measure currents varying over orders of magnitude. (The current changes by roughly one order of magnitude for every angstrom of vertical motion of the tip relative to the surface.)

The second advantage is that the resulting image will, under certain conditions, correspond closely to a "topograph" of the surface [14.1]. Examples are given in Sect. 14.2.2–4 below, both of cases where the image corresponds closely to a topograph, and cases where interpreting the image as a topograph would be grossly misleading.

It is sometimes convenient, with very flat surfaces, to operate in a "constant height" mode [14.14]. Often, this mode is used when tunneling under "dirty" conditions, in air or some other fluid, especially with a graphite surface. While such studies can be useful, a *quantitative* interpretation of STM images under these conditions is not yet possible. One of the problems which may arise under these conditions is discussed in Sect. 14.4.

Because of the general superiority of the constant-current imaging mode for well-controlled quantitative investigations, only that mode is specifically considered in the rest of this section. However, most of the results can be directly carried over to the constant-height mode.

14.2.2 Imaging of Metals

For simple metals, there is typically no strong variation with energy of the local density of states or wave functions near the Fermi level. For purposes of STM, the same is presumably true for noble and even transition metals, since the d shell apparently does not contribute significantly to the tunneling current [14.15]. It is, therefore, convenient in the case of metals to ignore the voltage dependence, and consider the limit of small voltage, (14.6). (Effects of finite voltage are discussed in Sects. 14.2.3 and 3.1 below.)

This is particularly convenient, since we then require only the calculation of $\varrho(\boldsymbol{r}_t, E_F)$, a property of the bare surface. Nevertheless, even this calculation is quite demanding numerically. In fact, I know of only one case of the STM image being calculated for a real metal surface, and compared with experiment. That is the case of Au$\{110\}$ 2×1 and 3×1, discussed in [14.11].

The real strength of STM is that, unlike diffraction, it is a local probe, and so can be applied even to disordered surfaces, or to isolated features such as defects. In order to interpret the resulting images quantitatively, it is often necessary to calculate the image for a proposed structure or set of structures, and compare with the actual image. However, while the accurate calculation of $\varrho(\boldsymbol{r}_t, E_F)$ is difficult even for Au$\{110\}$ 2×1, it is out of the question for surfaces with large unit cells, and *a fortiori* for disordered surfaces or defects. It is, therefore, highly desirable to have a method, however approximate, for calculating STM images in these important but intractable cases.

Such a method has been suggested and tested in [14.11]. It consists of approximating (14.6) by a superposition of spherical atomic-like densities. This approach is expected to work very well for simple and noble metals, and was tested in detail [14.11] for Au$\{110\}$. The success of the method relies on the fact that the model density, by construction, has the same analytical properties as the true density, so that if the model is accurate near the surface, it will automatically describe accurately the decay with distance.

An example of where this approach can be useful has also been presented [14.11]. The image expected for Au$\{110\}$ 3×1 was calculated for two plausible models of the structure, differing only in the presence or absence of a missing row in the second layer. The similarity of the model images at distances of interest suggested that the structure in the second layer could not be reliably inferred from experimental images. Quantifying the limits of valid interpretation in this way is an essential part of the analysis of STM data.

While this method is intended primarily for metals, *Tromp* et al. [14.16] applied it to Si{111} 7×7 with remarkable success. They simulated the images for a number of different proposed models of this surface, and compared them with experimental images. The so-called "Dimer-Adatom-Stacking fault" model gives an image which agrees almost perfectly with experiment, while most models lead to images with little similarity to experiment. Thus, the usefulness of such image simulations must not be underestimated, although few such applications have been made to date.

14.2.3 Imaging of Semiconductors

At very small voltages, the s-wave approximation for the tip led to the very simple result (14.6). At larger voltages, one might hope that this could be easily generalized to give a simple expression such as

$$J \sim \int_{E_{\mathrm{F}}}^{E_{\mathrm{F}}+V} \varrho(\boldsymbol{r}_{\mathrm{t}}, E) dE \, . \tag{14.7}$$

This is not strictly correct for two reasons. First, the matrix elements and the tip density of states are at least slightly energy dependent, and any such dependence is neglected in (14.7). Second, the finite voltage changes the potential, and hence the wave functions, outside the surface. A more careful discussion, especially of the latter effect, is given in Sect. 14.3. Nevertheless, there is considerable evidence that (14.7) is a reasonable approximation for many purposes [14.17], as long as the voltage is much below the workfunction. We shall, therefore, use (14.7) in discussing qualitative aspects of STM images of semiconductors at modest voltages.

Unlike metals, semiconductors show a very strong variation of $\varrho(\boldsymbol{r}_{\mathrm{t}}, E_{\mathrm{F}} + V)$ with voltage. In particular, this quantity changes discontinuously at the band edges. With negative sample voltage, current tunnels out of the valence band, while for positive voltage, current tunnels into the conduction band. The corresponding images, reflecting the spatial distribution of valence and conduction-band wave function, respectively, may be qualitatively different.

A particularly simple and illustrative example, which has been studied in great detail, is GaAs(110). There, it was proposed [14.11] that since the valence states are preferentially localized on the As atoms, and the conduction states on the Ga atoms, STM images of GaAs{110} at negative and positive bias should reveal the As and Ga atoms, respectively. Such atom-selective imaging was confirmed by direct calculation of (14.7), [14.11] and was subsequently observed experimentally [14.18].

In a single image of GaAs{110}, whether at positive or negative voltage, one simply sees a single "bump" per unit cell. In fact, the images at opposite voltage look quite similar. It is, therefore, crucial to obtain both images *simultaneously*, so that the dependence of the absolute position of the "bump" on voltage can be determined. Combining the two images, the zig-zag rows of the {110} surface are clearly seen [14.18].

Even in this simple case, however, the interpretation of the voltage-dependent images as revealing As or Ga atoms directly is a bit simplistic. A detailed analysis [14.18] shows that the apparent positions of the atoms in the images deviate significantly from the actual positions. This deviation could be viewed as an undesirable complication since it makes the image even less like a topograph.

Alternatively, it is possible to take advantage of this deviation. The apparent position of the atom turns out to be rather sensitive to the degree of buckling associated with the {110} surface reconstruction, so that it is possible to infer the surface buckling quantitatively from the apparent atom positions [14.18]. Thus, the images are actually quite rich in information, but the quantitative interpretation requires a more detailed analysis than is often feasible.

Even for semiconductors, there may be cases where the image (at least at some voltage) corresponds fairly closely to a simple topograph of the surface. A striking example of this is Si{111} 7×7, discussed above [14.16]. Similarly, for Si{111} 2×1, the image at low voltage is rather peculiar [14.19], for reasons discussed in Sect. 2.4; but at higher voltage, the image begins to look a bit more like a simple topograph [14.20].

In tunneling to semiconductors, there is an added complication not present in metals: there may be a large voltage drop associated with band-bending in the semiconductor, in addition to the voltage drop across the gap [14.21]. This means that the tunneling voltage may be substantially less than the applied voltage, complicating the interpretation. Moreover, local band-bending associated with defects or adsorbates on the surface can lead to striking non-topographic effects in the image [14.21].

14.2.4 Imaging Band-Edge States

A particularly interesting situation can arise in tunneling to semiconductors at low voltages [14.19, 22]. At the lowest possible voltages, only states at the band edge participate in tunneling. These band edge states typically (though not necessarily) fall at a symmetry point at the edge of the surface Brillouin zone. In this case, the states which are imaged have the character of a standing wave on the surface.

This standing-wave character leads to an image with striking and peculiar properties [14.22, 23]. The corrugation is anomalously large, and unlike the normal case, it does not decrease rapidly with distance from the surface. This gives the effect of unusually sharp resolution. For example, in the case of graphite (a semimetal which also satisfies these conditions), the unit cell is easily resolved despite the fact that it is only 2 Å across. This effect was also seen on Si{111} 2×1 [14.19].

In most cases, the image can be described by a universal form, [14.22] consisting of an array of sharp dips with the periodicity of the lattice. (The dips, however, are broadened by a variety of effects, and in any case may not be well resolved because of instrumental response time [14.22].) Such an image can be extremely misleading. Specifically, when there is one "bump" per unit cell, it is tempting to infer that there is one topographic feature per unit cell. However, the image may in fact carry no information whatever regarding the distribution of atoms within the unit cell in this case. It is strange indeed that the sharpest image with the best apparent resolution sometimes carries the least structural information.

14.3 Spectroscopy

14.3.1 Qualitative Theory

Tunneling spectroscopy in planar junctions was studied long before STM [14.24]. However, the advent of spatially-resolved STM spectroscopy has led to a resurgence of interest in this area. Because of the difficulty of calculating $J(\boldsymbol{r}_t, V)$ in general, most detailed analyses have instead focused simply on $J(V)$, without regard to its detailed spatial dependence [4.17]. Moreover, the important issue of inelastic tunneling spectroscopy is only beginning to be discussed theoretically in the context of STM [4.25].

Selloni et al. [14.26] suggested that the results of *Tersoff* and *Hamann* [14.11] for small voltage could be *qualitatively* generalized as

$$J(V) \propto \int_{E_F}^{E_F+V} \varrho(E) T(E, V) dE \,, \tag{14.8}$$

where $\varrho(E)$ is the local density of states (14.6) at or very near the surface, and assuming a constant density of states for the tip. This is similar to (14.7), except that the qualitative effect of the finite voltage on the surface wave functions is included through a barrier transmission coefficient $T(E, V)$.

Unfortunately, despite suggestions in [14.26], this simple qualitative model still does not lend itself to a straightforward interpretation of the tunneling spectrum [14.17]. In particular, the derivative dJ/dV has no simple relationship to the density of states $\varrho(E_F + V)$, as might have been hoped. At best, one can say that a sharp feature in the density of states of the tip or sample, at an energy $E_F + V$, will lead to a feature in $J(V)$ or its derivatives at voltage V.

Even this rather weak statement may prove unreliable in practice, where spectral features have considerable widths. The reason for this problem is that $T(E, V)$ is very strongly V-dependent when the voltage becomes an appreciable fraction of the work function. Thus the V-dependence of $T(E, V)$ may distort features in the spectrum [14.17].

Stroscio, Feenstra and coworkers [14.19, 27] proposed a simple but effective solution to this problem. They normalize dJ/dV by dividing it by J/V [14.19]. This yields $d\ln J/d\ln V$, and so effectively cancels out the exponential dependence of $T(E, V)$ on V. At semiconductor band edges, where the current goes to zero, a slight smoothing of J/V eliminates the singular behavior at the band edge [14.27].

This normalization is, however, both unnecessary and undesirable at small voltages; in that case, J/V is well behaved, whereas $(dJ/dV)/(J/V)$ is identically equal to unity for ohmic systems, and so carries no information. Thus, the appropriate way of displaying spectroscopy results depends on the problem at hand, and a variety of approaches for collecting and displaying data have been considered [14.19,21,27].

14.3.2 Quantitative Theory

A proper treatment of the tunneling spectrum in STM requires calculation of the wave functions of surface and tip at finite voltage. This is a difficult problem, which is not yet tractible except in simplified models.

A natural approximation [14.17] is, therefore, to use (14.5) with the zero-voltage wave functions, but to shift all the surface wave functions in energy relative to the tip, by an amount corresponding to the applied voltage V. Unfortunately, the result then depends on the position of the surface of integration for (14.3).

Lang [14.17] showed that, by positioning the surface of integration half-way between the two planar electrodes, the resulting error is second order in the voltage, and rather small as long as the voltage is much less than the work function. This result assumes that surface and tip have equal work functions, and is derived for a one-dimensional model only. Nevertheless, it seems safe to assume that the conclusion is more generally valid.

With this approximation, calculation of the tunneling spectrum becomes relatively straightforward. Such calculations [14.17] confirm the qualitative applicability of simple models such as (14.8). Moreover, they confirm that the spectrum $(dJ/dV)/(J/V)$ mimics the density of states reasonably well, as proposed by *Stroscio* et al. [14.19].

14.4 Mechanical Interactions Between Tip and Sample

Ideally, in STM the tip and surface are separated by a vacuum gap, and are mechanically non-interacting. However, sometimes anomalies are observed, which are most easily explained by assuming a mechanical interaction between the tip and surface.

In particular, since the earliest vacuum tunneling experiments of *Binnig* et al. [14.28], it has been observed that for dirty surfaces, the current varies less rapidly than expected with vertical displacement of the tip. *Coombs* and *Pethica* [14.29] pointed out that this behavior can be explained by assuming that the dirt mediates a mechanical interaction between surface and tip.

The current is expected to vary with tip height z as $J \propto \exp(-2\kappa z)$, where $\hbar^2\kappa^2/2m = \phi$, ϕ being the work function. Thus, in principle ϕ can be determined from $d\ln J/dz$. For dirty surfaces, this dependence is weaker than expected, leading to an inferred work function which is unphysically small. In liquids, this "effective work function" is often 0.1 eV or less.

The explanation proposed by Coombs and Pethica is that some insulating dirt (e.g. oxide) is squeezed between the surface and tip, acting in effect as a spring. Tunneling might take place from a nearby part of the tip which is free of oxide, or from a "mini-tip" poking through the dirt. As the tip is lowered, the dirt becomes compressed, pushing down the surface or compressing the tip if they are sufficiently soft. As a result, the surface-tip separation does not really decrease as much as expected from the nominal lowering of the tip, and so the current variation is correspondingly less.

This issue gained renewed importance with the observation of huge corrugations in STM of graphite [14.30, 31]. While theoretical calculations [14.26] suggested corrugations of at most 1 Å or so, ridiculously large corrugations were sometimes observed, up to 10 Å or more vertically within the 2 Å wide unit cell of graphite.

Soler et al. [14.30] at first attributed these corrugations to direct interaction of the tip and surface, but a detailed study by *Mamin* et al. [14.31] suggests that in fact the interaction is mediated by dirt, consistent with the earlier proposal of *Coombs* and *Pethica* [14.29]. Clean UHV measurements in fact gave corrugations of under 1 Å [14.32].

For the model of direct interaction, [14.30] there is a complex nonlinear behavior. But for the dirt-mediated interaction model, [14.29] a linear treatment is appropriate. In this case, assuming the mechanical interaction has negligible corrugation (e.g., because of the large interaction area), the image seen is simply the ideal image, with the vertical axis distorted by a constant scale factor.

It seems surprising that it is possible to obtain any image at all, when there is dirt between surface and tip, since some scraping might be expected as the tip is scanned. For graphite, one could imagine that the dirt (e.g., metal oxide) moves with the tip, sliding nicely along the inert graphite surface. However, any model here is based on indirect inference, and this phenomenon must be considered as not really understood at present.

This lack of a complete understanding is most dramatically pointed out by the recent observation of close-packed metal surfaces, in which individual atoms were resolved [14.33, 34]. It is possible [14.34] that these cases, like graphite, may represent an enhancement of the corrugation by mechanical interactions between tip and surface. However, no dramatic lowering of the "effective work function" was observed, [14.34] so the model of Coombs and Pethica for the interaction may not be directly applicable. Moreover, it is hard to imagine how mechanical contact could fail to disrupt the metal.

14.5 Conclusion

STM images at low voltage are relatively well understood at present, although the detailed dependence on tip shape has been considered only qualitatively. Tunneling spectra are also understood in principle, although they bear only a qualitative resemblance to the density of states.

However, this rather satisfactory state of affairs applies only to clean surfaces in ultra-high vacuum. A great deal of STM work is actually done in air or liquids, because of the tremendous reduction in labor and expense relative to vacuum. In this case, the understanding is much less complete.

In particular, mechanical interactions between surface and tip, mediated either by solid dirt or by the surrounding fluid, appear to play a crucial role. Such interactions are not understood at a microscopic level. They may not even be reproducible, depending upon their origin.

References

14.1 G. Binnig, H. Rohrer, Ch. Gerber, E. Weibel: Phys. Rev. Lett. **49**, 57 (1982). For reviews, see G. Binnig, H. Rohrer: Rev. Mod. Phys. **59**, 615 (1987); P.K. Hansma, J. Tersoff: J. Appl. Phys. **61**, R1 (1987); R.J. Behm, W. Hösler: in *Chemistry and Physics of Solid Surfaces VI*, ed. by R. Vanselow, R. Howe, Springer Ser. Surf. Sci., Vol. 5 (Springer, Berlin, Heidelberg 1986)

14.2 J.K. Gimzewski, R. Möller: Phys. Rev. B **36**, 1284 (1987)

14.3 N.D. Lang: Phys. Rev. B **36**, 8173 (1987)

14.4 M.D. Stiles, D.R. Hamann: Phys. Rev. B **38**, 2021 (1988)

14.5 N. Garcia, C. Ocal, F. Flores: Phys. Rev. Lett. **50**, 2002 (1983); E. Stoll, A. Baratoff, A. Selloni, P. Carnevali: J. Phys. C **17**, 3073 (1984)

14.6 J. Bardeen: Phys. Rev. Lett. **6**, 57 (1961)

14.7 For a unique exception, see Y. Kuk, P.J. Silverman, H.Q. Nguyen: J. Vac. Sci. Technol. A **6**, 524 (1988)

14.8 N.D. Lang: Phys. Rev. Lett. **56**, 1164 (1986)

14.9 N.D. Lang: Phys. Rev. Lett. **55**, 230 (1985)

14.10 J. Tersoff, D.R. Hamann: Phys. Rev. Lett. **50**, 1998 (1983)

14.11 J. Tersoff, D.R. Hamann: Phys. Rev. B **31**, 805 (1985)

14.12 R.J. Hamers, R.M. Tromp, J.E. Demuth: Phys. Rev. Lett. **56**, 1972 (1986)

14.13 R.M. Feenstra, W.A. Thompson, A.P. Fein: Phys. Rev. Lett. **56**, 608 (1986); J.A. Stroscio, R.M. Feenstra, D.M. Newns, A.P. Fein: J. Vac. Sci. Technol. A **6**, 499 (1988)

14.14 A. Bryant, D.P.E. Smith, C.F. Quate: Appl. Phys. Lett. **48**, 832 (1986)

14.15 N.D. Lang: Phys. Rev. Lett. **58**, 45 (1987)

14.16 R.M. Tromp, R.J. Hamers, J.E. Demuth: Phys. Rev. B **34**, 1388 (1986)

14.17 N.D. Lang: Phys. Rev. B **34**, 5947 (1986)

14.18 R.M. Feenstra, J.A. Stroscio, J. Tersoff, A.P. Fein: Phys. Rev. Lett. **58**, 1192 (1987)

14.19 J.A. Stroscio, R.M. Feenstra, A.P. Fein: Phys. Rev. Lett. **57**, 2579 (1986)

14.20 R.M. Feenstra, J.A. Stroscio: Phys. Rev. Lett. **59**, 2173 (1987)

14.21 R.M. Feenstra, J.A. Stroscio: J. Vac. Sci. Technol. B **5**, 923 (1987); J.A. Stroscio, R.M. Feenstra: J. Vac. Sci. Technol. B **6**, 1472 (1988); R.M. Feenstra, P. Mårtensson: Phys. Rev. B **39**, 7744 (1989)

14.22 J. Tersoff: Phys. Rev. Lett. **57**, 440 (1986)

14.23 J. Tersoff: Phys. Rev. B **39**, 1052 (1989)

14.24 C.B. Duke: *Tunneling in Solids*, Suppl. 10 of *Solid State Physics*, ed. by F. Seitz, D. Turnbull (Academic, New York 1969), p. 1

14.25 B.N.J. Persson, A. Baratoff: Phys. Rev. Lett. **59**, 339 (1987); B.N.J. Persson, J.E. Demuth: Solid State Comm. **57**, 769 (1986); G. Binnig, N. Garcia, H. Rohrer: Phys. Rev. B **32**, 1336 (1985)

14.26 A. Selloni, P. Carnevali, E. Tosatti, C.D. Chen: Phys. Rev. B **31**, 2602 (1985)
14.27 R.M. Feenstra, P. Mårtensson: Phys. Rev. Lett. **61**, 447 (1988)
14.28 G. Binnig, H. Rohrer, Ch. Gerber, E. Weibel: Appl. Phys. Lett. **40**, 178 (1982)
14.29 J.H. Coombs, J.B. Pethica: IBM J. Res. Develop. **30**, 455 (1986)
14.30 J.M. Soler, A.M. Baro, N. Garcia, H. Rohrer: Phys. Rev. Lett. **57**, 444 (1986)
14.31 H.J. Mamin, E. Ganz, D.W. Abraham, R.E. Thomson, J. Clarke: Phys. Rev. B **34**, 9015 (1986)
14.32 R.J. Hamers: unpublished
14.33 V.M. Hallmark, S. Chiang, J.F. Rabolt, J.D. Swalen, R.J. Wilson: Phys. Rev. Lett. **59**, 2879 (1988)
14.34 J. Wintterlin, J. Wiechers, H. Brune, T. Gritsch, H. Höfer, R.J. Behm: Phys. Rev. Lett. **62**, 59 (1989)

15. Proximal Probes: Techniques for Measuring at the Nanometer Scale

James S. Murday and Richard J. Colton

Chemistry Division, Naval Research Laboratory,
Washington, DC 20375–5000, USA

Innovation in analytical surface science has evolved a panoply of tools which are sensitive to nanometer scale lengths in one dimension (typically depth). A listing of those techniques would include Auger and photoelectron spectroscopies, low energy electron diffraction, secondary ion mass spectrometry, etc.; these tools have been under development since the late 1960s and are well covered in many textbooks and reviews [15.1–5]. They are surface specific, but probe relatively large areas (i.e., > 50 nm diameter). The continued scientific/technological evolution toward smaller structures raises the need for a different class of surface analytical tools which specifically probe nanometer volume elements. Even for larger structures, the analysis of defects requires the capability to examine properties of small volume elements.

There is a cornucopia of science/technology research opportunities at the nanometer scale. The impact will be revolutionary in scope and character:

Low Dimensioned Structures. Many coherence lengths have dimensions in the 1–100 nm range. As we grow structures of this scale, one can expect brand new phenomena to be discovered with revolutionary scientific and technological impact. Superconductivity, magnetism, quantum electronic states, non-linear optics, and other phenomena will show new, unexplored behavior [15.6, 7]. The major problem inhibiting progress in low dimension structures is fabrication technology. Lithographic techniques used in present electronics technology work principally above the 100 nm level, with some advanced laboratories at smaller dimensions.

Nucleation/Growth. The study of nucleation phenomena has long been hampered by the lack of analytical tools to address structures of nanometer scale. This is an area of rapidly expanding work in the STM community with some spectacular results [15.8–11], especially when combined with RHEED and LEED [15.12]. The growth of atomically precise nanometer structures is already important to many devices. The nucleation events and growth mechanisms, essential to produce atomically precise structures, are prime areas for proximal probe investigation.

Site Specific Reactions. Corrosion and catalytic reactions are frequently determined by site specific events on a surface. The new analytical tools allow these processes to be visualized directly, and hold promise for studies of dynamics. Present work in the field is limited to the reaction of a couple of molecules on Si [15.13] and some metal surfaces [15.14, 15].

Elastic/Plastic Deformation. The operating assumption behind theories of friction, wear, polishing and machining has been asperity mediated contact. Similarly, the transition from elastic to plastic deformation involves the initiation and propagation of dislocations at nanometer scales. Slow progress in understanding these phenomena is attributable to the inability to probe directly their nanoscale behavior; the appropriate tools and surface finishes are now available to attack the problem [15.7].

Crack Initiation/Propagation. The initiation and propagation of cracks is determined by the atomic behavior in a very small volume element around a crack tip; frequently chemical influences at that point are also important. The development of proximal probes may help in the investigation of this phenomena, but the constrained geometry is still a problem.

Interfacial Electromagnetic Waves. The exchange of electromagnetic energy between two bodies is usually treated in the far-field approximation. The ability to position two bodies close compared to the infra-red, visible and ultraviolet wavelengths allows one to work in the near field. Imaging with resolution better than diffraction limits is possible. Image contrast due to energy exchange in the evanescent wave fields will provide new insights. There is also some evidence to suggest that potential differences due to Fermi level misalignment (thermoelectric effect as one example) can provide a contrast mechanism for a scanning proximal probe. New classes of analytical tools will be developed from these investigations [15.16] and they also might be exploited in the detection of electromagnetic radiation (sensors).

Solid-Fluid Boundary Layers. The structure and dynamics of the solid-fluid boundary are critical to such diverse fields as hydrodynamic flow, electrode processes, tribology, corrosion and liquid crystal displays. However, very little is known at the molecular level about how or when the solid imparts order to adjacent "fluid" species [15.17]. The availability of in-situ analysis at the atomic scale portends important progress for these technologies.

Macromolecular Configuration. Imaging the structure of organic materials at the nanometer scale will be a major advance for biomolecular and polymer sciences [15.18]. The structure of these complex molecules and self-organized systems (membranes) is determined by a fine balance of many weak forces and, further, is strongly influenced by hydration. Understanding the chemical/physical behavior of these molecules will require the development of tools to help define their structure and to watch their dynamics. As a witness to the importance,

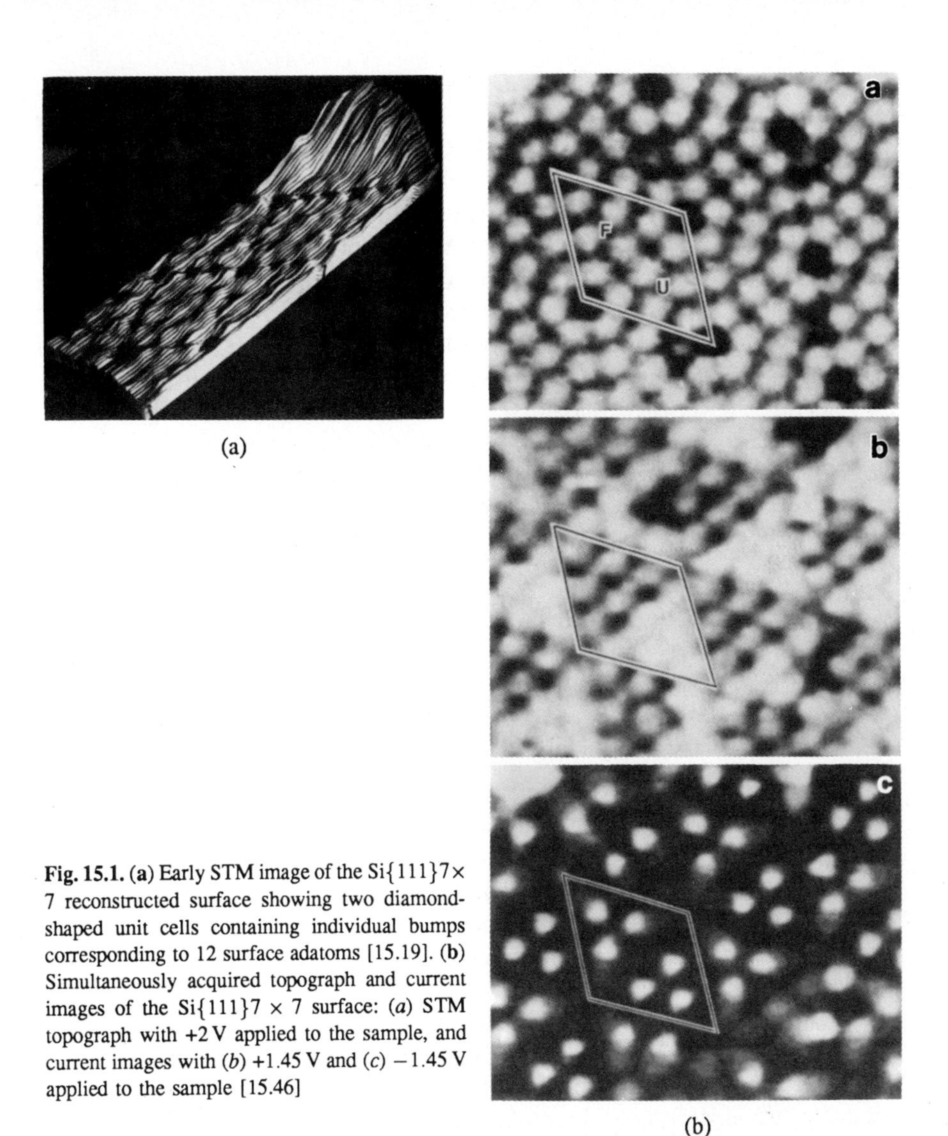

Fig. 15.1. (**a**) Early STM image of the Si{111}7× 7 reconstructed surface showing two diamond-shaped unit cells containing individual bumps corresponding to 12 surface adatoms [15.19]. (**b**) Simultaneously acquired topograph and current images of the Si{111}7 × 7 surface: (*a*) STM topograph with +2 V applied to the sample, and current images with (*b*) +1.45 V and (*c*) −1.45 V applied to the sample [15.46]

approximately 25 groups have instituted proximal probe investigations in this area during the past three years.

In 1982, *Binnig*, *Rohrer*, *Gerber* and *Weibel*'s [15.19, 20] striking picture of the Si{111}7 × 7 reconstructed surface (Fig. 15.1a) fired the imagination of the scientific community. If atoms could be resolved, so could structures of 1–100 nanometers. The subsequent progress has been nothing short of spectacular. A whole new class of analytical tools [15.21–23], *Proximal Probes* (so called for the dependence on probe-to-sample proximity for their analytical capabilities), is under development. These tools have been shown not only to identify the composition/structure of surfaces with nanometer scale resolution, but also to determine site specific chemical/physical properties.

The following section presents an overview of the various manifestations of proximal probes, organized about the physical phenomena by which the probe derives its capabilities. Section 15.1.1 reviews tools dependent on electronic tunneling phenomena between a probe and a surface. In Sect. 15.1.2, we examine the analytical possibilities as the bias between tip and surface is increased to work function values ($\sim 4\,\mathrm{eV}$) where the electron transport changes character into field emission. In this mode (i.e., higher primary beam energies) secondary and core electron microscopies/spectroscopies become possible. Sampling areas are small (but not atomic) due to the proximity of the tip and surface, without need of focussing elements. Section 15.1.3 discusses a different class of proximal probe, the force apparatus, where surface features are probed by their ability to bend a cantilever beam. Finally, Sect. 15.1.4 examines several approaches to nanometer probes based on near-field electrical and magnetic phenomena. The last section covers progress in fabricating structures using proximal probe techniques.

15.1 Proximal Probes

15.1.1 Tunneling

a) Scanning Tunneling Microscopy (STM) is based on the observation that electrons, because of their wave-like nature, can tunnel through the potential barrier established when two surfaces under appropriate bias are positioned about a nanometer from each other. While the concept of vacuum tunneling is quite simple, the experimental realization of scanning tunneling proved to be quite elusive. Electron tunneling and tunneling spectroscopy are mature disciplines in physics [15.24–28]. Tunneling through solid insulating barriers was first demonstrated in 1957. Early studies used metal-insulator-metal (MIM) tunnel junctions which consist of a thin oxide layer between planar conducting electrodes. The thin insulating oxide layer defines a rigid – hence controllable – potential barrier through which electrons could tunnel under appropriate potential bias [15.29–32]. The problems associated with maintaining a vacuum gap between two electrodes less than a nanometer apart were addressed by *Young* at NBS in the late 1960s [15.33] and overcome by *Binnig*, *Rohrer* and co-workers at IBM Zurich in the late 1970s [15.19, 20, 34]. Progress to date in STM instrumentation is covered in several reviews [15.14, 16, 21, 35–38], conference proceedings [15.39], and a bibliography [15.40]. More complete descriptions of STM theory and use can be found in Chaps. 14 and 16 by *Tersoff* and *Avouris*.

The scanning tunneling microscope consists of a sharp tip that can be positioned on the order of a nanometer from a surface. Electrons tunnel between the tip and the surface when a small bias voltage is applied. An image of the surface is acquired by scanning the tip or surface, usually by piezoelectric actuators. As the tip approaches a high (or low) spot on the specimen surface, the tunneling current increases (or decreases) in response to the change in separation between the tip and the surface. In a common manifestation of STM, a simple electronic feedback circuit is used to adjust the position of the tip to maintain a constant tunneling current. Scanning the tip in the plane of the surface produces a map or contour of the surface. The real power of the technique comes from its extremely high resolution – which can approach 0.2 nm in the lateral dimension and 10^{-3} nm in the vertical direction. That resolution originates in the sensitivity of the tunneling current to the electron wavefunction overlap between the proximal surfaces. For small voltages ($V \ll \varphi$) the tunneling current I, across a vacuum gap, can be expressed as [15.21, 28, 41]:

$$\mathrm{I} \propto (V/s)\exp\{-2\kappa s\} , \tag{15.1}$$

where V is the tunneling voltage, s is the distance between the electrodes, and κ is the decay constant of the electron wavefunction in the barrier where $\kappa = 2\pi\sqrt{2m\varphi}/h \approx 0.1\,\mathrm{nm}^{-1}$ for electron mass m and work function $\varphi = 4\,\mathrm{eV}$. When the separation is around 1 nm, the exponential term in (15.1) increases one order of magnitude for 0.1 nm change in s, while the preexponential decreases by 10%. The exponential dependence of the tunneling current on distance gives STM its high sensitivity.

The lateral resolution of the STM derives from a combination of the vertical sensitivity and spatial localization of the electronic wavefunctions whose overlap govern the tunneling. *Tersoff* and *Hamann* [15.42] (see also Chap. 14) have modelled the process and provide an estimate of an instrumental resolution function rms width W as:

$$W = \left[\frac{s+R}{2\kappa}\right]^{1/2} , \tag{15.2}$$

where R is the radius of curvature of a hemispherical tip. For a single atom tip, one could find $(s + R)$ as small as 0.5 nm, in which case $W \approx 0.2\,\mathrm{nm}$. *Kuk* et al. [15.43] prepared and characterized a tip with Field Ion Microscopy, then verified the STM image was consistent with (15.2) by measuring the change in the corrugation of Au$\{100\}$ and $\{110\}$ surfaces as a function of tip size. However, a satisfactory understanding of lateral resolution is not in hand; there is convincing evidence that, under the influence of strong tip-surface forces, mechanical effects are also playing a role (Sect. 15.1.3) [15.44, 45].

The improvements in atomic imaging with STM since 1982 are illustrated in Fig. 15.1b for Si$\{111\}7 \times 7$ where not only each adatom is delineated, but different kinds of surface atoms are identified through contrast dependence on tip-surface bias [15.46]. The improved images are the result of several advances in state-of-the-art, not the least of which is image processing. Improvements in the design of piezoelectric actuators responsible for tip-sample displacement has increased the total scan range to distances approaching 0.1 mm [15.47, 48] and has augmented the speed of data acquisition (scanning rates up to 40 kHz). To achieve these higher scan rates without crashing the tip on surface protuberances, the tip is generally not scanned in constant current mode, but at constant height, with modulation in tunneling current providing the contrast mechanism [15.35].

STM has been demonstrated viable over a wide range of conditions. Low temperature tunneling is prevalent both to quench electronic, vibrational, and thermal problems as well as to probe low temperature phenomena. Early experiments simply submerged the tip-sample in a liquid cryogen. Several examples of variable temperature devices have now been published; great care must be taken in their design to minimize thermal drift effects [15.49]. Tunneling is not constrained to vacuum barriers but also happens through dielectrics. Operation in fluid media is possible [15.50, 51]; this is being exploited to study in-situ electrochemical phenomena [15.52–54] (where careful shielding of the tip from Faradaic currents must be exercised [15.55]) and to image macromolecules adsorbed from solution [15.56, 57]. The presence of a liquid environment has the advantage of damping macroscopic vibrational motions, but may add to tunneling noise via diffusional processes. This sensitivity to fluctuations in the tunneling gap has been utilized to examine the diffusion of oxygen on Ni up to 680 K [15.58] and to characterize the trapping/detrapping of electrons [15.59]. The STM has high resolution in both the lateral and vertical directions, making it more useful than SEM where the vertical scale is hard to quantify.

Perhaps the least understood component of the STM apparatus is the tip [15.60]. A variety of prescriptions [15.61] have been developed to prepare a sharp tip including fracture, sputtering, electrochemical etch, field emission/evaporation, and field enhanced chemical attack. Electrochemical techniques are the most utilized in preparing a tip. But in order to get the atomic resolution demonstrated in the best semiconductor work, it is necessary to scan the tip for several hours or to operate it at higher voltages for a short period of time. Even then, the high resolution mode can come and go, apparently subject to changes of very small (atomic) asperities at the tip apex.

Meaningful characterization of surface features does not have to be on the atomic scale to be useful. Numerous groups have begun to utilize the STM for characterization of high technology nanometer structures such as machined surfaces [15.62, 63], fracture surfaces [15.64], Vicker imprints [15.65], x-ray optical multilayer films [15.66], microbridges of SQUID devices [15.67], microfabricated electronic patterns [15.68], and optical disks [15.69].

b) Scanning Tunneling Spectroscopy (STS). In addition to the imaging capability of STM, it is possible to hold a tip over a specific location and measure its tunneling current-voltage characteristics [15.21, 35]. A tip-sample potential range approaching ± 4 V is possible before the tunneling approximations are violated. When the tip is biased negative, the tunneling electrons originate in filled states of the tip and end in empty states of the surface (and vice versa); current-versus-bias data therefore yields a convolution of the tip and surface electron density-of-states.

Feenstra, Stroscio, and *Fein* [15.70, 71] showed that a plot of $(dI/dV)/(I/V) = d\ln(I)/d\ln(V)$ vs eV approximates the sample electron density-of-states $\varrho(\mathrm{eV})$

for a metal tip/semiconductor pair:

$$\frac{d\ln(I)}{d\ln(V)} \propto \varrho(\mathrm{eV}) + \ldots \,. \tag{15.3}$$

The interpretation of the spectral plots must be done carefully since several factors (wavefunction spatial extent as well as occupation) are important. *Lang* [15.72] has modelled the effect of local electronic wavefunctions (chemical state) on the tunneling current (see Chap. 14 for a more detailed discussion).

The spectroscopic capability has been used to investigate the site specific surface state character of numerous semiconductor surfaces [15.71, 73–75] (see also Chap. 16), and metal [15.76] surfaces, to differentiate between elements on coumpound semiconductor surfaces [15.77, 78], to investigate the superconducting band gap [15.79, 80], including the new high T_c oxides [15.81–86], and to image magnetic flux vortices in Type II superconductors [15.80, 87].

c) Ballistic Electron Emission Microscopy. As described above, STM/S is effective at a solid/fluid interface (vacuum being the low density limit of a fluid). *Kaiser* et al. [15.88] have shown that in selected cases the tunneling tip can be used to image buried interface structure as well; they call their technique Ballistic Electron Emission Microscopy (BEEM). When a solid-solid interface forms a barrier to electron transmission, such as a Schottky barrier, the electron current injected into the buried layer will depend on the energy of the electron approach-

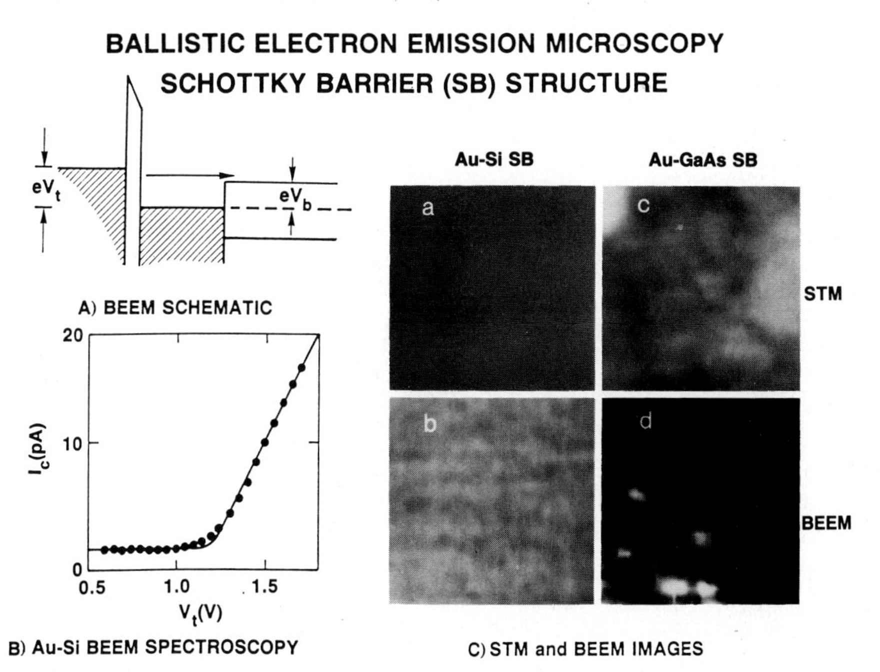

Fig. 15.2a–c. Study of Shottky Barrier (SB) structures using Ballistic Electron Emission Microscopy (BEEM): (**a**) schematic of BEEM experiment, (**b**) BEEM I-V spectrum at a single site of the Au/Si SB, and (**c**) STM and BEEM images of the surfaces and barrier interfaces, respectively, of the Au/Si and Au/GaAs SB [15.88]

ing the barrier. In Fig. 15.2a a flat band barrier is illustrated for a thin (< 10 nm, the length scale determined by the electron inelastic mean free path) metal film on a semiconductor. For a low biased tunneling tip, electrons have insufficient energy to penetrate into the semiconductor; when the tip is scanned, contrast in the current collected from the metal film is derived from the STM mechanisms described above. As the tip bias is raised, the electron energy becomes sufficient to surmount the barrier. The current voltage characteristic in Fig. 15.2b shows the expected behavior. When the tip is biased at the knee of the I-V curve, the current transmitted into the semiconductor depends critically on the local structure. Images of Au/Si (known to grow good interfaces) and Au/GaAs (known to have interface problems), taken with the bias in the I-V knee, show contrast patterns different from the STM images (Fig. 15.2c). Image resolution appears to be 1 nm. This can be explained through momentum conservation for electrons penetrating the interface; only those with a high forward angle get into the semiconductor [15.88].

15.1.2 Field Emission

Significant tunneling current occurs only when the tip and surface are close, ~ 0.2–2.0 nm and at low bias conditions (< 4 V). As the surfaces are withdrawn to larger distances and/or the bias raised, electron transfer occurs under the field emission approximation (see Fig. 15.3). The Fowler and Nordheim model for the field emission current predicts [15.28]:

$$I \propto C(\varepsilon_{\mathrm{F}}, \varphi) V^2 \exp\left[-2\kappa s \frac{2\varphi}{3\mathrm{eV}}\right] , \tag{15.4}$$

where ε_{F} is the Fermi energy and the other symbols are as defined above. After penetrating the tunneling barrier, the electrons propagate across the vacuum

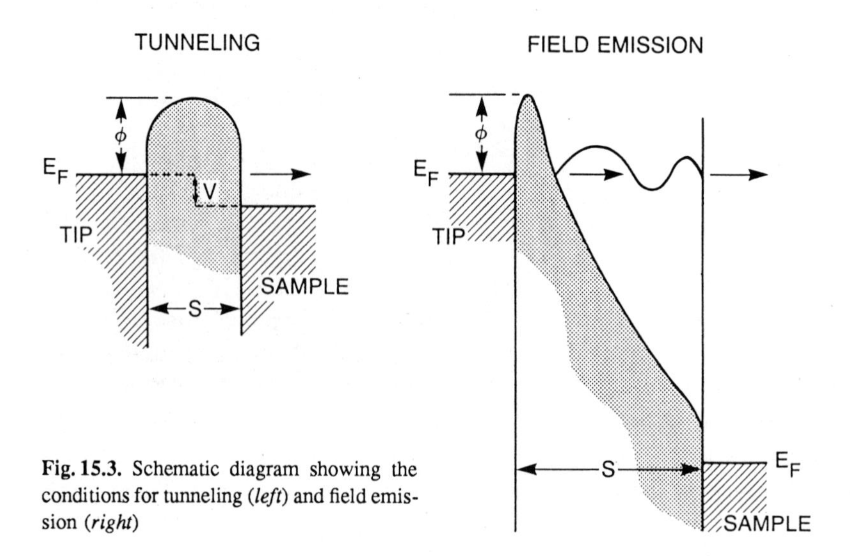

Fig. 15.3. Schematic diagram showing the conditions for tunneling (*left*) and field emission (*right*)

region. Under the right conditions, the electron wavefunction can form standing waves in this region, leading to field emission resonances [15.89] which can be sensitive probes of sample surface conditions.

Field emission tips are commonly used in scanning electron microscopes where their high brightness and small source volume enable high resolution work. Spot sizes of 0.5–5 nm are achieved with working potentials in the range 0.5–50 kV. The work of *McCord* and *Pease* (spherical tip, [15.90]) and especially *Fink* (atomically sharp tips, [15.91]) has shown that if the tips used for field ion microscopy are used as the electron source, then the field emission half-angle can be very small. Electron beams from these sharp tips placed close to the surface (~ 10 nm) can have spot diameters in the tens of nanometer size. In contrast to conventional SEM, the proximal tip-sample potentials for small sample spot diameters are much lower ~ 10–100 eV. Some benefits of the lower potential are greater surface sensitivity, diminished beam damage, and the possibility of varying the electron energy across chemical bond strengths. Another approach to low electron energy, high lateral resolution microscopy is the low energy electron microscopy (LEEM) technique developed by *Bauer* (Chap. 12).

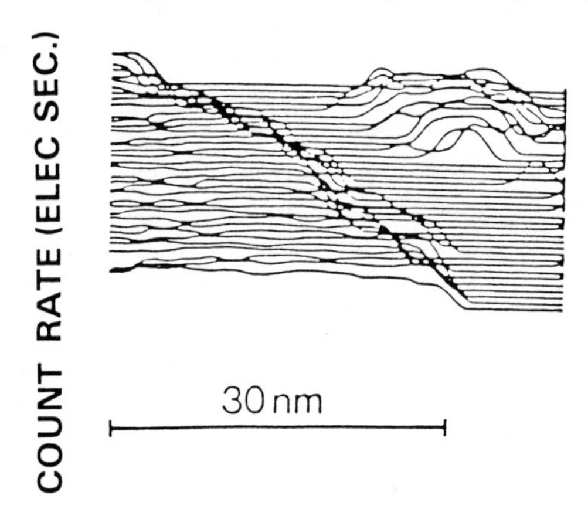

Fig. 15.4. Secondary electron micrograph of a polycrystalline gold sample showing features with a lateral resolution of ~ 3 nm. The image was generated with a 15 V electron beam from a tip positioned close to the sample surface [15.91]

The line image in Fig. 15.4 shows an SEM picture of an Au surface with 3 nm resolution [15.91] generated with a proximal field emission tip biased at 15 V to the substrate. The detection of secondary electrons is influenced by the presence of the electric field applied by the tip; however, *Allenspach* and *Bischof* [15.92] conclude that the field near the tip apex falls off rapidly enough for properly shaped tips to collect those electrons.

As the field emitted electron energies grow sufficiently large to ionize core electrons, Auger [15.93] and inverse photo emission [15.94] processes provide spectroscopic means of probing the chemical state of the surface. Of course, as the tip-surface bias increases, the tip must be retracted to limit the current [see (15.4)], thereby spoiling the lateral resolution. *Coombs* et al. [15.94] show an example of repeatable emission of electron stimulated light intensity with lateral features on a 0.5 nm scale for an evaporated silver film. Spin-polarized secondary electron emission is used to characterize the magnetic characteristics of thin films and surfaces; a proximal probe version of that experiment has been demonstrated to work [15.92].

15.1.3 Force

a) Atomic Force Microscopy. A telling limitation to STM/S and its field emission cousins is the necessity for accepting significant electron currents into very small areas; the substrate needs to be reasonably conducting. To defeat this limitation, *Binnig* et al. [15.95] developed the concept for atomic force microscopy (AFM). Figure 15.5 shows a schematic diagram of an AFM configured for constant force imaging using a tunneling tip to measure the displacement of the cantilevered beam. With the sample at a large distance from the tip of the cantilever, the cantilever rest position is determined by advancing the tunneling tip until a tunneling current is established. The tunneling tip is then withdrawn a small distance, determined by the amount of force desired on the sample. The z-feedback circuit on the sample piezo is then enabled, and the sample is moved towards the tip until the cantilever is pushed back to the point when tunneling begins again. The z-feedback circuit adjusts the z-position of the sample so as to keep the deflection (or force) of the cantilever beam constant while scanning the sample in the x and y direction. The z-feedback voltage as a function of x and y is plotted by a computer, thereby creating a constant force image or map of the sample surface. Alternatively, the AFM can be configured in a variable force imaging mode. The tunneling tip piezo (now under feedback control) is

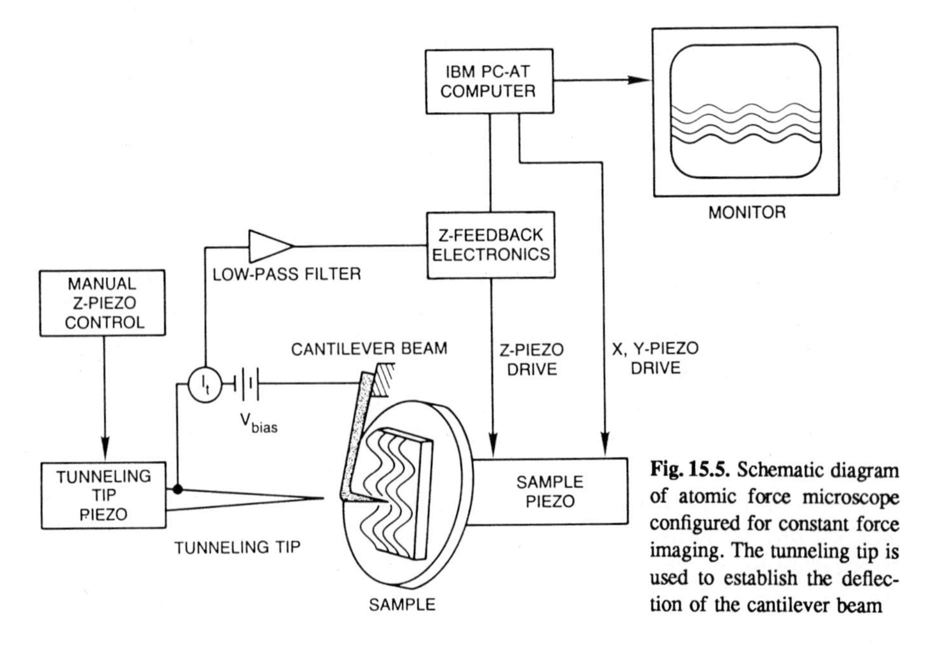

Fig. 15.5. Schematic diagram of atomic force microscope configured for constant force imaging. The tunneling tip is used to establish the deflection of the cantilever beam

used to measure the deflection of the cantilever beam as it moves up and down with sample corrugation while scanning the sample.

Because the tunneling tip is very sensitive to small displacement, it can be used to detect sub-Ångstrom deflections of the cantilever beam. The force is determined by multiplying the measured displacement by the calibrated spring constant of the cantilever. If the beam is either microfabricated or a fine wire, then very small forces are sufficient to bend it. Table 15.1 gives some examples of the force sensing capability of cantilever beams made from different-sized and -shaped materials. These AFM measurable forces are comparable to the forces associated with chemical bonding, e.g., $\sim 10^{-7}$ N for an ionic bond, $\sim 10^{-11}$ N for a van der Waals bond, and $\sim 10^{-12}$ N for surface reconstruction.

Table 15.1. Physical characteristics of some cantilever beams used in AFM

Material	*Dimensions* [mm]	ν [kHz]	k [N/m]	*Force Sensitivity* [N][a]
W wire	5×0.05 dia.	5	105	$\sim 10^{-9}$
Au wire	5×0.05 dia.	2	25	$\sim 10^{-10}$
SiO_2 bar	$0.2 \times 0.02 \times 0.002$	40	0.2	$\sim 10^{-12}$
Si bar	$2\mu m \times 0.5\mu m \times 0.1\mu m$	33 Mhz	2.5	$\sim 10^{-11}$

[a] Assuming a cantilever displacement of 0.1 Å

The resonant frequency ν and spring constant k for cantilever beams are two important properties of the cantilever that must be known when designing the force sensing cantilever. The principle design goal is to maximize ν while minimizing k. These properties of the beam are dependent on its physical size and on the materials from which it is made. The resonant frequency should be high to increase the immunity of the beam to vibration; and the spring constant should be small to make the beam sensitive to small forces, but not too small such that it is overly susceptible to thermal vibration. (ac modulation of the beam improves its sensitivity and performance, especially for measuring weak attractive forces.) The general guidelines, therefore, for choosing a cantilever beam are that the beam should be *soft* (low modulus), *light* (low mass and density), *thin* and *short.* In Table 15.1, substituting gold wire for tungsten (of the same dimension) reduces the spring constant of the cantilever, but also its resonant frequency. Microfabricating a cantilever from silicon or silicon dioxide also reduces its spring constant and raises its resonant frequency due primarily to the smaller dimensions of the microfabricated beams.

The nature of the surface forces can be determined using the AFM; both attractive and repulsive forces can be measured as well as the adhesive force necessary to separate the cantilever tip and sample surface, once in contact. Figure 15.6 illustrates the force measurement. The curve depicts the net forces acting between the cantilever tip and sample as a function of the separation s between cantilever tip and sample. The arrows are used to guide the eye through the full interaction cycle. The cycle starts with the sample far away and the

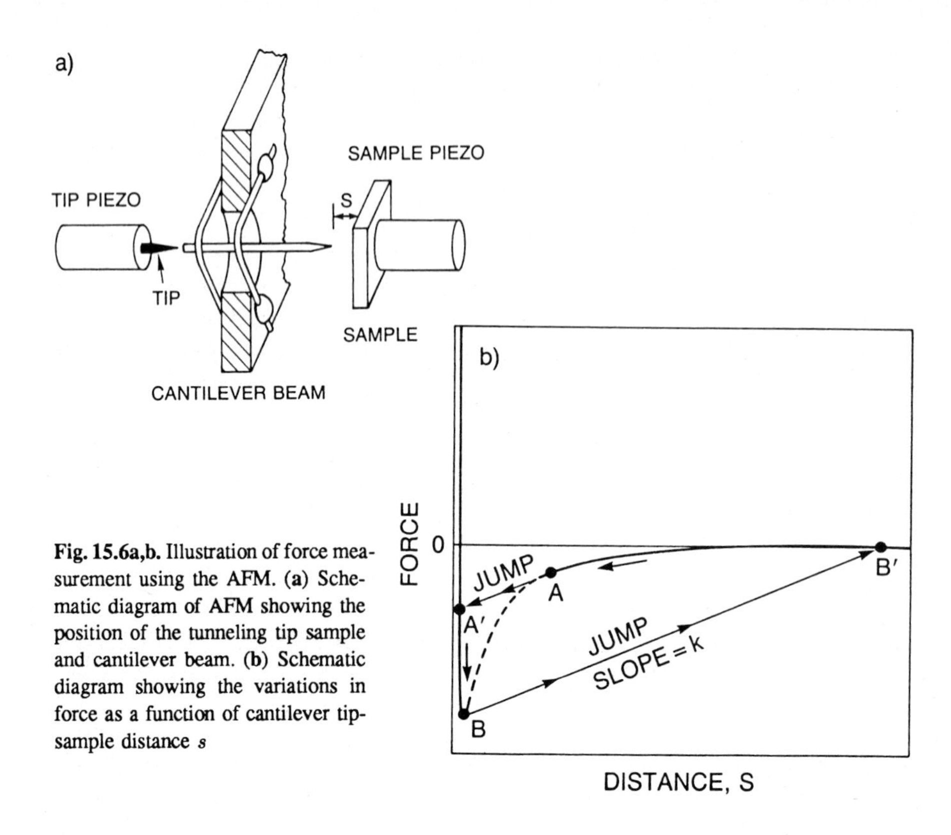

Fig. 15.6a,b. Illustration of force measurement using the AFM. (**a**) Schematic diagram of AFM showing the position of the tunneling tip sample and cantilever beam. (**b**) Schematic diagram showing the variations in force as a function of cantilever tip-sample distance s

cantilever in its rest position. As s decreases, the cantilever bends towards the sample such that at any equilibrium separation s, the attractive force exactly balances the restoring force of the cantilever defined by its spring constant k. However, once the gradient of the attractive force dF/ds exceeds k (point A), an instability occurs causing the cantilever to jump into contact with the sample (point A'). On reversing the direction of the sample, the cantilever will jump apart at B to some point B'. The force of adhesion (the force at point B) is found by simply multiplying the $B-B'$ jump distance by k. The force of attraction, on the other hand, is much more difficult to determine since its actual magnitude and distance dependence is masked once the surfaces jump into contact. The net force measured at A' has both attractive and repulsive components. In addition, we are unable to measure the separation s directly with the AFM, but can only measure the change in s during attraction up to the point of the jump into contact. Once in contact, the tip-sample distance can be taken to be $s_0 \sim 0.2\,\mathrm{nm}$, the interfacial separation or distance of closest approach.

Binnig et al. [15.95] demonstrated that this approach to stylus profilometry (i.e., tip in contact with surface) could produce surface images with atomic resolution of graphite. Since then, several other groups have reproduced that result and extended the results to other lamellar compounds [15.96–99]. Images of non-lamellar material surfaces have also been made, but with resolution nearer to 1–2 nm [15.100–102]. Most of these images were obtained with the cantilever tip in contact with the sample surface and under a repulsive force of typically 10^{-5} to 10^{-7} N. Under these loads, the tip-surface contact pressure is so high that single atom contact is impossible [15.103, 104]. A considerable amount of tip-sample deformation (elastic and possibly plastic) can be expected, resulting in a large, but finite, area of contact [15.105, 106].

A method employed by *Hansma* and coworkers [15.107] permits the imaging of delicate biological samples, even though the tip is in repulsive contact with the sample. They have observed that the repulsive force associated with cantilever tip-sample contact in air can be reduced substantially by placing the cantilever and sample in water. For example, they have been able to reduce the repulsive force from 10^{-7} N in air to 10^{-9} N in water, and believe that the forces can be further reduced by using cantilevers with low spring constants, by using more thermally stable microscopes, and by using other fluid systems. Calculations show that the force should not exceed 10^{-11} N when imaging biological surfaces.

As a means of measuring cantilever displacement, the tunneling tip is exquisitely sensitive to a very small area on the beam; this can lead to unwarranted sources of noise. Optical [15.102, 108–110] and capacitance [15.111] techniques are also capable of sensing small displacements; they are frequently easier to implement and sample a large beam area. Many new AFM designs utilize them.

If the atomic force microscope tip is a conductor, then one can perform STM and AFM simultaneously [15.112, 113]. This has a powerful advantage. The tunneling tip measures displacements very sensitively, but does not provide an absolute calibration of the tip-surface separation. In particular, if an insulating film covers a surface, say an oxide on a metal, then tunneling current may not be acquired until the tip is poked through part of the oxide. Physical contact with the surface is readily detected by the AFM and can be used to benchmark the position of contact.

Several groups [15.113–117] interested in adhesive forces and mechanical behavior of surfaces are exploiting the force microscope ideas. The AFM can be used to examine elastic/plastic flow of a surface with true nanometer resolution. This is important to surface modification/thin film technologies where the nanometer thickness encompasses the entire zone of interest. The atomic force microscope can be used to examine shear as well as normal forces. *Mate, McClelland* and coworkers [15.97, 106, 118] have shown that friction coefficients can be measured with the AFM. In their work on graphite and mica surfaces, they have observed the very interesting behavior shown in Fig. 15.7. Stick-slip phenomena are well known in tribology; contacting surfaces bind, freezing motion until the building force finally breaks the surfaces free. Just this kind of behavior is observed with the AFM, except that the slip distances are multiples of the lamellar substrate unit cell. Whether the AFM stick-slip is in any way related to its macroscopic cousin is unknown. It is clear, however, that the AFM is a powerful new approach to the study of tribology (Chap. 18), especially boundary lubrication and asperity behavior.

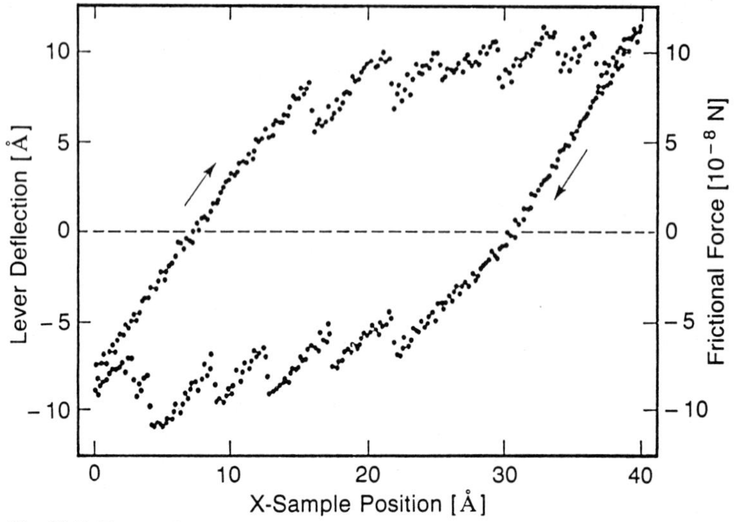

Fig. 15.7. Force microscope lever deflection in the x (in plane) direction and the corresponding frictional force as a Muscovite mica surface is moved first in one direction, then in the reverse direction, under a tip with normal force loading estimated to be 10^{-6} N [15.118]

b) Force Microscopy. The AFM described above makes contact with the surface; it is also possible to detect small cantilever displacements caused by attractive forces between substrate and cantilever [15.119–122]. As it is usually practiced, a tip at the end of a wire cantilever is vibrated at the resonant frequency of the cantilever. A laser interferometer (or other method) measures the amplitude of the ac vibration. The gradient of the force between the tip and sample modifies the compliance of the lever, hence inducing a change in vibration amplitude due to the shift of the cantilever resonance frequency. Measuring the vibration amplitude as a function of the tip-sample separation allows one to deduce the gradient of the force and the force itself. However, in order to avoid tip-sample contact, the attractive forces are measured at a tip-sample separation of 3 to 20 nm. As a result, the image resolution is not atomic. Several variants have been demonstrated, including the mapping of magnetic structure [15.120, 123–127] and surface potentials [15.119, 128, 129]. This capability provides powerful new approaches to the investigation of magnetic films [15.130–132] and integrated circuits.

c) Surface Force Apparatus. There is a different sort of proximal probe, the surface force apparatus [15.17, 133, 134] that has been developed independently. In this technique (Fig. 15.8) two crossed cylinders are brought close together. One cylinder is cantilevered; contact forces are measured by the amount it bends. Optical interferometry can be used to measure directly the separation of the two mating surfaces. This force microscope variant does not allow for imaging with nanometer resolution. It does however provide a more direct measure of the

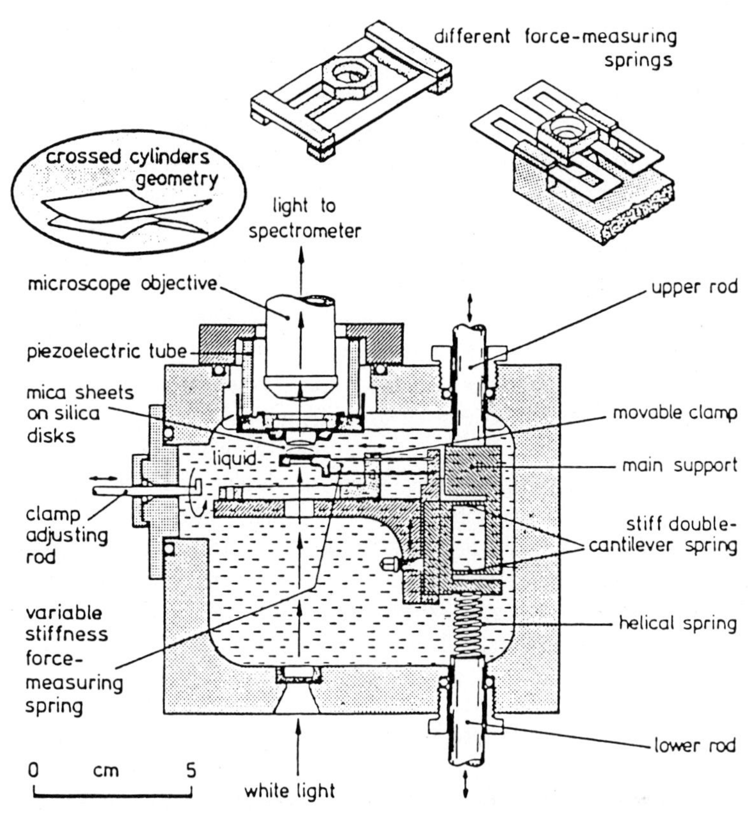

Fig. 15.8. Schematic diagram of the surface force apparatus that employes the crossed cylinders geometry [15.133]

separation, something that the tip-flat configuration of the force microscope does not allow.

Results from the surface force apparatus have been as equally startling as from the AFM. The attractive forces between the two surfaces have been measured in vacuum and found to conform to classical dispersive force models. However, when liquid molecules are juxtaposed, differences occur [15.133]. Large oscillations in the surface forces appear. The oscillations are presently attributed to structure in the "liquid" molecule boundary layers. The surface force apparatus has also been modified to examine shear forces [15.134, 135], also with tantalizing results. When the mating surfaces are within several "liquid" molecule diameters of each other, the shear force seems to have discrete values determined by the surface separation. Structure in the boundary layer is also invoked to explain the result.

15.1.4 Near-Field

a) Radiation. When two surfaces are brought closer than the wavelength of a photon, the normal far-field treatments of electromagnetic radiation are invalid. For instance, conventional diffraction theory is a far-field phenomenon; images can be obtained below the diffraction limit of light if a small aperture is located close (near-field) to the surface [15.136, 137]. Present demonstrated resolution is < 50 nm; estimates suggest ultimate resolution of about 10 nm using visible light. Near Field Optical Microscopy (NFOM) is being explored with transmission [15.136], reflection [15.136, 137] and fluorescence [15.137] approaches.

The macroscopic description of radiation transport of energy, described by the Stefan-Boltzmann law, is also a far-field description. When surfaces are close (but not in contact) the energy flux may be considerably higher [15.138]. The physics of this process is believed to be due to evanescent waves and provides new mechanisms to map surface properties. A recent variant of NFOM makes use of this phenomenon to provide contrast via what can be called "optical tunneling" [15.139]. *Williams* and *Wickramasinghe* [15.16, 140] and *Dransfeld* [15.138] have shown that a microfabricated thermocouple tip can map local surface features when a heated tip is scanned over a surface. If the tip penetrates the evanescent wave field which extends about 0.1 micron above the surface, thermal accommodation is rapid.

b) Other Concepts. Surface temperatures can also be investigated through the thermoelectric effect. The tip and surface can form a thermocouple-like junction through the tunnel barrier. Tip-sample temperature differentials can be sensed as the tip is scanned over the surface. Near atomic resolution is anticipated [15.16]. There is a recent paper describing the imaging of a single paramagnetic electron spin site on an oxidized Si/Si interface with a tunneling tip [15.141]; the mechanism of contrast is not yet clear, but one suggestion is the interaction of the precessing spin field with the tunneling current. *Hansma* et al. [15.142] has developed yet another technique: Scanning Ion Conductance Microscopy, which depends on the transport of ions. A hollow tip with a small aperture is placed proximal to a surface to be imaged. Electrodes are placed inside the tip and in the fluid media surrounding it; ion conductance along the electric field lines between the two electrodes is modulated by the proximal surface constraining the diffusion path.

15.2 Nanoscale Fabrication Using Proximal Probes

The availability of analytical tools which measure site specific properties at the nanometer scale can be exploited only in-so-far as one can fabricate structures worthy of study. Traditional microfabrication techniques are commercially viable at the 1 micron scale; laboratory techniques are presently focussed on solving problems with fabricating 0.1–0.5 micron features. Below this scale, the physical principles governing present electronics devices are compromised by interface and size effects. From a short-term commercial perspective, that reduces the incentive to aggressively pursue smaller structures since it will likely be many years before we learn how to make them, characterize their new and novel properties, and learn to exploit them in a new generation of devices. On the other hand, from a scientific perspective, structures $< 100\,\mathrm{nm}$ are the key to opening exciting new fields of research, with the long term certainty of future generations of electronic, magnetic, optic, superconductive, and chemical devices. In addition, the same fabrication and characterization tools may make Feynman's dream of nanotechnology [15.143, 144] a reality. Microfabrication of STM [15.145] and AFM [15.107] is already in process.

Lithographic techniques (X-ray [15.146, 147] and lens focussed electron [15.6, 148] and ion [15.6, 149] beams) developed for present electronics technology have been extended below 0.1 micron. The lens focussed beams require reasonably high energies (1–50 keV) to reduce the effects of chromatic abberation. The higher energies also enable uniform (in depth) exposure of resist materials, but contribute problems of proximity effect and beam damage [15.6]. Some of the proximal probe advantages for characterizing surface properties were discussed above; there are parallel advantages for their use in fabricating nanometer structures. Proximal liquid metal ion beam sources are being explored for etching and deposition [15.150], but this work is not widespread and will not be examined in any more detail here. The use of proximal tips as an electron beam source or a mechanical tool has wider and growing interest. Proximal field emission tips can produce low energy (1–1000 eV) electron beams with large current densities in relatively small beam diameters [15.90, 91]. This has the advantages of reducing the proximity effects in a resist (fewer backscattered and secondary electrons with sufficient energy to affect bonds) [15.151], less beam damage in the substrate (electrons do not penetrate as far), and the capability to vary the beam energy through a range where bond breaking/making thresholds might be expected.

A principle concern with focussed electron beam fabrication of these small dimensions is writing speed. If a structure of $10^2\,\mathrm{nm}^2$ can be written in a second, an area of $1\,\mathrm{cm}^2$ will require $10^{12}\,\mathrm{s}$ ($\sim$ 32 centuries). Parallel writing heads can be projected to reduce this time; field emission arrays have microfabricated tip densities of 10^7–$10^8\,\mathrm{cm}^{-2}$. However, it is still clear that writing speeds will have to be very fast. This argues for the electron beam to be used as an activator (requiring few chemical bond alterations), followed by a chemical processing step where the activated sites are processed (in parallel) into the desired structure or resist pattern; this concept has yet to be pursued aggressively.

A synopsis of published research in nanofabrication via proximal probes is detailed below.

Lithography: PMMA [15.152] and CaF_2 [15.153] have been explored with film thicknesses on the order of 10 nm and greater; the resulting structures are on the same size scale. More recently, a Langmuir Blodgett deposited polydiacetylene material has been explored with initially promising results, but with structures still larger than 10 nm [15.154]. Langmuir Blodgett techniques have generally been more successful for the deposition of very thin resist films than has spin casting. In related work, several groups are exploring the modification of organic [15.155–158] and polymeric materials [15.159, 160] by voltage pulses on a tip; while their research is oriented toward understanding the phenomena (dielectric breakdown, quantum excitation, etc.), the results can be applied to lithography.

Direct Writing: The first example of electron beam direct writing with a proximal probe was the deposition of carbon patterns through contamination writing [15.161]. Subsequent work with OMCVD of metals [15.162–164] has shown features of $> 10\,\mathrm{nm}$ size, and the metals have been high in carbon content. Deposition by a voltage pulse on a tip submerged in a fluid phase has been observed [15.155], but the identification of the species is uncertain. There is one report which shows convincing evidence for the controlled deposition of single atoms [15.165].

Direct Etch: It is known that high energy electrons will stimulate chemical etching of carbon by water. Evidence for this has been demonstrated using the proximal electron sources as well [15.166–168]; feature sizes near 5 nm have been observed [15.166]. These features have been shown [15.168] to trap Au atoms deposited on a graphite surface. Evidence of electron enhanced chemical etching of Si by WF_6 (presumably by F) has been presented by *Ehrichs and de Lozanne* [15.169].

In electrolytic media one expects electrochemical etching to be viable. *Lin, Fan* and *Bard* [15.170] have demonstrated fine line etching in GaAs ($\sim 0.3\,\mu\mathrm{m}$); electrodeposition and etching have also been shown with solid electrolytes [15.171].

Surface Modification: Sufficient power may be deposited in the surface layer by a proximal probe to cause surface melting. In metallic glasses, where the electron inelastic mean free path is short, surface melting has been observed, followed by structural changes associated with Taylor cone formation [15.172–174]. Features as small as 3 nm have been observed and complex patterns written [15.174]. *Nagahara* et al. [15.175] and *Emch* et al. [15.176] have reported features induced on a gold surface by a voltage pulse; the origin of the effect is unknown.

Mechanical contact of the probe and surface is an ever present danger with proximal probes. A number of groups have sought to deliberately exploit this possibility to 'machine' the surface. CaF_2 was shown to be readily machined by *McCord* and *Pease* [15.177], who even obtained pictures of a shaving. Several groups have shown the ability to produce indents into soft metal surfaces [15.178–181]; the lifetime of these indents can be highly limited by surface diffusion [15.178]. *Packard* et al. [15.180] illustrate surface features characterized as 'sanded' and 'swept' following the raster of a tip in contact with a gold surface.

15.3 Conclusion

The proximal probes are a suite of analytical tools sensitive to nanometer scale structures. Their availability provides tremendous research and development opportunities. Nanometer scale structures are a key in defining the science base necessary to solve many venerable technological problems – corrosion, tribology, fracture, fatigue and adhesion most notably. These technologies have long histories of empirical progress (some dating back 4000 years in recorded history),

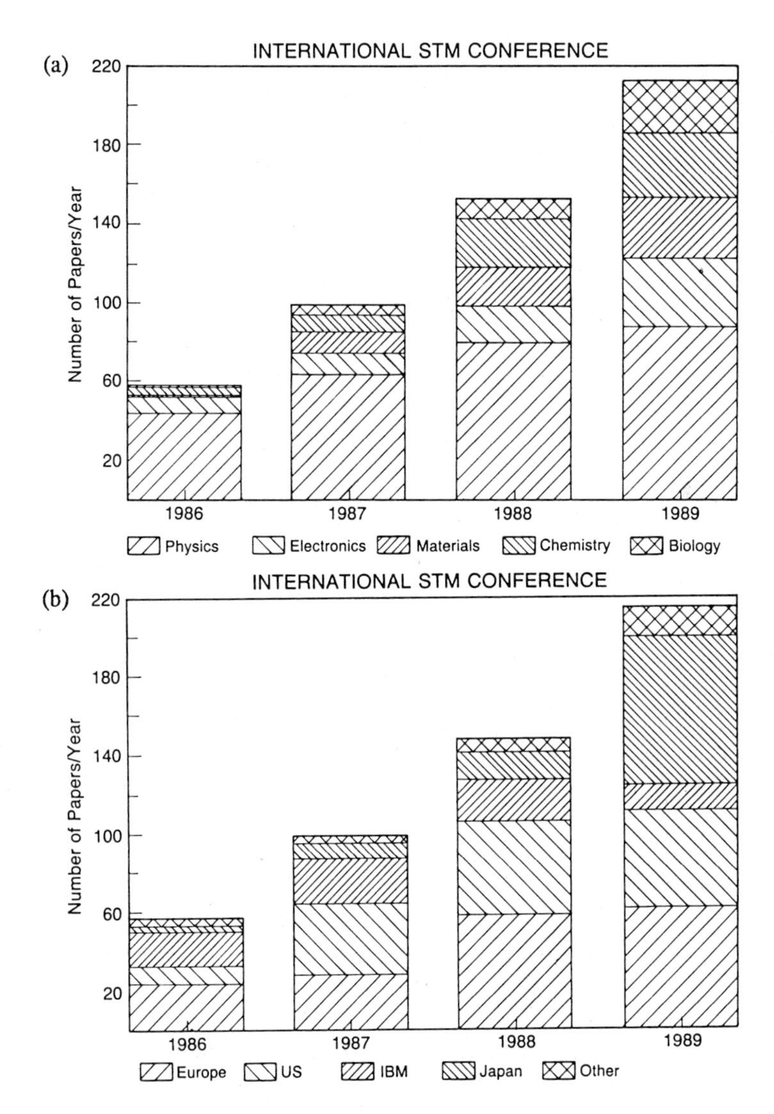

Fig. 15.9a,b. Graph of STM/AFM and related publications showing the rapid growth rate delineated by: (**a**) discipline and (**b**) country of origin. IBM is whimsically defined as a country to emphasize the major role its scientists have played in this field

empirical largely because one could not probe the phenomena on the relevant scale. Nanometer scale structures also tantalize us with the possibility of new science concepts and novel material properties: (1) Neither the force laws/models developed for atomic/molecular systems, nor those developed for solid states, are believed adequate to model the effects found at the nanometer scale; and (2) Structures of nanometer size go below critical scale lengths found in many phenomena: near-field electromagnetic radiation, superconductivity, ballistic electron transport, interfacial electronic states, dislocation motion, etc. Three dimensional nanometer structures thereby open intriguing possibilities for novel electronic, magnetic, optical, mechanical and chemical phenomena.

Progress in the development of analytical tools for the nanometer scale is moving at a breathtaking rate. Figure 15.9a shows the growth in the form of papers presented at the annual International Conference on Scanning Tunneling Microscopy/Spectroscopy. It is clear that the proximal probes are opening the door to an exciting new era in science. Many disciplines are and will be exploiting the opportunities to investigate the behavior of nanometer structures. Figure 15.9b is of interest because the proximal probe will be an essential part of technology development. Countries more proficient in that art will likely have distinct advantages in the commercial marketplace of the twenty-first century.

References

15.1 L.C. Feldman, J.W. Mayer: *Fundamentals of Surface and Thin Film Analysis* (North Holland, NY 1986)
15.2 A.W. Czanderna (ed.): *Methods of Surface Analysis* (Elsevier, NY 1975)
15.3 D. Briggs, M.P. Seah (eds.): *Practical Surface Analysis by Auger and X-ray Photoelectron Spectroscopy* (Wiley, NY 1983)
15.4 A. Benninghoven, F.G. Rüdenauer, H.W. Werner: *Secondary Ion Mass Spectrometry: Basic Concepts, Instrumental Aspects, Applications and Trends* (Wiley, New York 1987)
15.5 N.H. Turner: "Surface Analysis: X-ray Photoelectron Spectroscopy and Auger Electron Spectroscopy", Anal. Chem. **60**, 337R (1988)
15.6 T.G.P. Chang, D.P. Kern, E. Kratschmer, K.Y. Lee, H.E. Luhn, M.A. McCord, S.A. Rishton, Y. Vladimirsky: IBM Journal of Research and Development **32**, 462 (1988)
15.7 A. Franks: J. Phys. E. **20**, 1442 (1987)
15.8 U. Köhler, J.E. Demuth, R.J. Hamers: J. Vac. Sci. Technol. **A7**, 2860 (1989)
15.9 A.J. Hoeven, D. Dijkkamps, E.J. Van Loenen, J.M. Lenssinck, J. Dielemann: J. Vac. Sci. Technol. **A8**, 207 (1990)
15.10 St. Tosch, H. Neddermeyer: J. Microsc. **152**, 415 (1988)
15.11 E. Ritter, R.J. Behm, G. Potschke, J. Wintterlin: Surf. Sci. **181**, 403 (1987)
15.12 X.-S. Wang, R.J. Phaneuf, E.D. Williams: J. Microsc. **152**, 473 (1988)
15.13 Ph. Avouris, R. Wolkow: Phys. Rev. **B39**, 5091 (1989)
15.14 R.J. Behm, W. Hösler: In *Chemistry and Physics of Solid Surfaces VI*, ed. by R. Vanselow, R. Howe, Springer Ser. Surf. Sci. Vol. 5 (Springer, Berlin, Heidelberg 1986) p.361
15.15 J.P. Rabe: Angew. Chem. **101**, 117 (1989)
15.16 H.K. Wickramasinghe: Scientific American (October), 98 (1989)
15.17 J.N. Israelachvili: *Intermolecular and Surface Forces* (Academic, London 1985)
15.18 P.K. Hansma, V.B. Elings, O. Marti, C.E. Bracker: Science **242**, 157 (1988)
15.19 G. Binnig, H. Rohrer, Ch. Gerber, E. Weibel: Phys. Rev. Lett. **50**, 120 (1983)
15.20 G. Binnig, H. Rohrer: Helv. Physica Acta **55**, 726 (1982)
15.21 Y. Kuk, P.J. Silverman: Rev. Sci. Instrum. **60**, 165 (1989)

15.22 J.S. Murday, R.J. Colton: J. Mater. Sci. and Engn. **B**, **6**, 77 (1990)
15.23 J. Gimzewski: Physics World **2** (Aug), 25 (1989)
15.24 C.B. Duke: *Tunneling in Solids* (Academic, New York 1969)
15.25 E. Burstein, S. Lundqvist (eds.): *Tunneling Phenomena in Solids* (Plenum, New York 1969)
15.26 T. Wolfram (ed.): *Inelastic Electron Tunneling Spectroscopy*, Springer Series in Solid-State Science, Vol. 4 (Springer, Berlin, Heidelberg 1978)
15.27 P.K. Hansma (ed.): *Tunneling Spectroscopy* (Plenum Press, New York 1982)
15.28 E.L. Wolf: *Principles of Electron Tunneling Spectroscopy* (Oxford University Press, New York 1985)
15.29 L. Esaki: Phys. Rev. **109**, 603 (1958)
15.30 I. Giaver: Phys. Rev. Lett. **5**, 147 (1960)
15.31 R.C. Jaklevic, J. Lambe: Phys. Rev. Lett. **17**, 1139 (1966)
15.32 D.G. McDonald: Phys. Today **34**, 37 (1981)
15.33 R.D. Young: Phys. Today **21**, 42 (1972); R.D. Young, J. Ward, F. Scire: Phys. Rev. Lett. **27**, 922 (1971); R.D. Young, J. Ward, F. Scire: Rev. Sci. Instrum. **43**, 999 (1972)
15.34 G. Binnig, H. Rohrer: Rev. Modern Phys. **59**, 615 (1987)
15.35 P.K. Hansma, J. Tersoff: J. Appl. Phys. **61**, R1 (1987)
15.36 J.E. Demuth: In *Physics in a Technological World*, ed. by A.P. French, American Inst. of Phys., New York 1988, p. 141
15.37 R.M. Tromp: In *Chemistry and Physics of Solid Surfaces VII*, ed. by R. Vanselow, R. Howe, Springer Ser. Surf. Sci. Vol. 10 (Springer, Berlin, Heidelberg 1988) p.547
15.38 J.A. Golovchenko: Science **232**, 48 (1986)
15.39 Proc. of the First International Conference on Scanning Tunneling Microscopy, Surf. Sci. **181** (March, 1987); Proc. of the Second International Conference on Scanning Tunneling Microscopy, J. Vac. Sci. Technol. **A6** (March/April, 1988); Proc. of the Third International Conference on Scanning Tunneling Microscopy, J. Microscopy 152 (Oct/Nov/Dec, 1988); Proc. of the Fourth International Conference on Scanning Tunneling Microscopy, J. Vac. Sci. Technol. **A8** (Jan/Feb, 1990); Proc. of the Topical Conference on Probing the Nanometer Scale Properties of Surfaces and Interfaces, J. Vac. Sci. Technol. **A7** (July/Aug, 1989); Proc. of the Adriatico Research Conference on Scanning Tunneling Microscopy, Physica Scripta **38** (August 1988)
15.40 Y.-C. Cheng, P.J. Bryant: "Bibliography of Scanning Tunneling Microscopy/Spectroscopy and Atomic Force Microscopy", UMKC-Physics; 1110 E. 48th St.; Kansas City, MO 64110
15.41 J.G. Simmons: J. Appl. Phys. **34**, 2581 (1963)
15.42 J. Tersoff, D.R. Hamann: Phys. Rev. Lett **50**, 1998 (1983); Phys. Rev. **B31**, 805 (1985)
15.43 Y. Kuk, P.J. Silverman, H.Q. Nguyen: J. Vac. Sci. Technol. **A6**, 524 (1988)
15.44 J. Wintterlin, J. Wiechers, Th. Gritsch, H. Höfer, R.J. Behm: J. Microsc. **152**, 423 (1988)
15.45 U. Dürig, O. Züger, D.W. Pohl: J. Microsc. **152**, 259 (1988)
15.46 R. Hamers, R. Tromp, J. Demuth: Phys. Rev. Lett. **56**, 1972 (1986)
15.47 G. W. Stupian, M.S. Leung: J. Vac. Sci. Technol. **A7**, 2895 (1989)
15.48 N. Tsuda, H. Yamada, F. Ishida, M. Miyashita, R. Yamaguchi: Jap. J. Precis. Engn. **155**, 146 (1989)
15.49 J.W. Lyding, S. Skala, J.S. Hubacek, R. Brockenbough, G. Gammie: Rev. Sci. Instrum. **59**, 1897 (1988)
15.50 R. Sonnenfeld, P.K. Hansma: Science **232**, 211 (1986)
15.51 J. Schneir, O. Marti, G. Remmers, D. Glaser, R. Sonnenfeld, B. Drake, P.K. Hansma, V. Elings: J. Vac. Sci. Technol. **A6**, 283 (1988)
15.52 R.S. Robinson: J. Microsc. **152**, 541 (1988)
15.53 I. Otsuka, T. Iwasaki: J. Microsc. **152**, 289 (1988)
15.54 T. Twomey, J. Wiechers, D.M. Kolb, R.J. Behm: J. Microsc. **152**, 537 (1988)
15.55 M.J. Heben, M.M. Dovek, N.S. Lewis, R.M. Penner, C.E. Quate: J. Microsc. **152**, 651 (1988)
15.56 J.S. Foster, J.E. Frommer: Nature **333**, 542 (1988)
15.57 S.M. Lindsay, T. Thundat, L. Nagahara, U. Knipping, R.L. Rill: Science **244**, 1063 (1989)
15.58 G. Binnig, H. Fucks, E. Stoll: Surf. Sci. **169**, L295 (1986)
15.59 R.H. Koch, R.J. Hamers: Surf. Sci. **181**, 333 (1987)
15.60 J.E. Demuth, U. Köhler, R.J. Hamers: J. Microsc. **152**, 299 (1988)
15.61 J.P. Ibe, P.P. Bey, Jr., S.L. Brandow, R.A. Brizzolara, N.A. Burnham, D.P. DiLella, K.P. Lee, C.R.K. Marrian, R.J. Colton: J. Vac. Sci. Technol. **A8** July/August (1990)
15.62 M. Gehrtz, H. Strecker, H. Grimm: J. Vac. Sci. Technol. **A6**, 432 (1988)

15.63 R.A. Dragoset, R.D. Young, H.P. Layer, S.R. Mielezanek, E.C. Teague, R.J. Celotta: Opt. Lett. **11**, 560 (1986)
15.64 D.R. Denley: J. Vac. Sci. Technol. **A8**, 603 (1990)
15.65 Y. Miyazaki, Y. Koga, H. Hayashi: J. Vac. Sci. Technol. **A8**, 628 (1990)
15.66 M. Green, M. Richter, J. Kortright, T. Barbee, R. Carr, I. Lindau: J. Vac. Sci. Technol. **A6**, 428 (1988)
15.67 M. Anders, M. Thaer, M. Mück, C. Heiden: J. Vac. Sci. Technol. **A6**, 436 (1988)
15.68 S. Okayama, M. Komuro, W. Mizutani, H. Tokumoto, M. Okana, K. Shimizu, Y. Kobayashi, F. Matsumoto, S. Wakiyama, M. Shigeno, F. Sakai, S. Fujiwara, O. Kitamura, M. Ono, K. Kajimura: J. Vac. Sci. Technol. **A6**, 440 (1988)
15.69 B.A. Sexton, G.F. Cotterill: J. Vac. Sci. Technol. **A7**, 2734 (1989)
15.70 J.A. Stroscio, R.M. Feenstra, A.P. Fein: Phys. Rev. Lett. **57**, 2579 (1986)
15.71 R.M. Feenstra, J.A. Stroscio, A.P. Fein: Surf. Sci. **181**, 295 (1987)
15.72 N.D. Lang: Phys. Rev. **B34**, 5947 (1986); **B37**, 10395 (1988)
15.73 R.J. Hamers, U.K. Köhler: J. Vac. Sci. Technol. **A7**, 2854 (1989)
15.74 R.S. Becker, B.S. Swartzentruber, J.S. Vicker, M.S. Hybertsen, S.G. Louie: Phys. Rev. Lett. **60**, 116 (1988)
15.75 R.M. Tromp, R.J. Hamers, J.E. Demuth: Phys. Rev. **B34**, 1388 (1986)
15.76 W.J. Kaiser, R.C. Jaklevic: Surf. Sci. **181**, 55 (1987)
15.77 R.M. Feennstra, J.A. Stroscio: J. Vac. Sci. Technol. **B5**, 923 (1987)
15.78 J.A. Stroscio, R.M. Feenstra, D.M. Newns, A.P. Fein: J. Vac. Sci. Technol. **A6**, 499 (1988)
15.79 H.G. Le Duc, W.J. Kaiser, J.A. Stern: Appl. Phys. Lett. **50**, 1921 (1987)
15.80 R. Berthe, U. Hartmann, C. Heiden: J. Microsc. **152**, 831 (1988)
15.81 J.R. Kirtley, C.C. Tsuei, S.I. Pack, C.C. Chi, J. Rozen, M.W. Shafer: Phys. Rev. **B35**, 7216 (1987)
15.82 S. Pan, K.W. Ng, A.L. de Lozanne, J.M. Tarascon, L.H. Greene: Phys. Rev. **B35**, 7220 (1987)
15.83 M.E. Hawley, K.E. Gray, D.W. Capone II, O.G. Hinks: Phys. Rev. **B35**, 7224 (1987)
15.84 M. Naito, D.P.E. Smith, M.D. Kirk, B. Oh, M.R. Hahn, K. Chai, D.B. Mitzi, J.Z. Sun, D.J. Webb, M.R. Beasley, O. Fischer, T.H. Geballe, R.H. Hammond, A. Kapitulnik, C.F. Quate: Phys. Rev. **B35**, 7228 (1987)
15.85 H. Heinzelmann, D. Anselmetti, R. Wiesendanger, H.-R. Hidber, H.-J. Güntherodt, M. Düggelist, R. Auggenheim, H. Schmidt, G. Güntherodt: J. Microsc. **152**, 399 (1988)
15.86 R. Laiko, M. Aarnio, L. Heikkilä, H. Snellman, I. Kirschner: J. Microsc. **152**, 407 (1988)
15.87 H.F. Hess, R.B. Robinson, R.C. Dynes, J.M. Valles, Jr., J.V. Waszczak: Phys. Rev. Lett. **62**, 214 (1989)
15.88 W.J. Kaiser, L.D. Bell: Phys. Rev. Lett. **60**, 1406 (1988); **61**, 2368 (1988)
15.89 J.H. Coombs, J.K. Gimzewski: J. Microscopy **152**, 841 (1988)
15.90 M.A. McCord, R.F.W. Pease: J. Vac. Sci. Technol. **B3**, 198 (1985)
15.91 H.-W. Fink: IBM J. Res. Devel. **30**, 460 (1986); Physica Scripta **38**, 260 (1988)
15.92 R. Allenspach, A. Bischof: Appl. Phys. Lett. **54**, 587 (1989)
15.93 B. Reihl, J.K. Gimzewski: Surf. Sci. **189**, 36 (1987)
15.94 J.H. Coombs, J.K. Gimzewski, B. Reihl, J.K. Sass, R.R. Schlittler: J. Microscopy **152**, 325 (1988)
15.95 G. Binnig, C.F. Quate, Ch. Gerber: Phys. Rev. Lett. **56**, 930 (1986); G. Binnig, Ch. Gerber, E. Stoll, T.R. Albrecht, C.F. Quate: Surf. Sci. **189/190**, 1 (1987)
15.96 T.R. Albrecht, C.F. Quate: J. Appl. Phys. **62**, 2599 (1987); J. Vac. Sci. Technol. **A6**, 271 (1988)
15.97 C.M. Mate, G.M. McClelland, R. Erlandsson, S. Chiang: Phys. Rev. Lett. **59**, 1942 (1987)
15.98 O. Marti, B. Drake, P.K. Hansma: Appl. Phys. Lett. **51**, 484 (1987)
15.99 H. Heinzelmann, P. Grütter, E. Meyer, H. Hidber, L. Rosenthaler, M. Ringger, H.-J. Güntherodt: Surf. Sci. **189/190**, 29 (1987)
15.100 H. Heinzelmann, E. Meyer, P. Grütter, H.-R. Hidber, L. Rosenthaler, H.-J. Güntherodt: J. Vac. Sci. Technol. **A6**, 275 (1988)
15.101 H. Heinzelmann, E. Meyer, H.-J. Güntherodt: Surf. Sci. **221**, 1 (1989)
15.102 S. Alexander, L. Hellemans, O. Marti, J. Schneir, V. Elings, P. Hansma, M. Longmire, J. Gurley: J. Appl. Phys. **65**, 164 (1989)
15.103 J.B. Pethica: Phys. Rev. Lett. **57**, 3235 (1986)
15.104 J.B. Pethica, W.C. Oliver: Physica Scripta **T19**, 61 (1987)
15.105 H.J. Mamin, E. Ganz, D.W. Abraham, R.E. Thompson, J. Clarke: Phys. Rev. **B34**, 9015 (1986)

15.106 C.M. Mate, R. Erlandsson, G.M. McClelland, S. Chiang: Surf. Sci. **208**, 473 (1989)
15.107 A.L. Weisenhorn, P.K. Hansma, T.R. Albrecht, C.F. Quate: Appl. Phys. Lett. **54**, 2651 (1989)
15.108 Y. Martin, C.C. Williams, H.K. Wickramasinghe: J. Appl. Phys. **61**, 4723 (1987)
15.109 R. Erlandsson, G.M. McClelland, C.M. Mate, S. Chiang: J. Vac. Sci. Technol. **A6**, 266 (1988)
15.110 D. Rugar, H.J. Mamin, R. Erlandsson, J.E. Stern, B.D. Terris: Rev. Sci. Instrum. **59**, 2337 (1988)
15.111 G. Neubauer, S.R. Cohen, G.M. McClelland: In *Interfaces between Polymers, Metals, and Ceramics*, ed. by B.M. DeKoven, A.J. Gellman, R. Rosenberg, MRS Proceedings, 1989
15.112 U. Dürig, J.K. Gimzewski, D.W. Pohl: Phys. Rev. Lett. **57**, 2403 (1986)
15.113 P.J. Bryant, R.G. Miller, R. Yang: Appl. Phys. Lett. **52**, 2233 (1988)
15.114 J.B. Pethica, A.P. Sutton: J. Vac. Sci. Technol. **A6**, 2490 (1988)
15.115 N.A. Burnham, R.J. Colton: J. Vac. Sci. Technol. **A7**, 2906 (1989)
15.116 N.A. Burnham, D.D. Dominguez, R.L. Mowery, R.J. Colton: Phys. Rev. Lett. **64**, 1931 (1990)
15.117 U. Landman, W.D. Luedtke, M.W. Ribarsky: J. Vac. Sci. Technol. **A7**, 2829 (1989)
15.118 R. Erlandsson, G. Hadziioannou, C.M. Mate, G.M. McClelland, S. Chiang: J. Chem. Phys. **89**, 5190 (1988)
15.119 Y. Martin, C.C. Williams, H.K. Wickramasinghe: J. Appl. Phys. **61**, 4723 (1987)
15.120 Y. Martin, H.K. Wickramasinghe: Appl. Phys. Lett. **50**, 1455 (1987)
15.121 G.M. McClelland, R. Erlandsson, S. Chiang: In Review of progress in Quantitative Non-Destructive Evaluation, Vol. 6, Plenum, New York, NY 1987
15.122 U. Landman, W.D. Luedtke, A. Nitzan: Surf. Sci. **210**, L177 (1989)
15.123 P. Grütter, E. Meyer, H. Heinzelmann, L. Rosenthaler, H.-R. Hidber, H.-J. Güntherodt: J. Vac. Sci. Technol. **A6**, 279 (1988)
15.124 T. Göddenhenrich, U. Hartmann, M. Anders, C. Heiden: J. Microsc. **152**, 527 (1988)
15.125 U. Hartmann: J. Vac. Sci. Technol. **A8**, 411 (1990)
15.126 U. Hartmann, C. Heiden: J. Microsc. **152**, 281 (1988)
15.127 J.J. Saenz, N. Garcia, P. Grütter, E. Meyer, H. Heinzelmann, R. Wiesendanger, L. Rosenthaler, H.R. Hidber, H.-J. Güntherodt: J. Appl. Phys. **62**, 4293 (1987)
15.128 Y. Martin, D.W. Abraham, H.K. Wickramasinghe: Appl. Phys. Lett. **52**, 1103 (1988)
15.129 P. Muralt, H. Meier, D.W. Pohl, H.W.M. Salemink: Appl. Phys. Lett. **50**, 1352 (1987)
15.130 D.W. Abraham, C.C. Williams, H.K. Wickramasinghe: J. Microsc. **152**, 863 (1988)
15.131 H.J. Mamin, D. Rugar, J.E. Stern, B.D. Terris, S.E. Lambert: Appl. Phys. Lett. **53**, 1563 (1988)
15.132 Y. Martin, D. Rugar, H.K. Wickramasinghe: Appl. Phys. Lett. **52**, 245 (1988)
15.133 J.N. Israelachvili: J. Coll. Interf. Sci. **110**, 263 (1986)
15.134 J.N. Israelachvili, P.M. McGuiggan, A.M. Homola: Science **240**, 189 (1988)
15.135 J. Van Alsten, S. Granick: Phys. Rev. Lett. **61**, 2570 (1988)
15.136 U. Dürig, D.W. Pohl, F. Rohner: J. Appl. Phys. **59**, 3318 (1986); IBM J. Res. Develop. **30**, 478 (1986); D.W. Pohl, U.Ch. Fischer, U.T. Dürig: J. Microscopy **152**, 853 (1988)
15.137 A. Lewis, M. Isaacson, A. Harootunian, A. Muray: Ultramicroscopy **13**, 227 (1984)
15.138 K. Dransfeld: J. Microscopy **152**, 35 (1988)
15.139 U. Ch. Fischer, D.W. Pohl: Phys. Rev. Lett. **62**, 458 (1989)
15.140 C.C. Williams, H.K. Wickramasinghe: Appl. Phys. Lett. **49**, 1587 (1986)
15.141 Y. Manassen, R.J. Hamers, J.E. Demuth, A.J. Castellano Jr.: Phys. Rev. Lett. **62**, 2531 (1989)
15.142 P.K. Hansma, B. Drake, O. Marti, S.A.C. Gould, C.B. Prater: Science **243**, 641 (1989)
15.143 R.P. Feynman: "There's Plenty of Room at the Bottom", in *Miniaturization*, ed. by H.D. Gilbert (Reinhold, New York 1961), pp. 282–296
15.144 C. Schneiker, S. Hameroff, M. Voelker, J. He, E. Dereniak, R. McCuskey: J. Microsc. **152**, 585 (1988)
15.145 T.R. Albrecht, S. Akamine, M.J. Zdeblich, C.F. Quate: J. Vac. Sci. Technol. **A8**, 317 (1990)
15.146 A.D. Wilson: Solid State Technology, May, 249 (1986)
15.147 Nanostructure Technology in IBM Journal of Research and Development **32**, 440–514 (1988)
15.148 R.W. Devenish, D.J. Eaglesham, D.M. Maher, C.J. Humphreys: Ultramicroscopy **28**, 324 (1989)
15.149 P.G. Blauner, J.S. Ro, Y. Butt, J. Melngailis: J. Vac. Sci. Technol. **B7**, 609 (1989)
15.150 A.E. Bell, K. Rao, L.W. Swanson: J. Vac. Sci. Technol. **B6**, 306 (1988)
15.151 M.A. McCord, R.F.W. Pease: Surf. Sci. **181**, 278 (1987)
15.152 M.A. McCord, R.F.W. Pease: J. Vac. Sci. Technol. **B6**, 293 (1988)
15.153 M.A. McCord, R.F.W. Pease: J. Vac. Sci. Technol. **B5**, 430 (1987)
15.154 C.R.K. Marrian, R.J. Colton: Appl. Phys. Lett. **56**, 755 (1990)

15.155 J.S. Foster, J.E. Frommer, P.C. Arnett: Nature **331**, 324 (1988)
15.156 R. Kaneko, E. Hamada: J. Vac. Sci. Technol. **A8**, 577 (1990)
15.157 R.H. Bernhardt, G.C. McGonigal, R. Schneider, D.J. Thomson: J. Vac. Sci. Technol. **A8**, 667 (1990)
15.158 J.P. Rabe, S. Buchholz, A.M. Ritcey: J. Vac. Sci. Technol. **A8**, 629 (1990)
15.159 T.R. Albrecht, M.M. Dovek, C.A. Lang, P. Grütter, C.F. Quate, S.W.J. Kuan, C.W. Frank, R.F.W. Pease: J. Appl. Phys. **64**, 1178 (1988)
15.160 M.M. Dovek, T.R. Albrecht, S.W.J. Kuan, C.A. Lang, R. Emch, P. Grütter, C.W. Frank, R.F.W. Pease, C.F. Quate: J. Microscopy **152**, 229 (1988)
15.161 M.A. McCord, R.F.W. Pease: J. Vac. Sci. Technol. **B4**, 86 (1986)
15.162 R.M. Silver, E.E. Ehrichs, A.L. de Lozanne: Appl. Phys. Lett. **51**, 247 (1987)
15.163 E.E. Ehrichs, R.M. Silver, A.L. de Lozanne: J. Vac. Sci. Technol. **A6**, 540 (1988)
15.164 M.A. McCord, D.P. Kern, T.H.P. Chang: J. Vac. Sci. Technol. **B6**, 1877 (1988)
15.165 R.S. Becker, J.A. Golovchenko, B.S. Swartzentruber: Nature **325**, 419 (1987)
15.166 T.R. Albrecht, M.M. Dovek, M.D. Kirk, C.A. Lang, C.F. Quate, D.P.E. Smith: Appl. Phys. Lett. **55**, 1727 (1989)
15.167 J. Schneir, P.K. Hansma, V. Elings, J. Gurley, K. Wickramasinghe, R. Sonnenfeld: Proc. Spie. **892**, 16 (1988)
15.168 K. Terashima, M. Kondoh, T. Yoshida: J. Vac. Sci. Technol. **A8**, 581 (1990)
15.169 E.E. Ehrich, A.L. de Lozanne: J. Vac. Sci. Technol. **A8**, 571 (1990)
15.170 C.W. Lin, F.-R.F. Fan, A.J. Bard: J. Electrochem. Soc. **134**, 1038 (1987)
15.171 O.E. Hüsser, D.H. Craston, A.J. Bard: J. Vac. Sci. Technol. **B6**, 1873 (1988)
15.172 U. Staufer, R. Wiesendanger, L. Eng, L. Rosenthaler, H.R. Hidber, H.-J. Güntherodt, N. Garcia: Appl. Phys. Lett. **51**, 244 (1987)
15.173 U. Staufer, R. Wiesendanger, L. Eng, L. Rosenthaler, H.R. Hidber, H.-J. Güntherodt: J. Vac. Sci. Technol. **A6**, 537 (1988)
15.174 U. Staufer, L. Scandella, R. Wiesendanger: Z. Phys. B **77**, 281 (1989)
15.175 L. Nagahara, S.M. Lindsay, T. Thundat, U. Knipping: J. Microscopy **152**, 145 (1988)
15.176 R. Emch, J. Nogami, M.M. Dovek, C.A. Lang, C.F. Quate: J. Microscopy **152**, 129 (1988)
15.177 M.A. McCord, R.F.W. Pease: Appl. Phys. Lett. **50**, 569 (1987)
15.178 R.C. Jaklevic, L. Elie: Phys. Rev. Lett. **60**, 120 (1988)
15.179 G. Van de Walle, H. Van Kempen, P. Wyder: Surf. Sci. **167**, L219 (1986)
15.180 W.E. Packard, Y. Liang, N. Dai, J.D. Dow, R. Nicolaides, R.C. Jaklevic, W.J. Kaiser: J. Microscopy **152**, 715 (1988)
15.181 J. Schneir, R. Sonnenfeld, O. Marti, P.K. Hansma, J.E. Demuth, R.J. Hamers: J. Appl. Phys. **63**, 717 (1988)

16. Studying Surface Chemistry Atom-by-Atom Using the Scanning Tunneling Microscope

Phaedon Avouris and In-Whan Lyo

IBM Research Division, T.J. Watson Research Center
Yorktown Heights, NY 10598, USA

A solid surface provides a great variety of sites for a possible chemical reaction. A perfect surface may have several strucutrally inequivalent sites while a real life surface may have steps, a multitude of point and extended defect sites, adsorbed foreign atoms, etc. It is clear then that in order to even begin discussing the chemistry of a particular surface we have to have a good idea about its structure. This recognized need has led to the development of a great variety of surface structural techniques [16.1]. Most of these techniques provide information averaged over a macroscopic area of the sample defined by the size of the probe beam. Moreover, diffraction-based structural probes require long-range order which is not present in many cases of interest. Even when the structure of the surface is known, the understanding of the reactivities of the various sites requires that a correlation be made between the chemical behavior of a particular site and its local electronic structure. Thus, ideally one needs a surface technique or techniques which will allow the study of the topography and valence electronic structure of surfaces with atomic resolution.

In this chapter we will try to show that the techniques of scanning tunneling microscopy (STM) [16.2] and atom-resolved tunneling spectroscopy (ARTS) can, in favorable cases, go a long way towards achieving these goals. We will demonstrate these capabilities using examples taken from our own work on silicon surface chemistry. Chemical studies on metal surfaces using the STM are also beginning to appear [16.3, 4]. We will first review briefly our STM and ARTS methodology. Then we will discuss the different ways by which semiconductor surface chemistry can be studied on an atom-by-atom basis. Examples from two types of surface reactions will be discussed in some detail: (a) Reactions that preserve the original surface structure. Our goal in this case is to use STM and ARTS to determine site-selectivity in such reactions, and (b) Reactions that destroy the original structure. In this case we will show how STM and ARTS can be coupled with conventional techniques and theory to provide information about intermediate disordered phases and solve complex chemical problems.

16.1 Topography and Spectroscopy with the STM

The basic principles on which STM is based are reviewed in Chaps. 14 and 15 by *Tersoff* and by *Murday* and *Colton*. We will only discuss briefly how the data presented here were obtained and analyzed. Our experimental set up is shown

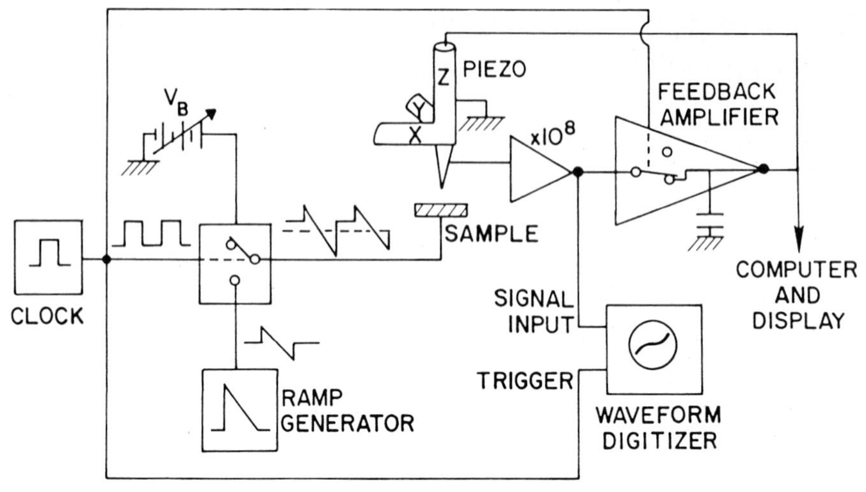

Fig. 16.1. Schematic diagram of the experimental setup used to obtain constant-current STM topographs and atom-resolved tunneling spectra. For details see text

diagrammatically in Fig. 16.1. The STM tip is moved in three dimensions using three orthogonal piezoelectric translators labeled x, y and z in Fig. 16.1. The z translator varies the tip-surface distance while the x, y translators scan the tip over the surface. The tunneling current depends exponentially on the tip-surface distance, tyically varying by about an order of magnitude for only 1 Å change in distance. Therefore, as the tip is scanned over a surface, topographic changes appear as changes in the tunneling current. In general, however, the surface scan is performed keeping the tunneling current constant. This is accomplished by the use of a feedback-loop (see Fig. 16.1) which maintains a predetermined constant tunneling current. By applying an appropriate bias on the z piezo the tip-surface distance is changed so as to maintain this constant current. The STM topographs are maps of the voltage applied to the z piezo so that a constant current of, typically, 1 nA is maintained. As was first pointed out by *Tersoff* and *Haman* [16.5], under certain conditions (see Chap. 14) constant current topographs represent the contours of a surface of equal local density-of-states (LDOS).

To obtain spectroscopic information using the STM, instead of applying a *dc* bias to the sample, a waveform composed of a *dc* part and a voltage-ramp generated by a clock and a ramp-generator (as shown in Fig. 16.1) is applied to the sample. During the *dc* part, the feedback loop is active, the tunneling current is kept constant and thus a constant current topograph is obtained. When the ramp-generator is triggered, the feedback-loop is inactivated, the tip-surface distance is fixed and the tunneling current is measured as a function of sample bias. In this way both topographic and spectroscopic information is obtained for the same area of the sample. An example is given in Fig. 16.2 which shows a topograph of the unoccupied states of a Si$\{111\}$-7×7 surface and a section of 14 Å$\times$ 14 Å on this surface for which tunneling spectra (in the form of dI/dV vs eV) have been obtained at 1.4 Å intervals.

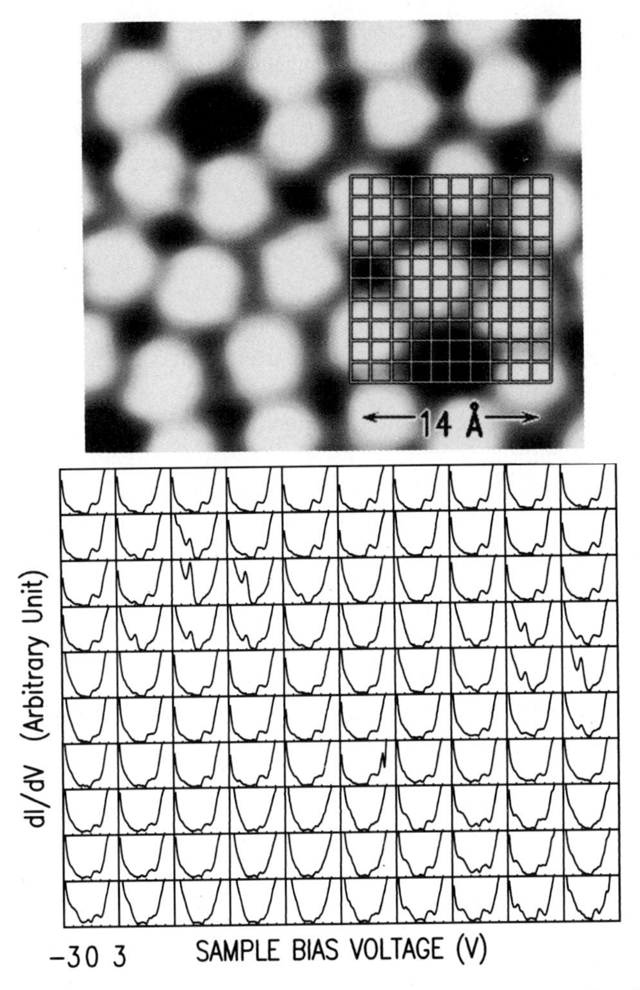

Fig. 16.2. (*Top*:) STM topograph of the unoccupied states of a Si{111}-7 × 7 surface (sample bias = +2 V). The atoms imaged are the top-layer Si adatoms. The grid encompasses a 14 Å× 14 Å area of this surface for which tunneling spectra have been obtained. (*Bottom*:) The 100 tunneling spectra obtained in the above area are plotted in the dI/dV form. Such spectral maps allow one not only to obtain the energies of the occupied (negative bias) and unoccupied (positive bias) states of particular atoms, but also to obtain information on the spatial extent of their wave functions

The interpretation of STM topographs and particularly of tunneling spectra is not, in general, straightforward. Within the Bardeen approximation [16.6] of weak tip-sample interaction the tunneling current can be written as:

$$I \propto \int_{E_{\mathrm{F}}}^{E_{\mathrm{F}}+\mathrm{eV}} \mathrm{DOS_S}(E, \boldsymbol{r}) \mathrm{DOS_T}(E - \mathrm{eV}, \boldsymbol{r}), T(E, \mathrm{eV}, \boldsymbol{r}) dE \;, \tag{16.1}$$

where $\mathrm{LDOS_S}$ and $\mathrm{LDOS_T}$ are the local density-of-states of sample and tip at the appropriate energies and tunneling position. The current depends on a convolution of both sample and tip properties. In STM studies, however, one is interested in the properties of the sample while the structure and electronic states of a tip are in general not known. *Tersoff* and *Haman* [16.5] have argued that constant-current topographs can be interpreted as contours of constant local density-of-states of the sample, provided that the tip wave function can be adequately represented by an s-wave. In this case the tip position $\boldsymbol{r}$ should be interpreted as the effective center of curvature of the tip; i.e., the origin of the s-wave that best approximates the tip wave function. In most cases such a simple picture is adequate. In general, however, the structure and nature of the tip may play an important role affecting not only the resolution but also the corrugation in STM topographs.

The interpretation of spectroscopy data requires careful consideration. As shown by (16.1) the tunneling current depends not only on $\mathrm{LDOS_S}$ and $\mathrm{LDOS_T}$ but also on the bias-dependent electron transmission coefficient $T(E, V, \boldsymbol{r})$. The derivative dI/dV is also a function of $T(E, V, \boldsymbol{r})$, and the exponential dependence of $T(E, V, \boldsymbol{r})$ on V can distort the positions of the peaks in the spectrum. A proposed solution to this problem involves dividing dI/dV by I/V. This procedure effectively cancels out the exponential dependence of $T(E, V, \boldsymbol{r})$ on V. The first-principles calculations of *Lang* [16.7] support the validity of this approach. Thus, a plot of $(dI/dV)/(I/V)$ vs eV corresponds to roughly a density-of-states spectrum on top of a usually smoothly varying background. It is usually assumed that the tip LDOS distribution is structureless and thus does not distort significantly the spectrum of the sample. In many cases this is true. This is verified by comparing the obtained tunneling spectra with conventional photoemission and inverse-photoemission spectra of the same surface. Such comparisons for the Si$\{111\}$-7×7 surface, for example, have shown that the same spectral peaks are present in both kinds of spectra [16.8, 9]. Such a comparison is not always meaningful, however, because of matrix element effects and the fact that ARTS measures only the z-component of $\mathrm{LDOS_S}$. There are cases, particularly involving contaminated tips, where the tip LDOS has sharply-peaked features. These features distort the spectra and in some cases give rise to interesting I-V curves showing regions of negative differential resistance, i.e., regions where increasing voltage leads to decreasing current [16.10, 11]. One must always be on the lookout for such "abnormal" tips.

16.2 Imaging Semiconductor Surface Chemistry Atom-by-Atom Using the STM

The termination of a solid at its surface produces surface dangling-bonds. Surface reconstructions reduce their numbers, but semiconductor surfaces, in general, are characterized by the presence of dangling-bonds. Electrons in surface dangling-bond states have low binding energies lying in the gap between the bulk valence and conduction bands. These electrons are the most chemically active, so that, at not too high temperatures and not very reactive species, surface dangling-bond sites can be considered as the chemically "active sites" of the semiconductor

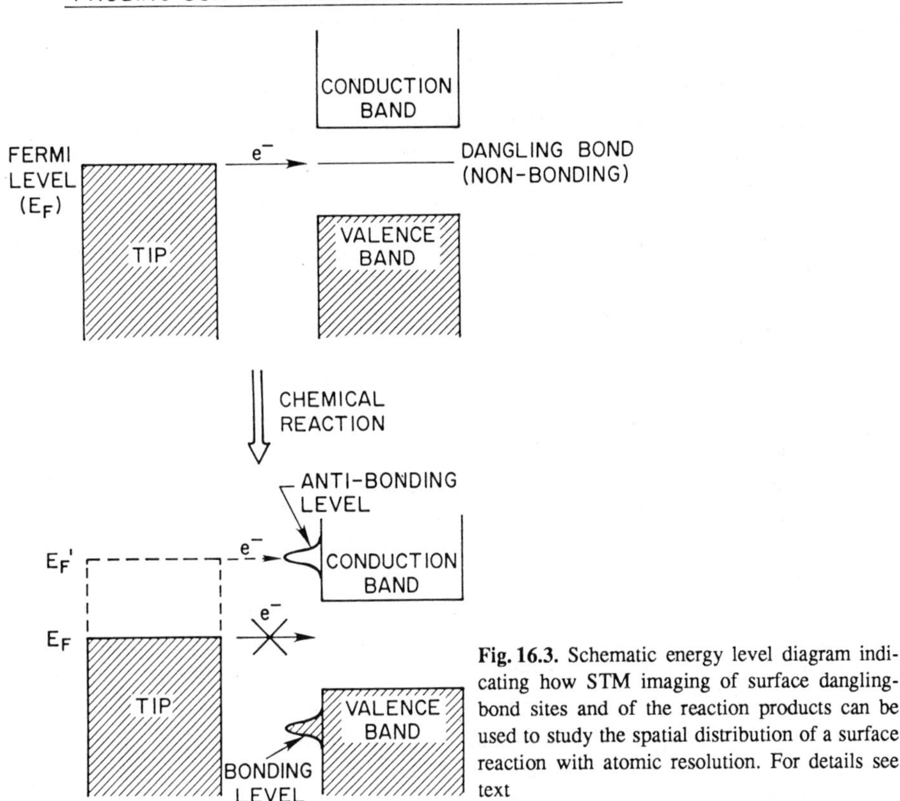

Fig. 16.3. Schematic energy level diagram indicating how STM imaging of surface dangling-bond sites and of the reaction products can be used to study the spatial distribution of a surface reaction with atomic resolution. For details see text

surface. This property provides a simple way for observing, with atomic resolution, the spatial distribution of a surface reaction [16.9]. This is illustrated schematically in Fig. 16.3. The sample bias is adjusted so that tunneling occurs in or out of the surface dangling-bond states. In this way, the spatial distribution of the active sites is obtained. If reaction now takes place at a particular site, its dangling-bond state is eliminated, while new bonding and anti-bonding levels are generated. If the sample bias is not changed, the reacted dangling-bond site will appear dark, due to the drastic reduction in the LDOS. In this way, the spatial distribution of the "dark sites" reflects the spatial distribution of the surface reaction. In some cases, such as in molecular chemisorption processes, the attached group may have low-lying occupied or unoccupied states and can lead to an increase of LDOS at the reaction site so that reacted sites appear brighter than unreacted sites. The fact that STM measures not only LDOS but also topography, (i.e., it provides height information), can be used to image the spatial distribution of the products of the surface reaction. As illustrated in Fig. 16.3, one can, in favorable cases, adjust the sample bias so as to be in an energy region where the reaction products have a finite LDOS. The finite LDOS, plus the fact that products generally lie above the surface plane, can in many cases provide enough contrast to allow the imaging of their spatial distribution. This is important when more than one type of product is formed.

At high temperatures, processes that involve the breaking of bonds and which induce large scale atomic rearrangements are activated. In some cases, at the end of the reaction, a new superstructure is formed which can be studied using conventional diffraction-based techniques. In other cases, the surface remains disordered and thus is not amenable to conventional structure analysis. Perhaps even more interesting are the intermediate steps of a reaction where no long-range order exists. It is in the study of such disordered phases that the STM offers unique advantages. These intermediate steps can be studied by quenching the surface reaction by rapid cooling. In this case the STM measurement is performed at the lower (usually room) temperature. It is expected, however, that drift compensation may allow STM measurements to take place at high reaction temperatures.

Atom-resolved tunneling spectroscopy (ARTS) is indispensable in surface chemistry studies. Certain dangling-bond sites are not readily imaged in STM topographs. This is, for example, the case of the rest-atoms in the 2nd atomic layer of the Si{111}-7×7 surface. Such sites, however, are readily recognizable by their spectra. Thus, the presence or absence of characteristic dangling-bond features in the spectra of a site can be used to determine if reaction has taken place at that site. However, one has to be alert to the possibility that sites are coupled, in the sense that reaction at one site can affect the spectrum of a neighboring site. The most important contribution of ARTS is in permitting a correlation of chemical reactivity with local electronic structure. Such a correlation is essential for the microscopic understandig of chemistry.

Besides its unique strenghts, the STM technique has a major drawback – namely, its lack of chemical selectivity. This drawback can be partially offset by careful consideration and correlation of the results of ARTS measurements, of electronic structure calculations, and of results of conventional surface science techniques. We will discuss such a case in Sect. 16.6.

16.3 The Structure of the Si{111}-7×7 Surface

We will illustrate the above possibilities using as examples reactions of the Si{111}-7×7 surface. The current, generally accepted model of the 7×7 reconstruction is the so-called dimer-adatom-stacking fault (DAS) model proposed by *Takayanagi* et al. [16.12]. The DAS model is illustrated in Fig. 16.4, which shows a top view and a side view along the long axis of the unit cell. In this picture, atoms at increasing depth from the surface are represented by circles of decreasing diameter. There are two triangular sub units surrounded by Si dimers. These sub units are rendered inequivalent by a stacking fault in the left sub unit which helps to eliminate surface dangling-bonds. A large reduction in the number of surface dangling-bonds occurs when Si atoms ejected from the original

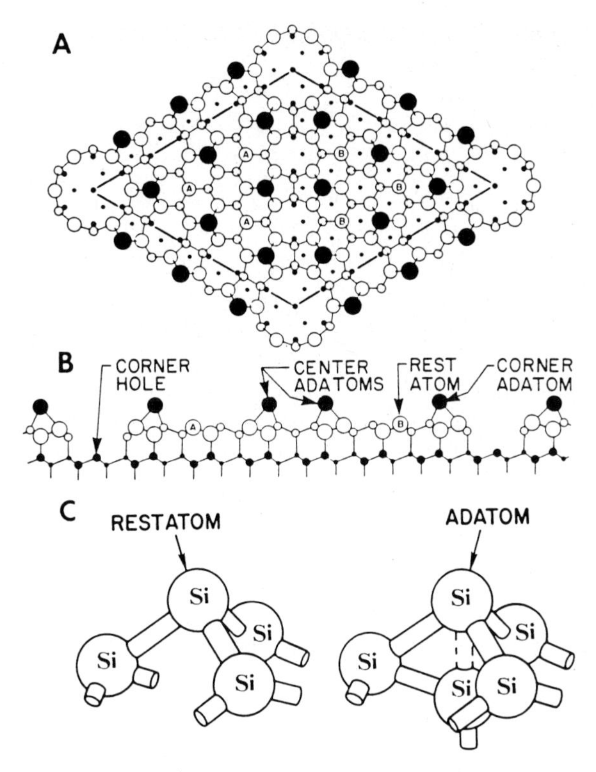

Fig. 16.4. (**a**) Top view of the dimer-adatom-stacking fault model of the Si{111}-7 × 7 reconstruction. The 7 × 7 unit cell containing the stacking fault (*left side*) is outlined. Atoms at increasing distances from the surface are indicated by circles of decreasing size. The large solid circles denote the 12 adatoms. The circles labeled A and B represent the rest-atoms in the faulted and unfaulted halves, respecitvely. Small open circles along the boundary of the unit cell represent the Si atoms forming the dimers. (**b**) Side view along the long diagonal of the unit cell. (**c**) Local structure and bonding at rest-atom and adatom sites

surface become adatoms on top of what is now the second atomic layer, each one of them eliminating 3 dangling-bonds on the layer while introducing a new one. Finally, at the corners of the unit cell there are Si vacancies. As a result of this reconstruction, from the 49 original dangling-bonds only 19 survive in the 7 × 7 unit cell. Six of them are located on the now 2nd layer, triply-coordinated Si atoms (A and B in Fig. 16.4) – the so-called rest-atoms. Twelve are on the adatoms, and finally one dangling-bond is located at the atom on the bottom of the corner vacancy, the so-called corner-hole. For reasons that will become apparent later in the discussion, we further separate the adatoms in twc groups: the 6 located near to a corner-hole are called corner-adatoms while the 6 not neighboring corner-holes are called center-adatoms.

In Fig. 16.5 we show topographs of the unoccupied (left) and occupied (right) states of the 7 × 7 surface. A unit cell is outlined in both cases. The 12 adatoms are clearly resolved. In the topograph of the occupied states, the two triangular

Fig. 16.5. (*Left*:) STM topograph of the occupied states of the Si{111}-7 × 7 (sample bias = −1.5 V). (*Right*:) Topograph of the unoccupied states (sample bias = +1.5 V). A unit cell is outlined in each case

sub units forming the unit cell appear inequivalent because of the presence of the stacking fault. Moreover, the same topograph shows that corner-adatoms appear higher than center-adatoms.

16.4 Site-Selective Reactions of Si{111}-7 × 7

Our initial examples of surface chemistry studies using the STM involve room temperature reactions which preserve the basic 7 × 7 reconstruction. The reaction of Si{111}-7 × 7 with NH_3 is one such reaction [16.9]. This reaction has been studied by a variety of techniques including photoemission [16.13, 14] and vibrational [16.15] spectroscopies, which indicate that NH_3 dissociates on the 7 × 7 surface producing Si-H and Si-NH_2 groups. Some NH groups may also be present [16.16]. In Fig. 16.6 we show a topograph of the unoccupied states of a 7 × 7 surface after exposure to NH_3 gas. We find that a large number of the Si adatoms have become dark. According to our previous discussion, this is the result of the elimination of the dangling-bonds at those sites as a result of the chemical reaction. This is confirmed by the relation between the number of "dark" sites and increasing NH_3 exposure. Topographs, such as Fig. 16.6, allow us to directly observe the adatom sites of the surface. It is clear from an inspection of Fig. 16.6 that the majority of the unreacted sites are corner-adatoms, indicating that these sites are less reactive than the center-adatom sites [16.9]. This is an important piece of chemical information which cannot be obtained by conventional surface science techniques. Information about reaction products can be obtained from topographs at high sample bias [16.9]. In Fig. 16.7 we show a 3d topograph of the occupied states of six adatom sites surrounding a corner-hole. Studies as a function of sample bias indicate that sites A and B in Fig. 16.7

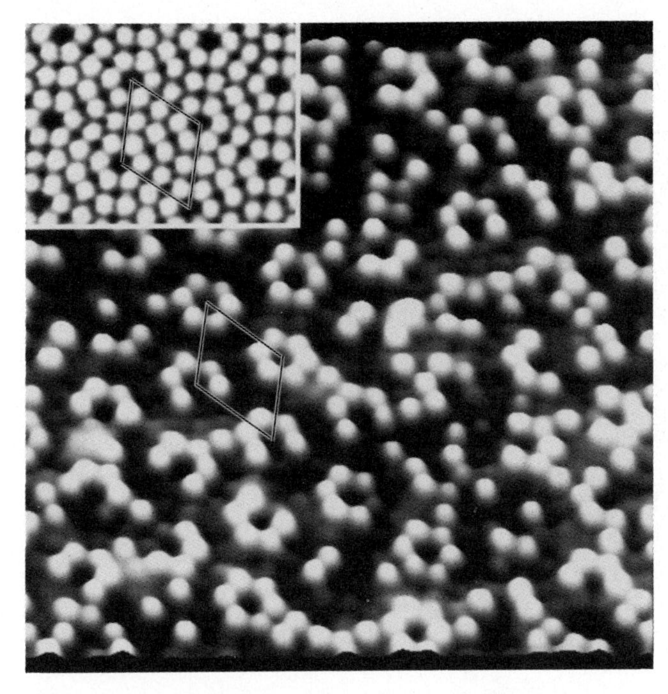

Fig. 16.6. STM topograph of a Si{111}-7 × 7 surface exposed to ~ 3 L of NH_3 at 300 K (sample bias = +2 V). Note the preferential reaction of center-adatoms. (*Inset:*) The unit cell of a clean Si{111}-7 × 7 surface

are product sites. The high intensity of these features is the combined result of the increased LDOS of product states at −3 eV, and the fact that these products lie at some height above the surface plane. Another interesting feature of such topographs is that such product features appear to come in two "sizes" (size A larger than B). The exact nature of these two types of features is not known, but it may be that they are the Si-NH_2 and Si-H surface groups detected by high resolution electron energy loss spectroscopy [16.15]. The Si-NH_2 group should appear larger because it contributes more than the Si-H group to the LDOS at −3 eV, and because it extends further out from the surface plane.

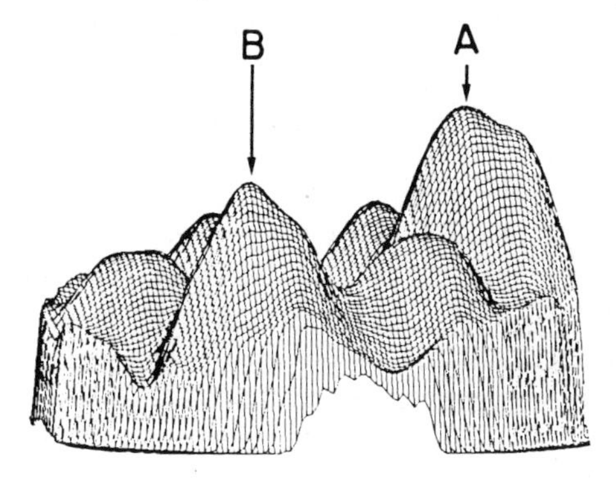

Fig. 16.7. Three-dimensional STM topograph of a group of six adatoms surrounding a corner-hole. Sites A and B are reaction products while the remaining sites are unreacted

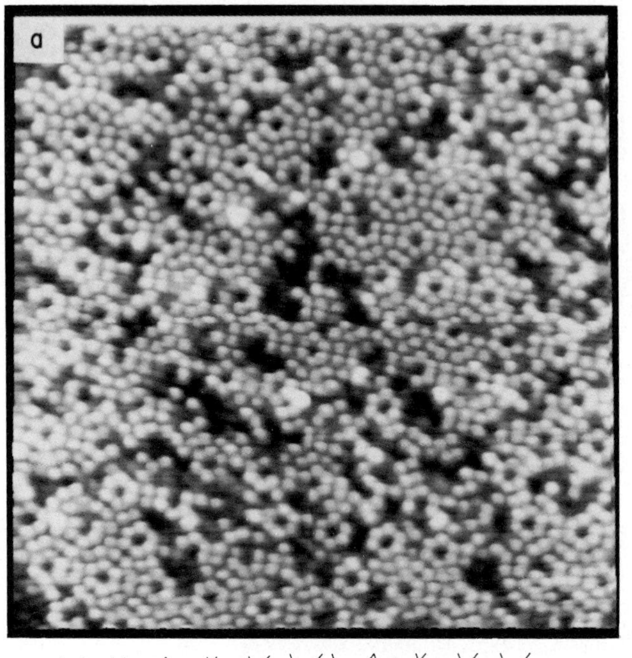

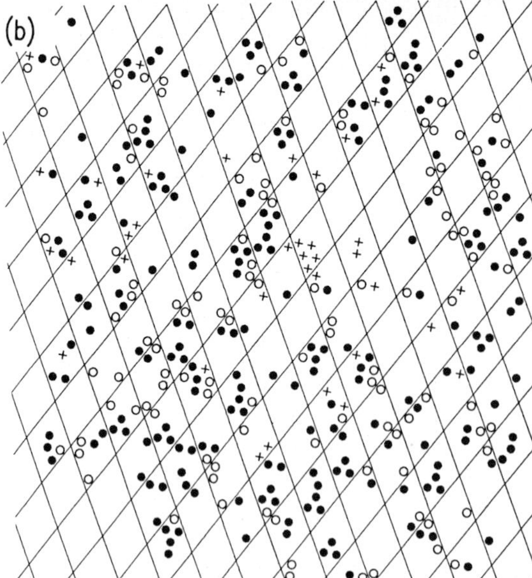

Fig. 16.8. (**a**) STM topograph of a Si{111}-7 × 7 surface after exposure to a few Langmuir of H_2O. (**b**) Schematic diagram showing the location of the reacted sites. Center-adatom sites are denoted by filled circles and corner-adatom sites by open circles. The location of defects on the original 7 × 7 surface before exposure to H_2O is indicated by pluses

As another example, we use the reaction of the 7 × 7 surface with H_2O. As in the case of the reaction with NH_3, reaction with H_2O leads to the appearence of dark areas on the 7 × 7 surface (Fig. 16.8a). In Fig. 16.8b we show a schematic diagram indicating the positions of the reacted atoms on the surface and which of them are corner- and which of them are center-adatoms. We find out that out of 325 reacted adatoms, 212 are center-adatoms and 113 corner-adatoms.

Center-adatom sites are fovored by H_2O, 2 to 1. It is worth noting that simple "reaction maps" such as that shown in Fig. 16.8b can be used to determine the role of the other factors which may influence reactivity such as defects. The positions of defects on the original surface are indicated on the reaction map of Fig. 16.8b by pluses. We can then, by direct observation, determine if these defect sites act as nucleation centers for the surface reaction. In the present case, no such correlation is seen, but in other reactions such a relation may exist [16.17]. It is also clear that accurate sticking coefficients thus can be obtained and if the sticking coefficient is known from other types of measurements, information about molecular fragmentation can be obtained.

Information about surface electronic structure and the ractivity of rest-atom sites can be obtained through ARTS studies. In Fig. 16.9a we show spectra of

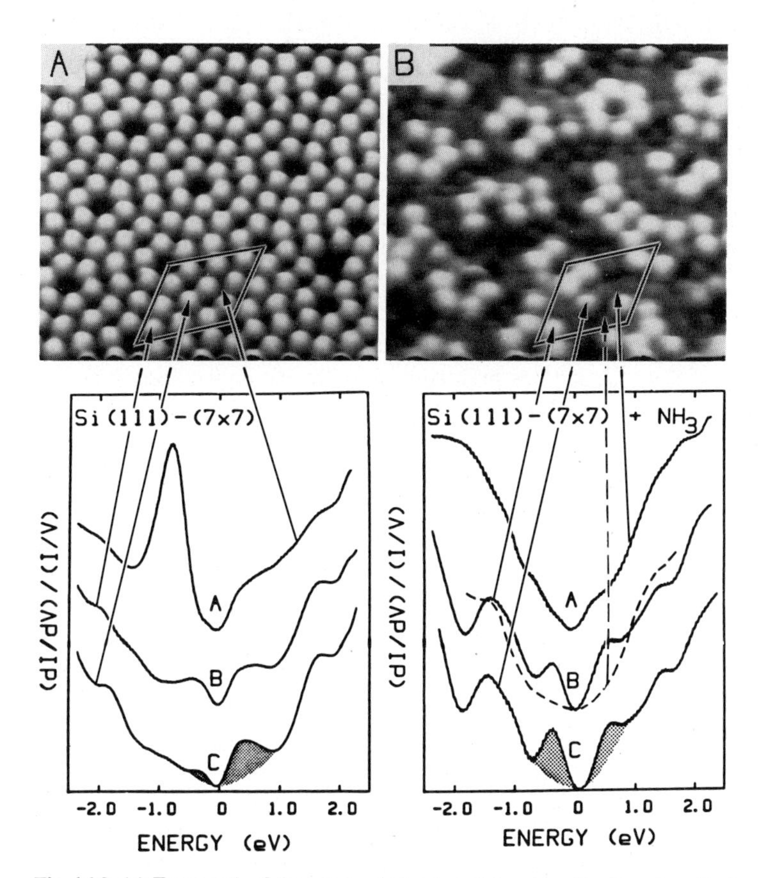

Fig. 16.9. (**a**) Topograph of the unoccupied states of the clean 7 × 7 surface (*top*) and atom-resolved tunneling spectra (*below*). Curves *A*, *B* and *C* give the spectra over rest-atom, corner-adatom and center-adatom sites, respectively. (**b**) Topograph of the unoccupied states (*top*) and atom-resolved tunneling spectra (*below*) of an NH_3-exposed surface. Curve *A* gives the spectrum over a reacted rest-atom site, curve *B* (*dashed*) gives the spectrum over a reacted corner-adatom, while curve *B* (*solid line*) and *C* give the spectra over unreacted corner- and center-adatoms, respectively

the clean 7×7 surface [16.9]. Spectrum (A) is obtained over a rest-atom site, spectrum (B) over a corner-adatom site and spectrum (C) over a center-adatom site. In Fig. 16.9 negative energies correspond to occupied states, while positive energies correspond to unoccupied states. The rest-atom spectrum (A) shows a strong occupied state at $\sim 0.8\,\mathrm{eV}$ below E_F. This peak is characteristic of the rest-atom dangling-bond. The corresponding dangling-bond states of the adatoms appear near $-0.5\,\mathrm{eV}$ (B and C). The energies of all these surface states are in very good agreement with the results of conventional, i.e., surface area averaged photoemission (occupied states), and inverse photoemission (unoccupied state) spectroscopies [16.18]. The magnitude of the electron-electron Coulomb repulsion is about 0.4 eV [16.19]. Thus, the high binding energy ($-0.8\,\mathrm{eV}$) rest-atom dangling-bond state should be fully occupied. Theoretical studies [16.19] have suggested that there is charge-transfer from the adatom to the rest-atom sites. The systematically higher intensity of corner adatom sites than that of center-adatoms supports this view [16.9]. According to electronic structure calculations [16.19, 20], the weak feature is about $-2\,\mathrm{eV}$ observed over adatom sites is due to a surface-resonance with adatom back-bond character.

In Fig. 16.9b (right) we show spectra obtained after the 7×7 surface has been exposed to a few L ($1L = 1 \times 10^{-6}$ TORR S of NH_3 and roughly half of the adatom dangling-bonds have been eliminated [16.9]. To determine the state of rest-atoms on this surface, we look at the spectra of the corresponding sites. If the characteristic $-0.8\,\mathrm{eV}$ peak is present, the rest-atom has not reacted. If, on the other hand, a structureless spectrum such as that shown in curve (A) is observed, then this is taken as indication that a chemical reaction has taken place at this site. More systematic studies involve spectral maps similar to that shown in Fig. 16.2 but encompassing large areas of the surface. In this way, we find that rest-atoms are more reactive than adatoms, reacting faster than one would predict on the basis of their relative numbers on the 7×7 surface.

Finally, spectra (B) and (C) of unreacted corner- and center-adatoms on the partially reacted surface (Fig. 16.9b) illustrate the effects of the reaction of the electronic structure of still unreacted sites. Curves B, C (Fig. 16.9B) show that after the neighboring rest-atoms have reacted, the electronic spectra of unreacted corner- and center-adatoms become virtually indistiguishable.

The fact that the reactivity of the rest-atoms is higher than that of the adatoms we find to be true in other reactions of Si{111}-7×7, including the technologically important reactions with phosphine (PH_3) [16.21] and disilane (Si_2H_6) [16.22].

The reactivity differences between the various dangling-bond sites of the 7×7 surface revealed by the STM reflect differences in the local structure of these sites. According to the DAS reconstruction model, rest-atom sites involve triply coordinated Si atoms with dihedral-angles between bonds very close to tetrahedral [16.12]. Reaction at these sites should not induce strain. Moreover, as we discussed above, the rest-atom dangling-bond should be fully occupied. The situation at adatom sites is different. Both theory [16.20, 23] and experiment [16.24] show that adatoms induce significant strain. According to the DAS model

[16.12], adatoms are menbers of three strained 4-member Si rings. These ring structures bring the adatoms close to the Si atoms directly below them in the 3rd atomic layer (see Fig. 16.4c). This proximity leads to repulsion, and distortion of the structure with adatom dihedral bond angles close to 90°. To reduce the local strain, a weak bond is formed between the adatom and the Si atom directly below it [16.9, 25] as is indicated in Fig. 16.4c. One would expect that on the basis of the large deviation from tetrahedral geometry and the delocalization of the dangling-bond density, adatom sites would be less favored in reactions which lead to dangling-bond saturation by the formation of the covalent bond. Adatom dangling-bond states are less than half-occupied. Even more intriguing is the fact that corner-adatoms appear to be less reactive that center-adatoms. By reference to the DAS model [16.12] we see that there are two obvious differences between the two kinds of adatoms: i) center-adatoms have two rest-atom neighbors while corner-adatoms have only one, and ii) corner-adatoms are bonded to two Si-dimers while center adatoms are bonded only to one dimer. The first difference between center- and corner-adatoms can be important in the case where a concerted reaction involves similtaneously both a rest-atom and a adatom site. These two sites are only about 4 Å apart. We can imagine, for example, a H_2O molecule dissociating in a concerted manner leaving an OH group at the adatom site and a H atom at the rest-atom site. Since there are twice as many atom/center-adatom pairs as there are rest-atom/corner-adatom pairs, a 2:1 reactivity ratio of center to corner-adatoms can be accounted for by such a mechanism. In the case of the reaction with NH_3, however, the reactivity ratio is much larger than 2:1, indicating that other differences between the two kinds of adatoms must be considered. Calculations have shown [16.20, 23] that the strain interaction of two Si-dimers mediated by a corner-adatom is large, leading to a distortion of the corner-adatom site which now has bond dihedral angles more removed from tetrahedral than a center-atom site. It may be this distortion or the initial differences in occupation of the two types of adatom sites that lead to the large reactivity ratios.

Most of the systems we have studied follow the above reactivity trends, that is, rest-atoms appear more reactive than adatoms and center-adatoms are more reactive than corner-adatoms. In the case of the reactions of 7×7 with the open-shell systems O_2 and NO, however, we have encountered the opposite behavior, namely adatom sites appear to be more reactive that rest-atom sites. We note in this respect that reactions at adatom sites may not only involve the dangling-bonds but also the weak and strained adatom back-bonds. The strain can be relieved by breaking the back-bond and attaching atoms at either end, or by inserting atoms so as to enlarge the size of the original 4-member Si ring.

We have also observed a most interesting and peculiar behavior in the reaction of the Si{111}-7×7 surface with N_2O [16.26]. Nitrous oxide is known to oxidize Si at room temperature [16.27]. It has been suggested that the oxidation is selective, taking place only at dangling-bond sites. In our STM studies of this system, we observe two reaction channels. One channel leads to the appearance of dark areas as in the case of H_2O or NH_3, while another, low yield channel leads to the appearance of bright spots almost exclusively on corner-adatom sites

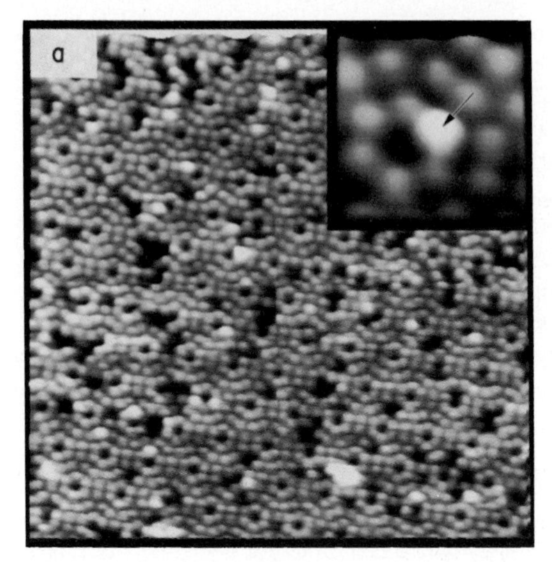

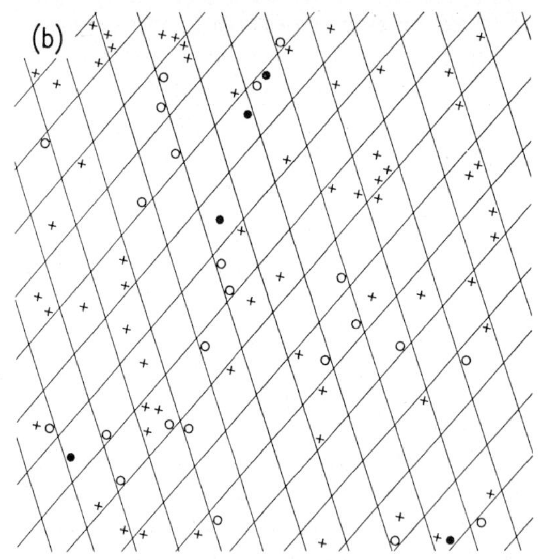

Fig. 16.10. (**a**) STM topograph of the Si{111}-7 × 7 surface after exposure to ~ 35 L of N_2O. The reaction leads to the appearance of dark areas at the surface and also to the appearance of bright spots, almost exclusively on corner-adatom sites. (**b**) Schematic diagram showing the spatial distribution of the "bright spots" on corner-adatom sites (*open circles*) and center-adatom sites (*filled circles*). The distribution of defects on the clean Si{111}-7 × 7 surface is indicated by pluses

(see Fig. 16.10). We do not know yet the nature of the reaction at the corner-adatom site. It may involve molecular N_2O adsorption by charge-donation to the adatom, or a reaction in which an oxygen atom is inserted in the backbond to relieve the strain at this most strained site.

Finally, it is interesting to note that the stacking-fault present in the 7 × 7 unit cell, although a sub-surface feature, can influence surface chemistry. Although we have not observed any significant reactivity differences between the two halves of the 7 × 7 unit cell in the various dissociation reactions we have studied, interesting effects were observed in the early stages of metal deposition on Si{113}-7 × 7. Specifically it was found that Pd [16.28], Ag [16.29] and Li [16.30] prefer the faulted-half of the unit-cell. Most likely, this difference in behavior reflects the interaction of the rest-atoms and through them of the adatoms with the Si atoms in the 4th atomic layer directly below the rest-atoms [16.31]. This interaction is present only in the faulted-half of the unit cell.

Thus the STM can provide us with new and unique information about the chemistry of semiconductor surfaces. As it is true with other surface science techniques, a complete understanding of this chemistry requires input from other experimental techniques and particularly from theory.

16.5 Molecular Adsorption on Si{111}-7 × 7

As an example of molecular adsorption of Si{111}-7 × 7 we consider the case of nido-decarborane ($B_{10}H_{14}$, DB). We have used this molecule as a source of B in our studies of B adsorption on Si{111} to be discussed in Sect. 6. Figure 16.11 shows the 7 × 7 surface after exposure to 0.2 L DB at room temperature. Unlike NH_3 or H_2O, chemisorption of DB does not lead to the formation of dark sites, but is characterized by the appearance of bright, disk-like features. The size of these features (about 1 nm) is roughly the size of the DB molecule. As in the previous studies we can study topographs such as that shown in Fig. 16.11 to determine if there is a preferred adsorption site for the DB molecules. We find that defect sites along with center-adatoms are preferred over corner-adatom sites.

16.6 Reactions That Involve Extensive Atomic Rearrangements

To exemplify reactions that destroy the 7 × 7 reconstruction and produce a new surface structure, we discuss the reaction of Si{111}-7 × 7 with boron. From

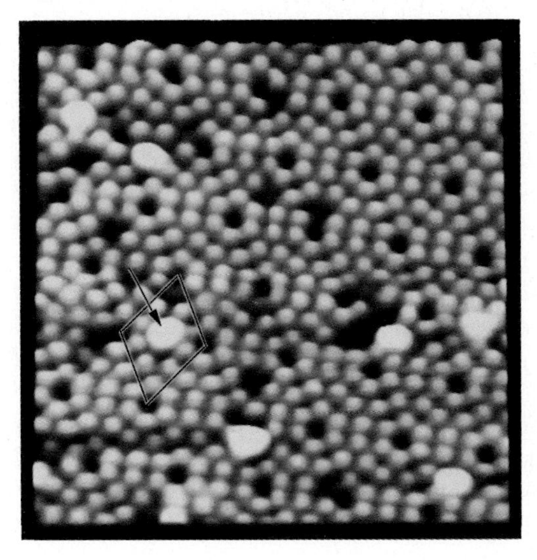

Fig. 16.11. STM topograph of the Si{111}-7 × 7 surface after exposure at 300 K to 0.2 L of nido-decaborane. The decaborane molecules appear as bright roundish disks

Fig. 16.12. STM topograph of the occupied states of the surface produced by exposing Si{111}-7 × 7 to 0.2 L of decaborane and briefly annealing to 600° C. Sample bias = -1 V

LEED studies [16.32, 33] it is known that interaction with B with the 7 × 7 surface at high temperatures leads to the formation of a $(\sqrt{3}\times\sqrt{3})\,R30°$ structure. Although from such studies the detailed nature of the $\sqrt{3}\times\sqrt{3}$ structure could not be deduced, it is clear that an extensive atomic rearrangement must take place. In our STM studies [16.34] we used decaborane (DB) as the source of B. As we saw in Fig. 16.11, at room temperature, DB adsorbs on Si{111}-7 × 7 in a molecular form (the exact structure in the chemisorbed state is not known). Heating the surface to temperatures above 500° C leads to the decomposition of the DB molecules and the desorption of hydrogen. In Fig. 16.12 we show a topograph of the occupied states of a 7 × 7 surface after exposure to 0.2 L DB followed by brief annealing to 600° C. Because B, when bonded to three Si atoms, does not have any occupied dangling-bond states, this type of site is not imaged in Fig. 16.12, which primarily reflects the spatial distribution of Si atoms. In the bottom right the unperturbed area of 7 × 7 can be seen, while the remainder of the surface shows a disordered distribution of Si atoms. Figure 16.13 shows another topograph, this time of the unoccupied states of a 7 × 7 surface exposed to 0.2 L DB and more thoroughly annealed at 700° C. We see that the reaction proceeds again in a non-homogeneous manner; there are large areas of unperturbed 7 × 7 surface, while, particularly in the top part of the picture, a new $\sqrt{3}\times\sqrt{3}$ structure appears. In this part of the surface there was a line defect which acted as a trap of DB molecules and a nucleation point for the $\sqrt{3}\times\sqrt{3}$ structure. The topograph in this area also shows globular clusters of displaced atoms. At higher DB exposure (0.4 L) and annealing temperature (800°C), the surface has the structure shown in Fig. 16.14. Now most of the surface shows the $\sqrt{3}\times\sqrt{3}$ structure. However, it is clear that there are at least two types of sites, bright (high) and dark (low), participating in the formation of the $\sqrt{3}\times\sqrt{3}$ surface structure. ARTS studies [16.34] show that the spectra over bright sites

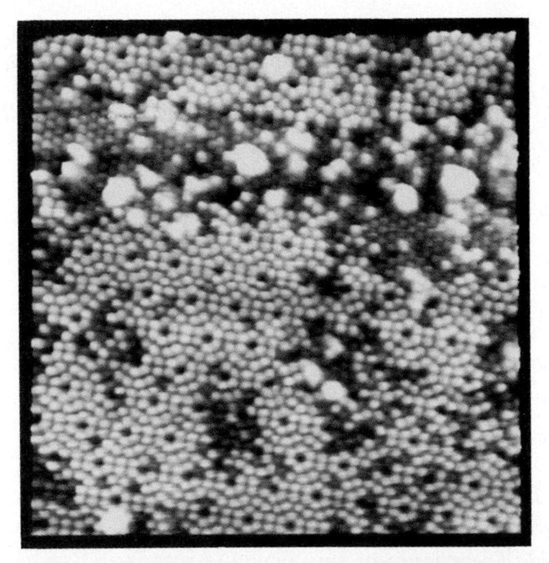

Fig. 16.13. STM topograph of the unoccupied states of the surface produced by exposing Si{111}-7 × 7 to 0.2 L of decaborane and annealing to 700°C. Sample bias = +2 V

Fig. 16.14. STM topograph of the unoccupied states of the mostly $(\sqrt{3}\times\sqrt{3})R30°$ surface produced by exposing the 7 × 7 surface to 0.4 L of decaborane and annealing briefly to 800°C

(Fig. 16.15a) are essentially identical to those of Si adatoms (Fig. 16.9a, local structure in Fig. 16.16a) on the clean 7 × 7 surface. The majority of the darker sites are due to B adatoms in a T_4 site (local structure in Fig. 16.16b). The B adatoms appear darker because, according to electronic structure calculations [16.34], they are $\sim$ 0.4 Å lower than the Si adatoms, and their wavefunction is more contracted.

Although it appears that B behaves very much like other Group III elements, such as Al [16.35], Ga [16.36] or In [16.37], which occupy T_4 adatom sites, local density functional slab-calculations [16.34, 38] suggest that this configuration is not the most stable one. We have performed total energy calculations on a variety of structures. We found that the lowest energy structure is that shown in

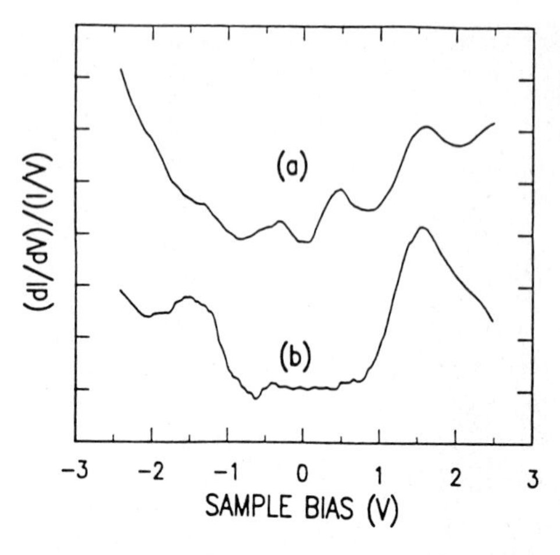

Fig. 16.15. Atom-resolved tunneling spectra obtained over (*a*) "bright"-atom sites in Fig. 14, (*b*) over majority "dark" sites of Figs. 16.17 (**a**) and (**b**)

Fig. 16.16c. This structure, which we denote as B-S_5, involves B in a substitutional site at the 2nd Si layer and a Si adatom in a T_4 site directly over it. In this site B is 5-fold coordinated. The energy of this structure was found to be about 1 eV per $\sqrt{3} \times \sqrt{3}$ unit cell lower than that of the B-T_4 adatom configuration. Because of the proximity of the B atom and the Si adatom ($\sim$ 2.2 Å) and the higher electronegativity of B, the dangling-bond state of the Si adatom is emptied by charge-transfer to B. Although the substitutional site is most stable, its formation involves the breaking of bonds and for this reason the B-T_4 configuration is kinetically favored. If the sample is properly annealed, however, the substitutional configuration (Fig. 16.16c) should be formed. By annealing the sample of Fig. 16.16 to 1000°C, we obtain the surface shown in Fig. 16.17a. The spectra of the dark sites of this surface (Fig. 16.15b) are different from those of the dark sites in Fig. 16.14. If the exposure to decarborane is increased to 1 L and the sample is annealed again to 1000°C, the uniform surface shown in Fig. 16.17b is obtained. Low energy ion scattering shows that this surface is

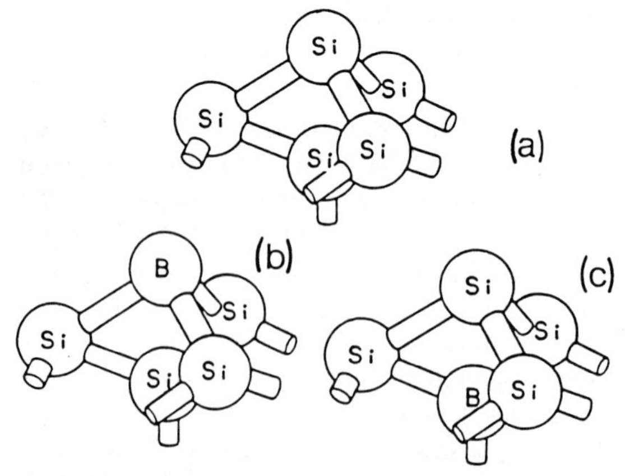

Fig. 16.16a–c. The local structures of: (**a**) a Si-T_4, (**b**) a B-T_4 site, and (**c**) a B-S_5 site

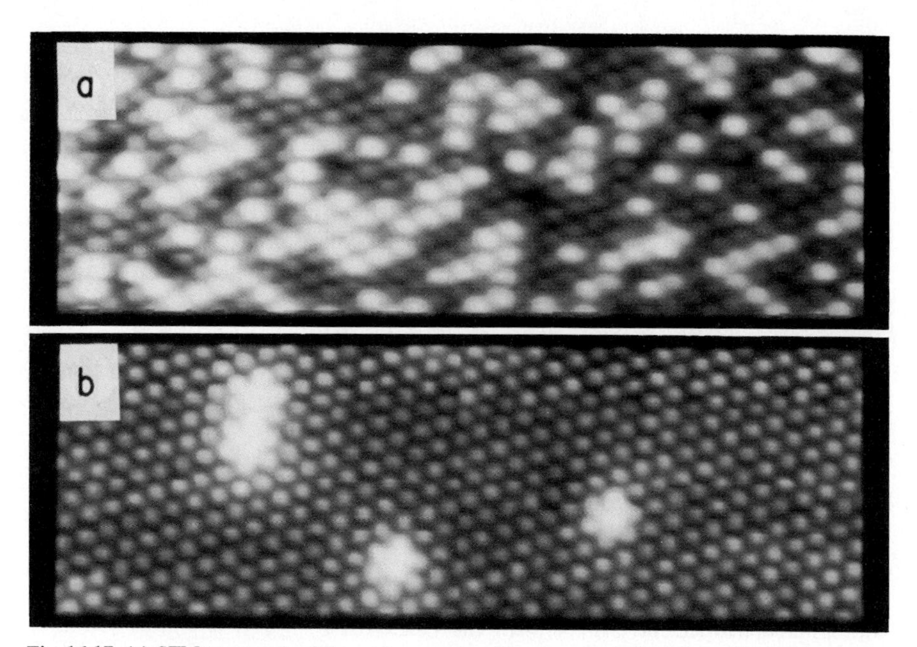

Fig. 16.17. (a) STM topograph of the surface produced by exposing Si{111}-7 × 7 to 0.4 L of decaborane and annealing to 1000° C. (b) The surface which results after exposure to 1 L of decaborane and annealing to 1000° C

composed of Si atoms. The B is still very near the surface, however, as shown by the fact that while the energy of the B(1s) XPS peak is changed upon high temperature annealing, its intensity is little affected. A similar behavior is seen in the Auger spectrum. Further evidence in favor of a substitutional (B-S_5) sub-surface B site is provided by the ARTS spectrum shown in Fig. 16.15b. We see that Si-adatoms on this equilbrium structure have a spectrum very different from that of Si-adatoms on 7 × 7 (Fig. 16.9a). As a result of the interaction with the sub-surface B layer, the Si top-layer is insulating. It has an unoccupied state with dangling-bond character at about 1.5 eV above E_F and an occupied state with back-bond character at about 1.5 eV below E_F. These findings are in accord with theoretical predictions [16.34].

The conclusions regarding the equilibrium adsorption site of B on Si{111} are further supported by independent studies [16.38, 39] of B surface segregation from heavily B-doped bulk Si. In these studies, the same substitutional (B-S_5) configuration of B was identified as the most stable structure. Although intermediate, metastable configurations may be different depending on the way B is introduced (by adsorption from the gas-phase or diffusion from the bulk), the most stable configuration should be the same in both types of experiments. The stability of the B-S_5 configuration can be traced to two factors. First, by placing the B atom in the sub-surface site no sub-surface strain is generated. The B-T_4 configuration on the other hand induces substantial sub-surface strain. Second, the surface dangling-bonds are saturated by the Si adatoms with each of them introducing a new dangling-bond. However, the Si-to-B charge-transfer empties these dangling-bonds, further reducing the energy of the system. We find that, in general, minimization of lattice strain and of the numbers of dangling-bonds determines the equilibrium adsorption configuration. For example, we find that phosphorus, the other common dopant of Si, does not give rise to a $\sqrt{3} \times \sqrt{3}$ structure, but instead a disordered 1×1 structure is formed [16.21]. Moreover, ion scattering shows that P is present at the surface of the sample [16.21]. Phosphorus, unlike B, has essentially the same covalent radius as Si (1.06 Å vs 1.11 Å) and thus can occupy a surface substitutional site without inducing lattice strain. This P configuration is stable because no dangling-bonds are introduced. Phosphorus has the valence electronic configuration $3s^2 3p^2$ so that a three-fold coordinated surface P atom has no dangling-bonds but simply introduces a doubly-occupied lone pair orbital.

It is clear that STM and ARTS with the help of electronic structure calculations and conventional surface spectroscopic measurements can indeed successfully tackle complex surface phenomena.

16.7 Doping Effects on Silicon Surface Chemistry

From our discussion in Sect. 4, it is clear that the chemistry of Si surfaces depends crucially on the local structure and strain considerations at the reactive site. As we have shown above, the equilibrium structure of the B/Si{111}-$\sqrt{3} \times \sqrt{3}$ surface involves a Si-adatom top layer, with the B-dopants below the Si adatoms in substitutional sites. Because of the Si-to-B charge-transfer, the Si top layer of the B/Si{111}-$\sqrt{3} \times \sqrt{3}$ system has no occupied dangling-bond levels. This drastic change in the dangling-bond level occupancy and the absence of rest-atom

Fig. 16.18. The B/Si{111}-$\sqrt{3} \times \sqrt{3}$ surface after exposure to $\sim$ 400 L of NH_3 at 300 K

sites should be reflected in the chemical properties of the surface. Indeed, we find that the top Si layer of this B/Si{111}-$\sqrt{3} \times \sqrt{3}$ surface has very different chemical properties than those of clean Si surfaces [16.40]. We illustrate this using the reaction with NH_3 as an example. Exposure of a clean Si{111}-7×7 surface to a few L of NH_3 (Fig. 16.6) leads to extensive reaction [16.9] which involves dissociation to form Si-H and Si-NH_2 surface species. In contrast to this behavior, STM topographs such as that in Fig. 16.18 show that exposure of the B/Si{111}-$\sqrt{3} \times \sqrt{3}$ surface at room temperature to even hundreds of L of NH_3 leads to very little reaction. When the surface temperature is lowered, the sticking coefficient increases drastically. However, photoemission studies show [16.40] that the reaction that is taking place on the B/Si{111}-$\sqrt{3} \times \sqrt{3}$ surface is different from that on the Si{111}-7×7 surface. While NH_3 dissociates on the 7×7 surface, it adsorbs molecularly and reversibly on the $\sqrt{3} \times \sqrt{3}$ surface. This is a novel doping effect on chemical reactivity where not only the reaction rate is affected by doping but the products of the reaction are also determined by it. Moreover, this doping effect is one that involves direct dopant-reaction site interactions.

Long-range effects may also be important, and may be responsible for the observed stabilization by B-doping of Si-based $\sqrt{3} \times \sqrt{3}$-islands. In this case, a shift of the Fermi level position towards the valence band might lead to a decreased population of the Si-adatom dangling-bonds and thus lower the energy of the Si{111}-$\sqrt{3} \times \sqrt{3}$ reconstruction.

16.8 Conclusions and Prospects for the Future

In the above we have discussed some of the ways in which STM and ARTS can be used in the study of surface chemistry. We have shown that STM and ARTS can provide unique information about chemical systems not readily obtainable by conventional surface analysis techniques. The spatial distribution of reactions on semiconductor surfaces can be studied on an atom-by-atom basis and the observed reactivity can be correlated with local electronic structure. The products of the reaction can be imaged and the role of the defects, steps, etc. can be directly determined. We also have seen that there are still important unanswered questions, particularly involving the role of the tip, in both topographs and spectra.

The examples discussed above do not in any way exhaust the possible kinds of chemical information that can be obtained or the types of chemical systems to which such techniques can be applied. In the future, we should expect not only static STM studies like those discussed here, but also dynamical studies such as studies of surface diffusion. Tunneling or field-emitted electrons can be used as active ingredients to induce local surface chemistry through local electronic excitations or through dissociative electron attachment type of reactions [16.41] and more extensive use of this capability is to be expected. Moreover, not only metals and semiconductors, but even insulators can be studied with the STM under appropriate conditions [16.42, 43]. Finally, one major drawback of STM is the difficulty of identifying the chemical nature of surface species. Currently, this can only be done on the basis of parallel measurements with conventional analytical techniques, and, of course, with the help of ARTS which, however, measures valence spectra which are not very chemically specific. The development of vibrationally inelastic tunneling spectroscopy which is currently being undertaken in many laboratories will provide the necessary "fingerprint" information for chemical identification.

References

16.1 See for example: D.P. Woodruff, T.A. Delchar: Modern Techniques of Surface Science (Cambridge University Press, Cambridge 1986); P.J. Estrup: Low Energy Electron Diffraction. In *Modern Diffraction and Imaging Techniques in Materials Science*, ed. by Amelinckx, Remaut and Van Landuyt (North-Holland, Amsterdam 1970); W.M. Gibson: Determination by Ion Scattering of Atomic Positions at Surfaces and Interfaces, in *Chemistry and Physics of Solid Surfaces*, ed. by Vanselow and Howe (Springer, Berlin, Heidelberg 1984)
16.2 G. Binnig, H. Rohrer, Ch. Gerber, E. Weibel: Phys. Rev. Lett. **49**, 57 (1982); G. Binnig, H. Rohrer: Rev. Mod. Phys. **57**, 615 (1977)
16.3 F.M. Chua, Y. Kuk, P.J. Silverman: Phys. Rev. Lett. **63**, 386 (1989)
16.4 T. Gritch, D. Coulman, R.J. Behm, G. Ertl: Phys. Rev. Lett. **63**, 1086 (1989)
16.5 J. Tersoff, D.R. Haman: Phys. Rev. B **31**, 805 (1985)
16.6 J. Bardeen: Phys. Rev. Lett. **6**, 57 (1961)
16.7 N.D. Lang: Phys. Rev. B **34**, 5947 (1986); Comments Cond. Matter Physics **14**, 253 (1989)
16.8 R.J. Hamers, R.M. Tromp, J.E. Demuth: Phys. Rev. Lett. **56**, 1972 (1986)
16.9 Ph. Avouris, R. Wolkow: Phys. Rev. B **34**, 5091 (1989); R. Wolkow, Ph. Avouris: Phys. Rev. Lett. **60**, 1049 (1988); Ph. Avouris, J. Phys. Chem. **94**, 2246 (1990)
16.10 I.-W. Lyo, Ph. Avouris: Science **245**, 1369 (1989); Ph. Avouris, I.-W. Lyo, F. Bozso, E. Kaxiras: J. Vac. Sci. Technol. A, to be published
16.11 P. Bedrossian, D.M. Chen, K. Mortensen, J.A. Golovchenko: Nature **342**, 258 (1989)
16.12 K. Takayanagi, Y. Tanishiro, M. Takahashi, M. Motoyoshi, K. Yagi: J. Vac. Sci. Technol. A **3**, 1502 (1985)
16.13 J. Kubler, E.K. Hlil, D. Bolmont, G. Gewinner: Surf. Sci. **183**, 503 (1987)
16.14 F. Bozso, Ph. Avouris: Phys. Rev. B **38**, 3943 (1988)
16.15 S. Tanaka, M. Onchi, M. Nishijima: Surf. Sci. **191**, L756 (1987)
16.16 B.G. Koehler, P.A. Coon, S.M. George: J. Vac. Sci. Technol. B **7**, 1303 (1989)
16.17 F.M. Leibsle, A. Samsavar, T.-C. Chiang: Phys. Rev. B **38**, 5780 (1988)
16.18 F.J. Himpsel, Th. Fauster: J. Vac. Sci. Technol. A **2**, 815 (1984)
16.19 J.E. Northrup: Phys. Rev. Lett. **57**, 154 (1986)
16.20 G.X. Qian, D.J. Chadi: J. Vac. Sci. Technol. A **5**, 906 (1987)
16.21 F. Bozso, I.-W. Lyo, Ph. Avouris: to be published
16.22 Ph. Avouris, F. Bozso: J. Phys. Chem. **94**, 2243 (1990)
16.23 G.X. Qian, D.J. Chadi: Phys. Rev. B **35**, 1288 (1988)
16.24 I.K. Robinson, W.K. Waskiewicz, P.H. Fuoss, L.J. Norton: Phys. Rev. B **37**, 4325 (1988)
16.25 W. Daun, H. Ibach, J.E. Muller: Phys. Rev. Lett. **59**, 1593 (1987)
16.26 I.-W. Lyo, Ph. Avouris: to be published
16.27 E.G. Keim, A. Van Silfhout: Surf. Sci. **152/153**, 1096 (1985)
16.28 U.K. Koehler, J.E. Demuth, R.J. Hamers: Phys. Rev. Lett. **60**, 2499 (1988)
16.29 St. Tosch, H. Neddermeyer: Phys. Rev. Lett. **61**, 349 (1988)
16.30 Y. Hasegawa, I. Kamiya, T. Hashizume, T. Sakurai: J. Vac. Sci. Technol. **A8**, 238 (1990)
16.31 Ph. Avouris: J. Phys. Chem. **94**, 2246 (1990)
16.32 V.V. Korobtsov, V.G. Lifshits, A.V. Zotov: Surf. Sci. **195**, 466 (1988)
16.33 H. Hirayama, T. Tatsumi, N. Aizaki: Surf. Sci. **193**, L47 (1988)
16.34 I.-W. Lyo, E. Kaxiras, Ph. Avouris: Phys. Rev. Lett. **63**, 1261 (1989)
16.35 R.J. Hamers, J.E. Demuth: Phys. Rev. Lett. **60**, 2527 (1988)

16.36 J.M. Nicholls, B. Reihl, J.E. Northrup: Phys. Rev. B **35**, 4137 (1987)
16.37 J. Nogami, S. Park, C.F. Quate: Phys. Rev. B **36**, 6221 (1987)
16.38 P. Bedrossian, R.D. Meade, K. Mortensen, D.M. Chen, J.A. Golovchenko: Phys. Rev. Lett. **63**, 1257 (1989)
16.39 R.L. Headrick, I.K. Robinson, E. Vlieg, L.C. Feldman: Phys. Rev. Lett. **63**, 1253 (1987)
16.40 Ph. Avouris, I.-W. Lyo, F. Bozso, E. Kaxiras: J. Vac. Sci. Technol. A, to be published
16.41 See for example: M.A. McCord, R.F.W. Pease: J. Vac. Sci. Technol. B **5**, 430 (1987); E.E. Ehrichs, S. Yoon, A.L. de Lozanne: Appl. Phys. Lett. **53**, 2287 (1988)
16.42 G.P. Kochanski: Phys. Rev. Lett. **62**, 2285 (1989)
16.43 Ph. Avouris, R. Wolkow: Appl. Phys. Lett. **55**, 1074 (1989)

17. Bonding and Structure on Semiconductor Surfaces

S.Y. Tong, H. Huang and C.M. Wei

Department of Physics and Laboratory for Surface Studies
University of Wisconsin-Milwaukee, Milwaukee, WI 53201, USA

Since the early days of surface science and vacuum technology, it has been known that semiconductor surfaces exhibit a rich variety of reconstructed structures. Through the years, an assortment of surface analytical tools and theoretical methods have been developed to probe the arrangement and bonding of atoms on semiconductor surfaces [17.1].

Now, after a considerable amount of work, some of the most important reconstruction structures on semiconductor surfaces have been solved. A few important trends have emerged. In this chapter we shall discuss some of the semiconductor reconstructed structures, and the nature of bonding that forms on these surfaces.

17.1 Basic Mechanism Driving Surface Reconstruction

The primary reason for surface reconstruction is the lowering of surface electronic energies. Bulk-terminated semiconductor surfaces are invariably unstable because of the large number of unpaired dangling bonds which result from the creation of a sharp, idealized vacuum-solid interface (i.e., an ideal surface). To reduce the number of such unpaired bonds, surface atoms often rearrange in position to form new bonds having a higher electron density per bond. In many instances, elastic stress is introduced in the surface region. The balance between electronic and stress energies in the near surface region gives rise to the array of superlattices found on reconstructed semiconductor surfaces of different materials: e.g., (2×1), (2×2), $(\sqrt{3} \times \sqrt{3})$ R 30°, (7×7), $c(2 \times 8)$, etc.

To reduce the number of unpaired surface bonds, two processes are commonly found to occur. These are:

i) The formation of novel bonds via adatoms or dimer pairs, and
ii) through bond-bond interaction, surface rehybridization takes place lowering the density of unpaired bonds by filling the lower surface band and leaving the higher surface band empty.

In the following, we shall use the surface reconstructions of GaAs$\{110\}(1 \times 1)$, GaAs$\{111\}(2 \times 2)$, Si$\{111\}(7 \times 7)$ and Ge$\{111\}c(2 \times 8)$ to illustrate the occurence of the above two processes.

17.2 The Geometric and Electronic Structures of GaAs{110}

Dangling bonds of the sp^3 type are unstable on the surfaces of III-V compound semiconductors. The orbitals of surface atoms of the group III species rehybridize to form three planar sp^2-type bonds plus an empty dangling hybrid. The group V species rehybridize to have a doubly occupied dangling hybrid and three p-type bonds separated by approximately 90°. In both cases, the rehybridized orbitals resemble more closely the atomic configurations than the sp^3 configuration found for bulk atoms. Chemically, the rehybridization lowers the energy of the As surface bands, which are doubly occupied. It also raises the energy of the Ga surface bands, but these are empty. Thus, the relaxation process is energetically favorable. Generally, the amount of charge transfer is less for the surface atoms than for bulk atoms.

The orbital rehybridization is most favorable if there are equal numbers of dangling bonds on the nearest Ga and As neighbors. On the (110) surface, the bulk-terminated surface already satisfies this condition, so complete relaxation can take place without altering the (1×1) periodicity. The reconstructed structure has Ga surface atoms receding towards the bulk, while As surface atoms rotate outwards. Thus, the reconstruction of this surface is driven by process (ii) of Sect. 17.1.

The angle of tilt between the surface As and Ga atoms, projected along the $(1\bar{1}0)$ mirror plane, was determined by dynamical low-energy electron-diffraction to be $27° \leq \omega \leq 31°$ [17.2, 3]. A detailed re-evaluation of this projected tilt angle was done by *Puga* et al. [17.4], who determined the optimal value as $\omega = 30° \pm 2°$. Figure 17.1 shows schematically the ideal and relaxed surfaces. In Fig. 17.1, the y-axis is along the [001] azimuth, and the x- axis is along the $[1\bar{1}0]$ azimuth. Figure 17.2 shows the values of three R-factors as a function of the projected tilt angle. The optimal structure also has the Ga atoms in the second layer raised vertically by 0.03 Å, while the As atoms in the second layer are displaced vertically down by an equal amount, to further relieve the surface stress. Figure 17.3 shows a side-view of the fully-relaxed GaAs{110} surface, with multilayer relaxations of atoms in the near surface region. The reader is referred to Table 4 of the article by *Puga* et al. [17.4] for a listing of the atomic coordinates of this surface.

The structural and electronic properties of the GaAs{110} surfaces have been extensively studied by a host of surface analytical techniques [17.2–17]. Besides dynamical LEED [17.2–4], there were results of photoemission [17.8–10], ion-scattering [17.13], STM [17.16] and RHEED [17.14, 15], as well as total energy calculations [17.11, 12, 17]. Surface rehybridization of near surface bonds results in the lowering of the total energy by $\simeq 0.35$ eV/atom [17.17]. Figure 17.4 shows a band structure of the relaxed surface [17.7]. The As-derived surface bands A_4, A_5 and A_6 are filled, while the Ga-derived surface bands C_3 and C_4 are empty [17.7]. No intrinsic surface states exist in the band gap.

Thus, on the GaAs{110} surface, the major lowering of surface energy is through the electronic energy gained by filling the surface As-derived bands.

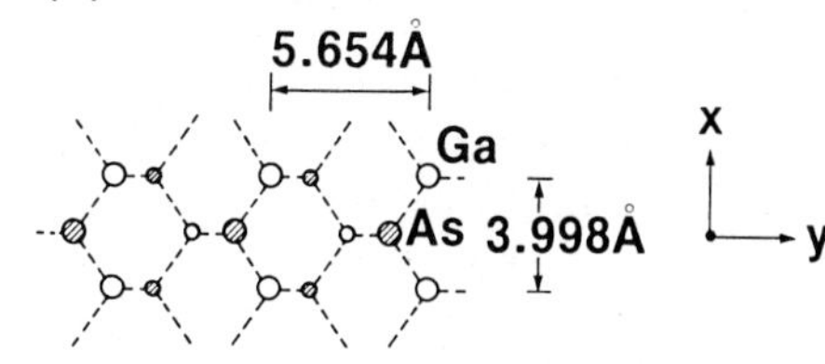

Fig. 17.1a–d. Schematic diagrams of ideal (a, b) and relaxed (d) structures of GaAs{110}. Large circles denote surface atoms, small circles denote atoms in second atomic plane. The reciprocal-lattice beams circled in (c) are predicted to be weak for the ideal surface structure but are not weak in a relaxed structure

(b) PROJECTED SIDE VIEW

As Ga
1.999Å
y
z

(c) RECIPROCAL SPACE

g_x
(30)
(20)
(hk) = ($\bar{h}$k)
(10)(11)(12)
mirror plane
(00)(01)(02)
g_y

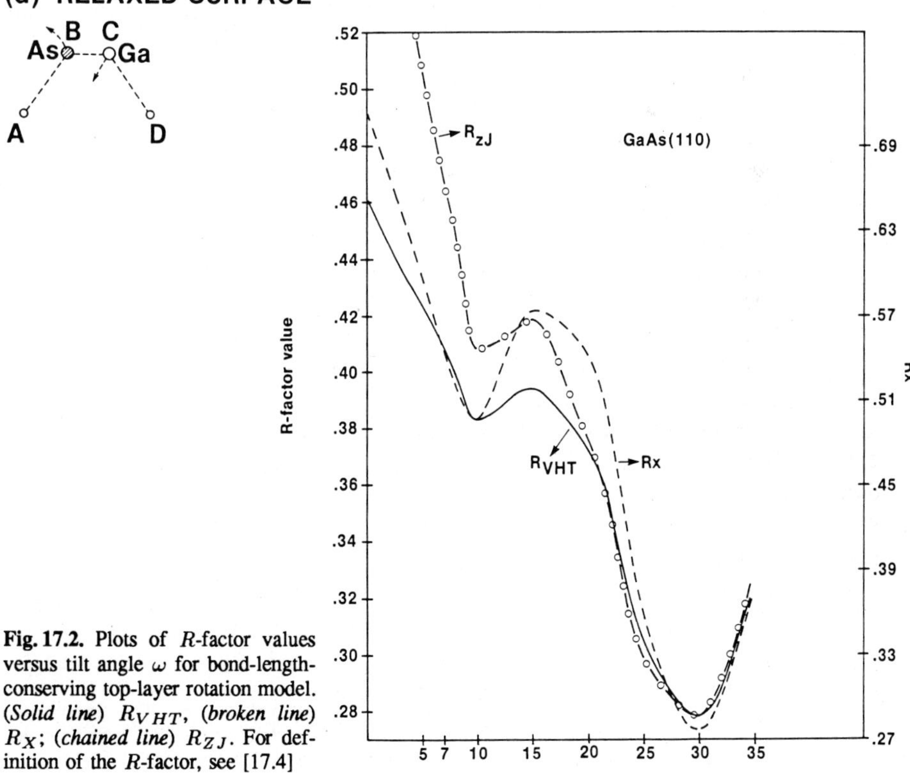

Fig. 17.2. Plots of R-factor values versus tilt angle ω for bond-length-conserving top-layer rotation model. (*Solid line*) R_{VHT}, (*broken line*) R_X; (*chained line*) R_{ZJ}. For definition of the R-factor, see [17.4]

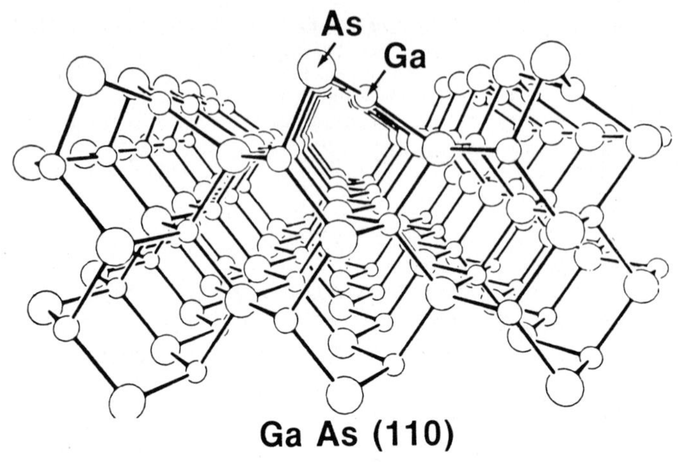

Fig. 17.3. Schematic view of the relaxed GaAs{110} structure

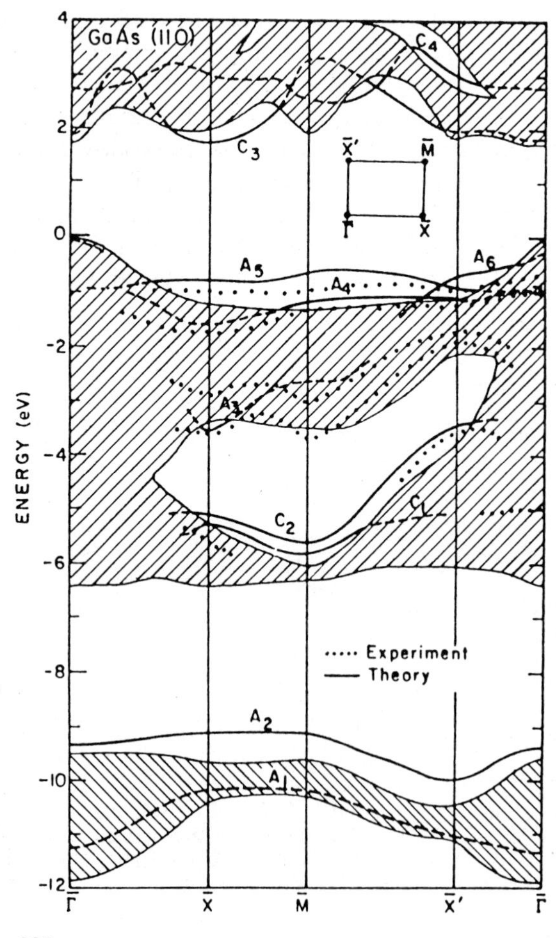

Fig. 17.4. Projected band structure for the GaAs relaxed {110} surface. Bulk features are indicated by the shaded regions. Calculated surface bands are shown by solid lines. If the surface band becomes resonant, then a dashed line is displayed. The experimental features are indicated by a dotted line. The experimental work is from [17.8] and this diagram is taken from [17.7]

The surface bond lengths are within a few percent of bulk bond lengths [17.2–4]. Thus, the surface region is only under a moderate elastic stress. The electronic structure of surface atoms are more atomic-like than that of the bulk, resulting in less charge transfer in the surface region. The geometric relaxation decays very rapidly as it enters the bulk, with second layer relaxations ~ 0.03 Å.

17.3 The GaAs{111}(2 × 2) Surface Reconstruction

On a bulk-terminated GaAs{111} surface, there is one Ga-derived dangling bond per surface atom. The surface is polar and metallic. This configuration is unstable and the surface atoms will relax to reduce the Coulomb forces between them. Consider now that a Ga vacancy is formed on the {111} surface. Then, three As dangling bonds are created. To achieve equal numbers of Ga and As dangling bonds on nearest neighbors, the new unit cell must contain three surface Ga atoms. A (2 × 2) periodicity satisfies this condition – each unit cell has a Ga vacany. There remain three surface Ga atoms, each of which is bonded to an As atom in the layer below.

In Fig. 17.5, we show a top view of the vacancy-buckling model that was proposed by *Tong* et al. [17.18]. In this model, each unit cell, bounded by the broken arrows of Fig. 17.5, has a Ga atom missing. The remaining three surface Ga atoms, denoted by A, B, and C, recede towards the bulk. By symmetry, they stay on a single plane parallel to the surface. These Ga atoms are bonded to three As atoms in the layer below; the latter (As atoms labeled a, c, and d) also form a single plane, by symmetry. Each of these Ga and As atoms has a dangling bond.

The fourth As atom (labeled b) is tetrahedrally bonded to Ga atoms above and below. This As atom is displaced deeper towards the bulk, relative to the other

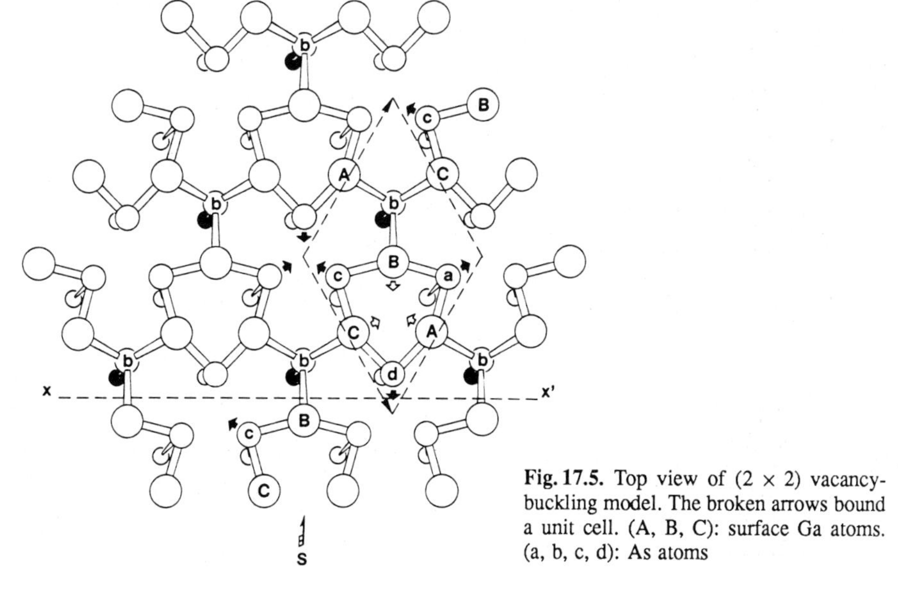

Fig. 17.5. Top view of (2 × 2) vacancy-buckling model. The broken arrows bound a unit cell. (A, B, C): surface Ga atoms. (a, b, c, d): As atoms

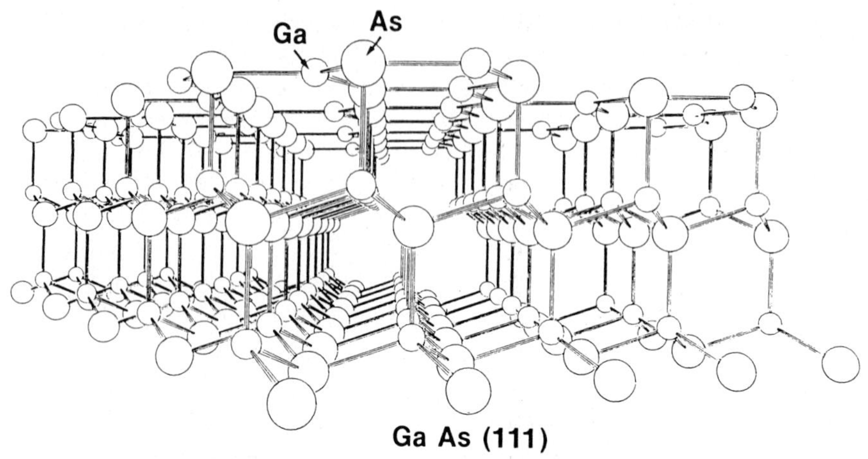

Fig. 17.6. Side view of (2×2)GaAs$\{111\}$, showing an almost flat surface layer, with $\frac{1}{4}$ layer Ga atoms missing

three As atoms in the unit cell. The Ga atoms in the third layer are also displaced to form two planes; three Ga atoms (smallest empty circles in Fig. 17.5) are raised towards the surface while the fourth Ga atom (solid circles) is depressed towards the bulk. Deeper layers have essentially the (1×1) bulk structure.

The detailed structure of the GaAs$\{111\}(2 \times 2)$ surface was determined by dynamical LEED analysis and the coordinates of atoms in the near surface region was tabulated in Table 1 of [17.18]. The structural characteristic of this surface is that the first GaAs bilayer distance is very small, ~ 0.07 Å compared to the bulk value of 0.816 Å. Figure 17.6 shows a side-view of this reconstructed surface. One could view the almost planar first bilayer as a Ga_3As_4 compound layer. Electronically, this compound layer is almost atomic-like, resulting in a non-polar termination of the $\{111\}$ surface. Figure 17.7 shows a plot of the LEED R-factor vs. the first bilayer distance. We note the deepest minimum is in the 0.05–0.08 Å range. Within this almost flat compound layer, the Ga orbitals are

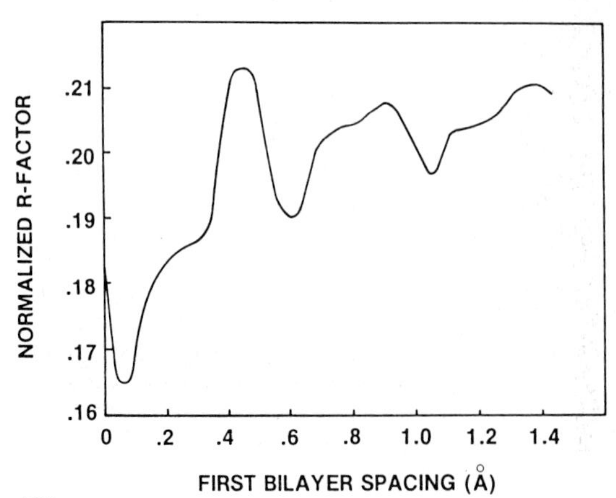

Fig. 17.7. Plot of R-factor versus first bilayer spacing

rehybridized into planar sp^2 orbitals and the orbitals on three As atoms in the unit cell are rehybridized into prismatic s^2, p^3 orbitals. The fourth As atom in the unit cell remains four-fold bonded with Sp^3 orbitals.

The vacancy model for the {111} group III atom terminated surface of III-V compounds was also found to be valid by quantitative Auger intensity analysis on GaAs{111} [17.19], angle-resolved photoemission (also on GaAs{111}) [17.20], by surface X-ray scattering (on InSb{111}) [17.21] and by another dynamical LEED analysis (on GaP{111}) [17.22]. Total-energy calculations showed that the Ga-vacancy model is $\simeq 3.1$ eV per (2×2) unit cell lower in energy than the bulk-terminated structure [17.23,24]. Based on such calculations, *Kaxiras* et al. [17.23] suggested that the Ga-vacancy model was the most likely structure that would form under normal annealing in high-vacuum conditions. They also found an As-triangle structure which is even 2 eV per (2×2) unit cell lower in energy than the Ga-vacancy model. *Kaxiras* et al. postulated that the As-triangle structure would form under an As-rich environment [17.23]. So far, such a surface has not been reproduced by experiment.

To make sure that the LEED IV spectra were taken on a surface with the Ga-vacancy structure and not the As-triangle structure, we calculated the IV spectra from a surface with the optimized As-triangle model. Figure 17.8 shows the comparison of the $(\frac{1}{2}, 0)$ beam for the two models with the experimental spectra. The normalized individual beam r-factors for the two models are in-

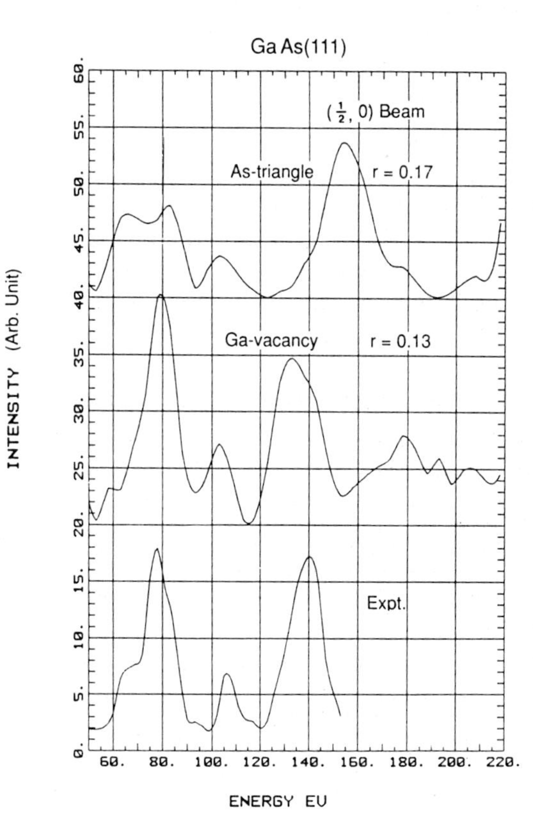

Fig. 17.8. Comparison between calculated (upper two curves) and measured IV spectra for the optimized As-triangle and Ga-vacancy models. The r-factor for the individual beam is shown

dicated in the figure. For this beam, the r-factor of the Ga-vacancy model is 30% better than that of the optimizied As-triangle model. Averaging over the 10 beams used in the analysis, the overall R-factor of the Ga-vacancy model is 31% better than that of the As-vacancy model. Based on this analysis, we conclude that the GaAs$\{111\}(2 \times 2)$ surface that was studied by dynamical LEED has the Ga-vacancy model. We believe that this conclusion also holds true for the IbSb$\{111\}(2 \times 2)$ structure (studied by surface X-ray scattering [17.21] and the GaP$\{111\}(2 \times 2)$ structure (again studied by LEED) [17.22].

17.4 Structure of the Si$\{111\}7 \times 7$ Surface

The determination of the Si$\{111\}(7 \times 7)$ reconstruction structure must rank among the top achievements in the 25 year history of surface crystallography. Perhaps it is also fitting that this most complicated surface was also the very first reconstructed structure that was ever reported (i.e., by LEED in 1959) [17.25]. Since the first experiment, for over 25 years the atomic geometry of this complicated surface has been hidden from investigators. However, important clues to the correct geometry have been revealed by investigators using a combination of theoretical deductions and experimental observations. The concept of adatoms in the outermost layer was proposed by *Harrison* [17.26] many years ago. Key experiments using constant momentum transfer averaging (CMTA) LEED [17.27] and ion-scattering spectroscopy (ISS) [17.28, 29] revealed non-cubic stacking in the selvedge. *Binnig* et al. [17.30] in 1983 obtained the first scanning-tunneling microscopy (STM) picture of this surface, which showed triangular units with deep holes at the vertices and smaller holes along the boundaries of the triangles. This led *Himpsel* [17.31], *McRae* [17.32], and *Bennett* et al. [17.33] to propose independent models in which the (7×7) unit cell is divided into two equal triangles with different stacking. A key contribution was made by *Himpsel* [17.31] and *McRae* [17.32], who showed that rebonding of Si atoms across the boundary between the faulted and unfaulted triangular regions results in three dimers, a ring of 12 atoms and two rings of 8 atoms, in exact agreement with features in the earlier STM image. Finally, *Takayanagi* et al. [17.34] using transmission electron diffraction correctly pieced together a model with dimers, adatoms and stacking faults (DAS). Thus, the unraveling of this highly complicated structure: one that embodies ingredients of adsorbed atoms, vacancies, stacking faults and rebonding dimerization, is a remarkable demonstration of the ability of surface analytical tools for structural analysis.

Complete atomic coordinates of the DAS model have been determined by dynamical LEED analysis [17.35, 36]. The Cartesian coordinates of the first 200 surface atoms of this structure were tabulated in Table 1 of [17.36]. Atomic coordinates parallel to the surface have also been determined by surface X-ray scattering [17.37]. The structure of this surface has also been studied recently by RHEED [17.38] and high-energy ion scattering [17.39].

The dynamical LEED analysis [17.35, 36] uses symmetries in real [17.40] and reciprocal spaces [17.41] to reduce the number of independent scatterers and the number of beams needed for the calculation. The basic features of the DAS model consist of an adatom layer and a (7×7) bilayer below, whose unit cell is made up of faulted and unfaulted triangular halves. These layers have C_{3v} symmetry. If the atoms in the faulted and unfaulted halves are assumed to have the same height, then the adatom layer and the bilayer below both have C_{6v} symmetry. Without using symmetry, the number of atoms/unit cell are 12 and 90, respectively, for the adatom layer and the bilayer below. With full real and reciprocal spaces symmetry, the corresponding numbers become 2(4) and 12(21), respectively, for $C_{6v}(C_{3v})$ symmetries. If we assume the next bilayer to relax into the (7×7) unit cell, its number of atoms/cells is reduced from 98 to 24, using C_{3v} symmetry.

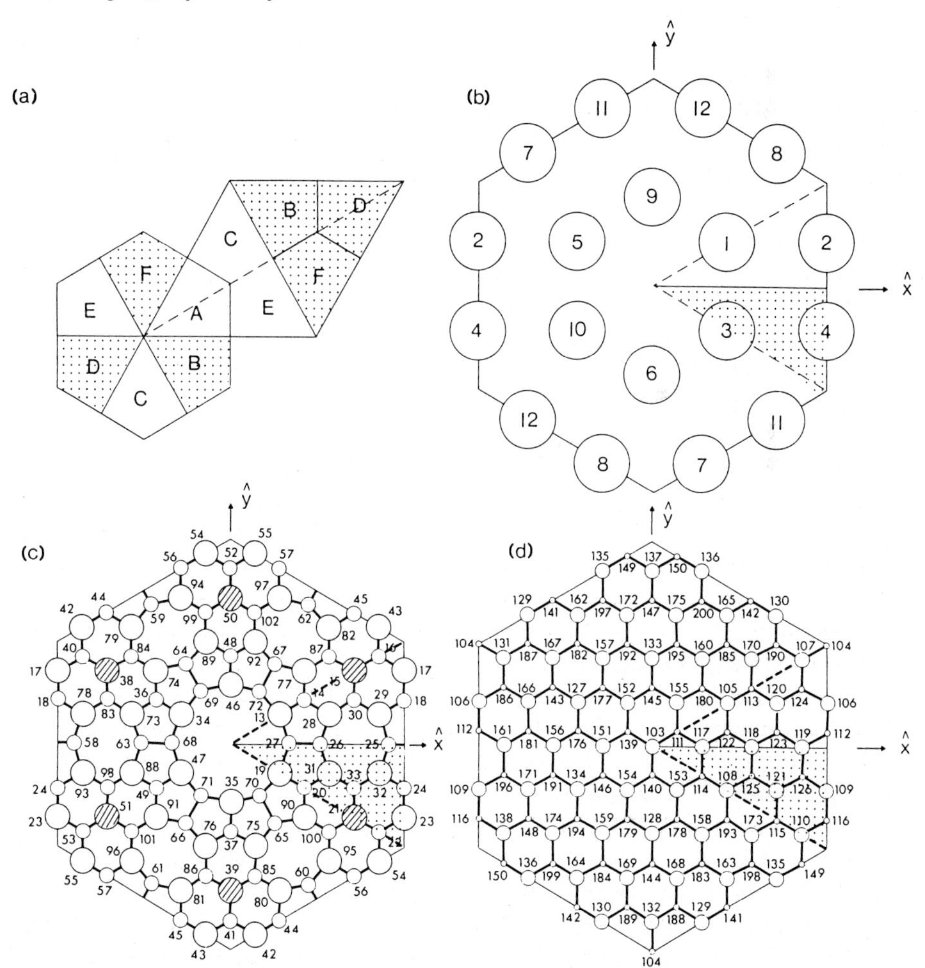

Fig. 17.9. (**a**) Symmetrized areas in (7×7) real-space unit cell. Dotted areas represent faulted regions. Ordering of atoms in (**b**) adatom layer, (**c**) first bilayer, and (**d**) second bilayer. Broken lines represent mirror planes

Figure 17.9 shows a plan view of the atoms in a (7×7) unit cell for the adatom layer (b), the bilayer below (c) where half of the atoms are faulted, and the atoms in the second bilayer (d). Due to C_{3v} symmetry, the irreducible area in the unit cell is the wedge bounded by the two broken lines (these are also the mirror planes). A structural feature determined by LEED, and supported by recent calculations of the total energy [17.42, 43], indicates that the atoms in the faulted and unfaulted sections of the unit cell have different vertical heights. This result is very clear in the dynamical LEED analysis. If the boundaries between the faulted and unfaulted sections are mirror planes [e.g., the x-axis of Fig. 17.9b and (c)], then the DAS structure should have C_{6v} symmetry for the adatom layer and the first bilayer below. Since the main contribution to the fractional order spots comes from diffraction off these layers, we expect that the IV curves of fractional order beams related by this new mirror plane (which exists only for the adatom and first bilayer below) will be rather similar. The intensity of these beams should not be identical due to contributions from deeper layers, which have the lower C_{3v} symmetry.

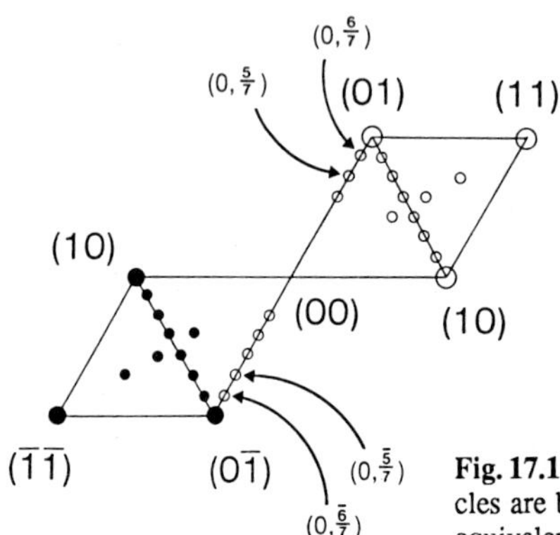

Fig. 17.10. Reciprocal space of the (7×7) structure. Open circles are beams whose IV curves are measured. Solid circles are equivalent beams by symmetry at normal incidence

In Figs. 17.10–12, we show the reciprocal lattice of the (7×7) surface and the IV spectra for the C_{6v} related beams: $(0\,5/7)$, $(0\,\bar{5}/7)$; and $(0\,6/7)$, $(0\,\bar{6}/7)$. From Fig. 17.11, the measured IV spectra for the $(0\,5/7)$ and $(0\,\bar{5}/7)$ pair of beams are very different. The same is true for the $(0\,6/7)$ and $(0\,\bar{6}/7)$ pair, as shown in Fig. 17.12. This is strong evidence that the C_{6v} symmetry does not exist for the adatom layer and the bilayer below. The calculated IV curves, using the atomic coordinates of Table 1 of [17.36], show good agreement with the data. In the structural model, the adatoms in the faulted half are raised by $\simeq 0.08$ Å relative to those in the unfaulted side. The higher positions of adatoms in the faulted half of the unit cell are consistent with the interpretation of STM images in which these atoms show the greatest brightness [17.44, 45].

We now turn to the energetics of the DAS structure. Using the criteria stated in Sect. 17.1, we look for the reduction in the number of unpaired dangling bonds in the surface region. The formation of the stacking-fault, with the creation of 3 dimers, a ring of 12 atoms and two rings of 8 atoms reduces the number of

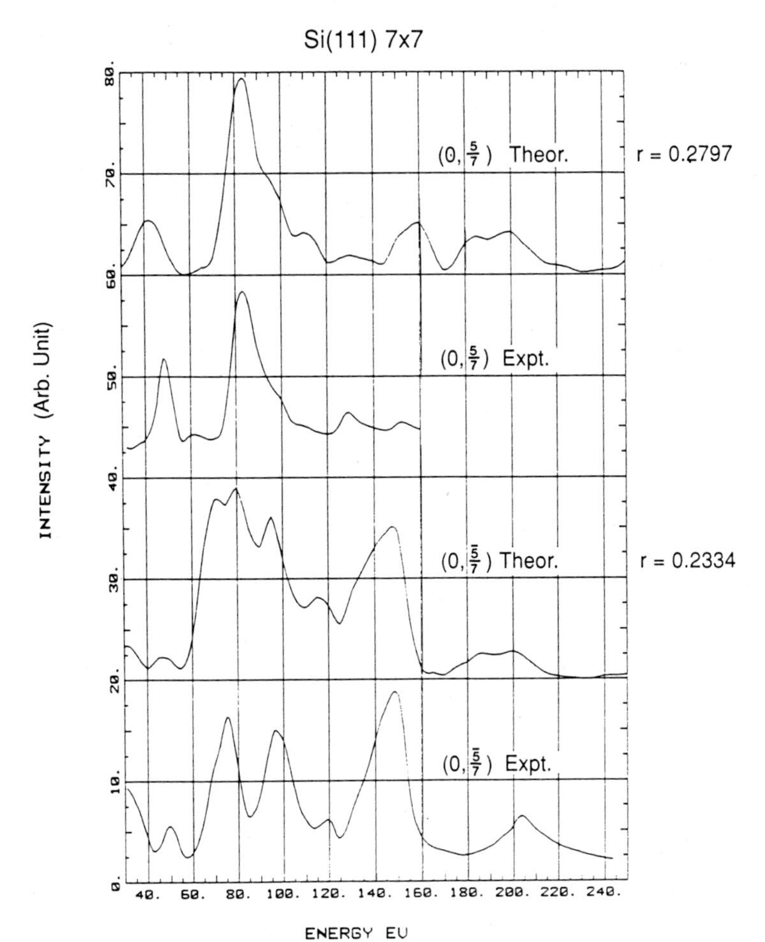

Fig. 17.11. Comparison between theory and experiment of the pair of beams (0 5/7) and (0 $\bar{5}$/7). The individual beam r-factor is indicated

unpaired dangling bonds from 49 to 43 per unit cell (Fig. 17.9c). Of these, 42 are dangling bonds on atoms in the first bilayer, and the remaining one is on atome '103' in the second bilayer (see Fig. 17.9d). The atoms forming the dimers (e.g., atoms 26 and 27 of Fig. 17.9c) are 4-fold coordinated, although the bond angles are severly distorted from the ideal tetrahedral angle of 109°28′. Thus, the dimer atoms and their neighbors are under stress. The 12 adatoms at three-fold sites, each would pair-up three dangling bonds but adds one unpaired dangling bond on the adatom itself, further reducing the unpaired dangling bond population to 19 in a unit cell. Of these, 12 dangling bonds are on adatoms, 6 on the rest atoms in the first bilayer (cross-hatched circles in Fig. 17.9c) and one atom '103' in the second bilayer.

A further reduction in the unpaired dangling bond population is possible through bond-bond interaction between the dangling bonds on the rest atoms and

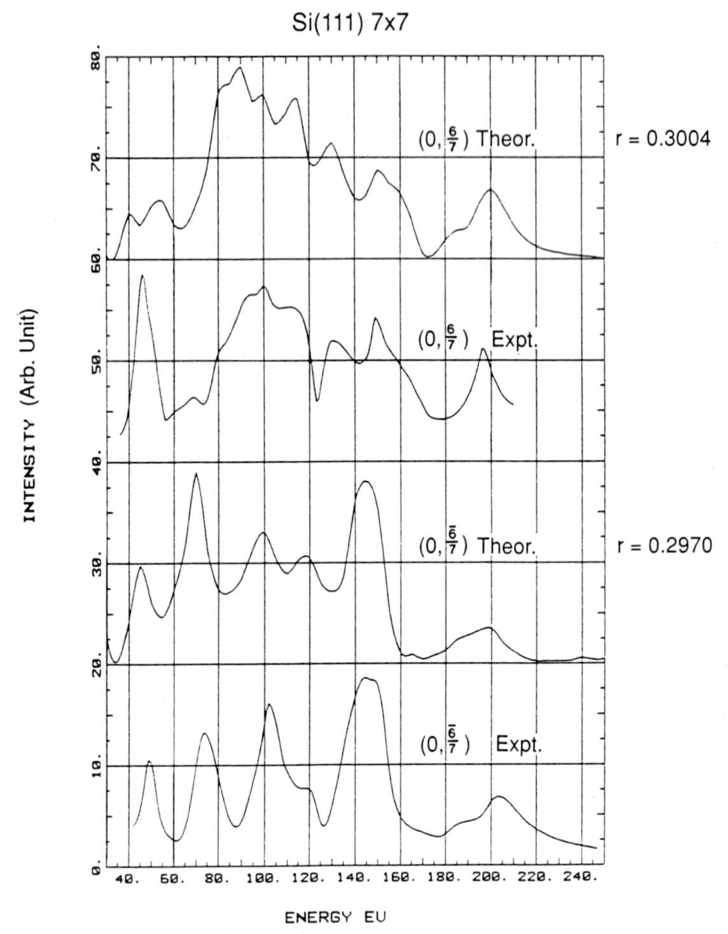

Fig. 17.12. Same as in Fig. 17.11 except for the pair of beams (0 6/7) and (0 $\bar{6}$/7)

adatoms. *Meade* and *Vanderbilt* [17.43] have carried out a first-principles total energy calculation to show that rehybridization of this type does take place. Two surface bands are created in which the lower (filled) band S_2 is centered primarily on top of the rest atoms, while the higher (empty) band S_1 is centered above and around the adatom. Figure 17.13 shows the charge density contour plots of these two bands. Although the calculation of *Meade* and *Vanderbilt* [17.43] was done for a (2×2) unit cell, the phenomenon it predicted should also hold true for the (7×7) surface. This interaction between adatom and rest atom bonds effectively further reduces the number of unpaired dangling orbitals to 7 (i.e., 6 on adatoms and 1 on atom '103'). An unresolved discrepancy in this description is that in the total energy pseudopotential calculation of *Meade* and *Vanderbilt* [17.43], the rest atoms are raised (due to doubly filled surface orbitals) much higher than the values which were determined by dynamical LEED in the (7×7) structure. A possible explanation is that in the (7×7) unit cell, because of the

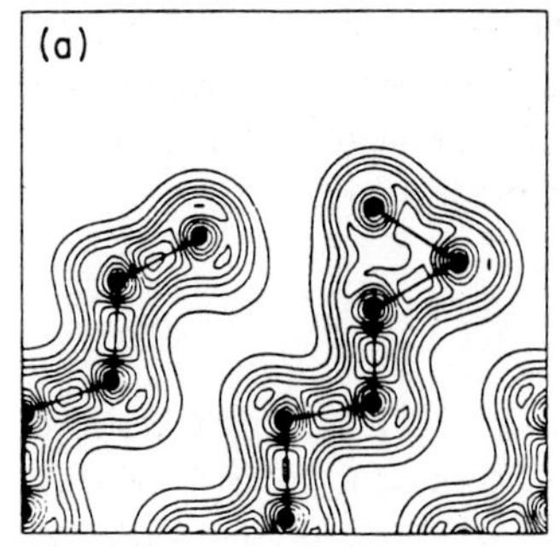

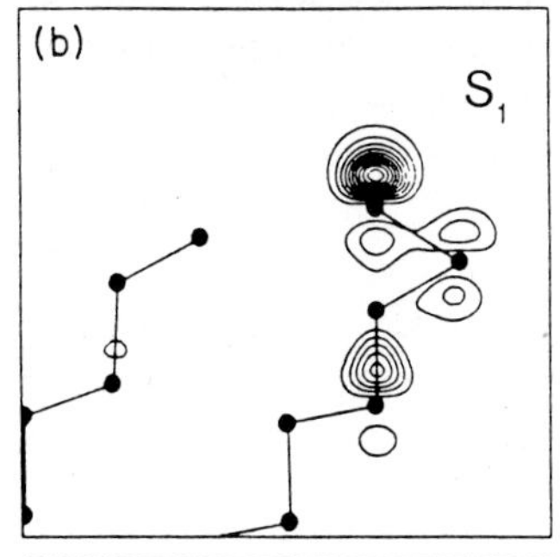

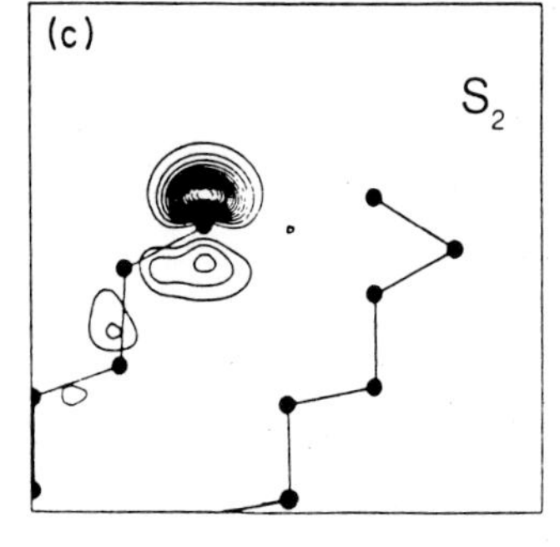

Fig. 17.13a–c. Contour plots of the charge density of (2 × 2) adatom-covered Si{111}. The figures show a cross section of the charge density on a plane passing through both the adatom and the rest atom. **(a)** is the total charge density. **(b)** and **(c)** show the average charge density of all states in bands S_1 and S_2, respectively. From [17.43]

C_{3v} symmetry and the unequal numbers of adatoms to rest atoms in the unit cell, the rehybridization may not be as complete as that for a (2 × 2) surface.

It is useful to compare, then, the number of unpaired dangling bonds in a totally relaxed DAS structure with that of a saturated adatom model. In a saturated adatom model, the unit cell is $(\sqrt{3} \times \sqrt{3})$R 30°. The most efficient way for adatoms to reduce dangling bonds is to absorb at 3-fold sites. On a saturated adatom (at 3-fold sites) surface, the number of unpaired dangling bonds equals 1/3 ML, all of these are on top of the adatoms. Therefore, for 49 atoms in a (7 × 7) cell, the number of unpaired dangling bonds would be $49/3 \sim 16$. Each unpaired dangling bond is on top of an adatom. Due to the C_{3v} symmetry of adatoms, these dangling bonds are unlikely to go through a Jahn-Teller distortion. The fully relaxed DAS model, on the other hand, has only 7 unpaired dangling bonds per (7 × 7) unit cell. By comparison, the DAS structure is thus more stable than an all adatom $\sqrt{3}$-surface.

In summary, the Si{111} DAS model is stabilized via processes (i) and (ii) described in Sect. 17.1. The number of unpaired dangling bonds is reduced from 49 per unit cell to 7. This is achieved via a combination of adatoms, dimerization (resulting in a stacking fault), and surface bond rehybridization.

17.5 The Ge{111} c(2 × 8) Reconstruction

The Ge{111} c(2 × 8) surface, together with the Si{111}(7 × 7) surface, rank among the most difficult surfaces for structural determination because of their large unit cells. The LEED pattern, i.e., reciprocal unit cell of the Ge{111} c(2 × 8) surface, is shown schematically in Fig. 17.14a for one domain and (b) for an average of three domains. *Chadi* and *Chiang* [17.46], *Yang* and *Jona* [17.47] have suggested basic building blocks for the c(2 × 8) unit cell, although the number of atoms in each unit cell was not specified. In 1985, *Becker* et al. [17.48] obtained the first scanning tunneling microscope (STM) image for this surface. The STM pictures are consistent with a surface unit cell with two adatoms in an arrangement shown in Fig. 17.15. The adatom model has also been suggested by ion-scattering [17.49], photoemission [17.50] and X-ray diffraction [17.51] studies.

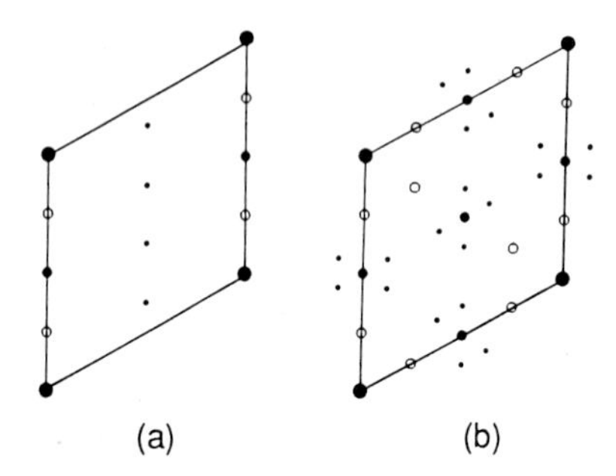

Fig. 17.14a,b. Reciprocal space of c(2 × 8) structure: (**a**) a single domain, (**b**) three domains

The complete structural coordinates of this surface have recently been determined by dynamical LEED analysis [17.52]. Again, in order to overcome the difficulty posed by the large unit cell, we used a vectorized LEED code that employed symmetries in real and reciprocal spaces. The structural analysis by LEED is summarized as follows: (i) The Ge{111} c(2 × 8) reconstruction consists of adatoms on a relaxed (1 × 1) substrate. The adatoms occupy the T_4 site. The H_3 site was tested and ruled out. (ii) The Ge atoms in the (1 × 1) substrate undergo substantial relaxations in directions normal and parallel to the surface. Appreciable sub-surface relaxations run as deep as five atomic layers. (iii) The dimer-chain mode, which consists of stacking faults and chains of dimers, constructed by *Takayanagi* et al. [17.53] in a manner similar to that of the dimer–adatom–stacking fault (DAS) model [17.34] which successfully explained the Si{111}(7 × 7) reconstruction, was found to be inconsistent with the IV data. (iv) The quasi-dynamical (QD) method [17.54], introduced earlier to treat large unit cells, was found to work very well for the Ge{111} c(2 × 8) reconstruction.

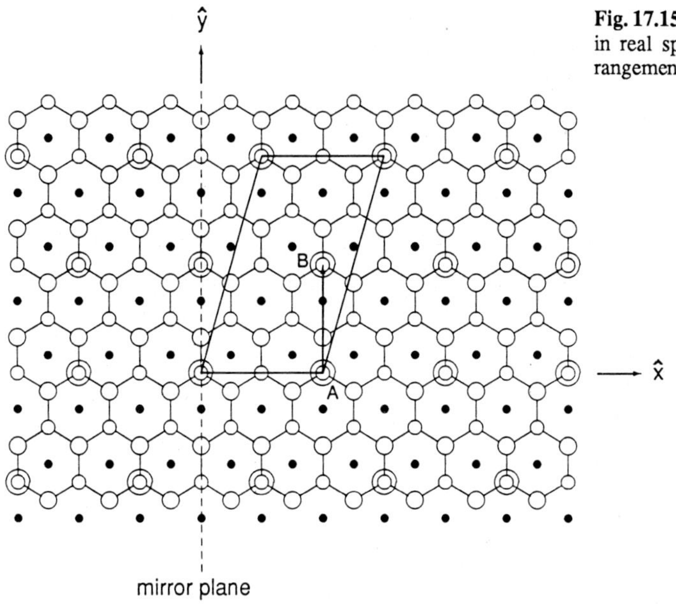

Fig. 17.15. c(2 × 8) unit cell in real space, showing the arrangement of adatoms at T_4 sites

This may be due to the fact that the surface structure [adatoms on a relaxed (1 × 1) surface] is relatively open and the data used in the analysis were taken at normal incidence.

The Ge{111} c(2 × 8) adatom structure has a mirror plane denoted in Fig. 17.15 by the broken line. For analysis of normal incidence data (or data with the incident direction in the mirror plane), the number of atoms/unit cell for the adatom layer is 2. For the bilayer below (i.e., atomic planes 2 and 3 in Fig. 17.16a), because the vertical distance between the atoms are < 1 Å, the two atomic planes are treated as a single "composite" layer in the multiple scattering formulation. For this composite layer, the number of atoms/unit cell is 12 (16 if real-space symmetry is not used). Similarly, atomic planes 4 and 5 in Fig. 17.16a are treated as a single composite layer and there are 12 atoms in the c(2 × 8) unit cell. Thus, the structural search involved relaxing the coordinates of atoms in the first 5 atomic planes, while holding atoms in deeper layers fixed at the bulk positions. Plan views of atomic layers made up of atomic planes 1, 2, 3 and 4, 5 and 6, 7 are shown in Fig. 17.16b–d respectively.

Using the available data, the LEED analysis considered the following models in the structural search: the dimer-chain model of *Takayanagi* et al. [17.53], adatom models with adatoms at H_3 sites, and adatoms at T_4 sites. For each mode, the atoms in the first five atomic planes are relaxed in directions normal and parallel to the surface, to obtain the best fit within each model. We show in Figs. 17.17 and 18 the comparisons for two selected beams. In these figures, panel (a) is for the optimized dimer-chain model, (b) for the optimized adatom model with adatoms at H_3 sites and (d) is for the optimized adatoms model with adatoms at T_4 sites. Panel (c) is for adatoms at T_4 sites sitting on an

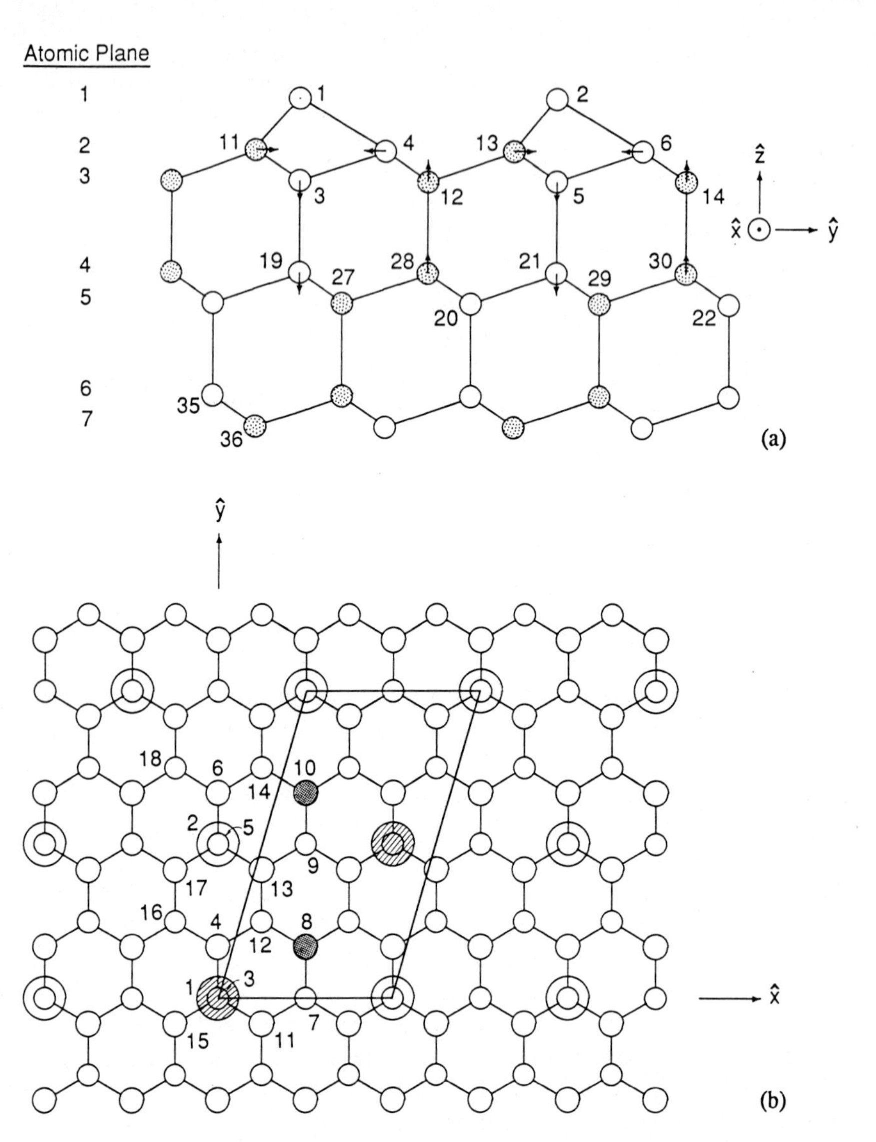

Fig. 17.16. (**a**) Side view of 7 layers of Ge{111} c(2 × 8) structure. The 36 atoms correspond to the coordinates listed in Table 17.1. (**b**) Top view of atomic planes 1, 2 and 3. (**c**) Top view of atomic planes 4 and 5. (**d**) Top view of atomic planes 6 and 7. In (b–d), the same 36 atoms are numbered

unrelaxed substrate. The best dimer-chain model [i.e., panel (a)] produced an overall R-factor (averaged over seven beams) of 0.34. This is 26% worse than the averaged R-factor of the optimized T_4 structure (shown in panel (d)]. The best model for adatoms at H_3 sites produced a R-factor of 0.31 [15% worse than (d)]. Our results show that substrate relaxation is *more important* than the adatom location. Without substrate relaxation, putting adatoms at T_4 sites [panel

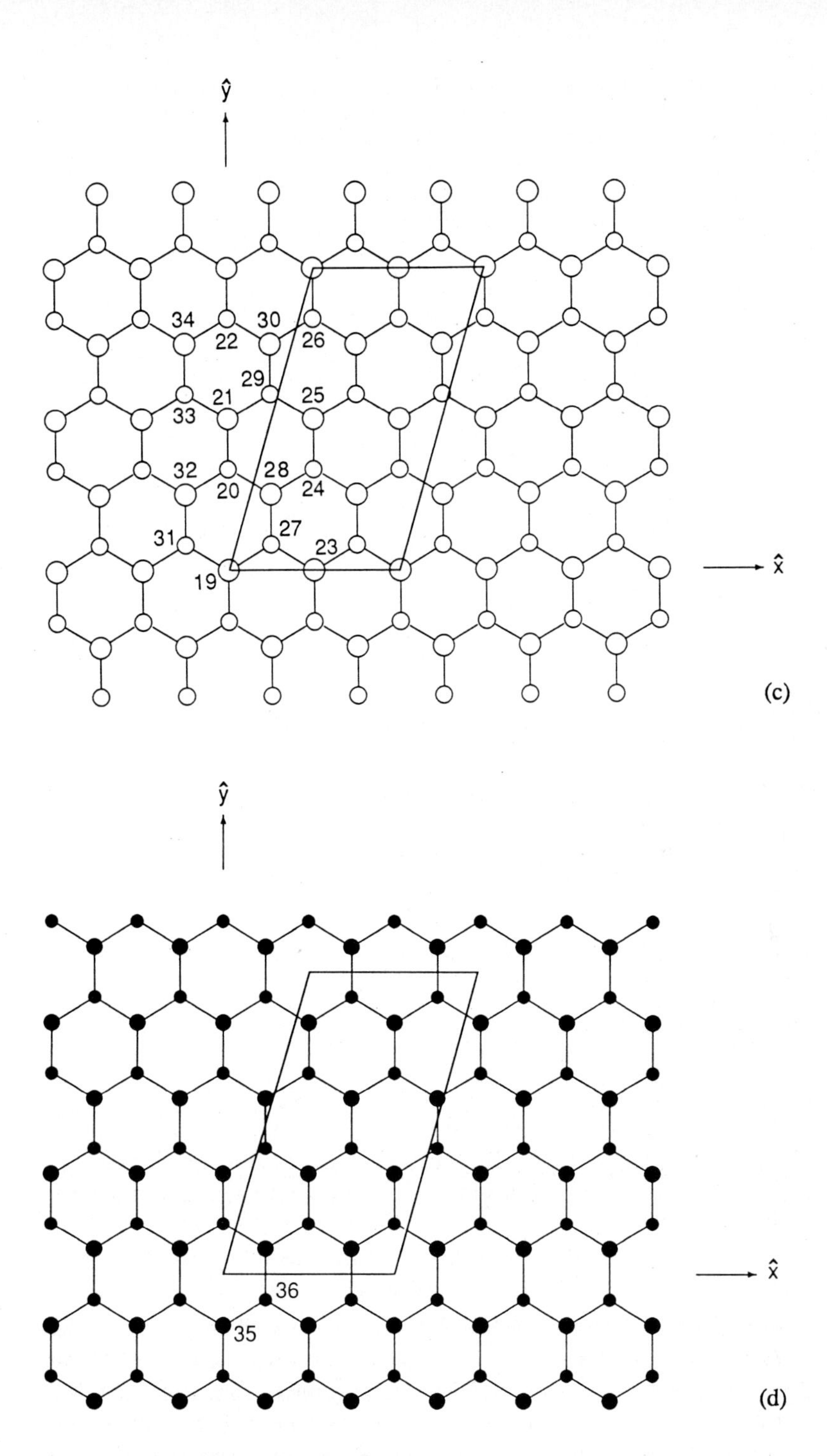

Fig. 17.16c,d. Caption see opposite page

Table 17.1. Atomic coordinates of the 36 surface atoms shown in Fig. 17.16 are listed. The $z = 0$ point is the bulk location of atomic plane 7. The $x = y = 0$ points is at the bulk location of atom 3. Δx, Δy, Δz indicate displacements from corresponding bulk position.

atom no.	$\hat{x}$	$\hat{y}$	$\hat{z}$	Δx	Δy	Δz (Å)
1	0.000	0.000	8.506			
2	0.000	6.932	8.506			
3	0.000	0.000	5.919			−0.620
4	0.000	2.011	7.136		−0.300	−0.220
5	0.000	6.932	5.919			−0.620
6	0.000	8.943	7.136		−0.300	−0.220
7	4.002	0.120	6.639		+0.120	+0.100
8	4.002	2.311	7.356			
9	4.002	7.052	6.639		+0.120	+0.100
10	4.002	9.243	7.356			
11	1.741	−1.005	7.136	−0.260	−0.150	−0.220
12	2.105	3.406	6.639	+0.104	−0.060	+0.100
13	1.741	5.927	7.136	−0.260	−0.150	−0.220
14	2.105	10.338	6.639	+0.104	−0.060	+0.100
15	−1.741	−1.005	7.136	+0.260	−0.150	−0.220
16	−2.105	3.406	6.639	−0.104	−0.060	+0.100
17	−1.741	5.927	7.136	+0.260	−0.150	−0.220
18	−2.105	10.338	6.639	−0.104	−0.060	+0.100
19	0.000	0.000	3.746			−0.340
20	0.000	4.621	3.199			−0.070
21	0.000	6.932	3.746			−0.340
22	0.000	11.553	3.199			−0.070
23	4.002	0.000	4.126			+0.040
24	4.002	4.621	3.199			−0.070
25	4.002	6.932	4.126			+0.040
26	4.002	11.553	3.199			−0.070
27	2.001	1.155	3.199			−0.070
28	2.001	3.466	4.126			+0.040
29	2.001	8.087	3.199			−0.070
30	2.001	10.398	4.126			+0.040
31	−2.001	1.155	3.199			−0.070
32	−2.001	3.466	4.126			+0.040
33	−2.001	8.087	3.199			−0.070
34	−2.001	10.398	4.126			+0.040
35	0.000	−2.311	0.816			
36	2.001	−1.155	0.000			

(c)] produced a R-factor of 0.35, which is 13% worse than the best adatom model with H_3 sites [i.e., panel (b)]. The LEED optimized structure of panel (d) is shown schematically in Fig. 17.16a and Table 17.1 lists the atomic coordinates for this surface structure.

The adatom model we used assumed that the local environments of adatoms A and B are identical (see Fig. 17.15). This structural symmetry leads to the extinction of the $\frac{1}{4}$ order spots in the kinematical scattering model. However, by multiple scattering, the $\frac{1}{4}$ order spots are non-zero. The calculated intensities of the $\frac{1}{4}$ order spots are two orders of magnitude smaller than those of the $\frac{1}{2}$ order spots, and the same ratio was found for the corresponding experimental beams. From this, we conclude that the difference in the local environments of the two adatoms in the unit cell is indeed very small.

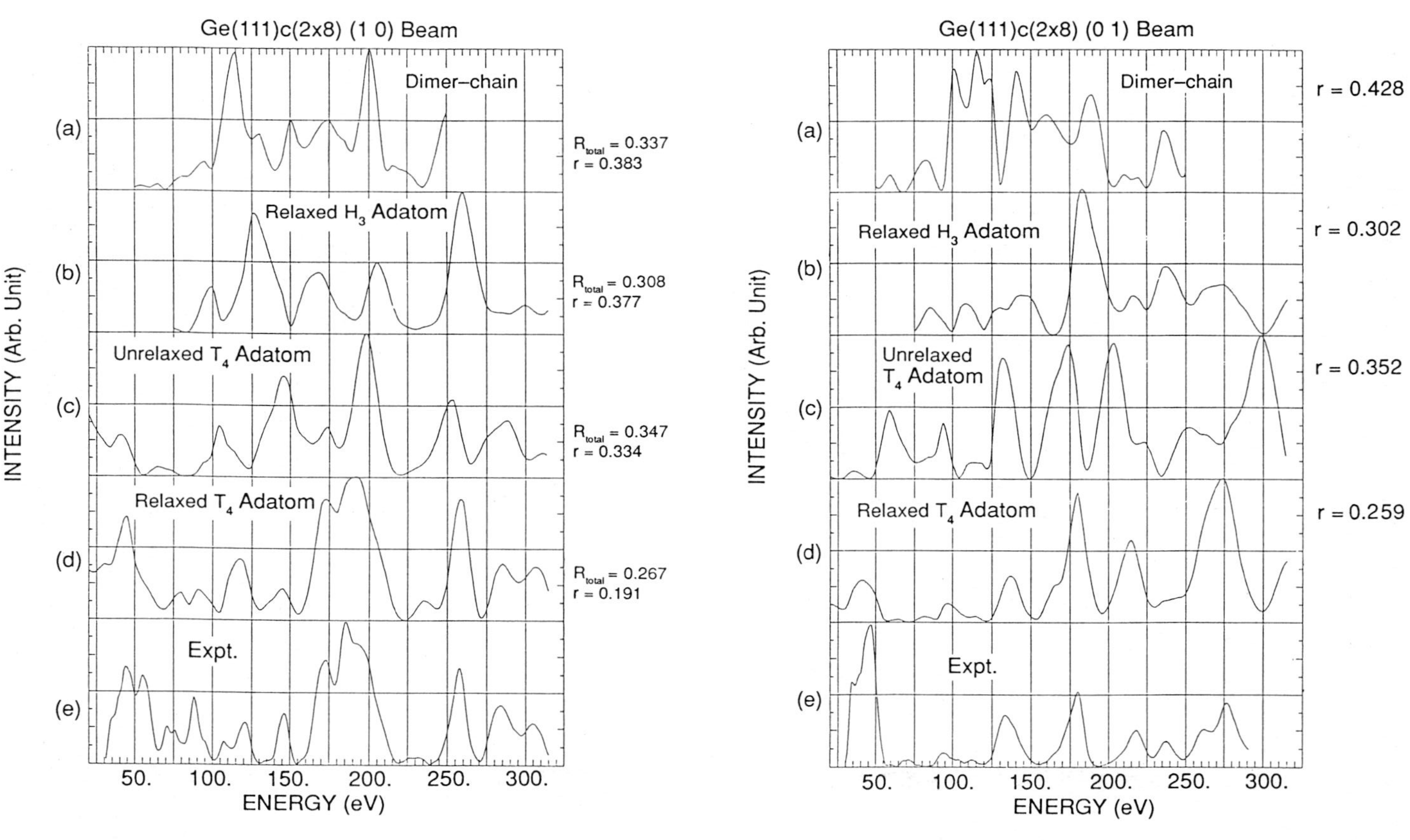

Fig. 17.17. Comparison of LEED IV curves for the different models indicated. Top four curves are calculated. Individual beam r-factor and average R-factor over all beams are indicated for each model. The (10) beam is shown

Fig. 17.18. Same as in Fig. 17.18, but now for the (01) beam

For the dimer-chain model [17.53], the calculated intensities of the $\frac{1}{4}$ order beams are of the same order of magnitude as those of $\frac{1}{2}$ order beams. This is a direct consequence of the lack of quasi-translational symmetry for the model. Furthermore, the dimer-chain model with atomic relaxations does not have a mirror plane, whereas the LEED data indicate that there is one. Therefore, based on symmetry consideration alone, the dimer-chain model can be ruled out as unlikely.

In the quasi-dynamical (QD) approximation [17.54], multiple scattering *within* a layer is neglected while multiple scatterings among layers are kept to all orders. Here, a *layer* may be a single plane of atoms or a composite layer defined earlier. The QD approximation works best at normal incidence because multiple scattering within a layer involves at least one 90° (or near 90°) scattering. For a 100 eV or higher energy electron, a 90° scattering is usually small. The QD approximation works very well for the Ge{111} c(2 × 8) structure. Figures 17.19 and 20 show the comparison between the QD and full-dynamical calculations with experiment for the optimal T_4 site model. The agreement between the QD and full-dynamical calculation, especially above 100 eV, is extremely good. Since the QD approximation is an order of magnitude faster and requires a much smaller core-space, it is often used in the initial stages of a search process.

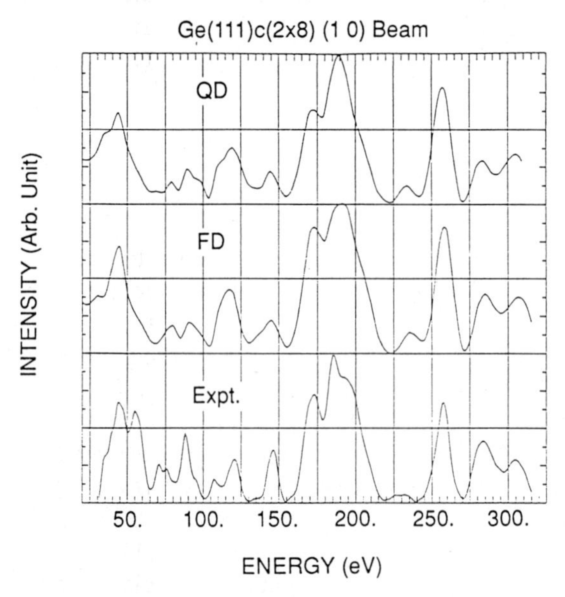

Fig. 17.19. Comparison of quasidynamical (QD) and fully dynamical (FD) calculations with experiment. The (10) beam is shown

We now consider the energetics of this surface. On the Ge{111} c(2 × 8) surface, the number of unpaired dangling bonds is reduced by adatoms. If we consider Fig. 17.16b, in a c(2 × 8) unit cell, there remain 4 dangling bonds. Of these, two are above the adatoms (cross-hatched) and two above the rest atoms (dotted). Further rehybridization in a manner similar to the Si{111}(7 × 7) surface results in (mainly) filled bands on top of the rest atoms and (mainly) empty bands on top of the adatoms. If this rehybridization is complete, there

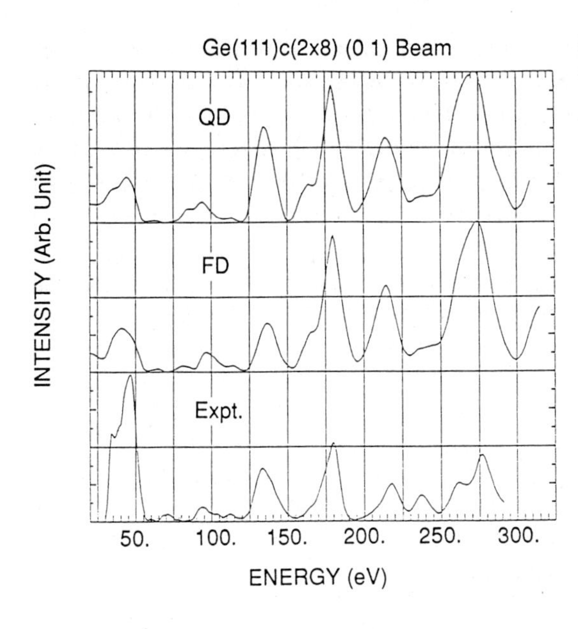

Fig. 17.20. Same is in Fig. 17.19, but now for the (01) beam

will be no unpaired dangling orbitals left on the surface. Such an arrangement should be more stable than the saturation adatom $(\sqrt{3} \times \sqrt{3})\mathrm{R}\,30°$ structure. In a $\sqrt{3}$-structure, a $c(2 \times 8)$ unit cell contains $8/3 \sim 2.7$ unpaired dangling bonds, all centered above the adatoms.

17.6 Summary

We have shown that semiconductor surfaces reconstruct to reduce the number of *unpaired* dangling bonds in the surface region. The common processes are orbital rehybridization and the formation of novel bonds involving adatoms and/or dimers. For the four reconstructed surfaces considered, the remaining unpaired orbital density are 0, 0, 0.14 and 0 per (1×1) surface atom for GaAs$\{110\}(1 \times 1)$, GaAs$\{111\}(2 \times 2)$, Si$\{111\}(7 \times 7)$ and Ge$\{111\}$ $c(2 \times 8)$, respectively. On a bulk-terminated surface, the number of unpaired orbital density is 1 per surface atom. While the particular reconstruction structure assumed by a surface is a delicate balance between electronic and stress energies (this balance would be affected by external pressure, temperature, preparation procedure, etc.: e.g. Ge$\{111\}$ $c(2 \times 8) \rightarrow$ DAS (7×7) under external compression), the general trend discussed in this review suggests that an equilibrium structure should have a low density of unpaired surface orbitals.

We have shown that both the (7×7) DAS model and the $c(2 \times 8)$ adatom model have lower unpaired dangling bond density (i.e., 0.14 and 0, respectively) compared to that of a saturation adatom $\sqrt{3}$-structure (whose unpaired orbital density is 0.33). It is interesting to note that recent studies [17.55–58] suggest that an all adatom $\sqrt{3}$-structure could form an Si$\{111\}$, if the near surface region is doped with B-atoms. The B-atoms are found to occupy substitutional sites (called B_5) [17.55–60]. Due to its smaller size, its presence relieves elastic stress in the adatom cluster. More importantly, the B atom removes the unpaired electron from the adatom dangling bond via charge transfer from the adatom to the B atom. This process further reduces the density of unpaired orbitals on the $\sqrt{3}$-surface, thus making its formation more favorable.

Acknowledgements. The authors acknowledge support of this work by the National Science Foundation, Grant No. DMR-8805938, the Department of Energy, Grant No. De-FG02-84ER45076, the Office of Naval Research and the Petroleum Reserach Fund, Grant No. 1154-AC5,6.

References

17.1 See, for example, S.Y. Tong: Physics Today **37**, 50 (1984)
17.2 S.Y. Tong, A.R. Lubinsky, B.J. Mrstik, M.A. Van Hove: Phys. Rev. B **17**, 3303 (1978)
17.3 A. Kahn, G. Cisneros, M. Bonn, P. Mark, C.B. Duke: Surf. Sci. **71**, 387 (1978)
17.4 M.W. Puga, G. Xu, S.Y. Tong: Surf. Sci. **164**, L789 (1985)
17.5 S.Y. Tong, G. Xu, W.Y. Hu, M.W. Puga: J. Vac. Sci. Technol. B **3**, 1076 (1985)
17.6 D.J. Chadi: Phys. Rev. **18**, 1800 (1978)
17.7 J.R. Chelikowsky, M.L. Cohen: Phys. Rev. B **20**, 4150 (1979)
17.8 A. Huijser, J. van Laar, T.L. van Rooy: Phys. Lett. **65A**, 337 (1978)
17.9 W.E. Spicer, I. Lindau, D.E. Gregory, C.M. Garner, P. Pianetta, P.W. Chye: J. Vac. Sci. Technol. **13**, 780 (1976)
17.10 W. Gudat, D.E. Eastman, J.L. Freeouf: J. Vac. Sci. Technol. **13**, 250 (1976)
17.11 D.J. Chadi: Phys. Rev. Lett. **41**, 1062 (1978)
17.12 K.C. Pandey: Phys. Rev. Lett. **49**, 223 (1982)
17.13 L. Smit, T.E. Derry, J.F. van der Veen: Surf. Sci. **150**, 245 (1985)
17.14 S.Y. Tong, T.C. Zhao, H.C. Poon, K.D. Jamison, D.N. Zhou, P.I. Cohen: Phys. Lett. A **128**, 447 (1988)
17.15 K.D. Jamison, D.N. Zhou, P.I. Cohen, T.C. Zhao, S.Y. Tong: J. Vac. Sci. Technol. A **6**, 611 (1988)
17.16 R.M. Feenstra, J.A. Stroscio, J. Tersoff, A.P. Fein: Phys. Rev. Lett. **58**, 1192 (1987)
17.17 G.X. Qian, R.M. Martin, D.J. Chadi: Phys. Rev. B **37**, 1303 (1988)
17.18 S.Y. Tong, G. Xu, W.N. Mei: Phys. Rev. Lett. **52**, 1693 (1984)
17.19 M. Alonso, F. Soria, J.L. Sacedon: J. Vac. Sci. Technol. A **3**, 1598 (1985)
17.20 R.D. Bringans, R.Z. Bachrach: Phys. Rev. Lett. **53**, 1954 (1984)
17.21 J. Bohr, R. Feidenhans'l, M. Nielsen, M. Tonney, R.L. Johnson, I.K. Robinson: Phys. Rev. Lett. **54**, 1275 (1985)
17.22 G. Xu, W.Y. Hu, M.W. Puga, S.Y. Tong, J.L. Yeh, S.R. Wang, B.W. Lee: Phys. Rev. B **32**, 8473 (1985)
17.23 E. Kaxiras, K.C. Pandey, Y. Bar-Yam, J.D. Joannopoulos: Phys. Rev. Lett. **56**, 2819 (1986)
17.24 D.J. Chadi: Phys. Rev. Lett. **52**, 1911 (1984)
17.25 R.E. Schlier, H.E. Farnsworth: J. Chem. Phys. **30**, 917 (1959)
17.26 W.A. Harrison: Surf. Sci. **55**, 1 (1976)
17.27 P.A. Bennett, M.B. Webb, unpublished
17.28 R.J. Culbertson, L.C. Feldman, P.J. Silverman: Phys. Rev. Lett. **45**, 2043 (1980)
17.29 R.M. Tromp, E.J. van Loenen, M. Iwami, F.W. Saris: Solid State Commun. **44**, 971 (1982)
17.30 G. Binnig, H. Rohrer, Ch. Gerber, E. Weibel: Phys. Rev. Lett. **50**, 120 (1983)
17.31 F.J. Himpsel: Phys. Rev. B **27**, 7782 (1983)
17.32 E.G. McRae: Phys. Rev. B **28**, 2305 (1983)
17.33 P.A. Bennett, L.C. Feldman, Y. Kuk, E.G. McRae, J.E. Rowe: Phys. Rev. B **28**, 3656 (1983)
17.34 K. Takayanagi, Y. Tanishiro, M. Takahashi, S. Takahashi: J. Vac. Sci. Technol. A **3**, 1502 (1985)
17.35 H. Huang, S.Y. Tong, W.E. Packard, M.B. Webb: Physics Lett. **130**, 166 (1988)
17.36 S.Y. Tong, H. Huang, C.M. Wei, W.E. Packard, F.K. Men, G. Glander, M.B. Webb: J. Vac. Sci. Technol. A **6**, 615 (1988)

17.37 I.K. Robinson, W.K. Waskiewicz, P.H. Fuoss, L.J. Norton: Phys. Rev. B **37**, 4325 (1988)
17.38 A. Ichimiya: Surface Sci., to be published
17.39 J. Yanagisawa, A. Yoshimori: Surface Sci., in print
17.40 M. Moritz: J. Phys. C **17**, 353 (1984)
17.41 M.A. Van Hove, J.B. Pendry: J. Phys. C **8**, 1362 (1975)
17.42 M.Y. Chou, M. L. Cohen, S.G. Louie: Phys. Rev. B **32**, 7979 (1985)
17.43 R.D. Meade, D. Vanderbilt: Phys. Rev. B **40**, 3905 (1989)
17.44 R.M. Tromp, R.J. Hamers, J.E. Demuth: Phys. Rev. B **34**, 1388 (1986)
17.45 R.S. Becker, J.A. Golovchenko, E.G. McRae, B.S. Swartzentruker: Phys. Rev. Lett. **55**, 2028 (1985)
17.46 D.J. Chadi, C. Chiang: Phys. Rev. B **23**, 1843 (1981)
H. Huang, S.Y. Tong, W.E. Packard, M.B. Webb: Phys. Lett. A **130**, 166 (1988)
17.47 W.S. Yang, F. Jona: Phys. Rev. B **29**, 899 (1984)
17.48 R.S. Becker, J.A. Golovchenko, B.S. Swartzentruker: Phys. Rev. Lett. **54**, 2678 (1985)
17.49 P.M.J. Maree, K. Nakagawa, J.F. van der Veen, R.M. Tromp: Phys. Rev. B **38**, 1585 (1988)
17.50 J. Aarts, A.J. Hoeven, P.K. Larsen: Phys. Rev. B **38**, 3925 (1988)
17.51 R. Feidenhans'l, J.S. Pedersen, J. Bohr, M. Nielsen, F. Grey, R.L. Johnson: Phys. Rev. B **38**, 9715 (1988)
17.52 H. Huang, C.M. Wei, T.C. Zhao, H. Li, B.P. Tonner, S.Y. Tong, F.K. Men, M.B. Webb: Phys. Rev., to be published
17.53 K. Takayanagi, Y. Tanishiro: Phys. Rev. B **34**, 10324 (1986)
17.54 S.Y. Tong, M.A. Van Hove, B.J. Mrstik: in *Proceedings of the 7th International Vacuum Congress and 3rd International Conference on Solid Surfaces*, ed. by. R. Dobrozemsky et al. (Berger, Vienna 1977), Vol. 3, p. 2407
17.55 P. Bedrossian, R.D. Meade, K. Mortensen, D.M. Chen, J.A. Golovchenko, D. Vanderbilt: Phys. Rev. Lett. **63**, 1257 (1989)
17.56 I.W. Lyo, E. Kaxiras, Ph. Avouris: Phys. Rev. Lett. **63**, 1261 (1989)
17.57 I.W. Lyo, Ph. Avouris: Science **245**, 1369 (1989)
17.58 H. Huang, S.Y. Tong, J. Quinn, F. Jona: Phys. Rev. B, to be published
17.59 L. Headrick, I.K. Robinson, E. Vlieg, L.C. Feldman: Phys. Rev. Lett. **63**, 1253 (1989)
17.60 E. Kaxiras, K.C. Pandey, F.J. Himpsel, R.M. Tromp: Phys. Rev. B, to be published

18. Tribology at the Atomic Scale

*Gary M. McClelland and Sidney R. Cohen**

IBM Research Division, Almaden Research Center
San Jose, CA 95120–6099, USA

Tribology is the study of contacting surfaces rubbing against each other. It encompasses the topics of friction, lubrication, and wear, and historically has received the most attention from mechanical engineers. The study of tribology has generally been motivated by the need to optimize the efficiency, endurance, and precision of mechanical devices. There has been a gradually increasing appreciation that tribology presents interesting fundamental problems to materials scientists, physicists, chemists and particularly surface scientists, and that solving these fundamental problems could have striking ramifications for applied problems.

This paper reviews recent experimental and theoretical developments involving tribology at the atomic scale. After a brief summary of key concepts in classical tribology, experimental results will be reviewed from instruments whose resolution approaches the atomic scale, including the field ion microscope, the atomic force microscope, and the surface force apparatus. Finally, theoretical models are discussed in which the motion of individual atoms are followed. This chapter includes little discussion of the extensive previous work which has shown *indirectly* the role of atomic scale phenomena, such as the effect of monolayers on friction and adhesion.

For a compilation of many research articles dealing with atomic scale and near atomic scale tribology, the reader is referred to a recent proceedings volume [18.1]. An excellent monograph concerning surface forces (but not frictional forces) has appeared [18.2], and two recent monographs highlight the role of surface physics in tribology [18.3, 4]. A collection of articles outlines new directions in tribology [18.5].

18.1 Concepts in Classical Tribology

Since many readers of this chapter may be unfamiliar with classical tribology, a summary of key concepts follows; many books on the subject are available [18.6]. The major goal of tribology is to reduce wear of rubbing mechanical parts, but wear may also be the desired outcome during machining and polishing. For a given system, wear may occur through one or more of several mechanisms, including abrasion, adhesion, chemical processes, and fatigue. Even for a given

---

* Chaim Weizmann Postdoctoral Fellow

load and contact area, wear rates can vary tremendously, over 10 orders of magnitude. Another goal in most tribological applications is to reduce friction, although in devices such as brakes, high friction is desirable. In contrast to wear rates, for different materials, friction coefficients are found to vary over a range of 0.01 to 3, with the vast majority of systems falling between 0.1 and 1.

Because of the complexity of sliding interfaces, there are few universal laws in tribology. One important empirical relationship, Amontons' Law, holds that for a given sliding system, the ratio of the frictional force to load is, to a good approximation, constant over several orders of magnitude variation in load. This ratio, the friction coefficient, is independent of the apparent contact area and depends only weakly on the surface roughness. Qualitatively, Amontons' Law arises from the fact that for real surfaces the true contact area is generally much less than 1% of the apparent contact area. Most surfaces can be described as a collection of mildly sloped asperities (bumps) with various heights and curvatures arranged across the surface (Fig. 18.1a,b). When the load increases from zero, at first, only the tips of a very few asperities come into contact. As the load increases further, the true contact area increases because existing contacts broaden and more asperities come into contact. Qualitatively, it is this increasing contact area which leads to the increase in friction with load described by Amontons' Law.

For the particular case of plastic asperity contact, Amontons' Law arises in a straightforward fashion. The frictional force is the product of the material shear strength and the contact area, which is in turn proportional to the load divided by the yield strength. Comparing different materials, the shear and yield strengths tend to vary together, leading to the relatively small range of friction coefficients observed. For many systems, individual asperities interact not by plastic, but rather by elastic contact, for which the contact area is proportional to the 2/3 power of the load [18.7]. However, it has been shown that for almost any type of asperity contact, the total contact area summed over all asperities is

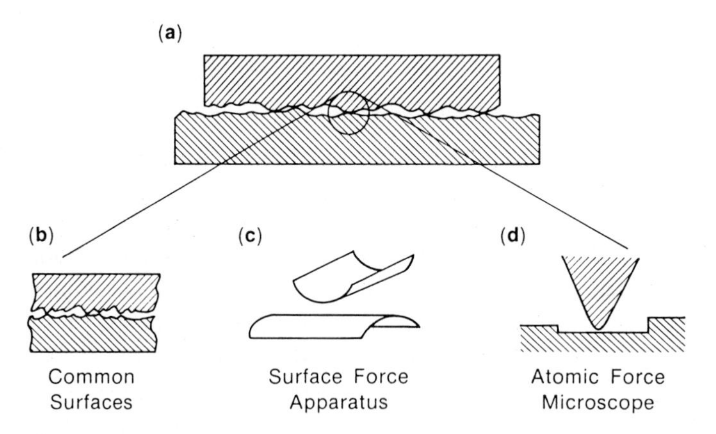

Fig. 18.1. Two macroscopic rubbing surfaces in contact (*a*), and in enlargement (*b,c,d*). For common surfaces, the contact is heterogeneous even on an atomic scale (*b*). In model experiments, contact may be simplified by using a surface force apparatus (*c*) or by using a single asperity technique, such as the atomic force microscope (*d*)

proportional to the load, if a distribution in heights of the asperities is taken into account [18.8].

Lubricants decrease friction and wear by separating sliding surfaces by a layer of low shear strength. Hydrodynamic lubrication occurs when the two moving surfaces are separated by a liquid film. The Reynolds Equation predicts that a liquid film is maintained quite naturally in very simple systems, such as that of a cylinder rotating with respect to another concentric cylinder which contains a liquid lubricant. As long as rotation is maintained, the load will not squeeze out the film. Hydrodynamic lubrication is an ideal subject for engineering analysis, because the performance of bearings can be predicted very accurately by a numerical analysis of fluid flow. Under some conditions, the viscosity of very thin films can be increased by the high pressure at the contact, resulting in elastohydrodynamic lubrication.

In boundary lubrication, the friction and wear of two surfaces is reduced by an adsorbed film as thin as a monolayer on one or both surfaces. If the layer does not bind strongly to the opposing surface, the interfacial shear strength is small, resulting in low friction and wear. Through this scheme, low friction can be achieved by using a hard bearing material for low contact area together with a soft overlayer of low shear strength.

Stick-slip friction is a commonly observed phenomenon, in which an increasing lateral force on an object produces no motion (stick) until the object suddenly leaps forward (slip) before coming to rest again. This behavior can occur when the force required to initiate sliding from zero velocity (the static friction) is greater than the force required to maintain sliding at a non-zero velocity (the kinetic friction). The static friction can be higher because slow processes such as chemical reactions have more time to occur at a static interface than at a moving one.

18.2 Experimental Approaches

18.2.1 Surface Force Apparatus

The intricate role in tribology of multiasperity contact, outlined above, highlights the importance of making measurements under conditions where asperities are well-defined or absent, even at the atomic level. An atomically flat surface is an ideal sample, but generally such surfaces can be prepared only in vacuum and with a size only up to tens of nanometers. A virtually unique exception is mica, which can be cleaved step-free over areas of several square centimeters (Fig. 18.2). It is the sample employed in the surface force apparatus (SFA) [18.9, 10] which has been successfully used to measure the forces between microscopic surfaces at distances ranging from several angstroms to hundreds of nanometers. In this instrument, cleaved mica sheets are are fashioned into a cylindrical shape and oriented with the axes perpendicular to create a "contact" zone of several tens of square microns. A delicate spring attached to one of the cylinders allows detection of forces as low as 10^{-8} N, while the distance between

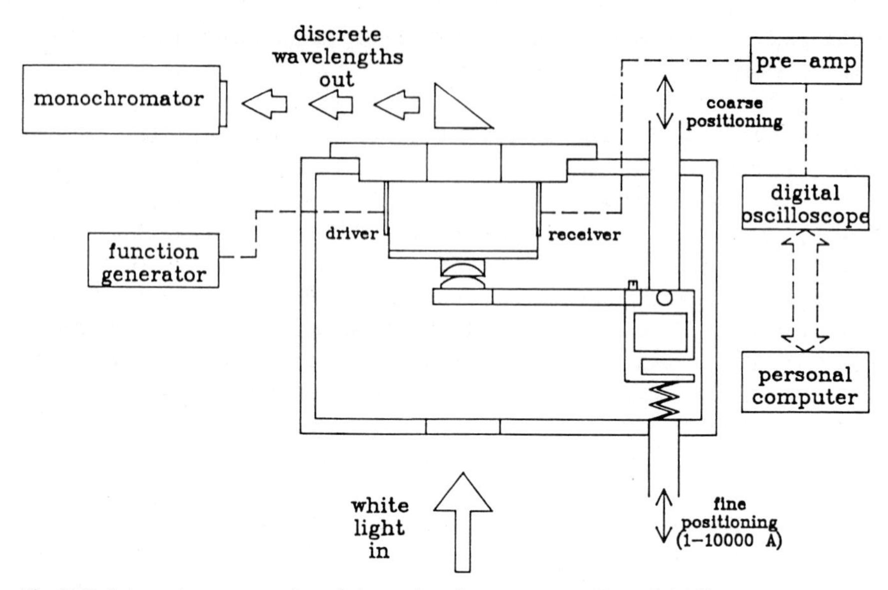

Fig. 18.2. Schematic representation of the surface force apparatus. From [18.33]

the interfaces is monitored using an optical technique; by silvering the back of the mica and shining a white light into the interface, multiple beam interferometry and fringes of equal chromatic order are used to observe separations to an accuracy of 1 Å.

If the surfaces of the SFA are immersed in a liquid and their separation decreased to near molecular dimensions, the liquid layer separating them will, at some point, cease to exhibit bulk properties. In particular, the density in a liquid oscillates at short radial distances from a molecule or surface, as may also be deduced from X-ray studies [18.11, 12, 13]. A number of theoretical studies of liquids near surfaces have been able to model the oscillatory behavior [18.14, 15, 16], the envelope of which decays with distance from the surface [18.17, 18]. This change in liquid density gives rise to oscillatory solvation or structural forces, which have been observed experimentally using the SFA [18.2, 19, 20]. The effect is due to the disruption of ordered molecular layers upon the approach of a second surface. The oscillations, which vary alternately between attraction and repulsion, have a periodicity of the mean molecular diameter.

a) Dependence of Adhesive Energy on Lattice Orientation. Besides its use in monitoring forces for non-contacting surfaces during approach or retraction, the SFA can measure the adhesive force required to separate two surfaces. When two identically oriented crystals of the same material are brought into contact, a single perfect crystal can be formed, and the adhesion energy is twice the surface energy of a single crystal. However, if the crystals are misoriented, a grain boundary, such as exists in a polycrystalline sample, is produced when the surfaces are brought together. Perfect bonding is not achieved, and the adhesion is reduced.

McGuiggan and *Israelachvili* have recently determined how the adhesion force between two mica cylinders in water depends on their angular misorientation [18.21, 22]. Narrow angular minima in the adhesive force are found by SFA measurements (Fig. 18.3). The adhesion enhancement at perfect alignment is large ($\simeq$ 50%), with a full width at half maximum of only about 0.5°. The very narrow peak widths observed in the mica adhesion experiments arise from the relative strength of the *intersolid* bonds formed across the interface with respect to the *intrasolid* bonds near the surface within each solid. The weak intersolid bonds between the mica cylinders are able to laterally distort the lattices only slightly, so that only a slight angular mismatch can be accomodated.

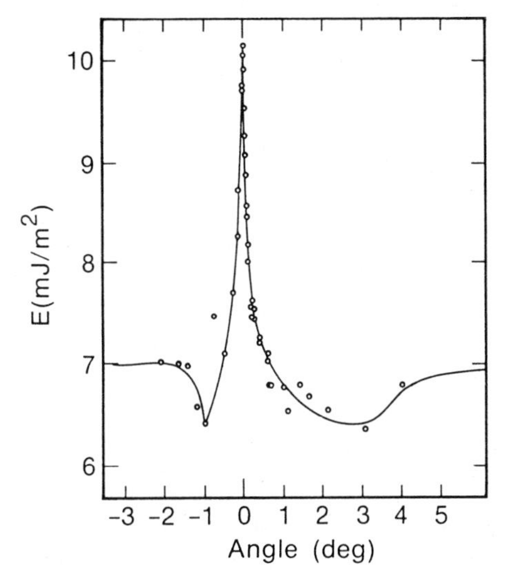

Fig. 18.3. Adhesion energy versus misorientation angle θ for two mica surfaces in water. From [18.21]

b) Thin Fluid Flow. In a lubricated system, an abrupt increase in friction and wear is expected when the lubricant layer decreases to a thickness of a few molecules. It is essential to understand to what extent continuum fluid dynamics based on the bulk viscosity can be extended into this regime. *Chan* and *Horn* [18.23] and *Israelachvili* [18.24] have explored this problem with the SFA, which in one experiment can probe fluid dynamics from the continuum micrometer level down to the single monolayer level. Figure 18.4 presents a "drainage" experiment with OMCTS (octamethylcyclotetrasiloxane, $[CH_3)_2\ SiO]_4$), described by *Chan* and *Horn* [18.23]. In this experiment a well-defined force pushes the mica plates of the SFA together, expelling the liquid between them. As the plates move closer, the approach rate decreases as the drainage path becomes more constrained; in the figure, this effect is manifested in the plot of cylinder separation vs. time. Since the experimental geometry is very well defined, the approach rate

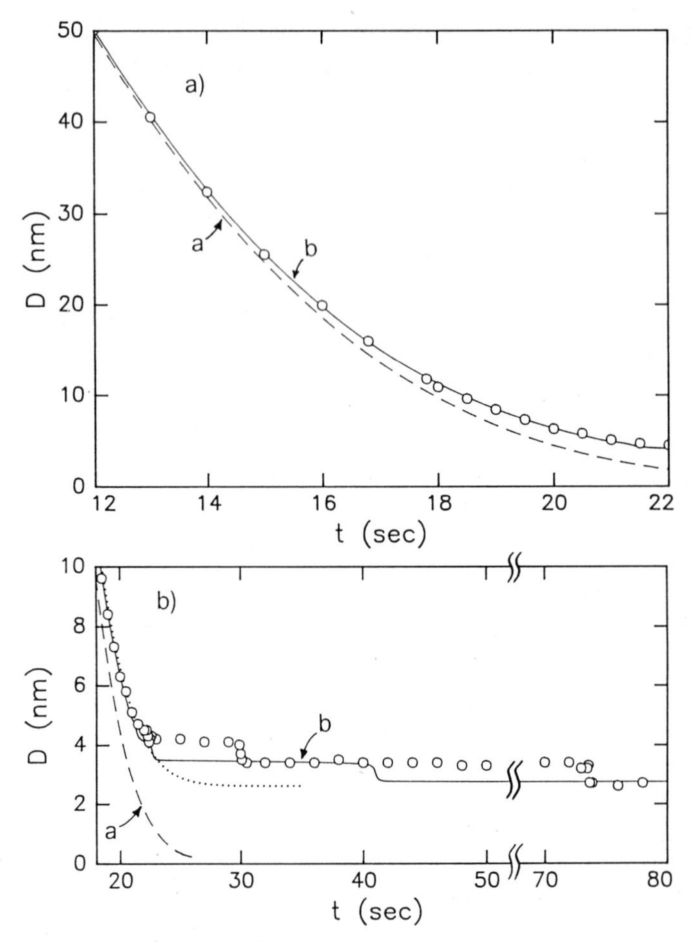

Fig. 18.4a,b. Drainage of OMCTS from between two crossed mica cylinders pushed together by a force. Distance D between the cylinders versus time. See text. From [18.23]

expected for a structureless liquid continuum with the viscosity of bulk OMCTS can be calculated, and is displayed in Fig. 18.4 with the experimental results. The continuum model, labelled "a" in Fig. 18.4 agrees quite well with the data until the liquid thickness decreases to below about 20 nm, at which point the experimental drainage rate is slower than predicted by the model.

The authors suggest that the deviation of their results from the continuum theory is due to a layer of liquid adjacent to each surface being immobilized by its interaction with the surface. They find that, down to a surface separation of about 4 nm, the experiment can be fit (curve "b") by assuming that the thickness of each immobile layer is about 1.3 nm, just less than twice the molecular diameter of OMCTS. The authors emphasize that, while this is a physically appealing interpretation, it is obviously an oversimplified picture.

At separations below about 4 nm, the discreteness of the flowing layers themselves becomes manifest in the abrupt changes of plate separation in steps of one molecular diameter. The change occurs at later times than can be accounted for from the measured static force alone, indicating that the effect is a dynamic one. Chan and Horn speculate that the retarded drainage of the discrete layers may be due to the effect of layering on molecular mobility or to the inability of the constricted molecules to move around each other.

Georges and his co-workers [18.25] have studied the dynamic and static forces between curved macroscopic surfaces other than mica. They forego the absolute distance calibration afforded by interferometry, instead using capacitance measurements to monitor the relative displacement of the surfaces. In a study of the flow of dodecane between two alumina cylinders with roughness of 2 nm, they found deviations from bulk behavior at separations less than 5 nm, where the observed irreversibility in the normal force was attributed to molecular packing [18.25]. They have also investigated a lubricant solution containing colloidal calcium carbonate particles and the effect of the adsorption of these particles on the mechanical interactions of surfaces < 100 nm apart [18.26].

Matthewson and *Mamin* have recently emphasized that the role of liquid films in determining meniscus and viscous forces involved in adhesion and friction depends qualitatively on the thickness of the films and on surface flatness [18.27].

c) Shear Motion. The use of curved mica plates to study surfaces in shear motion has a long history, starting in 1955 with a study by *Bailey* and *Courtney-Pratt* [18.28], who demonstrated that coating the mica surfaces with a monolayer of calcium stearate reduces the friction by a factor of forty. Later, *Briscoe* and *Evans* studied the tribological behavior of dry Langmuir-Blodgett films [18.29].

Several recent studies have employed the SFA to probe the shear behavior of thin liquid films constrained between mica plates. Because of the oscillating force behavior described above, even a multilayer of liquid can support a load. For the shear studies, the mica plates are attached to their supports by a flexible adhesive, which allows the plates to deform locally at the contact so as to become parallel. *Israelachvili* et al. described a study in which mica surfaces separated by cyclohexane and OMCTS were sheared hundreds of μm over each other while monitoring the shear force, load, and separation between the plates [18.30]. The observed response can be characterized by a critical shear stress; at shear stresses below this critical value, no sliding occurs, while at this shear stress, sliding commences. The frictional force during sliding is roughly independent of velocity. The novel feature of this experiment is the observation of "quantized friction". The measured shear stress varies over a set of discrete values, which are directly related to the number of liquid layers trapped between the plates. For cyclohexane, the shear stress is 2×10^7 N/m^2 for one layer, and decreases by about a factor of two for each additional layer up to four layers. The friction measured for OMCTS multilayers is considerably less, an effect which the authors attribute to the fact that the more gradually curved OMCTS molecules feel smoother than cyclohexane to the mica surfaces.

Israelachvili et al. have also reported stick-slip behavior for these systems [18.31], indicating that the static friction is greater than the kinetic friction under conditions where plastic deformation or chemical reactions with the surface are unlikely. The fact that the friction for static plates is greater than that for moving plates suggests a liquid ordering process occurs when the plates are not sliding. Water is found to act as a good lubricant under conditions when the K^+ ions are not removed from the mica surface, so that they exert a repulsion between the plates [18.32].

Van Alsten and *Granick* have reported a series of experiments [18.33, 34, 35] using an apparatus similar to that of Israelachvili et al. A major difference was that the amplitude of the oscillating shear applied was only on the order of hundreds of angstroms. For hexadecane, OMCTS, and other liquids, these workers found that at moderate pressures, there is no critical shear stress, but rather a viscous response is observed. In other words, shear motion occurs even in response to a very small shear force, and the shear rate is proportional to the force. The authors deduce an effective viscosity from their measured data, but caution that this number cannot be regarded as a true viscosity, because of the very thin anisotropic geometry involved. The fact that the effective viscosities measured are up to four orders of magnitude larger than those of the corresponding bulk liquids is probably attributable to the constrained quasi-two-dimensional geometry available for fluid flow between the mica plates. An increasing effective viscosity was found with decreasing film thickness. Also, the pressure could be varied over a factor of two without decreasing the film thickness, while the effective viscosity increased by about a factor of four. From such observations an effective activation volume of $3\,nm^3$ can be calculated. This is much larger than the volume of a single molecule, suggesting that the fundamental process for flow involves many molecules.

As the pressure is increased, Van Alsten and Granick find a sudden apparent "freezing" of the liquid film. At this point the viscous response disappears, and the sample appears to take on a critical shear stress. Sliding only occurs when this shear stress is exceeded. After the pressure is increased, the critical shear stress increases slowly with time, requiring about 15 minutes to attain its equilibrium value. Above the threshold for its onset, the shear stress increases very rapidly with pressure, exhibiting a very large pressure corresponding to an activation pressure of $10^7\,nm^3$ per molecule, implying a unit process for flow which is enormously larger than a single molecule.

It is important to understand what conditions determine whether a system exhibits a critical shear stress or viscous friction response to a shear. To do this, the full range of experimental conditions between those investigated by these two groups must be explored. The critical difference may be the amplitude of the shear motion [18.35] which in one study is much larger than the contact diameter, while in the other study is much smaller than this diameter. Also, very low level laboratory vibrations have been found to affect the frictional behavior [18.34]. Another topic which has not been addressed is the relationship of drainage experiments [18.23 24] to shear experimetns carried out on the SFA.

Although the external forces applied in the shear and drainage experiments are obviously very different, the motion of the entrapped molecules in both cases is constrained to be parallel to the interface. Perhaps the effective viscosities deduced from the two sorts of experiments can be related.

18.2.2 Single Asperity Techniques

Unfortunately, there are few materials which can be made smooth enough over a large enough surface area to be used in the SFA technique. Another approach to precisely defining a contact is to mimic the contact of a single asperity with a surface (Fig. 18.1d). The results can then be interpreted without the averaging implied in Amontons' Law. If the tip is small enough, contact with the surface can be limited to a region over which there are no defects on the surface.

There is a rather extensive literature concerning the contact of sharp tips with surfaces in vacuum, with and without sliding. In addition to normal and lateral force measurement, the contact resistance can be measured. Some measurements have been combined with a scanning electron microscope and transmission electron microscope to examine indentations and wear tracks, others with field ion and field emission spectroscopy to examine changes in the tip during contact. Among the interesting effects observed in these experiments are the development of strong metal-metal adhesion at zero loads, and friction at "negative loads".

When the contact area is reduced to that of several contacting atoms, and indentation depths are a few nanometers or less, inhomogeneities in the tip and surface may influence the contact mechanics. On this scale, a monolayer of contaminant will profoundly affect the attractive and adhesive forces. The behavior will also be influenced by the local crystal structure. Indeed, as the contact area between surfaces approaches atomic dimensions, the interface and its associated stress zone may be free of dislocations, leading to mechanical behavior approaching that of an ideal lattice. This idea has existed for some time [18.36, 37] although its careful examination has only been possible in recent years. *Maugis* et al. have related observed dislocation density to adhesion, friction, and microhardness of Al foils in contact with a fine tip [18.38]. The experiments indicate two different regimes exist, depending on the applied load. Below a critical load, the behavior is elastic so that the measured adhesion force is low and friction is near zero. As the load is increased, however, the range at which the load can act on a dislocation source increases, so that plastic deformation occurs, accompanied by increases in both the adhesive force and friction. When sliding, the threshold comes at a lower applied load, because the size of the shear stresses involved augments the effective applied load. This work and others [18.39, 40, 41] however, have shown that the elastic energy stored in dislocations represents 1% or less of the total frictional energy expended, and heat must be dissipated by other means.

The key role of dislocation density has been substantiated recently by *Pharr* and *Oliver*, who have reinvestigated the dependence of hardness on indentation depth into Ag {111} surfaces [18.42]. A rapid increase in hardness of about a factor of two is observed as the indentation depth is decreased from 50 to 20 nm. In this range of indentation depths, dislocations are not observed to nucleate around the indentation, as they do at larger indentation depths. Thus, the increased hardness for small indentations may be the result of the confinement of dislocation activity to sub-surface regions.

Skinner and *Gane* have also investigated the friction and microhardness for sharp tips on a flat surface under small, and even 'negative' loads. Some controversy exists in these works concerning the effect of surface cleanliness. In their indentation experiments, Gane and co-workers [18.43, 44, 45] found that under certain conditions, the surface would not yield until it "punched" through, leaving a large indentation as the load was increased. This behavior was correlated with the presence of a carbon-containing polymeric film at the surface. The action of this film was believed to be one of preventing the spread of dislocations to the surface. *Maugis* et al. [18.38] cite the role of the contaminant film as one of reducing the attraction which would exist between two clean metals. This attraction, for fine metal tips, is strong enough to initiate plastic deformation even at zero applied load. The presence of such a strong attraction, which exists due to the natural strong short-range attraction between clean metals, is widely recognized, and has been observed by others [18.46, 47]. Intentional contamination of a clean nickel surface by oxidation has no effect on adhesion of a tungsten stylus until the oxide layer is several nm thick [18.47]. This topic has recently been reviewed [18.48] although it is worthwhile noting that determination of the hardness of a clean surface relative to a contaminated one is not yet fully understood.

a) The Nanoindenter. Many of the measurements cited above have utilized some kind of nanoindentor. Nanoindentation is a well-developed method for studying the elastic and plastic properties of surface materials to normal loads resulting in indentation depths between 10^{-8}–10^{-3} m [18.49]. Here, a standard indenter, usually a diamond pyramid, is loaded and unloaded from a surface while measuring the forces involved. For accurate measurements, the point of the indenter must be sharper than the diameter of the indent. Through these experiments, both the adhesive force and the microhardness of the materials can be investigated. Simultaneous measurement of contact resistance is useful for the estimate it provides of the contact area, and as a measure of the onset of contact during approach between conducting indenters and surfaces. *Chen* and *Hendrickson* [18.50] have demonstrated that nanoindentation can be accompanied by etch pitting and electron microscopy to image individual dislocations generated by the indentation where they intersect the surface.

Pethica, Oliver and co-workers have extended the technique to apply a desired load by magnetic coils, and monitor the tip displacement [18.51, 52, 53]. They have thus been able to observe the surprising robustness of thin oxide films and their effect on the contact mechanics. By applying a small, ac modulation to the tip, it is possible to measure the surface stiffness, $S = \partial P/\partial x$, with P the load and x the displacement. The attractiveness of this measurement is that the stiffness is related to the radius of the contact area a by a simple power law relationship $S = k \cdot aE^*$, where k is $\simeq 2$, and E^* the reduced elastic modulus, independent of the tip shape [18.54]. The resolution of this instrument allows the evaluation of stiffness, and thus contact radius, for contact areas on the atomic scale. Using this technique, near theoretical lattice strength was observed for a diamond tip on Cu and sapphire when the tip radius was 500 Å.

b) Field Ion Microscopy. Because of its power to both clean and image sharp tips with atomic resolution, special mention is given to the field ion microscope [18.55] which was invented by *Müller* in 1951 [18.56]. The potential of the FIM for studying tip-surface contacts was first exploited by *Nishikawa* and *Müller* who characterized the damage done to tungsten tips during contact with several types of metals [18.55]. *Buckley* reports that when a tungsten tip contacts a gold surface, only enough gold is transferred to the surface to create a dense covering of gold trimers [18.3]. From a platinum surface, epitaxial transfer to an iridium tip is found except when the interface is subject to vibration, in which case platinum clusters are found on the tip. In a FIM study of contact of tungsten and platinum tips touched to platinum surfaces with $\simeq 10^{-6}$ N loads, *Walko* finds that tip damage extends only 5–15Å deep [18.57]. By determining as a function of load the area over which tip damage occurred, it was determined that the contact followed Hertz's equations for elastic contact.

c) Other UHV Work. The concern, mentioned above, of deviation from clean surface behavior due to a contaminant film, is particularly strong when high energy surfaces such as active metals are under investigation. In these cases, the experiment should be performed in UHV. Standard surface science instrumentation which is typically available in UHV chambers can be used to monitor the phenomena occurring. *Buckley* [18.3 58] and *Pepper* [18.59] have proven the effectiveness of such tools as LEED, Auger spectroscopy, field ion microscopy, and the atom probe in studies of adhesion, wear, and the associated transfer of material. *Pollock* has made a careful analysis of surface forces and the effect of nanometer-scale topography aided by a single asperity UHV experiment [18.46].

Pepper has correlated the friction of Cu on diamond in vacuum with the presence or absence of unoccupied surface states in the bandgap [18.60]. The friction coefficient for a polished surface, annealed at 750°C was $\simeq 0.1$. Annealing above 850°C or bombarding with 500 eV electrons, which presumably dehydrogenates the surface, results in an increased friction coefficient of 0.5. Exposure of this state to excited hydrogen reduces the friction coefficient. Energy loss spectra showed the presence of a feature at 284 eV (in the bandgap) could be correlated with the conditions in which high friction was observed. Formation of an interfacial bond involving the gap state is postulated. The combination of UHV surface techniques with tribological measurements is being actively pursued by several groups [18.61, 62, 63].

d) Scanning Tunneling and Atomic Force Microscopy. The demonstration of single atom resolution by scanning tunneling microscopy (STM) in 1983 greatly expanded the horizons for the single-asperity technique [18.64]. The reader is referred to Chap. 16 (Avouris) and Chap. 14 for reviews of STM experiments and theory [18.66] and to scanning probe techniques in general (Chap. 15). Obviously, the STM shows great potential for determining the topography of surfaces of tribological interest [18.65]. Early on, it was recognized that tip-surface forces could play a role in the imaging process [18.66], an effect thought to be responsible for the anomalously large atomic corrugations observed on graphite [18.67, 68, 69]. Tip-surface forces are among the many effects demonstrated to achieve surface modification in the STM, which has been promoted as a technique for very high density recording of information. The scanning tip can be used both to modify the surface and to image the modification and any spontaneous changes at the surface. This approach was used on a gold {111} surface by *Jaklevic* and *Elie* [18.70]. They created craters $\simeq$ 10 Å deep and $\simeq$ 100 Å wide by tip contact, and, over a period of about an hour, observed the healing of the craters by surface diffusion. Other examples of such surface modifications, which are prototypes for both ultrahigh resolution lithography and wear at the atomic scale, are discussed in Chap. 15.

As its inventors, *Binnig, Quate* and *Gerber*, have emphasized, the atomic force microscope (AFM) is simply a stylus profilometer updated with technology developed for the STM [18.71]. In the atomic force microscope, the forces normal to the surface exerted by the surface on the tip are measured by the deflection of a cantilever on which the tip is mounted. As the surface is scanned past the tip, the variation in tip-surface forces reveals the topography of the surface. The most critical aspect of the AFM is reliably and accurately sensing the sub-angstrom deflection of the cantilever. Although this tunneling method offers unsurpassed sensitivity, the tunneling junction exerts an extraneous force on the cantilever, and is unreliable. Soon after the first AFM was demonstrated, simple [18.72] and heterodyne [18.73] optical interferometry were introduced in place of tunneling to detect the cantilever deflection. Fiber interferometry [18.74], laser deflection [18.75], and capacitance [18.76, 77, 78] methods have also been developed.

The force-sensing sensitivity of an AFM is ultimately limited by the noise level of the deflection sensing technique and by the unavoidable thermal excitation of the lever. However, 0.01 Å noise levels are achievable over a 100 Hz bandwidth. Although low lever force constants k_l give high sensitivity, they also allow the tip to "snap-in" to the surface under the influence of the attractive force of the tip. A good compromise value is k_l = 20 N/m, for which 0.01 Å gives a noise level of 2×10^{-11} N (1 N $\simeq$ 100 gram weight). For comparison, the force required to break a single chemical bond is about 10^{-8} N, while a single van der Waals bond breaks at about 10^{-10} N. Thus the AFM is sensitive enough to image individual atoms without breaking chemical bonds and to accurately measure forces (including frictional forces) exerted on a single atom of the tip.

Many AFM images showing atomic features have now been reported, but only on layered compounds such as graphite or boron nitride. Since almost all of the AFM work to date has been in air, such compounds are used because they can be cleaved to give unreactive surfaces. There have been near-atomic resolution images recorded on surfaces of polymers, organic crystals, and biopolymers [18.79]. To obtain a high resolution image, the instrument must be operated in the "respulsive mode", with a force of 10^{-9} to 10^{-7} N between the tip and the surface. The simple fact that two-dimensional atomic resolution images can be repeatedly scanned on the same region of a surface is very significant from a tribological perspective because the tip must be sliding across the surface with no wear. These AFM images represent the most direct evidence to date that wearless sliding is possible. Since a force of 10^{-7} N is much larger than required to break a single chemical bond, it seems surprising that such a large load can lead to atomic resolution images. This question has been the subject of some modeling calculations of the tip-surface interaction [18.80] which we will discuss in more detail below. These calculations have emphasized that, because of the layered structure of graphite, a deep and laterally extended deformation of the surface occurs at modest loads, so that tip-surface contact occurs over more than just one atom. It may be that, while imaging in air, most of this contact is lubricated by a mobile adsorbed layer with only a single tip atom "poking through" it to do the imaging [18.68]. While imaging a layered compound, it is also possible for a flake to break off of the surface and adhere to the tip. If the flake remains oriented with respect to the surface, the normal force of the surface on the tip will vary with the periodicity of the surface [18.81].

Many AFM images show anomalous features which are apparently the result of forces parallel to the surface. A featureless region appearing immediately after the surface changes direction is likely due to sticking of the tip to the surface [18.82]. "Grooving" is sometimes observed, in which two successive x line scans taken at slightly differing y values are very different [18.82]. It has been suggested that this occurs when the tip which has been stuck in a row of the surface lattice suddently slips to an adjacent row. Frictional forces have also been invoked to explain lattice distortion observed at the beginning of line scans in repulsive mode imaging [18.83]. The effect of frictional forces would be more clearly elucidated by comparing line scans obtained in opposite scanning direction while the sample is oscillated under the tip.

The first AFM study of frictional forces at atomic-level resolution was reported by *Mate* et al. and *Erlandsson* et al. for the 2000 Å radius tip of a tungsten wire sliding on graphite [18.84] and mica [18.85] surfaces in air. This work used an AFM based on optical interferometry in which the sample and tip were rotated by 90° from the usual orientation used for mapping topography (Fig. 18.5a), so that lever deflections detected parallel to the surface measure the frictional force. The frictional force on both the graphite and mica surfaces shows a clear pattern with the period of the sample surface (Fig. 18.5b), an effect most easily seen in two-dimensional scans [18.84, 85].

The force varies in a stick-slip pattern, which appears to be very much like that discussed above for the surface force apparatus experiments. Here, the origin of this effect is not the velocity dependence of the frictional force, but rather,

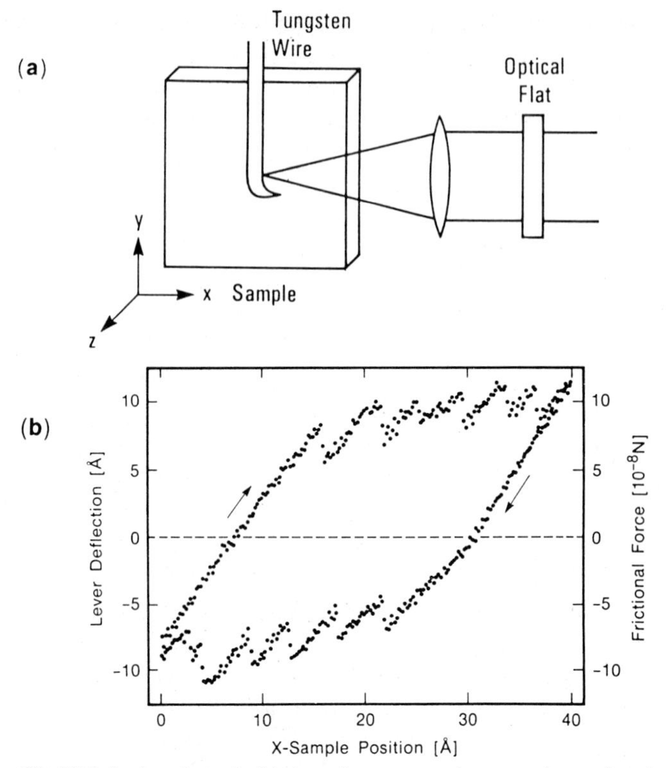

Fig. 18.5a,b. Atomic scale friction of a tungsten tip on a mica surface in air (**a**) Schematic representation of tip, sample, and interferometer geometry. (**b**) The lever deflection and corresponding frictional force in the x direction as a function of sample position as the sample is scanned back and forth under the tip. From [18.85]

its dependence on position. To understand this effect and other aspects of AFM experiments, the tip and cantilever cannot be considered simply as a probe of tip-surface forces. Rather, the dynamics of the tip moving in response to forces exerted by the surface and the cantilever must be considered. Because surface features are encountered at a frequency smaller than the lever's resonant frequency, the tip always comes to an instantaneous mechanical equilibrium (total force zero) in response to the combined force of the lever and sample. Considering only displacements of the sample and tip along the scanning direction x, and noting that the tip-sample force F_s depends only on the instantaneous relative position of the tip and sample, the total force F on the tip can be written as a function of the tip position x_t and the sample position x_s

$$F = -k_l x_t + F_s(x_t - x_s) .$$

The first term represents the force exerted by the cantilever on the tip, and has a negative slope with respect to x_t, pushing more towards negative x_t as x_t increases. If F_s is not constant, it will have regions of both positive and negative F_s'. For any region where $dF/dx_t > 0$, the tip will not be at a stable equilibrium. If the tip encounters such regions, it will slip until it finds a region where $dF/dx_t < 0$. The effect can be likened to pulling a stick across successive boards of a picket fence. This equation predicts that, for a given lever, if the frictional force is made low enough so that dF/dx_t is always less than k_l, no slipping will occur, and the tip position will vary smoothly with the sample position, while still reflecting the periodic character of the lattice. This change between discontinuous to continuous behavior has been observed on graphite [18.84].

An elastic analysis of the graphite and mica friction experiments indicates that the tip-surface contact covers many unit cells of the surface, and for the larger loads used for graphite, is as much as 1000 Å across [18.84]. In view of this fact, it seems surprising that the surface periodicity is directly manifested in the frictional force, because the large tip contact area would average any variations across the unit cell. However, the contact is not uniform and undoubtedly consists of the contact of many nm-scale asperities, which are probably responsible for the linearity of the frictional force with load. These contacts will not be spaced uniformly, and so will not completely average out the lateral dependence of the frictional force.

Another possibility which could explain the large force modulation is that flakes of the flat surface break off and remain on the tip, remaining in orientational registry with the surface. Such flakes have been postulated to be present during STM imaging of graphite [18.81]. In some recent low-resolution AFM experiments, care was taken to measure the frictional force before extensive sliding had occurred which might transfer a flake to the tip [18.76]. In this case the measured friction coefficient (0.28) was much larger than the value of 0.01 that had been observed after extensive contact.

Using the AFM in the nonscanning, nanoindenter mode, *Burnham* and *Colton* [18.86] and *Neubauer* et al. [18.76] have studied the elastic indentation of a graphite surface with hard metal tips of several thousand angstrom radii. The apparent surface stiffness is much lower than is to be expected from the known graphite elastic constants, but for indentation depths greater than 100 Å, the stiffness approaches its expected value [18.76]. This effect may be due to the partial delamination of graphite flakes, which are pushed back into the surface by the applied load. This sensitive technique was recently extended to surfaces consisting of Langmuir-Blodgett (LB) films, where it was shown that substituting the exposed methyl group of stearic acid with the trifluoromethyl group led to a noticeable variation in the tip-surface forces [18.87].

Blackman et al. [18.88] have recently used the AFM to study the tribology in air of a tungsten tip ($\simeq$ 1000 Å radius) on LB films of cadmium arachidate deposited on the oxidized Si$\{111\}$ surface. From friction loops of the type displayed in Fig. 18.5b, they found that for a monolayer film, friction was not proportional to load, but that the derivative of friction vs. load began at 0.1 for small loads, increased to 0.6 at intermediate loads, and fell back to about 0.3 for the maximum load of 6×10^{-8} N. Using the AFM tip to image by attractive

VDW forces, the monolayer was found to be undamaged by the friction although it was swept free of "molecular debris" adhering to the surface. In contrast, when a thicker film of five layers was used as a lubricant, the first four layers were easily worn away. Using the AFM tip as an indenter, the Meyer's hardness of a 7 layer film was determined to be 0.6 MPa.

Albrecht has studied self-assembled films which, in contrast to LB films, are attached to the surface through a covalent bond [18.82]. Using a microfabricated SiO_2 tip, and forces up to 10^{-6} N, it was possible to wear away a monolayer of octadecyl phosphate, but not a monolayer of octadecyltrichlorosilane. The latter undergoes chain polymerization (cross-linking) at the base, in addition to the chain-substrate bond. The AFM results may be compared with macroscopic studies, performed on both LB [18.89, 90, 91] and self-assembled [92, 93] films, which show that one layer provides all of the lubricating action. Interestingly, the strength of the monolayer-substrate bond plays no obvious role in the microscopic studies, whereas the intramolecular cohesion, altered by varying the chain substituents, does. More work needs to be done at the atomic level to elucidate the importance of substrate bonding and intramolecular forces in these lubricants.

For tribological studies of well-characterized surfaces, *Neubauer* et al. have recently incorporated an AFM into a UHV system equipped with sample and tip transfer airlocks and standard surface analysis equipment [18.78]. This AFM is the first to sense the lever deflection simultaneously both parallel and perpendicular to the surface. This enables direct measurement of the load during a friction experiment, and allows tip-surface dynamics to be followed in great detail. Furthermore, the surface can be imaged by repulsive tip-surface forces after measuring friction loops. To sense the lever deflection both parallel and perpendicular to the surface, a dual capacitance sensor is employed (Fig. 18.6a). In this capacitance method, a radio frequency oscillating voltage is applied to the sensor plates, which are positioned about 1000 Å from the lever. Angstrom-scale deflections of the lever cause $\simeq 10^{-17}$ F variations of the capacitance between the lever and plates, which are sensed by a detector locked onto the radio frequency.

This new AFM was used to characterize the indentation and friction of an iridium tip on a Au$\{111\}$ surface in UHV [18.97]. To obtain these data, the sample is oscillated back and forth parallel to the sample direction while being advanced toward and retracted from the sample. As the sample is advanced toward the tip, the tip snaps into the surface when the derivative of the attractive force exceeds the lever force constant (Fig. 18.6b). The sample is advanced until the load is about 10^{-7} N, and then is retracted. The curve of lever deflection vs. sample position is reversible and the slope is near one, indicating that the tip-surface contact undergoes very little elastic or plastic deformation. The snap-out of the lever establishes that the adhesive force is about 3×10^{-7} N. The frictional force computed from the height of friction loops of the sort shown in Fig. 18.5b is also displayed. Interestingly, the friction shows a very strong hysteresis, changing much more gradually when the load is decreased than when it is increased. While the dependence of friction on load is not strictly linear, the approximate friction coefficient upon increasing the load is 1.0, while upon de-

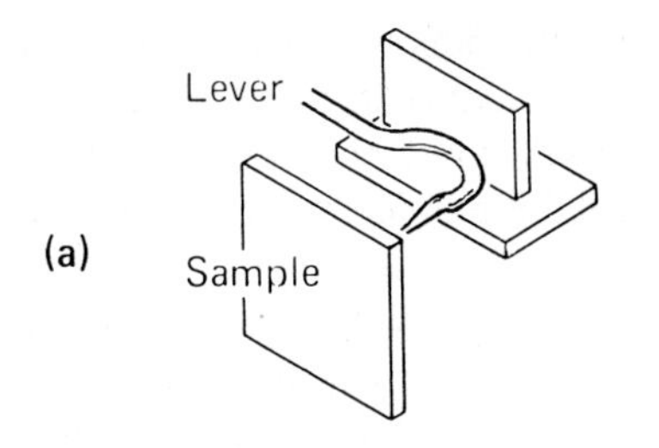

Fig. 18.6a,b. Friction of an iridum tip on a Au {111} surface in vacuum. (a) Schematic representation of geometry used to simultaneously measure friction and load by capacitance measurement. (b) Variation in frictional (lateral) and loading (normal) forces as the sample is pushed into the tip with simultaneous oscillation in the sample plane. From [18.18]

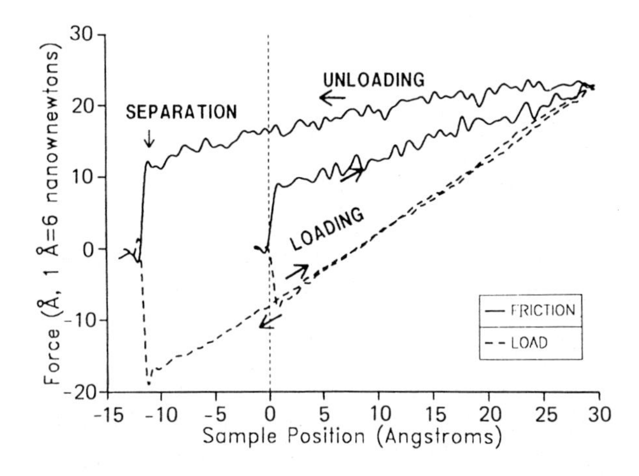

creasing the load it is 0.4. Considering that very little hysteresis is observed in the motion of the lever perpendicular to the sample, the changing frictional behavior is evidently associated with very subtle changes in the tip-surface interface. For these experiments, the surface was clean as verified by LEED and Auger analyses. The tip was cleaned by heating and ion bombardment before these experiments, but measurements of the tip-surface contact resistance indicate that some contaminant prevented full tip-surface contact.

In order to better understand the head-disk interaction in magnetic disk recording devices, the AFM has been used to characterize the friction and adhesion properties of magnetic disks [18.95]. In addition to its usefulness in probing the dynamics of tribological interactions, the AFM is a very useful surface analytical tool for characterization of tribologically interesting surfaces. For example, the AFM has been used to correlate the topography of hard carbon films with their tribological properties [18.83]. *Mate* et al. have used the AFM as a "dipstick" to profile the thickness of lubricant coatings [18.96, 97] and have profiled the liquid-air interface by sensing van der Waals forces [18.96].

Contact electrification refers to the exchange of charge which occurs between two surfaces. When electrification is caused by the repeated contact of asperities during rubbing, it is referred to as tribocharging. The role of rubbing is to repeatedly make and remake asperity contacts. Between metals, electrification results from the electron flow which must occur to equilibrate the fermi level of two contacting metals of different work functions. Tribocharging of insulators is not at all understood, and has recently been investigated on a small scale by force microscopy [18.98]. *Terris* et al. have developed a method for detecting force gradients from charges as small as 6, with a lateral resolution of 0.2 μm [18.99]. This resolution, which is 3–4 orders of magnitude better than previous tribocharging studies, enables the observation of a bipolar charge distribution: when a Ni tip of 0.1 μm radius is touched to a polymethyl methacrylate surface, a charged region of 10 μm diameter is formed. Within this region are both positive and negative charges which are stable on the surface for days before decaying away [18.98].

18.3 Theoretical Descriptions of Tribology

Because of the complexity of tribological interfaces and the difficulty of obtaining reliable data from a "buried" sliding interface, interpretation of experimental data is bound to be ambiguous. Theoretical molecular dynamics models, even if simplified, provide the advantage of being able to study a well-defined system, and to follow directly any desired variable of the system. The relation of directly measurable properties to the atomic dynamics can be determined directly. Using empirical or *ab initio* interatomic potentials, molecular dynamics calculations integrate the classical equations for motion of atomic nuclei. Although, ultimately a quantitative prediction of experiment requires realistic interatomic potentials, a great deal of phenomena can be elucidated with less accurate potentials.

18.3.1 Simplified Model of Wearless Friction

As discussed above in the general overview of tribology, in most cases, friction is associated with wear. For example, according to the adhesive theory of friction, shear occurs within the contacting asperities, and the energy is dissipated during the movement of dislocations involved in the plastic deformation of asperities. The AFM and SFA results obtained recently seem to show clearly that friction can occur without wear, and *McClelland* has recently presented a discussion of friction in this regime, using a highly simplified model [18.100]. The motivation of such models is not to provide a realistic comparison to experiment, but rather to clarify important mechanisms of interfacial energy dissipation which can be obscured in complex calculations. Figure 18.7a presents a schematic representation of two atomically smooth surfaces sliding over each other in such a way as to generate no wear. We assume that each solid is bound together by strong (chemical, metallic, or ionic) *intrasolid bonds*, but that, because the valency of each atom at the interface is satisfied, the *intersolid* interaction across the interface is weak, consisting only of van der Waals interactions. Then, when the lower solid A is slid across the upper solid B, the interfacial interaction is not

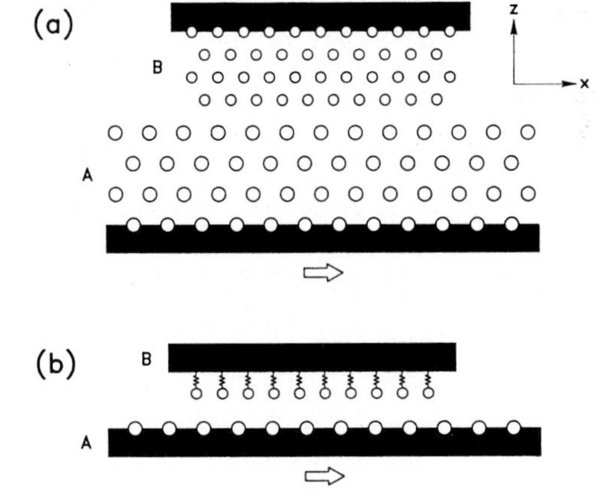

(c)
B
A

Fig. 18.7a–c. Models of wearless friction. (**a**) Solid A sliding across solid B in two dimensions. (**b**) Independent oscillator model. (**c**) Frenkel-Kontorova model. The black areas represent rigid supports of the atoms. From [18.100]

strong enough to break any of the bonds within either solid. However, it can, in principle, move the atoms within the solid, thus generating phonons and dissipating energy. For purposes of discussion, for each solid, a row of atoms far from the interface is attached to a rigid "handle". The top handle is held fixed while the bottom handle is translated slowly across the other solid.

To understand qualitatively how energy can be dissipated for this type of interface, a simplified "independent oscillator model" is analyzed, which is similar in spirit to a model described long ago by *Tomlinson* [18.101]. In this simplified model, solid A has been replaced by a single row of atoms anchored to a rigid handle while solid B has been replaced by non-interacting atoms, bound as damped harmonic oscillators to the B handle.

Suppose that the handle of the solids are held at absolute zero temperature during sliding. Noting that for reasonable sliding speeds, the time scale for sliding over a single atom is many orders of magnitude less than a vibrational period, we see that a particular B-atom, B_0, will at all times move toward achieving a position of lowest energy in the potential energy governing its motion. As illustrated in Fig. 18.8a, this potential V_S is the sum of two components: V_{BB}, a simple harmonic potential attaching it to its handle, and V_{AB} the periodic interfacial potential arising from the interaction with the A-atoms [18.103]. In the left column of the Fig. 18.8, V_{AB} is a rather strong hard-sphere interaction.

Following the left column of Fig. 18.8, as A is slid, B_0 at first moves gradually, following the minimum of V_s, until, at the point diagrammed in frame *e*, the atom is suddenly released because the minimum has disappeared. This disappearance occurs when the repulsive (concave downward) part of V_{AB} is overcome by the increasingly sloped concave upward part of V_{BB}. When the

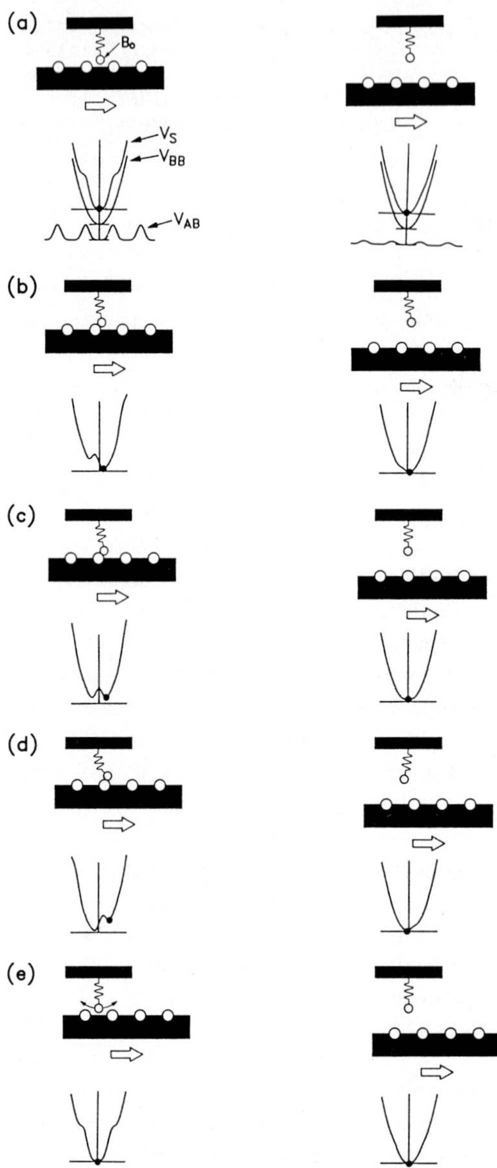

Fig. 18.8a–e. Motion of an atom B_0 of B in the independent oscillator model of friction. Left and right columns diagram strong and weak interfacial interactions, respectively. The top panel diagrams the relevant potentials, while subsequent panels illustrate the response of B_0 represented by a black dot on V_S to progressive sliding of A, as determined by the potential sum V_S plotted below each diagram. From [18.100]

B-atom is released, it falls suddenly down into the nearby minimum, becoming vibrationally excited in the process. This vibrational excitation is dissipated into the surrounding solid. The dissipation of energy results in friction.

A key feature of friction is that the frictional force opposes the direction of motion, so that it must reverse when the sliding direction is changed. How this reversal occurs in the independent oscillator model can be discovered in the left panel of Fig. 18.8c. Because solid A has been moving from left to right, the

B-atom sits in the right minimum of a symmetric V_S, and, therefore, exerts a force to the right on solid B.

One might expect that the frictional force would be roughly proportional to the interfacial interaction, but in fact the dynamics changes *qualitatively* as V_{AB} is reduced. The right column of Fig. 18.8 traces out the dynamics for a V_{AB} one fifth the size of that diagrammed in the left column. Following the sequence of frames, it is seen that there is no "plucking" step to release energy as there was for higher V_{AB}. Plucking can only occur when there is a disappearing local maximum in V_S, which happens only when a portion of V_{AB} has a downward curvature greater than the upward curvature of V_{BB} (the force constant of the B bond).

The situation where two solids interact by physical interactions while they are each held together by stronger bonds actually corresponds to the right column of Fig. 18.8. We conclude that in this case the friction vanishes. It should be emphasized that the origin of this frictionless behavior is the reversible adiabatic response of the system to the slow external perturbation. The conclusion of frictionless motion is not limited to the highly simple independent oscillator mode. As emphasized by *Sokoloff* [18.102], the same result arises in the Frenkel-Kontorova model, where the interaction between atoms is included, but all other forces on the interfacial atoms are neglected (Fig. 18.7c). In the Frenkel-Kontorova model, however, the effect of the relative lattice spacing can be seen. When the two interacting solids are commensurate (i.e., when their lattice spacings are related by a simple ratio of whole numbers) the force acts in phase on many atoms to produce a concerted motion of the interfacial atoms of each solid. The effective restoring force for such a concerted motion is very small, so that in this case energy dissipation occurs.

These concepts have been confirmed by molecular dynamics calculations [18.100]. In the simple model presented here, thermal excitation would reduce friction if the temperature were high enough to activate the B_0-atoms over the V_s barrier at a rate competitive to sliding. A similar model has been applied to discuss the effect of adsorbed molecules at a sliding interface [18.100].

18.3.2 Molecular Dynamics Studies of Shearing Solids

Recently, the Georgia Tech group has reported several large scale molecular dynamics simulations of interfacial shear using reasonable representations of the interatomic potentials. In one such simulation, two solids, each of which had 9 layers containing 70 atoms interacting by Leonard-Jones potentials, were placed in contact and sheared across each other [18.103–105]. One solid (the soft solid) was less weakly bound than the other, and three-dimensional periodic boundary conditions were employed. As an increasing shear stress was applied across the two solids, an elastic response was first observed, followed by stacking fault formation and eventually yield.

When the lattice constant parallel to the shear direction was set the same for the two solids, shear occurred not at the interface, but in the softer solid, an observation in general agreement with experimental experience. But when the lattice constant for the softer solid was set to be 50% larger than that of the hard solid, the yield stress was lowered because "pinning" no longer occurred at the interface. Yield occurred not within the soft solid but at the interface. In the context of the above discussion of frictionless sliding, we expect a non-zero friction in this case both because the intersolid bonds are not much weaker than the intrasolid bonds, and because the lattices are commensurate with a spacing ratio of 3 to 2.

Modeling of sharp tips sliding across surfaces is relevant both to asperity contact during sliding of two surfaces and to imaging in force microscopy. *Landman* et al. [18.104–107] have discussed the sliding of a Si {111} surface across a 16 atom {111} facet of a dynamical Si tip. This is clearly a "reactive" system, in which the surface can form strong bonds with the tip. Accordingly, when the tip is brought into contact with the surface and then withdrawn, a plastic "neck" is formed between the two, consisting mostly of tip atoms. On the other hand, if the tip is not withdrawn, but is simply slid over the surface, no damage occurs to surface or the tip. A striking feature of these results is the clear manifestation of abrupt behavior in the frictional force (Fig. 18.9d) of the sort discussed in the independent oscillator model above. The slips are not absolutely abrupt because the sliding velocity is so large, making the time interval in the figure quite short.

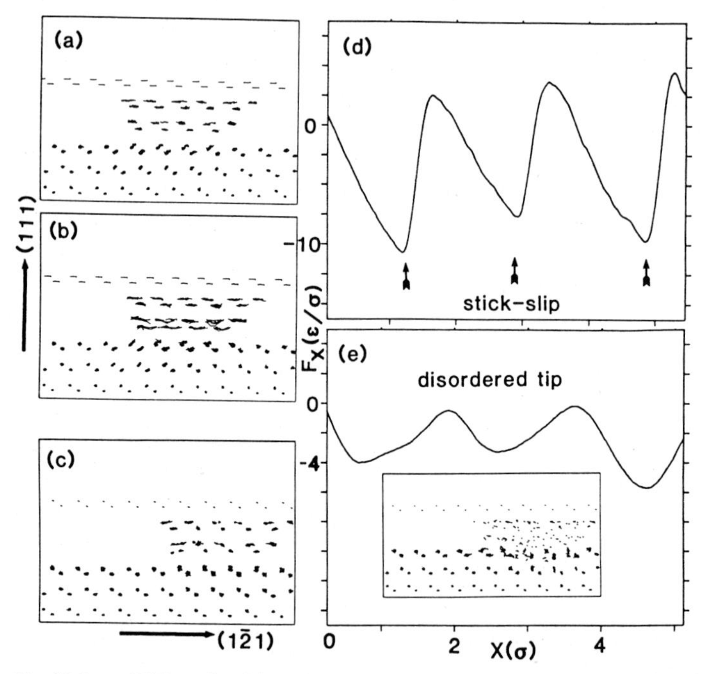

Fig. 18.9a–e. Sliding of a silicon tip across a silicon surface. (**a–c**) Particle trajectories viewed along the (10$\bar{1}$) direction just before (**a**) and after (**b**) an atomic scale stick-slip event, and (**c**) towards the end of the scan. (**d**) The frictional force. (**e**) The frictional force for an amorphous tip. From [18.106]

From the figure, it is clear that the interfacial atoms have become greatly excited during the slip process. Interestingly, a disordered tip shows a frictional force which also varies with the period of the lattice, but no distinct slips are visible (Fig. 18.9e) probably because the friction is composed of minor unsynchronized individual slips across the interface.

The relation of the STM and AFM imaging processes to tribology has been emphasized above. Normal forces between the tip and surface have been the subject of several theoretical calculations involving varying degrees of sophistication in their treatment of electronic structure and tip and surface deformation. Topics addressed include tip geometry [18.108, 109], elastic deformation of the substrate [18.80, 109, 110], detection of intercalation atoms [18.110], the effect of tip-surface forces on STM topographs [18.67, 111] and tip-induced modifications of the surface electronic structure [18.111, 112]. The possibility of jumping of a tip to a surface due to the flexibility not of a cantilever but of the tip itself has been discussed [18.113, 114], as have the universal aspects of interaction forces between metal tips and surfaces [18.115].

18.3.3 Molecular Dynamics Studies of Thin Fluid Flow

The importance of thin liquid films in lubrication and in recent SFA experiments has motivated theoretical studies of liquid flow in small channels. A recent molecular dynamics study addressed the shearing of a one to five atom thick Lennard-Jones fluid between two flat, rigid, ordered surfaces of atoms of the same type [18.116]. When the surfaces are held fixed, they induce freezing of the intervening fluid into a commensurate structure. Sliding of one surface over the other can occur only when a critical shear stress is exceeded, at which point the intervening solid becomes less dense and liquifies.

In the SFA experiments, it is the load, not the spacing, which is held fixed, and such a constraint was applied in the simulation of *Landmann* et al. of a three layer liquid [18.105]. For static surfaces, this liquid also orders. At the onset of shear, the thickness of the confined liquid is reduced by one layer, but residual liquid ordering is still manifested by oscillation of the frictional force during sliding.

The flow of a very narrow atomic liquid through a smooth pore of square cross-section about 6 molecular diameters wide was reported by *Bitsanis* et al. [18.117]. The viscosity was found to be less than that of the bulk liquid, but could be accounted for accurately by assigning a local viscosity equal to that of a bulk fluid with a density equal to that of the density found in the pore simulation.

18.4 Summary and Future Directions

During the last few years, tremendous advances have been made in the development of experimental and theoretical tools necessary for attaining a scientific understanding of tribology. Atomically resolved images of surfaces in real space can be generated by scanning tunneling and atomic force microscopy. Furthermore, the mechanical properties of the surfaces themselves can be determined with atomic resolution. Using the surface force apparatus, the angular dependence of adhesive forces has been revealed. The static and dynamic forces both normal and parallel to the surfaces have been determined as a function of the numbers of layers on the surface. Using the atomic force microscope, the periodic structure of a surface has been detected in the frictional force. Due to the increasingly widespread capability of doing large scale numerical computation, the experimental advances have been matched by realistic molecular dynamics simulations of sliding interfaces.

The possibilities are so new that they have barely been explored. For example, using an STM/AFM tip, it is now possible to image a surface (including its defects) with atomic resolution, measure the frictional force and load as a single apex atom of the tip slides across the surface, and then use the same tip to monitor atomic-scale defects in the surface. This experiment could be made perfectly well-defined if the tip were imaged at atomic resolution by field ion microscopy. Already reproducible manufacture of single atom tips has been demonstrated, and FIM-characterized tips have been used for STM [18.118]. Another approach for better characterized tribology experiments is to study surfaces of low specific surface free energy, such as LB and self-assembled monolayers, which do not require UHV to remain clean.

An important direction for the surface force apparatus experiments will be the extension of this technique to surfaces other than mica. The mica surfaces can be chemically modified, and recent results on sapphire [18.119] and glass [18.120] are promising.

There have been no direct experimental studies of the fate of energy dissipated at a sliding interface. Eventually, this issue may be addressed by spectroscopically detecting phonons created at the sliding interface.

At present, a major advantage of molecular dynamics approaches is that, in contrast to the experiments, the systems that are being studied are well-defined. In the future it will be important to use more accurate representations of inter-atomic potentials. A particular challenge will be to model slow phenomena which have a molecular basis, for example, stick-slip motion in the surface force apparatus.

Even with an accurate atomic scale picture of a tribological interface, a complete understanding requires a proper account of the very heterogeneous rough interface. Fractal treatments offer a possible approach to this problem [18.121].

The practical movitation for a fundamental understanding of tribology remains strong, with developing applications in micromechanics, space craft, medical implants, and information storage devices. We can look forward to a fruitful interplay of applied and fundamental studies during the next few years.

Acknowledgements. We are grateful to A. Homola, A. Gellman, S. Granick, R.G. Horn, J.N: Israelachvili, U. Landmann, and C.M. Mate for providing preprints of their work. Partial support of the ONR under contract N00014-88-C-0419 and the AFOSR under contract F49620-89-C-0068 is gratefully acknowledged. S.R.C. acknowledges the support of the Myron A. Bantrell Trust through a Chaim Weizmann postdoctoral fellowship.

References

18.1 L.E. Pope, L.L. Fehrenbacher, W.O. Winer (Eds): *New Materials Approaches to Tribology Theory and Applications* (Materials Research Society Symposium Proceedings **140**, Pittsburgh 1989)
18.2 J.N. Israelachvili: *Intermolecular and Surface Forces* (Academic, New York 1985)
18.3 D.H. Buckley: *Surface Effects in Adhesion, Friction, Wear, and Lubrication* (Elsevier, Amsterdam 1981)
18.4 G. Heinicke: *Tribochemistry* (Hanser, Munich 1984)
18.5 W.R. Loomis (Ed.): *New Directions in Lubrication, Materials, Wear, and Surface Interactions* (Noyes, Park Ridge 1985)
18.6 A.D. Sarkar: *Friction and Wear* (Academic, London 1980); J. Halling: *Introduction to Tribology* (Wykeham, London 1976); N.P. Suh: *Tribophysics* (Prentice Hall, New York 1986)
18.7 I.N. Sneddon: Int. J. Eng. Sci. **3**, 47 (1965)
18.8 J.A. Greenwood, J.H. Tripp: J. Appl. Mech. March 1979, p. 153; J.A. Greenwood: J. Lubrication Tech. Jan. 1967, p. 81; J.A. Greenwood, J.B.P. Williamson: Proc. Roy. Soc. Lond. **295**, 300 (1966)
18.9 J.N. Israelachvili, D. Tabor: Proc. Roy. Soc. Lond. A **331**, 19 (1972)
18.10 D. Tabor, R.H.S. Winterton: Proc. Roy. Soc. Lond. A **312**, 435 (1969)
18.11 J.A. Pryde: *The Liquid State* (Hutchinson Univ. Library, London 1966)
18.12 F. Kohler: *The Liquid State* (Verlag Chemie, Weinheim 1972)
18.13 P. Kruus: *Solids and Solutions: Structure and Dynamics* (Dekker, New York 1977)
18.14 W. van Mege, I.K. Snook: J. Chem. Phys. **74**, 1409 (1981)
18.15 N.I. Christou, J.S. Whitehouse, D. Nicholson, N.G. Parsonage: Symp. Faraday Soc. **16**, 139 (1981)
18.16 G. Rickayzen, P. Raymond: in *Thin Liquid Films*, ed. by I.B. Ivanov (Dekker, New York 1985) Chap. IV
18.17 F.F. Abraham: J. Chem. Phys. **68**, 3713 (1978)
18.18 I.K. Snook, W. van Megen: J. Chem. Phys. **70**, 3099 (1979)
18.19 R.G. Horn, J.N. Israelachvili: J. Chem. Phys. **75**, 1400 (1981)
18.20 H.K. Christenson, R.G. Horn: Chem. Phys. Lett. **98**, 45 (1983)
18.21 P.M. McGuiggan, J.N. Israelachvili: in *Characterization of Structure and Chemistry of Defects in Matter*, Mat. Res. Soc. Symp. p. 349 (1989)
18.22 P.M. McGuiggan, J.N. Israelachvili: Chem. Phys. Lett. **146**, 469 (1988)
18.23 D.Y.C. Chan, R.G. Horn: J. Chem. Phys. **83**, 5311 (1985)
18.24 J.N. Israelachvili: J. Colloid Interface Sci. **110**, 263 (1986)
18.25 A. Tonck, J.M. Goerges, J.L. Loubet: J. Colloid Interface Sci. **126**, 150 (1988)
18.26 J.M. Georges, J.L. Loubet, A. Tonck: Mat. Res. Soc. Symp. Proc. **140**, 67 (1989)
18.27 M.J. Matthewson, H.J. Mamin: Mat. Res. Soc. Symp. Proc. **119**, 87 (1988)
18.28 A.I. Bailey, J.S. Courtney-Pratt: Proc. R. Soc. London **A227**, 500 (1955)
18.29 B.J. Briscoe, B. Scruton, F.R. Willis: Proc. R. Soc. London **A333**, 99 (1973); B.J. Briscoe, D.C.B. Evans: Proc. R. Soc. London **A380**, 389 (1982)
18.30 J.N. Israelachvili, P.M. McGuiggan, A.M. Homola: Science **240**, 189 (1988)
18.31 P.M. McGuiggan, J.N. Israelachvili, M.L. Gee, A.M. Homola: in *New Materials Approaches to Tribology: Theory and Applications*, ed. by L.E. Pope, L.L. Fehrenbacher, W.O. Winer: Mat. Res. Soc. Symp. Proc. **140**, 51 (1989)
18.32 A.M. Homola, J.N. Israelachvili, M.L. Gee, P.M. McGuiggan: J. Tribology, to be published
18.33 J. Van Alsten, S. Granick: Phys. Rev. Lett. **61**, 2570 (1988)
18.34 J. Van Alsten, S. Granick: Tribology Trans, in press
18.35 J. Van Alsten, S. Granick: Langmuir, in press
18.36 H. Bückle: "Progress in Micro-Indentation Hardness Testing", Metallurgical Reviews (London, Institute of Metals), **4**, 49 (1959)
18.37 B.W. Mott: *Microindentation Hardness Testing* (Butterworths, London 1957)
18.38 D. Maugis, G. Desalos-Andarelli, A. Heurtel, R. Courtel: ASLE Trans. **21**, 1 (1976)
18.39 R. Feder, P. Chaudhari: Wear **19**, 109 (1972)
18.40 G. Andarelli, D. Maugis, R. Courtel: Wear **23**, 21 (1973)
18.41 N. Gane, J. Skinner: Wear **25**, 381 (1973)
18.42 G.M. Pharr, W.C. Oliver: J. Mater. Res. **4**, 94 (1989)
18.43 N. Gane, F.P. Bowden: J. Appl. Phys. **39**, 1432 (1968)

18.44 N. Gane: Proc. R. Soc. Lond. A **317**, 367 (1970)
18.45 N. Gane, J.M. Cox: Philos. Mag. **22**, 881 (1970)
18.46 Q. Guo, J.D.J. Ross, H.M. Pollock: Mat. Res. Soc. Symp. Proc. **140**, 51 (1989)
18.47 M.D. Pashley, J.B. Pethica, D. Tabor: Wear **100**, 7 (1984)
18.48 D. Maugis, H.M. Pollock: Acta Metall. **32**, 1323 (1984)
18.49 P.J. Blau, B.R. Lawn (Eds.): *Microindentation Techniques in Materials Science and Engineering* (ASTM, Ann Arbor 1986)
18.50 C.C. Chen, A.A. Hendrickson: J. Appl. Phys. **42**, 2208 (1971)
18.51 J.B. Pethica, R. Hutchings, W.C. Oliver: Philos. Mag. **A48**, 593 (1983)
18.52 J.B. Pethica: *Ion Implantation into Metals*, ed. by V. Ashworth et al. (Pergamon Press, Oxford 1982) p. 147
18.53 J.B. Pethica, W.C. Oliver: Physica Scripta **T19**, 61 (1987)
18.54 K. Kendall, D. Tabor: Proc. R. Soc. **A323**, 321 (1971)
18.55 E.W. Müller, T.T. Tsong: *Field Ion Microscopy* (Elsevier, New York 1969)
18.56 E.W. Müller: Z. Physik **131**, 136 (1951)
18.57 R.J. Walko: Surface Sci. **70**, 302 (1978)
18.58 D.H. Buckley: J. Vac. Sci. Technol. **13**, 88 (1976); Wear **46**, 19 (1978)
18.59 S.V. Pepper: J. Appl. Phys. **50**, 8062 (1979); ibid. **47**, 801 (1976); ibid. **45**, 2947 (1974)
18.60 S.V. Pepper: J. Vac. Sci. Technol. **20**, 643 (1982)
18.61 B.M. DeKoven, P.L. Hagans: Mat. Res. Soc. Symp. Proc. **140**, 357 (1989)
18.62 K.P. Walley, A.J. Gellman: preprint
18.63 K. Miyoshi: Mat. Res. Soc. Symp. Proc. **153**, 321 (1989)
18.64 G. Binnig, H. Rohrer, Ch. Gerber, E. Weibel: Phys. Rev. Lett. **50**, 120 (1983)
18.65 M.T. Dugger, Y.W. Chung, B. Bhushan, W. Rothschild: J. Tribology, to be published
18.66 J.H. Coombs, J.B. Pethica: IBM J. Res. Develop. **30**, 455 (1986)
18.67 J.M. Soler, A.M. Baro, N. Garcia, H. Rohrer: Phys. Rev. Lett. **57**, 444 (1986)
18.68 H.J. Mamin, E. Ganz, D.W. Abraham, R.E. Thomson, J. Clarke: Phys. Rev. B **34**, 9015 (1986)
18.69 C.M. Mate, R. Erlandsson, G.M. McClelland, S. Chiang: Surface Sci. **208**, 473 (1989)
18.70 R.C. Jaklevic, L. Elie: Phys. Rev. Lett. **60**, 120 (1988)
18.71 G. Binnig, C.F. Quate, Ch. Gerber: Phys. Rev. Lett. **56**, 930 (1986)
18.72 G.M. McClelland, R. Erlandsson, S. Chiang: in *Review of Progress in Quantitative Non-Destructive Evaluation*, ed. by D.O. Thompson, D.E. Chimenti (Plenum, New York 1987) p. 1307; R. Erlandsson, G.M. McClelland, C.M. Mate, S. Chiang: J. Vac. Sci. Technol. A **6**, 266 (1988)
18.73 Y. Martin, C.C. Williams, H.K. Wickramasinghe: J. Appl. Phys. **61**, 4723 (1987)
18.74 D. Rugar, H.J. Mamin, R. Erlandsson, J.E. Stern, B.D. Terris: Rev. Sci. Instrum. **59**, 2337 (1988); D. Rugar, H.J. Mamin, P. Günther: Appl. Phys. Lett. **55**, 2588 (1989)
18.75 G. Meyer, N.M. Amer: Appl. Phys. Lett. **53**, 1045 (1988); S. Alexander, L. Hellemans, O. Marti, J. Schneir, V. Elings, P.K. Hansma, M. Longmie, J. Gurley: J. Appl. Phys. **65**, 164 (1989)
18.76 G. Neubauer, S.R. Cohen, G.M. McClelland: In *Interfaces Between Polymers, Metals, and Ceramics*, ed. by B.M. DeKoven, A.J. Gellman, R. Rosenberg: Mat. Res. Soc. Symp. Proc. **153**, 307 (1989)
18.77 T. Göddenhenrich, H. Lemke, U. Hartmann, C. Heiden: to appear in J. Microscopy
18.78 G. Neubauer, S.R. Cohen, G.M. McClelland, D. Horne, C.M. Mate: Rev. Sci. Instrum. (1990), in press
18.79 S. Gould, O. Marti, B. Drake, L. Hellemans, C.E. Bracker, P.K. Hansma, N.L. Keder, M.M. Eddy, G.D. Stucky: Nature **332**, 332 (1988); P.K. Hansma, V.B. Elings, O. Marti, C.E. Bracker: Science **242**, 209 (1988)
18.80 F.F. Abraham, I.P. Batra: Surface Sci. **209**, L125 (1989)
18.81 J.B. Pethica: Phys. Rev. Lett. **57**, 3235 (1986)
18.82 T.R. Albrecht: Ph.D. Dissertation, Stanford University (1989)
18.83 E. Meyer, H. Heinzelmann, P. Grütter, Th. Jung, H.-R. Hidber, H. Rudin, H.-J. Güntherodt: Thin Solid Films, 180 (1989), to be published
18.84 C.M. Mate, G.M. McClelland, R. Erlandsson, S. Chiang: Phys. Rev. Lett. **59**, 1942 (1987)
18.85 R. Erlandsson, G. Hadziioannou, C.M. Mate, G.M. McClelland, S. Chiang: J. Chem. Phys. **89**, 5190 (1988)
18.86 N.A. Burnham, R.J. Colton: J. Vac. Sci. Technol. A **7**, 2906 (1989)
18.87 N.A. Burnham, D.D. Dominguez, R.L. Mowery, R.J. Colton: Phys. Rev. Lett. **64**, 931 (1990)

18.88 G.S. Blackman, C.M. Mate, M.R. Philpott, to be published
18.89 I. Langmuir: Trans Faraday Soc. **XV**, pt. 3, 62 (1920)
18.90 I. Langmuir: J. Franklin Inst. **218**, 143 (1934)
18.91 V. Novotny, J.D. Swalen, J.P. Rabe: Langmuir **5**, 485 (1989)
18.92 V. DePalma, N. Tillman: Langmuir **5**, 868 (1989)
18.93 O. Levine, W.A. Zisman: J. Chem. Phys. **61**, 1068 (1957)
18.94 S.R. Cohen, G. Neubauer, G.M. McClelland: to appear in J. Vac. Sci. Technol. A (July/August, 1990)
18.95 R. Kaneko, K. Nonaka, K. Yasuda: J. Vac. Sci. Technol. A **6**, 291 (1988); R. Kaneko: J. Microscopy **152**, 363 (1988)
18.96 C.M. Mate, M.R. Lorenz, V.J. Novotny: J. Chem. Phys. **90**, 7550 (1989)
18.97 C.M. Mate, M.R. Lorenz, V.J. Novotny: IEEE Trans. on Magnetics 1990 (in press)
18.98 B.D. Terris, J.E. Stern, D. Rugar, H.J. Mamin: Phys. Rev. Lett. **63**, 2669 (1989); J. Vac. Sci. Techn. A, in press
18.99 J.E. Stern, B.D. Terris, H.J. Mamin, D. Rugar: Appl. Phys. Lett. **53**, 2717 (1988)
18.100 G.M. McClelland: In *Adhesion and Friction*, ed. by M. Grunze, H.J. Kreuzer, Springer Series in Surface Science, Vol. 17 (Springer, Berlin, Heidelberg 1989), p. 1
18.101 G.A. Tomlinson: Philos. Mag. Series 7 **7**, 905 (1929)
18.102 J.B. Sokoloff: Surf. Sci. **144**, 267 (1984)
18.103 M.W. Ribarsky, U. Landman: Phys. Rev. B **38**, 9522 (1988)
18.104 U. Landman, W.D. Luedtke, M.W. Ribarsky: J. Vac. Sci. Technol. A **7**, 2829 (1989)
18.105 U. Landman, W.D. Luedtke, M.W. Ribarsky: Mat. Res. Soc. Symp. Proc. **140**, 101 (1989)
18.106 U. Landman, W.D. Luedtke, A. Nitzan: Surface Sci. **210**, L177 (1989)
18.107 U. Landman, W.D. Luedtke, R.N. Barnett: In *Many-Atom Interactions in Solids*, ed. by R. Nieminien (Springer, Berlin, Heidelberg 1989)
18.108 F.F. Abraham, I.P. Batra, S. Ciraci: Phys. Rev. Lett. **60**, 1314 (1988)
18.109 S.A.C. Gould, K. Burke, P.K. Hansma: Phys. Rev. B **40**, 5363 (1989)
18.110 D Tománek, G. Overney, H. Miyazaki, S.D. Mahanti, H.J. Güntherodt: Phys. Rev. Lett. **63**, 876 (1989)
18.111 S. Ciraci, A. Baratoff, I.P. Batra: to be published
18.112 I.P. Batra, S. Ciraci: J. Vac. Sci. Technol. A **6**, 313 (1988)
18.113 J.B. Pethica, A.P. Sutton: J. Vac. Sci. Technol. **6**, 2494 (1988)
18.114 J.R. Smith, G. Bozzolo, A. Bannerjea, J. Ferrante: Phys. Rev. Lett. **63**, 1269 (1989)
18.115 A. Bannerjea, J.R. Smith, J. Ferrante: Mat. Res. Soc. Symp. Proc. **140**, 89 (1989)
18.116 M. Schoen, C.L. Rhykerd, Jr., D.J. Diestler, J.H. Cushman: Science **245**, 1223 (1989)
18.117 I. Bitsanis, J.J. Magda, M. Tirrell, H.T. Davis: J. Chem. Phys. **87**, 1733 (1987)
18.118 H.-W. Fink: IBM J. Res. Develop. **30**, 460 (1986); Y. Kuk, P.J. Silverman: Appl. Phys. Lett. **48**, 1597 (1986); Y. Kuk, P.J. Silverman: Rev. Sci. Instrum. **60**, 165 (1989)
18.119 R.G. Horn, D.R. Clarke, M.T. Clarkson: J. Mater. Res. **3**, 413 (1988)
18.120 R.G. Horn, D.T. Smith, W. Haller: Chem. Phys. Lett. **162**, 404 (1989)
18.121 A. Majumdar, B. Bhushan: J. Trib., Trans. ASME (in press)

Subject Index

Contents of

Chemistry and Physics of Solid Surfaces IV

(Springer Series on Chemical Physics, Vol. 20)

Contents of
Chemistry and Physics of Solid Surfaces V

(Springer Series in Chemical Physics, Vol. 35)

Contents of
Chemistry and Physics of Solid Surfaces VI

(Springer Series in Surface Sciences, Vol. 5)

Contents of
Chemistry and Physics of Solid Surfaces VII

(Springer Series in Surface Sciences, Vol. 10)